INTERMEDIATE
STATISTICAL METHODS
AND APPLICATIONS

INTERMEDIATE STATISTICAL METHODS AND APPLICATIONS

A Computer Package Approach

MARK L. BERENSON
Baruch College

DAVID M. LEVINE
Baruch College

MATTHEW GOLDSTEIN
President
The Research Foundation
CUNY

Prentice-Hall, Inc., Englewood Cliffs, New Jersey   07632

Library of Congress Cataloging in Publication Data

BERENSON, MARK L. (date)
  Intermediate statistical methods and applications.

  Bibliography: p. **568**
  Includes index.
  1. Statistics—Data processing. 2. Mathematical
statistics—Data processing. I. Levine, David M., (date)
  II. Goldstein, Matthew, (date) III. Title.
QA276.4.B47 1983   519.5′028′54   82-12226
ISBN 0-13-470781-8

Editorial/production supervision: Kim Gueterman and Barbara Palumbo
Cover design: Diane Saxe
Manufacturing buyer: Ed O'Dougherty

Printed in the United States of America

10 9 8 7 6 5 4 3 2 1

ISBN 0-13-470781-8

Prentice-Hall International, Inc., *London*
Prentice-Hall of Australia Pty. Limited, *Sydney*
Editora Prentice-Hall do Brasil, Ltda., *Rio de Janeiro*
Prentice-Hall Canada Inc., *Toronto*
Prentice-Hall of India Private Limited, *New Delhi*
Prentice-Hall of Japan, Inc., *Tokyo*
Prentice-Hall of Southeast Asia Pte. Ltd., *Singapore*
Whitehall Books Limited, *Wellington, New Zealand*

To Rhoda B., Kathy B., and Lori B.

Marilyn L. and Sharyn L.

Brian G. and Seth G.

CONTENTS

PART II Linear Models: Experimental Design and Regression Analysis 51

PART III Multivariate Methods 419

PREFACE

Whenever a new textbook is being planned, the authors must resolve the issue of how the new text will differ from those already available and what contribution it will make to a field of study. Advances in computer technology over the past ten years have brought about a new dimension in statistical research: the capacity to efficiently and effectively analyze large-scale problems that heretofore could only be discussed theoretically. Thus it is not surprising that technological progress in both computer hardware and software capability has encouraged corresponding advances in statistical research endeavors. Therefore, when writing a textbook on the practical applications of statistical methods (beyond the introductory first-course level), we feel it has become essential to describe the use of computer packages throughout since, in reality, only limited meaningful applied research could be achieved today without them.

This text is characterized by its broad-based coverage of statistical procedures and applications in business and in the social and health sciences. It is divided into three parts. Part I comprises Chapters 1 and 2. The first chapter is intended as both a review of and a reference for fundamental statistical methods covered in a typical one-semester introductory course. In the second chapter we introduce the reader to the three most widely used computer packages: the Statistical Package for the Social Sciences (SPSS), the Statistical Analysis System (SAS), and Biomedical Programs (BMDP). Part II contains twelve chapters dealing with applied linear models. Experimental design models are covered in Chapters 3 through 7, while regression analysis is described in Chapters 8 through 14. Finally, Part III is concerned with the concepts and applications of such multivariate methods as: principal components and factor analysis; multidimensional scaling and cluster analysis; multiway contingency table analysis; and discriminant analysis. These are respectively presented in Chapters 15 through 18.

The broad coverage of topics in this text provides for much flexibility. The book can be used in four types of courses.

Material from Parts I and II (the first fourteen chapters) on applied linear models is designed for use in the second semester of a one-year introductory statistics course in business or in the social and health sciences wherein it is assumed that such introductory topics as basic descriptive statistics, probability, hypothesis testing, and simple regression and correlation were already covered. Thus the basic prerequisite background for this book would be the type of material contained in such texts as *Basic Business Statistics*: *Concepts and Applications* (Berenson, M., and Levine, D., Prentice-Hall, Englewood Cliffs, N.J., 1983), *Statistics for Management and Economics* (Mendenhall, W., and Reinmuth J., Duxbury Press, North Scituate, Mass., 1982), or *Applied Statistics* (Neter, J., Wasserman, W., and Whitmore, G., Allyn & Bacon, Boston, 1982).

Moreover, the book is also designed for use in an intermediate-level course in

Applied regression analysis (Chapters 1, 2, 8–14), or

Analysis of variance and experimental design (Chapters 1–7 and 12), or

Applied multivariate analysis (Chapters 1, 2, 8–10, 13, 15–18).

The major pedagogical features of this text are threefold:

1. The explanation and use of computer packages for every topic
2. The coverage of important multivariate procedures not previously considered for texts at this level
3. The emphasis on concepts and methods through both *real data* and *realistic-type* applications throughout.

It is our belief that knowledge of, and the ability to use (and interpret results from), computer packages has become essential for applied statistical research. Thus the *primary* pedagogical approach that can be offered for applied intermediate-level statistics courses in business as well as in the social and health sciences is to aid the researcher in data analysis through the use of computer packages. Hence the use of computer packages is explained and demonstrated for each statistical topic considered so that, in addition to developing a conceptual understanding of the technique involved, students will be able to utilize the computer to aid in the solution of problems. Well-known packages such as SPSS, SAS, and BMDP are used throughout the text in order to teach the reader (1) *how to access and use* one of the packages and (2) *how to interpret and use* the resulting computer output as an aid in the decision-making process.

The second unique feature of the text concerns the coverage of certain relevant, recently developed topics in multivariate analysis. Over the past decade, there has been a rapid development in the area of discrete multivariate methods. While many articles have appeared in various journals and some advanced texts have been published [see, for example, *Discrete Multivariate Analysis* (Bishop, Y., Holland, P., and Feinberg, S., M.I.T. Press, Cambridge, Mass., 1975), *Discrete Discriminant Analysis* (Goldstein, M., and Dillon, W., Wiley, New York, 1978), and *Applied Multidimensional Scaling* (Green, P., and Rao, V., Dryden Press, Hinsdale, Ill., 1972)], a practical treatment of this useful material has not previously appeared in intermediate-level texts. Thus a fundamental development of subjects such as multiway contingency tables, discrete discriminant analysis, and multidimensional scaling would make a contribution by exposing these important methods to additional researchers in business and in the social and health sciences.

The third unique pedagogical feature of the text is the emphasis on concepts and methods of data analysis through the use of numerous real-data and realistic-type applications. Thus, in examining the various linear models as well as multivariate techniques, the underlying principles will be discussed and illustrated by use of an actual (real-data) or realistic-type problem with but the minimal development of the necessary mathematical theory involved. For example, in describing the concepts of model development in multiple regression, the primary emphasis will focus on the analysis of results for a real-data problem obtained from a computer package rather than through the use of matrix algebra. In addition, as in the text *Basic Business Statistics: Concepts and Applications*, we again find it pedagogically essential to include a data base as a unifying theme when teaching the various subjects. Thus Appendix A contains a set of 335 responses regarding the sales of single-family houses in a small southwestern city. Questions pertaining to an analysis of this real estate data base are represented by the symbol ■ and are found among the end-of-chapter problems throughout the text. Also, certain problems are preceded by an asterisk. The solutions to these problems can be found at the end of the book in the section entitled *Answers to Selected Problems*.

Hence the scope of such a text should be appealing and useful to a wide level of audiences. It is intended for intermediate-level undergraduate or graduate students—depending on research interests, as well as professional researchers in business or in the social and health sciences. Since the primary emphasis is on the concepts and applications, it is our hope that this textbook will make the study of the various statistical methods more meaningful and comprehensible and, at the same time, provide the necessary tools to enable the reader to participate in actual research endeavors with the aid of computer packages.

We wish to express our thanks to the various individuals who provided substantial assistance to us in the development of this text. We are particularly grateful to Mr. David Stephan for assistance with the computer applications contained herein; and to Mrs. Ruth Meyer and Mrs. Ann Festa for their careful typing of the manuscript. We would also like to thank Professors Jack Friedman, Texas Real Estate Center, Texas A & M University, and William Dillon, University of Massachusetts, for providing data for use in the text. In addition, we would like to express our gratitude to SPSS, Inc., to the SAS Institute, and to BMDP for granting permission to use their syntax and output throughout the text. Furthermore, we are grateful to the Literary Executor of the late Sir Ronald A. Fisher, F.R.S., and to Dr. Frank Yates, F.R.S., and to Longman Group Ltd., London, for permission to reprint Tables III and IV from their book *Statistical Tables for Biological, Agricultural and Medical Research*, Sixth Edition (1974).

Moreover, we wish to express our thanks to the editorial staff at Prentice-Hall, Inc. for their continued encouragement—Mr. Doug Thompson, Ms. Kim Gueterman, and Ms. Barbara Palumbo. We also wish to thank Professors Wynn Abranovic, Applied Statistics and Computer Information Systems, University of Massachusetts; Beth Elaine Allen, Department of Economics, University of Pennsylvania; Charles Lamphear, School of Business Administration, University of Nebraska—Lincoln; Glen W. Milligan, Faculty of Management Sciences, Ohio State University; M. Anthony Schork, Department of Biostatistics, University of Michigan; and H. Tamura, Graduate School of Business Administration, University of Washington, for their ideas and constructive comments during the development stages. Finally we wish to express thanks to our families who patiently bore the pressures caused by our commitment to complete this book. To them we are most grateful.

Mark L. Berenson

David M. Levine

Matthew Goldstein

INTERMEDIATE
STATISTICAL METHODS
AND APPLICATIONS

Part I

PRELIMINARY CONCEPTS

In the first part of the text, consisting of Chapters 1 and 2, we shall present the background necessary for studying such topics as experimental design, regression, and multivariate methods. Chapter 1 serves as both a review of and a reference for the fundamental elements of descriptive and inferential statistics. In Chapter 2 we develop the rationale behind the use of computer packages in statistics and introduce the basic features of three well-known and widely used packages: the Statistical Package for the Social Sciences (SPSS), the Statistical Analysis System (SAS), and Biomedical Programs (BMDP). The application of these packages is integrated throughout all subsequent chapters in this text so that the reader not only becomes familiar with a particular statistical procedure but also learns how to utilize the computer to apply the procedure to actual sets of data.

1

FUNDAMENTAL
STATISTICAL METHODS:
A REFERENCE AND REVIEW

1.1 INTRODUCTION

This chapter is intended as a review of basic statistical methods. Moreover, it is intended to provide a common reference by highlighting those statistical ideas and procedures fundamental to the development and understanding of the concepts of experimental design, regression, and multivariate analysis. As such, the chapter may be skimmed, used as a reference, or omitted.

1.2 DATA AS OUTCOMES OF RANDOM VARIABLES

The basic resource necessary for any statistical endeavor is data. The data may be obtained from published sources, through survey research, or by designed experiment. However obtained, the data are the observed outcomes or responses of some phenomenon of interest or underlying *random variable*. As outlined in Figure 1.1, there are two types of random variables yielding two types of data: *qualitative* and *quantitative*. Moreover, the quantitative random variables may

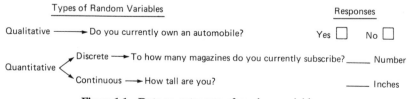

Figure 1.1 Data as outcomes of random variables.

be considered as either *discrete* or *continuous*; the former yield numerical responses arising from a counting process (e.g., to how many magazines do you currently subscribe?), while the latter yield numerical responses arising from a measuring process (e.g., how tall are you?).

1.3 DATA BASED ON MEASUREMENT SCALES

In addition, the resulting data may also be described in accordance with the scale of measurement attained.

This text primarily deals with procedures involving quantitative data. Thus, we shall be mainly concerned with procedures which require that a "fairly sophisticated" level of measurement be attained on the collected data. In the literature, such levels of measurement are known as *interval* or *ratio* scaling [see Table 1.1 for various examples as well as Conover (1980) or Glass and Stanley

TABLE 1.1   **Examples of Measurement Scales**

Scale	Type of Data	Examples
Nominal	Qualitative	Ethnic background. Type of life-insurance policy owned. Political party affiliation.
Ordinal	Rankings or Ratings	Basketball, boxing, or football ratings. Military rank. Bond ratings.
Interval	Quantitative	Fahrenheit versus Celsius temperature scales. Calendar dates. Utility scales.
Ratio	Quantitative	Length. Weight. Time.

(1970) for a detailed discussion of measurement and scaling concepts]. That is, the level of measurement attained on the data must be of "sufficient strength" to warrant such meaningful arithmetic operations as addition, subtraction, multiplication, and division. This, of course, implies that the observed data must be quantitative—not qualitative. Hence, for such procedures to be used the data cannot be attained by merely classifying the observations into various distinct categories (*nominal* scaling) such as "yes," "no," and "not sure"; nor can the data be attained by merely ranking the observations (*ordinal* scaling) based on some scheme such as highest to lowest.

On the other hand, some procedures based on qualitative data or rank data are also considered in this text. Procedures involving qualitative data applications are described in Chapter 12 (use of dummy variables), Chapter 17 (use of cross-classified categorical data), and Chapter 18 (use of discriminant analysis), while procedures involving ordinal data applications are presented in Chapter 16 (use of nonmetric multidimensional scaling).

1.4 DESCRIPTIVE STATISTICS

Descriptive statistics can be defined as those methods involving the collection, presentation, and characterization of a set of data in order to describe properly the various features of that set of data [see Berenson and Levine (1983), Chapters 2–4]. To characterize quantitative data several descriptive summary measures may be utilized.

The most commonly used measure of *central tendency* or *location* is the (arithmetic) *mean* $\bar{Y}$, which, for a sample of size n, is given by*

$$\bar{Y} = \frac{\sum_{i=1}^{n} Y_i}{n} \tag{1.1}$$

where $Y_i = i$th observation of the random variable Y and $\sum_{i=1}^{n} =$ "summation of" all values from 1 to n.

Other measures of central tendency are the *median* (i.e., the "middle" of an ordered sequence of observations), the *mode* (i.e., the most "typical" observation), and the *midrange* (i.e., the average of the smallest and largest observations).

The most commonly used measures of *dispersion* (i.e., variability or spread) are the *variance* S_Y^2 and (its square root) the *standard deviation* S_Y, which for a sample of size n are, respectively, expressed as

$$S_Y^2 = \frac{\sum_{i=1}^{n} (Y_i - \bar{Y})^2}{n-1} \tag{1.2}$$

and

$$S_Y = \sqrt{\frac{\sum_{i=1}^{n} (Y_i - \bar{Y})^2}{n-1}} \tag{1.3}$$

It is usually easier to compute S_Y^2 and S_Y from

$$S_Y^2 = \frac{\sum_{i=1}^{n} Y_i^2 - \left[\left(\sum_{i=1}^{n} Y_i\right)^2 \Big/ n\right]}{n-1} \tag{1.4}$$

*As in most introductory texts, Berenson and Levine (1983) use the symbol $\bar{X}$ to represent the *average* of the n observations of the random variable X. In this text, however, it is more convenient to represent the responses by the symbol Y.

and

$$S_Y = \sqrt{\frac{\sum\limits_{i=1}^{n} Y_i^2 - \left[\left(\sum\limits_{i=1}^{n} Y_i\right)^2 / n\right]}{n - 1}} \qquad (1.5)$$

rather than from (1.2) and (1.3), respectively.

Other measures of dispersion are the *range* (i.e., the difference between the largest and smallest observations) and the *interquartile range* (i.e., the difference between the third and first *quartiles*). See Berenson and Levine (1983), Chapter 4.

1.5 THE NORMAL DISTRIBUTION AND ITS ROLE IN STATISTICS

A quantitative random variable may be represented by a particular *mathematical model* or function pertaining to some phenomenon of interest. For discrete random variables the mathematical model is known as a *probability distribution function*. When such a mathematical model is available, the exact probability of occurrence of any particular outcome or value of the random variable can be computed, and the entire probability distribution can be enumerated.

On the other hand, for continuous random variables the mathematical expression representing some underlying phenomenon is known as a *probability density function*. When such a mathematical model is available, the probabilities that particular values of the random variable occur within certain ranges or intervals may be calculated. In this section the *normal probability density function* (the most important of the basic continuous mathematical models) will be reviewed. Its extensions to the multivariable case (i.e., the multivariate normal density function) will be considered in Chapter 18.

1.5.1 The Normal Probability Density Function

The normal distribution is vitally important for three main reasons. First, numerous continuous phenomena seem to follow the *bell-shaped* curve or could be approximated by it. Second, the normal distribution can be used to approximate various discrete probability distributions and thereby avoid much otherwise needed computational drudgery. Third, and perhaps most important, the normal distribution provides the basis for classical statistical inference because of its relationship to the *central limit theorem*.

Theoretically, the normal distribution has two interesting properties: It is bell-shaped and symmetrical so that the measures of central tendency (mean and median) are identical, and the observed outcomes of the continuous random variable of interest Y have an infinite range ($-\infty < Y < +\infty$). Despite the "theoretically" infinite range, the "practical" range for the observed outcomes of Y is obtained within approximately three standard deviations (distances) above and below the mean.

The mathematical expression representing the normal probability density function $f(Y)$ is

$$f(Y) = \frac{1}{\sqrt{2\pi}\sigma_Y} e^{(Y-\mu_Y)^2/2\sigma_Y^2} \tag{1.6}$$

where e = mathematical constant approximated by 2.71828
π = mathematical constant approximated by 3.14159
μ_Y = true mean
σ_Y = true standard deviation
Y = the continuous random variable, $-\infty < Y < +\infty$

Since e and π are mathematical constants, it is clear from (1.6) that the probabilities of occurrences within particular ranges of the random variable Y are dependent on the two parameters of the normal distribution, the mean μ_Y and the standard deviation σ_Y. That is, $Y \sim \mathcal{N}(\mu_Y, \sigma_Y)$. Every time a particular combination of μ_Y and σ_Y is specified, a different normal probability distribution may be generated.

1.5.2 Normally Distributed Random Variables

The form of the mathematical expression (1.6) used for generating a normal probability density function, and/or ranges of its particular probabilities, is computationally tedious. Fortunately, though, we may use the transformation formula

$$Z = \frac{Y - \mu_Y}{\sigma_Y} \tag{1.7}$$

to convert any normal random variable Y to a standardized normal random variable Z. This transformed variable Z has $\mu_Z = 0$ and $\sigma_Z = 1$, no matter what the values are for μ_Y and σ_Y. That is, $Z \sim \mathcal{N}(0,1)$ even though $Y \sim \mathcal{N}(\mu_Y, \sigma_Y)$. Thus we shall always be able to use the table of the standardized normal distribution (see Appendix B, Table B.1) to determine probabilities of interest. Such a table represents the probability or area under the standardized normal curve from the mean ($\mu_Z = 0$) to the transformed value of interest Z. As depicted in Figure 1.2, this of course corresponds to the probability or area under the normal curve from the *actual* mean μ_Y to the *actual* value of interest Y.

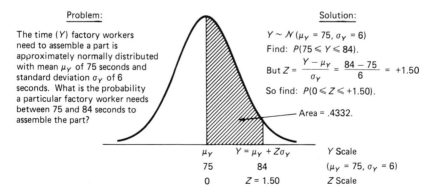

Problem:

The time (Y) factory workers need to assemble a part is approximately normally distributed with mean μ_Y of 75 seconds and standard deviation σ_Y of 6 seconds. What is the probability a particular factory worker needs between 75 and 84 seconds to assemble the part?

Solution:

$Y \sim \mathcal{N}(\mu_Y = 75, \sigma_Y = 6)$
Find: $P(75 \leqslant Y \leqslant 84)$.

But $Z = \dfrac{Y - \mu_Y}{\sigma_Y} = \dfrac{84 - 75}{6} = +1.50$

So find: $P(0 \leqslant Z \leqslant +1.50)$.

Area = .4332.

μ_Y	$Y = \mu_Y + Z\sigma_Y$	Y Scale
75	84	($\mu_Y = 75, \sigma_Y = 6$)
0	$Z = 1.50$	Z Scale

Figure 1.2   Determining areas by transformation of scales.

1.5.3 Approximating Various Discrete Probability Distributions

A second important application of the normal distribution occurs in the approximation of *exact* probabilities arising from various discrete probability models. As examples, when, under certain conditions (such as large sample size), it becomes too tedious to compute the exact probabilities arising from such common discrete probability distributions as binomial, hypergeometric, and Poisson, a normal approximation can be used [see Berenson and Levine (1983), Chapter 6] and much time and effort will be saved.

1.5.4 Relationship to Central Limit Theorem

A *sampling distribution of a statistic* is the probability distribution for that statistic. Theoretically, it can be obtained by drawing all possible samples of the same size n from some population and then grouping together the particular statistic computed from each one of these samples. What form should the sampling distribution of the statistic have? The importance of the normal distribution to classical inferential methods can largely be attributed to its relationship to the *central limit theorem*.

The theorem essentially states that, no matter what type of underlying population the random sample is drawn from (provided that it has a finite variance σ_Y^2), the statistic $\bar{Y}$ has a sampling distribution that is approximately normal with mean $\mu_{\bar{Y}} = \mu_Y$, the mean of the underlying population, and standard deviation $\sigma_{\bar{Y}} = \sigma_Y/\sqrt{n}$.

The sampling distribution of $\bar{Y}$ is exactly normal if the underlying population is normal, regardless of the sample size. In addition, if the underlying distribution does not differ markedly from normality, the distribution of $\bar{Y}$ will be approximately normal for fairly small sample sizes. Furthermore, even when the original population is far from normal, the sampling distribution of $\bar{Y}$ will be approximately normal in most instances if n (the sample size used to compute each $\bar{Y}$) is at least 30.

Therefore, when dealing with the statistic $\bar{Y}$, we have

$$Z = \frac{\bar{Y} - \mu_{\bar{Y}}}{\sigma_{\bar{Y}}} = \frac{\bar{Y} - \mu_{\bar{Y}}}{\sigma_Y/\sqrt{n}} \tag{1.8}$$

and since $Z \sim \mathcal{N}(0,1)$, Table B.1 in Appendix B may be used to determine the likelihood of obtaining values of $\bar{Y}$ which deviate from the true (or hypothesized) mean $\mu_{\bar{Y}}$ by particular amounts.

A detailed discussion of the development of the sampling distribution is presented in Berenson and Levine (1983, Chapter 7).

1.6 DERIVED SAMPLING DISTRIBUTIONS: χ^2, t, F

Another useful aspect of the normal probability density function is that certain derived sampling distributions—χ^2, t, and F—may easily be obtained. As will be demonstrated in Section 1.7, these distributions are particularly important to the development of classical inference procedures.

1.6.1 The χ^2 Distribution

Let $Z_i = (Y_i - \mu_Y)/\sigma_Y$ be the ith *independent* standardized normal random variable (where $i = 1, 2, \ldots, v$). The positive-valued random variable

$$\chi^2 = Z_1^2 + Z_2^2 + \cdots + Z_v^2 \tag{1.9}$$

is then distributed as *chi square* with v *degrees of freedom*. That is, $\chi^2 \sim \chi_v^2$.

As depicted in Figure 1.3, a χ^2 distribution is asymmetrical. It has mean v and variance $2v$. Table B.2 in Appendix B presents the percentiles $\chi_{1-\alpha;\,v}^2$ of χ^2 distributions for various values of v and $1 - \alpha$.

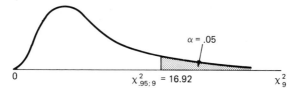

Figure 1.3 χ^2 distribution with nine degrees of freedom. (Source: Appendix B, Table B.2.)

Now suppose we take a random sample $(Y_1, Y_2, \ldots, Y_n)$ of size n from a normal distribution having mean μ_Y and variance σ_Y^2. Since for all $i = 1, 2, \ldots, n$, $Z_i = (Y_i - \mu_Y)/\sigma_Y$, from (1.9) it follows that

$$\chi^2 = \sum_{i=1}^{n} Z_i^2 = \frac{\sum_{i=1}^{n}(Y_i - \mu_Y)^2}{\sigma_Y^2} \sim \chi_n^2 \tag{1.10}$$

If, instead of $\sum_{i=1}^{n}(Y_i - \mu_Y)^2$, we take the sum of the squared differences between the observations and their *sample mean* [that is, $\sum_{i=1}^{n}(Y_i - \bar{Y})^2$], then from (1.2) and (1.10) it can be shown that*

$$\frac{\sum_{i=1}^{n}(Y_i - \bar{Y})^2}{\sigma_Y^2} = \frac{(n-1)S_Y^2}{\sigma_Y^2} \sim \chi_{n-1}^2 \tag{1.11}$$

*In (1.10) all n values of Y_i are independent of the population mean μ_Y. However, in (1.11) one "degree of freedom" has been "lost" because only $n - 1$ of the Y_i are independent of their own sample mean $\bar{Y}$.

The fact that the positive-valued random variable $(n - 1)S_Y^2/\sigma_Y^2$ has a χ^2 distribution with $n - 1$ degrees of freedom (when sampling from normal populations) is useful in estimation and testing (see Section 1.7).

1.6.2 The *t* Distribution

Let Z be a standardized normal random variable and let χ^2 be a chi-squared random variable with v degrees of freedom. If Z and χ^2 are *independent*, the random variable

$$t = \frac{Z}{\sqrt{\chi^2/v}} \tag{1.12}$$

is then distributed as t with v *degrees of freedom.* That is, $t \sim t_v$.

In practice, suppose we take a random sample $(Y_1, Y_2, \ldots, Y_n)$ of size n from a normal distribution (having mean μ_Y and variance σ_Y^2) and compute the statistics $\bar{Y}$ and S_Y^2. From (1.8) and (1.11), we may rewrite (1.12) as

$$t = \frac{\bar{Y} - \mu_Y}{\sigma_Y/\sqrt{n}} \div \sqrt{\frac{(n - 1)S_Y^2/\sigma_Y^2}{(n - 1)}} = \frac{\bar{Y} - \mu_Y}{S_Y/\sqrt{n}} \sim t_{n-1} \tag{1.13}$$

where the simplified expression is very useful in estimation and testing (see Section 1.7).

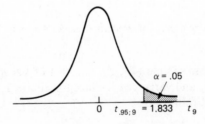

Figure 1.4 t distribution with nine degrees of freedom. (Source: Appendix B, Table B.3.)

As shown in Figure 1.4, a t distribution is symmetric with mean of zero. Table B.3 in Appendix B presents the percentiles $t_{1-\alpha;\,v}$ of t distributions for various values of v and $1 - \alpha$. From this table it is noted that as the number of degrees of freedom v increases, the t distributions gradually approach the normal distribution so that for very large samples the values are practically identical. That is, by letting $Z_{1-\alpha}$ represent the $1 - \alpha$ percentile point of the normal distribution,

$$t_{1-\alpha;\,n-1} \cong Z_{1-\alpha} \qquad \text{if } n \text{ is large} \tag{1.14}$$

since

$$\lim_{v \to \infty} t_{1-\alpha;\,v} \equiv Z_{1-\alpha} \tag{1.15}$$

1.6.3 The F Distribution

Let χ_1^2 be a *chi-squared* random variable with ν_1 degrees of freedom and let χ_2^2 be a *chi-squared* random variable with ν_2 degrees of freedom. If χ_1^2 and χ_2^2 are *independent*, the positive-valued random variable

$$F = \frac{\chi_1^2/\nu_1}{\chi_2^2/\nu_2} \tag{1.16}$$

is then distributed as F with ν_1 and ν_2 *degrees of freedom*. That is, $F \sim F_{\nu_1, \nu_2}$. As displayed in Figure 1.5, the F distribution is asymmetrical.

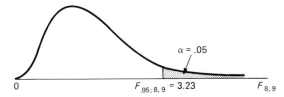

Figure 1.5 F distribution with $\nu_1 = 8$ and $\nu_2 = 9$ degrees of freedom. (Source: Appendix B, Table B.4.)

Table B.4 in Appendix B presents commonly selected percentiles $F_{1-\alpha; \nu_1, \nu_2}$ of F distributions for various combinations of ν_1 (*numerator* degrees of freedom), ν_2 (*denominator* degrees of freedom), and α (where $1 - \alpha = .95, .975,$ and $.99$). However, when $1 - \alpha < .50$, the desired *lower* percentile (i.e., *left-tail critical value*) is readily obtained from the reciprocal of the *complementary* percentile (i.e., *upper* or *right-tail critical value*) with degrees of freedom reversed. That is, for $1 - \alpha < .50$,

$$F_{1-\alpha; \nu_1, \nu_2} = \frac{1}{F_{\alpha; \nu_2, \nu_1}} \tag{1.17}$$

As an example, from Figure 1.5 and Appendix B, Table B.4,

$$F_{.05; 8, 9} = \frac{1}{F_{.95; 9, 8}} = \frac{1}{3.39} = .2950$$

In practice, the F statistic is both important and useful. Suppose, for example, we draw a random sample $(Y_{11}, Y_{21}, \ldots, Y_{n_1 1})$ of size n_1 from a normal population with mean μ_{Y_1} and variance $\sigma_{Y_1}^2$. In addition, suppose we take an *independent* random sample $(Y_{12}, Y_{22}, \ldots, Y_{n_2 2})$ of size n_2 from a normal population having mean μ_{Y_2} and variance $\sigma_{Y_2}^2$. We then compute the statistics $\bar{Y}_1$ and $S_{Y_1}^2$ from the first sample and $\bar{Y}_2$ and $S_{Y_2}^2$ from the second sample. Using (1.11), we may rewrite (1.16) as

$$F = \frac{(n_1 - 1)S_{Y_1}^2/\sigma_{Y_1}^2}{n_1 - 1} \div \frac{(n_2 - 1)S_{Y_2}^2/\sigma_{Y_2}^2}{n_2 - 1} = \frac{S_{Y_1}^2/\sigma_{Y_1}^2}{S_{Y_2}^2/\sigma_{Y_2}^2} \sim F_{n_1-1, \, n_2-1} \tag{1.18}$$

where, once again, the simplified expression is convenient for estimation and testing (Section 1.7).

1.6.4 Interrelationships Among Z, χ^2, t, and F

Throughout this section the relationships between the normal probability density function and the derived sampling distributions (χ^2, t, and F) have been expressed. In (1.9) and (1.10) χ^2 is given as a function of Z. Moreover, in (1.12) t is defined as a function of Z and χ^2. More specifically, the relationship between t and Z is stated in (1.15). In addition, F is presented as a function of χ^2 in (1.16).

Using Tables B.1–B.4 in Appendix B, we may express the following interrelationship among Z, χ^2, t, and F:

$$Z^2_{1-\frac{\alpha}{2}} = \chi^2_{1-\alpha;\,1} = (t_{1-\frac{\alpha}{2};\,\infty})^2 = F_{1-\alpha;\,1,\infty} \qquad (1.19)$$

As an example, if $\alpha = .05$, then

$$Z^2_{.975} = \chi^2_{.95;\,1} = (t_{.975;\,\infty})^2 = F_{.95;\,1,\infty} = 3.84$$

Other interesting relationships which may be verified from the tables are

$$F_{1-\alpha;\,1,\nu} = (t_{1-\frac{\alpha}{2};\,\nu})^2 \qquad \text{for all } \nu \qquad (1.20)$$

$$F_{1-\alpha;\,\nu,\infty} = \frac{\chi^2_{1-\alpha;\,\nu}}{\nu} \qquad \text{for all } \nu \qquad (1.21)$$

1.7 CLASSICAL INFERENCE: ESTIMATION AND TESTING

Inferential statistics involves the drawing of conclusions regarding population characteristics (i.e., *parameters*) based on information (i.e., *statistics*) obtained from a sample. Inference deals with both *estimating* population parameters and *testing* hypothesized values of the population parameters.

1.7.1 Estimation

Suppose a random sample $(Y_1, Y_2, \ldots, Y_n)$ of size n is drawn from a particular population having unknown parameter θ and some statistic $\hat{\theta}$ is computed from the sample. The numerical value that the statistic $\hat{\theta}$ possesses is then used as an estimate of the unknown parameter θ.

Point estimation.   The statistic $\hat{\theta}$ is a random variable since it would differ from sample to sample—depending on the observations Y_i (where $i = 1$, $2, \ldots, n$) in each sample. The random variable $\hat{\theta}$ is called a *point estimator* of θ, and the numerical value that the statistic possesses is considered a *point estimate* of that unknown parameter.

Properties of estimators.   Since different types of statistics can be used as estimators of the unknown parameter θ, what criteria should be considered in selecting from among the possible statistics? Classically, a "good estimator" is said to possess four properties: *unbiasedness, efficiency, consistency,* and *sufficiency.*

The statistic $\hat{\theta}$ is an *unbiased* estimator of the parameter θ if, over all possible samples of the same size n, the average (i.e., *expected*) value of the statistic $\hat{\theta}$ is equal to the value of the parameter θ.

The statistic $\hat{\theta}$ is an *efficient* (minimum variance) estimator of the parameter θ if, over all possible samples of the same size n, the variability in the sampling distribution of the statistic $\hat{\theta}$ is less than that of other estimators $\hat{\theta}'$ (for all $\hat{\theta}'$). That is, $\hat{\theta}$ is an efficient statistic provided that $\sigma_{\hat{\theta}}^2 < \sigma_{\hat{\theta}'}^2$ (for all $\hat{\theta}'$).

The statistic $\hat{\theta}$ is a *consistent* estimator of the parameter θ if, as the sample size n increases, the likelihood that $\hat{\theta}$ is "closer" to the true θ also increases.

The statistic $\hat{\theta}$ is a *sufficient* estimator of the parameter θ if, based on the sample of size n, it contains all the information in that sample.

One familiar statistic which possesses these properties is the sample mean $\bar{Y}$. That is, $\bar{Y}$ is an unbiased, efficient, consistent, and sufficient estimator of the parameter μ_Y, the true population mean.

Confidence interval estimation.   Since the *point estimate* rarely equals the true value of the parameter, the researcher constructs a *confidence interval estimate* that contains lower and upper limits ($\hat{\theta}_L$ and $\hat{\theta}_U$) such that the probability the obtained interval covers the parameter θ is given by $1 - \alpha$. We may also note that $1 - \alpha$ is called the *confidence coefficient,* and the interval between the two statistics $\hat{\theta}_L$ and $\hat{\theta}_U$ is known as the $100(1 - \alpha)\%$ *confidence interval.*

Like $\hat{\theta}$, the estimators $\hat{\theta}_L$ and $\hat{\theta}_U$ are random variables whose numerical values differ from sample to sample. Hence, if all possible samples of the same size n were drawn from a particular population and estimates $\hat{\theta}_L$ and $\hat{\theta}_U$ computed for each sample, $100(1 - \alpha)\%$ of all such *confidence intervals* would be expected to include the true parameter θ, while $100(\alpha)\%$ of them would not.

In practice, however, a sample of n observations is drawn only once. The statistics $\hat{\theta}_L$ and $\hat{\theta}_U$ are calculated, and the confidence interval is obtained. Either the true parameter θ is included within the particular interval or it is not. However, we can state that we have a specified degree of confidence [i.e., $100(1 - \alpha)\%$] that the particular interval covers θ.

A summary of useful confidence interval formulas is presented in Table 1.2.

1.7.2 Hypothesis Testing

Whereas confidence interval estimation involves the estimation of unknown parameter values, hypothesis testing is concerned with the testing of certain specified (i.e., *hypothesized*) values for those population parameters. A *null hypothesis,* H_0 (sometimes called the hypothesis of "no differences" or "status

TABLE 1.2 Summary of Useful Confidence Interval Formulas

Parameters Being Estimated	Conditions	Point Estimates	Confidence Interval Estimates
A. μ_Y, mean of a normal distribution	σ_Y is known; n is large (i.e., > 30) if the distribution is not $\mathcal{N}(\mu_Y, \sigma_Y)$	$\bar{Y}$	$\bar{Y} - Z_{1-\frac{\alpha}{2}}\dfrac{\sigma_Y}{\sqrt{n}} \leq \mu_Y \leq \bar{Y} + Z_{1-\frac{\alpha}{2}}\dfrac{\sigma_Y}{\sqrt{n}}$
B. μ_Y, mean of a normal distribution	σ_Y is unknown	$\bar{Y}$	$\bar{Y} - t_{1-\frac{\alpha}{2};\,n-1}\dfrac{S_Y}{\sqrt{n}} \leq \mu_Y \leq \bar{Y} + t_{1-\frac{\alpha}{2};\,n-1}\dfrac{S_Y}{\sqrt{n}}$
C. σ_Y^2, variance of a normal distribution		S_Y^2	$\dfrac{(n-1)S_Y^2}{\chi^2_{1-\frac{\alpha}{2};\,n-1}} \leq \sigma_Y^2 \leq \dfrac{(n-1)S_Y^2}{\chi^2_{\frac{\alpha}{2};\,n-1}}$
D. $\mu_{Y_1} - \mu_{Y_2}$, difference between the means of two *independent* normal distributions	σ_{Y_1} and σ_{Y_2} are known; $n_1 + n_2 > 30$ if the distributions are not normal	$\bar{Y}_1 - \bar{Y}_2$	$(\bar{Y}_1 - \bar{Y}_2) - Z_{1-\frac{\alpha}{2}}\sqrt{\dfrac{\sigma_{Y_1}^2}{n_1} + \dfrac{\sigma_{Y_2}^2}{n_2}} \leq (\mu_{Y_1} - \mu_{Y_2})$ $\leq (\bar{Y}_1 - \bar{Y}_2) + Z_{1-\frac{\alpha}{2}}\sqrt{\dfrac{\sigma_{Y_1}^2}{n_1} + \dfrac{\sigma_{Y_2}^2}{n_2}}$
E. $\mu_{Y_1} - \mu_{Y_2}$, difference between the means of two *independent* normal distributions	σ_{Y_1} and σ_{Y_2} are unknown but assumed equal	$\bar{Y}_1 - \bar{Y}_2$	$(\bar{Y}_1 - \bar{Y}_2) - t_{1-\frac{\alpha}{2};\,n_1+n_2-2}S_P\sqrt{\dfrac{1}{n_1} + \dfrac{1}{n_2}} \leq (\mu_{Y_1} - \mu_{Y_2})$ $\leq (\bar{Y}_1 - \bar{Y}_2) + t_{1-\frac{\alpha}{2};\,n_1+n_2-2}S_P\sqrt{\dfrac{1}{n_1} + \dfrac{1}{n_2}}$ where $S_P = \sqrt{\dfrac{(n_1-1)S_{Y_1}^2 + (n_2-1)S_{Y_2}^2}{n_1+n_2-2}}$

Parameter	Point estimator	Confidence interval
F. $\mu_{Y_1} - \mu_{Y_2}$, difference between the means of two *independent* normal distributions	$\bar{Y}_1 - \bar{Y}_2$	$(\bar{Y}_1 - \bar{Y}_2) - t_{1-\frac{\alpha}{2}; \nu'} \sqrt{\dfrac{S_{\bar{Y}_1}^2}{n_1} + \dfrac{S_{\bar{Y}_2}^2}{n_2}} \leq (\mu_{Y_1} - \mu_{Y_2})$ $\leq (\bar{Y}_1 - \bar{Y}_2) + t_{1-\frac{\alpha}{2}; \nu'} \sqrt{\dfrac{S_{\bar{Y}_1}^2}{n_1} + \dfrac{S_{\bar{Y}_2}^2}{n_2}}$ where $t_{1-\frac{\alpha}{2}; \nu'} \cong \dfrac{t_{1-\frac{\alpha}{2}; n_1-1}(S_{\bar{Y}_1}^2/n_1) + t_{1-\frac{\alpha}{2}; n_2-1}(S_{\bar{Y}_2}^2/n_2)}{(S_{\bar{Y}_1}^2/n_1) + (S_{\bar{Y}_2}^2/n_2)}$
σ_{Y_1} and σ_{Y_2} are unknown and not assumed to be equal (Behrens-Fisher problem)		
G. $\sigma_{Y_1}^2/\sigma_{Y_2}^2$, ratio of variances of two *independent* normal distributions	$\dfrac{S_{Y_1}^2}{S_{Y_2}^2}$	$\dfrac{S_{Y_1}^2}{S_{Y_2}^2 F_{1-\frac{\alpha}{2};\, n_1-1, n_2-1}} \leq \dfrac{\sigma_{Y_1}^2}{\sigma_{Y_2}^2} \leq \dfrac{S_{Y_1}^2 F_{1-\frac{\alpha}{2};\, n_2-1, n_1-1}}{S_{Y_2}^2}$
H. μ_D, difference between the means of two *related* normal distributions	$\bar{D}$	$\bar{D} - Z_{1-\frac{\alpha}{2}} \dfrac{\sigma_D}{\sqrt{n}} \leq \mu_D \leq \bar{D} + Z_{1-\frac{\alpha}{2}} \dfrac{\sigma_D}{\sqrt{n}}$ where D_j are considered a sample of *difference scores* $(D_1, D_2, \ldots, D_n)$ between related (i.e., *matched* or *paired*) observations $(Y_{1j} - Y_{2j})$
σ_D is known		
I. μ_D, difference between the means of two *related* normal distributions	$\bar{D}$	$\bar{D} - t_{1-\frac{\alpha}{2};\, n-1} \dfrac{S_D}{\sqrt{n}} \leq \mu_D \leq \bar{D} + t_{1-\frac{\alpha}{2};\, n-1} \dfrac{S_D}{\sqrt{n}}$
σ_D is unknown		

15

quo"), is tested against its complement, the *alternative* hypothesis, H_1 (sometimes called the "research" hypothesis of interest). Hypotheses are usually set up to determine if the data support a belief as specified by H_1.

Traditional hypothesis testing procedure.   Once H_0 and H_1 are stated, a *level of significance* α is selected so that it will be possible to assess the significance of the differences between the hypothesized parameter value and the corresponding statistic to be obtained from sample information. Since the level of significance represents the risk to the researcher of rejecting a true null hypothesis, α levels chosen either as .05 or .01 have been traditionally considered adequate for most testing purposes.

To decide whether or not there is a close agreement between the hypothesized value of the parameter θ_0 and the computed value of the statistic $\hat{\theta}$, the researcher compares the observed statistic against the (theoretical) sampling distribution of the statistic that would be attained had all possible samples of the same size been taken from a population having the parameter stated in the null hypothesis.

If there is close agreement, the likelihood of obtaining such a statistic as the one observed, or, for that matter, one even more extreme, will exceed α. In such cases the null hypothesis would not be rejected. The researcher would conclude that no *real* differences (between $\hat{\theta}$ and θ_0) exist and that the observed differences between the sample statistic and the parameter which it is estimating are attributable only to chance.

On the other hand, if the agreement between $\hat{\theta}$ and θ_0 is poor, the probability of obtaining such a result $(\hat{\theta})$ or one even more extreme (under the null hypothesis) will be very small (i.e., $\leq \alpha$). In such cases the null hypothesis would be rejected, and the researcher would conclude that the observed differences are *significant* at the α level (and cannot be attributed only to chance).

Hypotheses tests may be *two-tailed* or *one-tailed*, depending on the alternative H_1 of interest to the researcher. In a two-tailed test the researcher is merely looking for differences between the statistic and the hypothesized parameter and would reject H_0 if $\hat{\theta}$ were deemed significantly smaller or significantly larger than θ_0. However, in a one-tailed test the researcher specifies the particular direction of interest and would reject H_0 only if the observed differences were deemed significant in that corresponding direction (particular *decision rules* are presented in Table 1.3).

In making such decisions regarding the rejection or nonrejection of H_0, the appropriate test statistic (see Table 1.3) would be compared against the *critical* values for the particular sampling distribution of interest—Z, χ^2, t, or F (whose percentile points are given in Appendix B, Table B.1–B.4).

_P_-Value approach to hypothesis testing.   In recent years an approach to hypothesis testing which has increasingly gained acceptance deals with the concept of P value. The P value is often referred to as the *observed* *level of significance*—the smallest level at which H_0 can be rejected for a given set of data. We should realize that all the well-known computer packages now present the P values as an aid to decision making (i.e., drawing conclusions about θ). When read from computer printouts, the P values do not require the additional use of percentile points of such distributions as Z, χ^2, t, and F (see Appendix B, Table B.1–B.4) which are necessary for decision making when using traditional testing procedures.

In deciding whether or not there is close agreement between $\hat{\theta}$ and θ_0, the researcher determines the likelihood of obtaining such a statistic as the one observed or one even more extreme if the null hypothesis were true. If this probability is high, the null hypothesis is not rejected; if this likelihood is very small (traditionalists would say $\leq .05$), the null hypothesis is rejected.

Definition.   The P value is the probability of obtaining a test statistic equal to or more extreme than the result observed—given H_0 is true.

A summary of useful hypothesis testing formulas is presented in Table 1.3 along with corresponding decision rules. A case study is given in Section 1.7.3 which will enable the reader to demonstrate his/her understanding of the application of classical statistical inference.

1.7.3 Applications of Classical Inference: A Case Study

A financial analyst wishes to compare the earnings per share of two types of nonindustrial companies—commercial banks versus retailers—for potential investment purposes. Recorded in Table 1.4 are the earnings per share (for the years 1978 and 1979) for random samples of 9 commercial banking companies and 10 retailing companies. The summary computations and statistics for these data are displayed in Table 1.5.

Assuming each set of data is approximately normally distributed, suppose the financial analyst wanted to draw conclusions about

1. The population mean earnings per share for all commercial banks in 1979
2. The population variance (or standard deviation) in earnings per share for all commercial banks in 1979
3. The difference (i.e., *ratio*) in the population variances in earnings per share between commercial banks and retailers during 1979
4. The difference in the population mean earnings per share between commercial banks and retailers during 1979
5. The difference in the population mean earnings per share from 1978 to 1979 for all commercial banks
6. Whether or not there is evidence that the population mean earnings per share for all retailers in 1979 was *less than* $4.00

TABLE 1.3 Summary of Useful

Parameters Being Tested	Conditions	H_0	H_1
A. μ_Y, mean of a normal distribution	σ_Y is known; n is large (i.e., > 30) if the distribution is not $\mathcal{N}(\mu_Y, \sigma_Y)$	$\mu_Y = \mu_0$	$\mu_Y \neq \mu_0$
		$\mu_Y \leq \mu_0$	$\mu_Y > \mu_0$
		$\mu_Y \geq \mu_0$	$\mu_Y < \mu_0$
B. μ_Y, mean of a normal distribution	σ_Y is unknown	$\mu_Y = \mu_0$	$\mu_Y \neq \mu_0$
		$\mu_Y \leq \mu_0$	$\mu_Y > \mu_0$
		$\mu_Y \geq \mu_0$	$\mu_Y < \mu_0$
C. σ_Y^2, variance of a normal distribution		$\sigma_Y^2 = \sigma_0^2$	$\sigma_Y^2 \neq \sigma_0^2$
		$\sigma_Y^2 \leq \sigma_0^2$	$\sigma_Y^2 > \sigma_0^2$
		$\sigma_Y^2 \geq \sigma_0^2$	$\sigma_Y^2 < \sigma_0^2$
D. $\mu_{Y_1} - \mu_{Y_2}$, difference between the means of two *independent* normal distributions	σ_{Y_1} and σ_{Y_2} are known; $n_1 + n_2 > 30$ if the distributions are not normal	$\mu_{Y_1} = \mu_{Y_2}$	$\mu_{Y_1} \neq \mu_{Y_2}$
		$\mu_{Y_1} \leq \mu_{Y_2}$	$\mu_{Y_1} > \mu_{Y_2}$
		$\mu_{Y_1} \geq \mu_{Y_2}$	$\mu_{Y_1} < \mu_{Y_2}$
E. $\mu_{Y_1} - \mu_{Y_2}$, difference between the means of two *independent* normal distributions	σ_{Y_1} and σ_{Y_2} are unknown but assumed equal	$\mu_{Y_1} = \mu_{Y_2}$	$\mu_{Y_1} \neq \mu_{Y_2}$
		$\mu_{Y_1} \leq \mu_{Y_2}$	$\mu_{Y_1} > \mu_{Y_2}$
		$\mu_{Y_1} \geq \mu_{Y_2}$	$\mu_{Y_1} < \mu_{Y_2}$
F. $\mu_{Y_1} - \mu_{Y_2}$, difference between the means of two *independent* normal distributions	σ_{Y_1} and σ_{Y_2} are unknown and not assumed to be equal (Behrens-Fisher problem)	$\mu_{Y_1} = \mu_{Y_2}$	$\mu_{Y_1} \neq \mu_{Y_2}$
		$\mu_{Y_1} \leq \mu_{Y_2}$	$\mu_{Y_1} > \mu_{Y_2}$
		$\mu_{Y_1} \geq \mu_{Y_2}$	$\mu_{Y_1} < \mu_{Y_2}$
G. $\sigma_{Y_1}^2/\sigma_{Y_2}^2$, ratio of variances of two *independent* normal distributions		$\sigma_{Y_1}^2 = \sigma_{Y_2}^2$	$\sigma_{Y_1}^2 \neq \sigma_{Y_2}^2$
		$\sigma_{Y_1}^2 \leq \sigma_{Y_2}^2$	$\sigma_{Y_1}^2 > \sigma_{Y_2}^2$
		$\sigma_{Y_1}^2 \geq \sigma_{Y_2}^2$	$\sigma_{Y_1}^2 < \sigma_{Y_2}^2$
H. μ_D, difference between the means of two *related* normal distributions	σ_D is known; n is large (i.e., > 30) if the distribution of differences is not $\mathcal{N}(\mu_D, \sigma_D)$	$\mu_D = 0$	$\mu_D \neq 0$
		$\mu_D \leq 0$	$\mu_D > 0$
		$\mu_D \geq 0$	$\mu_D < 0$
I. μ_D, difference between the means of two *related* normal distributions	σ_D is unknown	$\mu_D = 0$	$\mu_D \neq 0$
		$\mu_D \leq 0$	$\mu_D > 0$
		$\mu_D \geq 0$	$\mu_D < 0$

Hypotheses Testing Formulas

Test Statistic Formulas	Decision Rules
$Z = \dfrac{\bar{Y} - \mu_0}{\sigma_Y/\sqrt{n}}$	Reject H_0 if $Z \geq Z_{1-\frac{\alpha}{2}}$ or if $Z \leq Z_{\frac{\alpha}{2}}$
"	Reject H_0 if $Z \geq Z_{1-\alpha}$
"	Reject H_0 if $Z \leq Z_\alpha$
$t = \dfrac{\bar{Y} - \mu_0}{S_Y/\sqrt{n}}$	Reject H_0 if $t \geq t_{1-\frac{\alpha}{2};\, n-1}$ or if $t \leq t_{\frac{\alpha}{2};\, n-1}$
"	Reject H_0 if $t \geq t_{1-\alpha;\, n-1}$
"	Reject H_0 if $t \leq t_{\alpha;\, n-1}$
$\chi^2 = \dfrac{(n-1)S_Y^2}{\sigma_0^2}$	Reject H_0 if $\chi^2 \geq \chi^2_{1-\frac{\alpha}{2};\, n-1}$ or if $\chi^2 \leq \chi^2_{\frac{\alpha}{2};\, n-1}$
"	Reject H_0 if $\chi^2 \geq \chi^2_{1-\alpha;\, n-1}$
"	Reject H_0 if $\chi^2 \leq \chi^2_{\alpha;\, n-1}$
$Z = \dfrac{\bar{Y}_1 - \bar{Y}_2}{\sqrt{(\sigma_{Y_1}^2/n_1) + (\sigma_{Y_2}^2/n_2)}}$	Reject H_0 if $Z \geq Z_{1-\frac{\alpha}{2}}$ or if $Z \leq Z_{\frac{\alpha}{2}}$
"	Reject H_0 if $Z \geq Z_{1-\alpha}$
"	Reject H_0 if $Z \leq Z_\alpha$
$t = \dfrac{\bar{Y}_1 - \bar{Y}_2}{S_P\sqrt{(1/n_1) + (1/n_2)}}$ where $S_P = \sqrt{\dfrac{(n_1 - 1)S_{Y_1}^2 + (n_2 - 1)S_{Y_2}^2}{n_1 + n_2 - 2}}$	Reject H_0 if $t \geq t_{1-\frac{\alpha}{2};\, n_1+n_2-2}$ or if $t \leq t_{\frac{\alpha}{2};\, n_1+n_2-2}$
"	Reject H_0 if $t \geq t_{1-\alpha;\, n_1+n_2-2}$
"	Reject H_0 if $t \leq t_{\alpha;\, n_1+n_2-2}$
$t = \dfrac{\bar{Y}_1 - \bar{Y}_2}{\sqrt{(S_{Y_1}^2/n_1) + (S_{Y_2}^2/n_2)}}$	Reject H_0 if $t \geq t_{1-\frac{\alpha}{2};\, v'}$ or if $t \leq t_{\frac{\alpha}{2};\, v'}$ where $t_{1-\frac{\alpha}{2};\, v'} \cong \dfrac{t_{1-\frac{\alpha}{2};\, n_1-1}(S_{Y_1}^2/n_1) + t_{1-\frac{\alpha}{2};\, n_2-1}(S_{Y_2}^2/n_2)}{(S_{Y_1}^2/n_1) + (S_{Y_2}^2/n_2)}$
"	Reject H_0 if $t \geq t_{1-\alpha;\, v'}$
"	Reject H_0 if $t \leq t_{\alpha;\, v'}$
$F = \dfrac{S_{Y_1}^2}{S_{Y_2}^2}$	Reject H_0 if $F \geq F_{1-\frac{\alpha}{2};\, n_1-1, n_2-1}$ or if $F \leq \dfrac{1}{F_{1-\frac{\alpha}{2};\, n_2-1, n_1-1}}$
"	Reject H_0 if $F \geq F_{1-\alpha;\, n_1-1, n_2-1}$
"	Reject H_0 if $F \leq \dfrac{1}{F_{1-\alpha;\, n_2-1, n_1-1}}$
$Z = \dfrac{\bar{D}}{\sigma_D/\sqrt{n}}$ where D_j are considered a sample of *difference scores* $(D_1, D_2, \ldots, D_n)$ between related (i.e., *matched* or *paired*) observations $(Y_{1j} - Y_{2j})$	Reject H_0 if $Z \geq Z_{1-\frac{\alpha}{2}}$ or if $Z \leq Z_{\frac{\alpha}{2}}$
"	Reject H_0 if $Z \geq Z_{1-\alpha}$
"	Reject H_0 if $Z \leq Z_\alpha$
$t = \dfrac{\bar{D}}{S_D/\sqrt{n}}$	Reject H_0 if $t \geq t_{1-\frac{\alpha}{2};\, n-1}$ or if $t \leq t_{\frac{\alpha}{2};\, n-1}$
"	Reject H_0 if $t \geq t_{1-\alpha;\, n-1}$
"	Reject H_0 if $t \leq t_{\alpha;\, n-1}$

TABLE 1.4 Comparison of Earnings Per Share (in dollars)

Commercial Banks				Retailers			
	Year				Year		
Company	1979	1978	Difference	Company	1979	1978	Difference
Manufacturers Hanover Corp.	6.42	5.59	+.83	J. C. Penney	3.52	4.12	−.60
First Chicago Corp.	2.83	3.29	−.46	Federated Department Stores	4.21	4.11	+.10
Crocker National Corp.	8.94	5.66	+3.28	American Stores	4.36	3.14	+1.22
First Bank System	6.80	6.01	+.79	Carter Hawley Hale Stores	2.67	2.52	+.15
Texas Commerce Bancshares	5.70	4.71	+.99	Wickes	3.49	3.14	+.35
Harris Bankcorp.	4.65	4.09	+.56	McDonald's	4.68	4.00	+.68
AmeriTrust Corp.	6.20	5.68	+.52	Zayre	3.30	2.71	+.59
Wachovia Corp.	2.71	2.25	+.46	Wal-Mart Stores	2.68	1.93	+.75
Republic New York Corp.	8.34	7.06	+1.28	Mercantile Stores	7.25	6.80	+.45
				Borman's	.16	2.00	−1.84

SOURCE: Peter D. Petre, "The Fortune Directory of the Largest Non-Industrial Companies," *Fortune* (July 14, 1980), pp. 146–159. (c) 1980 Time Inc. All rights reserved.

TABLE 1.5 Summary of Results from the Samples

Results	Commercial Banking Companies		
	1979	1978	Differences
Sample Size	9	9	9
Summed Observations	52.59	44.34	+8.25
Mean	5.843	4.927	+.917
Sum of Squared Observations	344.8411	236.3086	15.6971
Square of the Summed Observations	2765.7081	1966.0356	68.0625
Variance	4.6925	2.2325	1.0168
Standard Deviation	2.166	1.494	1.008

Results	Retailing Companies		
	1979	1978	Differences
Sample Size	10	10	10
Summed Observations	36.32	34.47	+1.85
Mean	3.632	3.447	+.185
Sum of Squared Observations	160.9960	137.2451	6.9645
Square of the Summed Observations	1319.1424	1188.1809	3.4225
Variance	3.2313	2.0474	.7358
Standard Deviation	1.798	1.431	.858

7. Whether or not there is evidence that the population standard deviation in earnings per share for all retailers in 1979 was *greater than* $1.45

8. Whether or not there is evidence of a *difference* in the variance in earnings per share between commercial banks and retailers during 1979

9. Whether or not there is evidence that mean earnings per share was *significantly higher* among all commercial banking companies than among all retailing companies in 1979

10. Whether or not there is evidence that an *increase* in mean earnings per share occurred from 1978 to 1979 among retailing companies

1.7.4 Results

Cases 1–5 involve estimating unknown parameters, while cases 6–10 deal with testing various hypotheses. In all situations traditional .95 *confidence coefficients* and .05 *levels of significance* are used where appropriate.

Case 1: From Table 1.2 (formula B), a 95% confidence interval estimate of the mean is given by

$$\bar{Y} - t_{.975;\,8} \frac{S_Y}{\sqrt{n}} \leq \mu_Y \leq \bar{Y} + t_{.975;\,8} \frac{S_Y}{\sqrt{n}}$$

$$5.843 - 2.306 \frac{2.166}{\sqrt{9}} \leq \mu_Y \leq 5.843 + 2.306 \frac{2.166}{\sqrt{9}}$$

$$4.178 \leq \mu_Y \leq 7.508$$

Case 2: From Table 1.2 (formula C), 95% confidence interval estimates of the variance and standard deviation are given by

$$\frac{(n-1)S_Y^2}{\chi_{.975;\,8}^2} \leq \sigma_Y^2 \leq \frac{(n-1)S_Y^2}{\chi_{.025;\,8}^2}$$

$$\frac{8(4.6925)}{17.53} \leq \sigma_Y^2 \leq \frac{8(4.6925)}{2.18}$$

$$2.1415 \leq \sigma_Y^2 \leq 17.2203$$

and

$$1.463 \leq \sigma_Y \leq 4.150$$

Case 3: From Table 1.2 (formula G), a 95% confidence interval estimate of the ratio in the population variances is obtained as

$$\frac{S_{Y_1}^2}{S_{Y_2}^2 F_{.975;\,8,9}} \leq \frac{\sigma_{Y_1}^2}{\sigma_{Y_2}^2} \leq \frac{S_{Y_1}^2 F_{.975;\,9,8}}{S_{Y_2}^2}$$

$$\frac{4.6925}{3.2313(4.10)} \leq \frac{\sigma_{Y_1}^2}{\sigma_{Y_2}^2} \leq \frac{4.6925(4.36)}{3.2313}$$

$$.3542 \leq \frac{\sigma_{Y_1}^2}{\sigma_{Y_2}^2} \leq 6.3316$$

Since the interval includes the number 1, it may be concluded with 95% confidence that there are no *real* differences in the two population variances.

Case 4: From the estimate in case 3 there is reason to assume *equality of variance* between the two populations. Hence, by using Table 1.2 (formula E), a 95% confidence interval estimate of the difference in the population means is

$$(\bar{Y}_1 - \bar{Y}_2) - t_{.975;\,17} S_P \sqrt{\frac{1}{n_1} + \frac{1}{n_2}} \leq \mu_{Y_1} - \mu_{Y_2}$$

$$\leq (\bar{Y}_1 - \bar{Y}_2) + t_{.975;\,17} S_P \sqrt{\frac{1}{n_1} + \frac{1}{n_2}}$$

$$2.211 - 2.110(1.980)\sqrt{\tfrac{1}{9} + \tfrac{1}{10}} \leq \mu_{Y_1} - \mu_{Y_2}$$

$$\leq 2.211 + 2.110(1.980)\sqrt{\tfrac{1}{9} + \tfrac{1}{10}}$$

$$.291 \leq \mu_{Y_1} - \mu_{Y_2} \leq 4.131$$

On the other hand, for situations in which the researcher cannot (or is not willing to) assume $\sigma_{Y_1}^2 = \sigma_{Y_2}^2$ (i.e., *homoscedasticity*), formula F may be used to give (approximate) 95% confidence interval estimates for $\mu_{Y_1} - \mu_{Y_2}$. Such situations in classical inference are known as the *Behrens-Fisher problem*, and over the years many researchers have devised formulas for dealing with it. Formula F utilizes an approximation developed by Cochran (1964). For illustrative purposes, using formula F an *approximate* 95% confidence interval estimate of the difference in the population means is given by

$$(\bar{Y}_1 - \bar{Y}_2) - t_{.975;\,v'}\sqrt{\frac{S_{Y_1}^2}{n_1} + \frac{S_{Y_2}^2}{n_2}} \le \mu_{Y_1} - \mu_{Y_2} \le (\bar{Y}_1 - \bar{Y}_2) + t_{.975;\,v'}\sqrt{\frac{S_{Y_1}^2}{n_1} + \frac{S_{Y_2}^2}{n_2}}$$

where

$$t_{.975;\,v'} \cong \frac{2.306(4.6925/9) + 2.262(3.2313/10)}{(4.6925/9) + (3.2313/10)} = 2.289$$

so that

$$.107 \le \mu_{Y_1} - \mu_{Y_2} \le 4.315$$

Case 5: From Table 1.2 (formula I), a 95% confidence interval estimate of the true mean difference over the years 1978–1979 is given by

$$\bar{D} - t_{.975;\,8}\frac{S_D}{\sqrt{n}} \le \mu_D \le \bar{D} + t_{.975;\,8}\frac{S_D}{\sqrt{n}}$$

$$.917 - 2.306\frac{1.008}{\sqrt{9}} \le \mu_D \le .917 + 2.306\frac{1.008}{\sqrt{9}}$$

$$.142 \le \mu_D \le 1.692$$

Case 6: From Table 1.3 (formula B),

$$H_0:\quad \mu_Y \ge 4.00$$

$$H_1:\quad \mu_Y < 4.00 \qquad \text{(one-tailed test)}$$

$$t = \frac{\bar{Y} - \mu_0}{S_Y/\sqrt{n}} = \frac{3.632 - 4.00}{1.798/\sqrt{10}} = -.647$$

Since t is not equal to or smaller than $t_{.05;\,9} = -1.833$, do not reject H_0. In using an .05 level of significance, there is insufficient evidence to reject the assertion that the mean was *at least* $4.00. The observed difference between $\bar{Y}$ and μ_0 is merely due to chance. In using the *P*-value approach, the probability that $t \le -.647$ given that the null hypothesis is true is slightly greater than .25 (see Appendix B, Table B.3). Such a result is too large to be deemed significant.

Case 7: From Table 1.3 (formula C),

$$H_0:\quad \sigma_Y \le 1.45$$

$$H_1:\quad \sigma_Y > 1.45 \qquad \text{(one-tailed test)}$$

$$\chi^2 = \frac{(n-1)S_Y^2}{\sigma_0^2} = \frac{9(3.2313)}{(1.45)^2} = 13.832$$

Since χ^2 is not equal to or greater than $\chi^2_{.95;\,9} = 16.919$, do not reject H_0. At the .05 level of significance, there is no evidence that the standard deviation was *greater than* $1.45. The observed differences are due to chance. In using the *P*-value approach, the probability that $\chi^2 \ge 13.832$ given that H_0 is true is between .10 and .20 (see the critical values of χ^2, Appendix B, Table B.2). Again, such a result is too large to be considered significant, and H_0 is not rejected.

Case 8: From Table 1.3 (formula G),*

$$H_0: \quad \sigma_{Y_1}^2 = \sigma_{Y_2}^2 \quad \text{or} \quad \frac{\sigma_{Y_1}^2}{\sigma_{Y_2}^2} = 1$$

$$H_1: \quad \sigma_{Y_1}^2 \neq \sigma_{Y_2}^2 \quad \text{or} \quad \frac{\sigma_{Y_1}^2}{\sigma_{Y_2}^2} \neq 1 \qquad \text{(two-tailed test)}$$

$$F = \frac{S_{Y_1}^2}{S_{Y_2}^2} = \frac{4.6925}{3.2313} = 1.452$$

Since F does not equal or exceed $F_{.975;\, 8,9} = 4.10$ nor is F equal to or smaller than $1/F_{.975;\, 9,8} = .229$, do not reject H_0. At the .05 level of significance, there is no evidence of a difference in the two population variances. In using the P-value approach, the probability of such a result as 1.452 or one even more extreme (given that H_0 is true) exceeds .50. Obviously, such a result is not significant, and H_0 is not rejected. It should be recalled that the same conclusions reached here were also made in case 3. Thus it is seen that the findings from the two-tailed hypothesis test are identical to the information obtained based on confidence interval estimation. This is true, in general, and highlights the relationship between estimation and testing procedures.

Case 9: Based on the results of case 3 or 8, there is reason to believe that the homoscedasticity assumption (i.e., $\sigma_{Y_1}^2 = \sigma_{Y_2}^2$) holds, and hence Table 1.3 (formula E) is used to test for differences in the population means. Therefore,

$$H_0: \quad \mu_{Y_1} \leq \mu_{Y_2} \quad \text{or} \quad \mu_{Y_1} - \mu_{Y_2} \leq 0$$

$$H_1: \quad \mu_{Y_1} > \mu_{Y_2} \quad \text{or} \quad \mu_{Y_1} - \mu_{Y_2} > 0 \qquad \text{(one-tailed test)}$$

$$t = \frac{\bar{Y}_1 - \bar{Y}_2}{S_p\sqrt{(1/n_1) + (1/n_2)}} = \frac{5.843 - 3.632}{1.980\sqrt{\frac{1}{9} + \frac{1}{10}}} = 2.430$$

Since t exceeds $t_{.95;\, 17} = 1.740$, reject H_0. At the .05 level of significance, there is sufficient evidence to indicate that mean earnings per share was significantly higher among commercial banks than among retailers. In using the P-value approach, similar results are found and H_0 is rejected. The probability that $t \geq 2.430$ given that H_0 is true is only between .01 and .025. Since this is unlikely if H_0 were really true, the result is deemed significant (and H_0 is rejected).

Case 10: From Table 1.3 (formula I),

$$H_0: \quad \mu_D \leq 0$$

$$H_1: \quad \mu_D > 0 \qquad \text{(one-tailed test)}$$

$$t = \frac{\bar{D}}{S_D/\sqrt{n}} = \frac{+.185}{.858/\sqrt{10}} = .682$$

Since t does not equal or exceed $t_{.95;\, 9} = 1.833$, do not reject H_0. At the .05 level of significance, there is no evidence that significant increases in mean earnings per share have occurred. The P value here is approximately .25.

---

*From (1.18) we note that $F = (S_{Y_1}^2/\sigma_{Y_1}^2)/(S_{Y_2}^2/\sigma_{Y_2}^2)$. However, in testing the null hypothesis, (1.18) reduces to $F = S_{Y_1}^2/S_{Y_2}^2$.

1.8 DATA TRANSFORMATIONS

The classical methods of inference, utilizing such statistics as χ^2, t, and F, all make the *assumption* that the underlying population from which the sample data are drawn is (at least approximately) a normal distribution.

With respect to this assumption, the t test is a *"robust"* test in that it is not sensitive to modest departures from normality—particularly with large-sized samples. On the other hand, the χ^2 and F tests are not robust with respect to violations in normality, and the accuracy of these procedures could be seriously affected.

Thus, to use classical methods properly, researchers have tested the assumption of normality [e.g., see Conover (1980)] and in its suspected absence have sought appropriate transformations which would permit the data to approximately follow the normal distribution.

One method of determining the correct transformation involves plotting the observed data on *normal probability paper*, using various functions of the original scale of measurement on the horizontal axis (abscissa), and determining whether such a plot appears to be *linear*. Other methods involve computing various summary measures and/or examining functional relationships.

1.8.1 Commonly Used Transformations

The most commonly used transformations which attempt to normalize the data and (when comparing two or more samples) stabilize the variances are: $\arcsin \sqrt{Y}$, $\sqrt{Y}$, $\ln Y$, and $1/Y$. Appropriate application of such transformations is identified in Table 3.7 (see Section 3.4.2).

A particularly interesting feature of the Behrens-Fisher problem (described in case 4 of Section 1.7.3) is that while the assumption of normality holds, the population variances may be unequal. Since transformations which stabilize variance would likely destroy the normality, the approximation methods used in such problems [e.g., see Table 1.2 (formula F) and Table 1.3 (formula F)] do not alter the observed data. Nevertheless, it is usually the case that both normality and homoscedasticity are violated, and transformations which stabilize one also aid the other.

For further discussion on transformations and "re-expressions," see Draper and Smith (1981), Dixon and Massey (1969), Mosteller and Tukey (1977), and Iman and Conover (1979).

1.8.2 A Warning

Nevertheless, finding the appropriate transformation often depends on the experience of the researcher. With small-sized samples it is both difficult to test for normality and to make statements as to the appropriateness of a particular transformation. In such instances it may be dangerous to assume normality without sufficient investigation as to the justification of the assumption.

Hence, when little is known about the underlying distribution being sampled and/or when the researcher is dealing with small-sized samples, *nonparametric* methods of analysis should be considered. The subject of nonparametric inference, however, is very broad and is not included within the scope of this text [see Conover (1980)].

2

INTRODUCTION TO
COMPUTER PACKAGES

2.1 COMPUTER PACKAGES IN STATISTICS

Over the past decade, revolutionary changes have taken place in our society due to the rapidly expanding application of computer technology. The computer has become increasingly intertwined with all aspects of our life from routine activities such as accounting for our credit card purchases to long-range activities involving the exploration of our solar system. This expansion of computer technology has also led to an information processing explosion in which massive amounts of data are being studied, analyzed, and published. Thus, from this perspective, researchers and professionals in business and the social and natural sciences have become increasingly aware of the potential for utilizing the computer in the application of statistical methods—particularly those that involve a high degree of computational complexity.

One approach toward using the computer for the solution of statistical problems involves the development of single-purpose computer programs. However, this approach suffers from high cost in terms of both the expense and time to develop and debug programs. These difficulties have provided the impetus for the development of standardized programs. Collections of standardized programs not only perform computations that otherwise would be cumbersome and time-consuming if done manually, but also are utilized for statistical analyses which are far too complex to ever be performed manually. Over the past several years, groups of these standardized programs have been assembled into a collection or *package* so that multiple users may share the programs with different data. Thus, for example, if we were to be concerned merely with computing a simple average, the computational algorithm would remain the same from problem to problem although the particular data and description would vary. Therefore, because data vary between users, the description of the data provides a large portion of the command structure of each particular package. The methods available in most packages range from the computation of simple descriptive statistics and the plotting of charts to the use of more complex procedures such as multiple regression (see Chapters 10–14), factor analysis (see Chapter 15), and discriminant analysis (see Chapter 18).

Although other computer packages (both special purpose and multi-purpose) are available [see Curtis (1976) and Ryan et al. (1976)], this text will focus upon three general-purpose, widely applicable computer packages: the Statistical Package for the Social Sciences (SPSS), the Statistical Analysis System (SAS), and Biomedical Programs (BMDP).

While older computer packages were designed especially for 80-column keypunch (Hollerith) card input, many present-day computer systems use *terminal devices for input*. In such systems the user of course does not utilize *punch cards* but instead *types lines of input*. **Thus, the terms card and line may be used interchangeably**.

Regardless of which computer package is called upon, three types of statements are needed so that the package may be used in a particular statistical analysis: system statements, control and procedure statements, and data.

System statements are those job control (JCL) statements that are needed in order to utilize a specific computer package at a particular computer installation. Although there are differences from one computer facility to another, these statements typically include a job statement that gives the name of the job, a procedure statement that calls the specific package being utilized, and delimiter statements that inform the computer system where the program begins and ends.

Once a package is initiated, *control and procedure statements* provide instructions for the computer package concerning the specific parts of the package to be used or accessed. These statements could include (depending on the computer package selected) commands that carry out the labeling of variable names as well as those that relate to the actual statistical procedure being performed. Since these control and procedure statements are the central component of each computer package, the explanation of their use for each appropriate statistical analysis is an important feature of this text.

If the data have not been previously placed on a remote storage device such as a tape or disk, the data may be included as part of the program. This is the method that is used for all examples throughout the text.

2.2 A SAMPLE DATA SET

To illustrate the use of the basic features of each computer package (SPSS, SAS, and BMDP), we shall refer to the questionnaire illustrated in Figure 2.1. Data for this questionnaire were obtained from the personnel records of a sample of 46 employees of a large company and are summarized in Table 2.1.

Cols.	EMPLOYEE QUESTIONNAIRE
	Respondent Number (for coding purposes only)
‾1‾ ‾2‾	
‾4‾	1. Sex: (0) Female (1) Male
‾6‾ ‾7‾	2. Age in years as of January 1, 1980: __ __
‾9‾	3. Education level: (1) Non-HS grad (2) HS grad (3) Post-HS education (4) Two-year college degree (5) Four-year college degree
‾11‾ ‾12‾ ‾13‾	4. Number of months employed as of January 1, 1980: __ __ __
‾15‾	5. Job classification: (1) Clerical (2) Technical (3) Managerial
‾17‾ ‾18‾ ‾19‾	6. Weekly salary as of January 1, 1980 ($): __ __ __

Figure 2.1 Employee questionnaire.

We may observe from Figure 2.1 that there are six questions to be answered. In addition, there is an identification number that is assigned to each employee. Once the data have been collected from the sample, we must decide on the way in which it is to be formatted or presented to the computer. Thus, we must determine the number of characters that are to be assigned for the responses to each question. In the interest of clarity, an arbitrary *blank* character or space (indicated here by a b̸) is inserted between the responses to each question. If we are using *keypunch cards*,* we may place the first character in column 1. The allocated column positions have been written in the left-hand margin next to each question in Figure 2.1. Since there are 46 employees, we note that only two digits (columns 1 and 2) need to be assigned as the employee respondent number. Moreover, since the variable "sex" has only two possible responses, only one character or column (column 4) is required for assignment, while the "weekly salary" variable requires three characters or columns (columns 17–19) because three-digit values are involved. Following this format, we observe that for each employee in the survey, a total of 19 characters or columns are required for the responses to the six questions. The responses of the first two employees are illustrated in Figure 2.2.

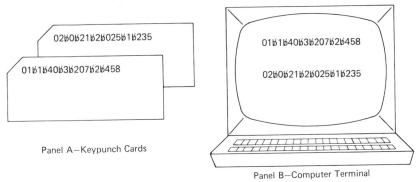

02b̸0b̸21b̸2b̸025b̸1b̸235

01b̸1b̸40b̸3b̸207b̸2b̸458

01b̸1b̸40b̸3b̸207b̸2b̸458

02b̸0b̸21b̸2b̸025b̸1b̸235

Panel A—Keypunch Cards

Panel B—Computer Terminal

Figure 2.2 Results for employees #01 and #02.

*If we are using *terminal devices*, column 1 corresponds to the first typeable space in an input *line*.

TABLE 2.1 Employee Questionnaire Results

Respondent Number	Sex	Age	Educ.	Months Employed	Job Class	Weekly Salary ($)
1	1	40	3	207	2	458
2	0	21	2	25	1	235
3	1	49	2	390	3	798
4	1	45	3	34	2	339
5	1	25	5	20	2	339
6	1	47	4	209	3	584
7	1	22	3	15	2	296
8	1	22	3	12	2	235
9	1	28	4	70	2	235
10	1	57	3	475	3	571
11	0	20	2	31	2	283
12	1	50	3	364	2	403
13	0	51	3	31	1	363
14	1	62	2	416	2	436
15	1	32	2	129	2	435
16	1	51	4	42	1	334
17	1	55	4	89	2	455
18	0	45	2	274	1	293
19	1	31	5	15	1	228
20	1	62	2	372	2	379
21	1	54	2	83	2	339
22	0	42	2	98	2	270
23	1	26	4	4	2	303
24	1	38	4	43	2	340
25	1	32	2	85	2	359
26	1	45	4	26	2	334
27	1	28	3	99	2	314
28	1	42	4	8	2	340
29	1	50	2	82	2	373
30	1	44	4	211	3	581
31	1	35	4	137	2	451
32	1	68	2	128	2	323
33	1	40	5	18	2	345
34	1	24	5	5	2	333
35	0	27	4	21	1	296
36	1	46	5	87	2	363
37	0	63	3	102	2	325
38	1	28	5	60	2	363
39	1	64	1	338	2	323
40	1	51	2	351	2	323
41	1	47	1	101	2	323
42	0	22	3	28	1	256
43	1	51	5	317	3	515
44	1	29	5	31	3	411
45	1	48	4	316	3	544
46	1	59	5	293	3	450

Cols.	EMPLOYEE QUESTIONNAIRE
__ __ 1 2	Respondent Number (for coding purposes only)

Cols.	EMPLOYEE QUESTIONNAIRE
__ __ ¯1¯ ¯2¯	Respondent Number (for coding purposes only)
__ ¯4¯	1. Sex: (0) Female (1) Male
__ __ ¯6¯ ¯7¯	2. Age in years as of January 1, 1980: __ __
__ ¯9¯	3. Education level: (1) Non-HS grad (2) HS grad (3) Post-HS education (4) Two-year college degree (5) Four-year college degree
__ __ __ 11 12 13	4. Number of months employed as of January 1, 1980: __ __ __
__ ¯15¯	5. Job classification: (1) Clerical (2) Technical (3) Managerial
__ __ __ 17 18 19	6. Weekly salary as of January 1, 1980 ($): __ __ __

Figure 2.1 Employee questionnaire.

We may observe from Figure 2.1 that there are six questions to be answered. In addition, there is an identification number that is assigned to each employee. Once the data have been collected from the sample, we must decide on the way in which it is to be formatted or presented to the computer. Thus, we must determine the number of characters that are to be assigned for the responses to each question. In the interest of clarity, an arbitrary *blank* character or space (indicated here by a ƀ) is inserted between the responses to each question. If we are using *keypunch cards*,* we may place the first character in column 1. The allocated column positions have been written in the left-hand margin next to each question in Figure 2.1. Since there are 46 employees, we note that only two digits (columns 1 and 2) need to be assigned as the employee respondent number. Moreover, since the variable "sex" has only two possible responses, only one character or column (column 4) is required for assignment, while the "weekly salary" variable requires three characters or columns (columns 17–19) because three-digit values are involved. Following this format, we observe that for each employee in the survey, a total of 19 characters or columns are required for the responses to the six questions. The responses of the first two employees are illustrated in Figure 2.2.

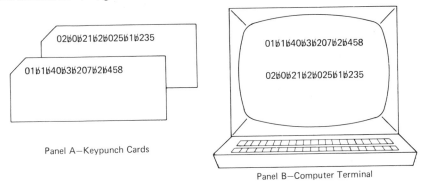

02ƀ0ƀ21ƀ2ƀ025ƀ1ƀ235

01ƀ1ƀ40ƀ3ƀ207ƀ2ƀ458

01ƀ1ƀ40ƀ3ƀ207ƀ2ƀ458

02ƀ0ƀ21ƀ2ƀ025ƀ1ƀ235

Panel A–Keypunch Cards

Panel B–Computer Terminal

Figure 2.2 Results for employees #01 and #02.

*If we are using *terminal devices*, column 1 corresponds to the first typeable space in an input *line*.

TABLE 2.1 Employee Questionnaire Results

Respondent Number	Sex	Age	Educ.	Months Employed	Job Class	Weekly Salary ($)
1	1	40	3	207	2	458
2	0	21	2	25	1	235
3	1	49	2	390	3	798
4	1	45	3	34	2	339
5	1	25	5	20	2	339
6	1	47	4	209	3	584
7	1	22	3	15	2	296
8	1	22	3	12	2	235
9	1	28	4	70	2	235
10	1	57	3	475	3	571
11	0	20	2	31	2	283
12	1	50	3	364	2	403
13	0	51	3	31	1	363
14	1	62	2	416	2	436
15	1	32	2	129	2	435
16	1	51	4	42	1	334
17	1	55	4	89	2	455
18	0	45	2	274	1	293
19	1	31	5	15	1	228
20	1	62	2	372	2	379
21	1	54	2	83	2	339
22	0	42	2	98	2	270
23	1	26	4	4	2	303
24	1	38	4	43	2	340
25	1	32	2	85	2	359
26	1	45	4	26	2	334
27	1	28	3	99	2	314
28	1	42	4	8	2	340
29	1	50	2	82	2	373
30	1	44	4	211	3	581
31	1	35	4	137	2	451
32	1	68	2	128	2	323
33	1	40	5	18	2	345
34	1	24	5	5	2	333
35	0	27	4	21	1	296
36	1	46	5	87	2	363
37	0	63	3	102	2	325
38	1	28	5	60	2	363
39	1	64	1	338	2	323
40	1	51	2	351	2	323
41	1	47	1	101	2	323
42	0	22	3	28	1	256
43	1	51	5	317	3	515
44	1	29	5	31	3	411
45	1	48	4	316	3	544
46	1	59	5	293	3	450

2.3 INTRODUCTION TO THE STATISTICAL PACKAGE FOR THE SOCIAL SCIENCES (SPSS)

The Statistical Package for the Social Sciences (SPSS)* [see Nie et. al. (1975) and Hull and Nie (1981)] was developed by researchers at the University of Chicago and the National Opinion Research Center primarily to aid in the analysis of social science data. SPSS represents a highly flexible package whose syntax is nontechnically oriented. SPSS is relatively easy to learn for individuals with limited statistical and computer backgrounds. Among its features, SPSS provides excellent capabilities for variable labeling and includes a wide variety of commonly used statistical procedures.

Figure 2.3 illustrates the SPSS program that has been written to compute various descriptive statistics (i.e., summary measures) for each of the quantitative variables "age," "length of employment," and "salary" and frequency distributions (i.e., summary tables) for the qualitative variables "sex" and "job classification."

Since the different types of statements have been explained in Section 2.1, we can focus upon the particular SPSS statements that are needed to obtain the required descriptive statistics and frequency distributions for our data. SPSS *control statements* provide information about the variables to be analyzed such as variable names, variable labels, etc. SPSS *procedure statements* provide the names of the SPSS statistical procedures that are to be utilized. We should note that all SPSS control statements *must precede* any SPSS procedure statements in the program sequence in order for the program to be processed correctly.

The format of both the SPSS control and procedure statements is rigidly specified. Regardless of whether the researcher is using an 80-column keypunch card or a terminal device, the *name* of the control or procedure being defined begins in *column 1*, and the *description* of the control or procedure begins in *column 16*:

1	16
{name of control or procedure}	{description of control or procedure}

The SPSS program provided in Figure 2.3 may be examined to obtain an understanding of the basic features of SPSS. The general format for each SPSS statement will be presented. This will be followed by an explanation of the statement as utilized in Figure 2.3:

1	16
RUNßNAME	{description of run}

The RUN NAME statement is intended to provide a description or title of the statistical methods being performed. The description begins in column 16

---

*SPSS is a trademark of SPSS, Inc., of Chicago, IL for its proprietary software. No materials describing such software may be produced or distributed without the written permission of SPSS, Inc.

31

```
{SYSTEM CARDS}
RUN NAME    ANALYSIS OF EMPLOYEE DATA
DATA LIST   FIXED(1)/1 RESPNO 1-2, SEX 4, AGE 6-7, EDUC 9
            EMPLOY 11-13, JOBCLASS 15, SALARY 17-19
VAR LABELS  RESPNO, RESPONDENT IDENTIFICATION NO/
            SEX, SEX OF EMPLOYEE/
            AGE, AGE IN YEARS AS OF JAN 1980/
            EDUC/EDUCATION LEVEL/
            EMPLOY, MONTHS EMPLOYED AS OF JAN 1980/
            JOBCLASS, JOB CLASSIFICATION/
            SALARY, WEEKLY SALARY AS OF JAN 1980/
VALUE LABELS  SEX(0)FEMALE(1)MALE/
            EDUC(1)NON HS GRAD(2)HS GRAD(3)POST HS EDUCATION
            (4)2 YR COLLEGE DEGREE(5)4 YR COLLEGE DEGREE/
            JOBCLASS(1)CLERICAL(2)TECHNICAL(3)MANAGERIAL/
READ INPUT DATA
01 1 207 2 458 1
02 0 025 1 235 3
...  ...  ...
46 1 593 5 450 3
END INPUT DATA
CONDESCRIPTIVE AGE, EMPLOY, SALARY
STATISTICS  1,5,9,10,11
FREQUENCIES  GENERAL=SEX, JOBCLASS
OPTIONS  8
FINISH
{SYSTEM CARDS}
```

Figure 2.3 SPSS program for analyzing the employee data.

32

and may continue up to column 80. Referring to Figure 2.3, we can observe that the particular job to be run is entitled "Analysis of Employee Data".

$$\underset{\text{DATA}\flat\text{LIST}}{\overset{1}{}} \qquad \underset{\text{FIXED}\flat}{\overset{16}{}} \left(\begin{array}{c} \text{number of} \\ \text{data cards per} \\ \text{respondent} \end{array} \right) /$$

$$\{\text{card number}\}\flat \left\{ \begin{array}{c} \text{variable} \\ \text{name} \\ \text{or list} \end{array} \right\} \flat \left\{ \begin{array}{c} \text{starting} \\ \text{column} \end{array} \right\} - \left\{ \begin{array}{c} \text{ending} \\ \text{column} \end{array} \right\}$$

The DATA LIST statement names the variables to be studied and indicates the columns on the data cards in which the responses are located (see Figures 2.1 and 2.2). The DATA LIST statement may be continued onto several cards or lines by merely continuing the description from the previous card or line into column 16 of the next card or line. The word FIXED indicates that the format of the data is the same for all respondents to be analyzed. This means that the same set of columns has been used to store the data for each respondent (see Table 2.1 and Figures 2.1 and 2.2). The number in parentheses following the word FIXED tells SPSS the number of separate data cards or lines needed for each respondent.* For our employee study, since the data are contained in only 19 columns of one card, this value is 1. The number after the slash indicates the particular data card which is to be read first by the computer. In our example this must also be 1 since the data for each employee are contained on only one card. After a blank (♭) space, the "names" of the variables to be analyzed appear next on the DATA LIST statement. Each variable name may contain up to *eight* alphabetic or numerical characters but must begin with a letter. The numbers which follow the variable names tell SPSS the columns on the data card in which the responses to the particular variables are entered. Thus, for our survey, the "respondent number" has been given the name RESPNO, and the information corresponding to this identification number appear in columns 1 and 2 of each data card. The "sex" variable has been named SEX, and the responses for this variable appear in column 4. Moreover, the variable "age" has been named AGE, and the responses for this variable appear in columns 6 and 7. In a similar manner we can observe that the responses for "educational level" (named EDUC) are located in column 9, responses for "length of employment" (named EMPLOY) are located in columns 11–13, responses for "job classification" (named JOBCLASS) are located in column 15, and responses for "weekly salary" (named SALARY) are located in columns 17–19 of the data record for each employee. We should also

---

*Since a card contains only 80 columns, large studies having many variables (questions) often require several cards in order to record the responses of each individual. In such studies, the number in parentheses after the word FIXED informs the computer how many cards per respondent to expect, while the number after the slash tells the computer what columns are being allocated for which variables on that particular data card. The DATA LIST statement continues until all variables are named.

note that a comma (,) separates the different variables on the DATA LIST statement.

1   16
VARβLABELS   variable name, variable label/

The VAR LABELS statement provides a descriptive label for each variable of interest. This label may contain up to 40 characters and ends with a slash. The characters)(and / *cannot* appear as part of the label. If we refer to Figure 2.3, we can observe that each of the variables in the study has been given its own label. Moreover, we should note that the VAR LABELS statement appears only on the first card of the group of variables. Since SPSS views the remaining variable label statements as a continuation, there are no other entries in columns 1–15.

1   16

$$\text{VALUE\textvisiblespace LABELS} \quad \left\{ \begin{array}{l} \text{variable} \\ \text{name} \\ \text{or list} \end{array} \right\} (\text{value}_1)\text{label}_1(\text{value}_2)\text{label}_2 \text{ etc./}$$

The VALUE LABELS statement assigns a descriptive label to each category of a qualitative variable. The label for each category may contain up to 20 characters but may not include the symbols / or) or (. A slash must be placed at the end of the label following the last category. Referring to Figure 2.3, we can observe that the variable SEX has two categories; 0 corresponds to female, and 1 corresponds to male. We also observe that the complete set of labels for the variable EDUC is continued from line 12 to line 13. Furthermore, we note that the name VALUE LABELS needs to be placed only on the first card or line since the set of labels has been grouped in succession.

1
READβINPUTβDATA

The READ INPUT DATA statement tells SPSS that the data should now be read from the particular input device utilized (cards, tape, disk). When the data are stored on punch cards these data cards immediately follow the READ INPUT DATA card. Thus, for our example, a set of 46 data cards would immediately follow the READ INPUT DATA statement.

1
ENDβINPUTβDATA

The END INPUT DATA statement informs SPSS that the end of the data deck has been reached. It can be utilized when the data are stored on punch cards. This completes the set of SPSS control statements that were needed to prepare the employee data for analysis. These are followed by the specific SPSS procedure statements or "subprograms" pertaining to the statistical methods the researcher wishes to use when analyzing the data.

1   16
CONDESCRIPTIVE $\left\{\begin{array}{c}\text{variable name or list}\\\text{or ALL}\end{array}\right\}$

The CONDESCRIPTIVE subprogram is used to compute descriptive statistics for quantitative variables. The variables listed in column 16 are separated by a comma. The word ALL is written beginning in column 16 if descriptive statistics are desired for all variables.

1   16

STATISTICS $\left\{\begin{array}{ll}\text{statistics numbers separated by commas:}\\1 = \text{mean} & 5 = \text{standard deviation}\\6 = \text{variance} & 8 = \text{skewness}\\9 = \text{range} & 10 = \text{minimum}\\11 = \text{maximum}\end{array}\right\}$

Each SPSS procedure statement has a set of OPTIONS and/or STATISTICS associated with the output obtained [see Nie et al. (1975) and Hull and Nie (1981)]. These statements, if desired, are placed after the particular procedure statement utilized. If each of the statistics is desired, the STATISTICS statement can be omitted from the CONDESCRIPTIVE subprogram.*

1   16
FREQUENCIES   GENERAL $= \left\{\begin{array}{c}\text{variable list}\\\text{or ALL}\end{array}\right\}$

OPTIONS   8

The FREQUENCIES subprogram is used to obtain frequency distributions for both qualitative and quantitative variables. The variables listed after the GENERAL = statement are separated by a space. As in the case of the CONDESCRIPTIVE subprogram, the word ALL would be utilized if frequency distributions for all variables were desired. Among the options for the FREQUENCIES subprogram is the plotting of a histogram for each variable when OPTIONS 8 has been specified.

1
FINISH

The FINISH statement signifies the end of processing for a particular computer job run. The FINISH statement can be considered as optional since SPSS will provide one if it is omitted from the program.

Now that we have explained the control and procedure statements involved in the SPSS program, we can examine the output obtained (see Figure 2.4). We observe that the requested statistics have been computed for the variables AGE, EMPLOY, and SALARY, while the frequency and percentage of the

---

*It is possible and, in fact, it is more efficient from a computer perspective for one SPSS procedure statement to be placed ahead of the READ INPUT DATA statement. Thus we alternatively could have moved the CONDESCRIPTIVE statement and its associated STATISTICS statement ahead of the READ INPUT DATA statement in Figure 2.3.

```
VARIABLE  EMPLOY     MONTHS EMPLOYED AS OF JAN 1980

MEAN         136.783                 STD DEV      136.547                    RANGE          471.000
MINIMUM        4.000                 MAXIMUM      475.000

VALID OBSERVATIONS -      46                  MISSING OBSERVATIONS -         0
- - - - - - - - - - - - - - - - - - - - - - - - - - - - - - - - - - - - - - - - -

VARIABLE  SALARY     WEEKLY SALARY AS OF JAN 1980

MEAN         373.826                 STD DEV      110.976                    RANGE          570.000
MINIMUM      228.000                 MAXIMUM      798.000

VALID OBSERVATIONS -      46                  MISSING OBSERVATIONS -         0
- - - - - - - - - - - - - - - - - - - - - - - - - - - - - - - - - - - - - - - - -

VARIABLE  AGE        AGE IN YEARS AS OF JAN 1980

MEAN          41.696                 STD DEV       13.522                    RANGE           48.000
MINIMUM       20.000                 MAXIMUM       68.000

VALID OBSERVATIONS -      46                  MISSING OBSERVATIONS -         0



      SEX        SEX OF EMPLOYEE

                                               RELATIVE   ADJUSTED    CUM
                                    ABSOLUTE      FREQ       FREQ      FREQ
      CATEGORY LABEL         CODE     FREQ       (PCT)      (PCT)     (PCT)

      FEMALE                  0.        8        17.4       17.4      17.4

      MALE                    1.       38        82.6       82.6     100.0
                                      ------     ------     ------
                            TOTAL      46        100.0      100.0


      JOBCLASS   JOB CLASSIFICATION

           CODE
              I
           1. ********* (     7)
              I  CLERICAL
              I
           2. ***************************** (     31)
              I  TECHNICAL
              I
           3. ********* (     8)
              I  MANAGERIAL
              I
              I........I........I........I........I........I
              0       10       20       30       40       50
           FREQUENCY
```

Figure 2.4 Partial output obtained from the SPSS subprograms CONDE-SCRIPTIVE and FREQUENCIES for the employee data.

respondents in various categories of the variables SEX and JOBCLASS have been summarized in tables and charts.

2.4 INTRODUCTION TO THE STATISTICAL ANALYSIS SYSTEM (SAS)

The Statistical Analysis System (SAS) is a computer package that was originally developed at North Carolina State University for use in statistical research [see SAS (1979)]. It has evolved over time into a widely utilized and extremely flexible package that is generally considered to be somewhat more statistically sophisticated than SPSS (see Section 2.3). Like SPSS, the syntax of SAS is relatively easy to learn for individuals with little statistical or computer background. As compared to SPSS, however, SAS unfortunately requires a more cumbersome approach for the labeling of variables as well as categories within variables. On the other hand, SAS has two distinct advantages over SPSS. First, SAS uses a less rigid format in syntax. Second, and even more important, SAS contains a more complete set of sophisticated statistical procedures.

Figure 2.5 illustrates the SAS program that has been written to compute various descriptive statistics for each quantitative variable ("age," "length of

```
{SYSTEM CARDS}
DATA EMPLOY;
    INPUT RESPNO 1-2 SEX 4 AGE 6-7 EDUC 9
    EMPLOY 11-13 JOBCLASS 15 SALARY 17-19;
LABEL RESPNO=RESPONDENT IDENTIFICATION NO
      SEX=SEX OF EMPLOYEE
      AGE=AGE IN YEARS AS OF JAN 1980
      EDUC=EDUCATION LEVEL
      EMPLOY=MONTHS EMPLOYED AS OF JAN 1980
      JOBCLASS=JOB CLASSIFICATION
      SALARY=WEEKLY SALARY AS OF JAN 1980;
CARDS;
01 1 40 3 207 2 458
02 0 21 2 025 1 235
 :  :  :  :  :   :  :
46 1 59 5 293 3 450
PROC FORMAT;
    VALUE SEXFMT 0=FEMALE 1=MALE;
    VALUE EDUCFMT 1=NON HS GRAD
                  2=HS GRAD
                  3=POST HS EDUCATION
                  4=2 YR COLLEGE DEGREE
                  5=4 YR COLLEGE DEGREE;
    VALUE JOBFMT 1=CLERICAL
                 2=TECHNICAL
                 3=MANAGERIAL;
PROC MEANS;
    VAR AGE EMPLOY SALARY;
PROC FREQ;
    FORMAT SEX SEXFMT.;
    TABLES SEX;
PROC CHART;
    FORMAT JOBCLASS JOBFMT.;
    HBAR JOBCLASS/DISCRETE;
{SYSTEM CARDS}
```

Figure 2.5   SAS program for analyzing the employee data.

employment," and "salary"). As we have described in Section 2.1, we observe that there are system statements, control and procedure statements, and data. In particular, the control and procedure statements in SAS can be considered to be of two types: those that relate to the DATA *step* and those that relate to the *procedure* (PROC) *step*. Although DATA and procedure (PROC) steps can be intermingled in some situations, all DATA steps should precede any statistical procedures that are to be performed.

We note that the syntax in SAS is quite flexible. The information does not have to begin in any particular column; it may continue through column 80 and, if desired, onto the following card or line. Each statement in SAS begins with a one-word name (that tells SAS what to do) and ends with a semicolon (;).

The program illustrated in Figure 2.5 can now be examined to obtain an understanding of the basic features of SAS. The general format for each statement will be provided and then followed by an explanation of the statement as it was used in Figure 2.5.

$$\text{DATAb} \begin{Bmatrix} \text{name of} \\ \text{data set} \end{Bmatrix};$$

The DATA step in SAS creates a data set with the particular name provided by the user. For our example (see Figure 2.5), we have named the data set EMPLOY.

$$\text{INPUTbvariable}_1 \text{b} \begin{Bmatrix} \text{starting} \\ \text{column} \end{Bmatrix}$$

$$- \begin{Bmatrix} \text{ending} \\ \text{column} \end{Bmatrix} \text{bvariable}_2 \text{b} \begin{Bmatrix} \text{starting} \\ \text{column} \end{Bmatrix} - \begin{Bmatrix} \text{ending} \\ \text{column} \end{Bmatrix} \text{b etc.;}$$

The INPUT statement names the variables to be studied and indicates the columns* in which the responses are located (see Figures 2.1 and 2.2). Each variable name can contain up to *eight* characters but must begin with a letter. If the response provided by a variable is alphabetic (such as the names of the employees), a dollar sign ($) must follow that particular variable name. If more than one data card or line is needed for each respondent, a $\#$ symbol followed by the card or line number must precede the names of any variables for which responses are to be provided on subsequent data cards or lines.

$$\text{LABELbvariable}_1 = \text{label}_1$$

$$\text{variable}_2 = \text{label}_2$$

$$\cdot$$
$$\cdot$$
$$\cdot$$

$$\text{variable}_i = \text{label}_i;$$

The LABEL statement in SAS provides a descriptive label for each variable desired. The label may include up to 40 characters. If the symbol $=$ is included

---

*If each data value is separated by one or more blank characters, the column numbers may be omitted from the INPUT statement.

as part of a label, it must be enclosed by quotation marks. If a list contains several variables to be labeled, a semicolon is needed only at the end of the list. Referring to Figure 2.5, we observe that a separate card or line has been provided for each variable label. However, we should note that this has been done only for convenience; in SAS several labels could be combined on one card or line.

CARDS;

The CARDS statement tells SAS that the data are punched on cards. Note in Figure 2.5 that 46 data cards immediately follow this statement in the program sequence.

PROCþFORMAT;

$$\text{VALUEþ} \begin{Bmatrix} \text{name} \\ \text{of format} \end{Bmatrix} \text{þvalue}_1 = \text{label}_1$$

$$\text{value}_2 = \text{label}_2$$

$$\text{value}_i = \text{label}_i;$$

The naming of the value labels for the qualitative variables is handled as a PROC (procedure) step in SAS. This procedure allows the researcher to define output formats that give categorical labels to the different numerical values of the variable. A name that has not previously been assigned to a variable must be given to the format followed by the set of values and their respective labels. Once this format has been defined, whenever the variable is to be used in a statistical procedure the format is associated with the variable by placing

$$\text{FORMATþ} \begin{Bmatrix} \text{variable} \\ \text{name} \end{Bmatrix} \text{þ} \begin{Bmatrix} \text{format} \\ \text{name} \end{Bmatrix} .;$$

after the particular PROC step. If we examine Figure 2.5, we can observe that the FORMAT named SEXFMT has been associated with the labels for the sex variable, EDUCFMT has been associated with the labels for the education variable, and JOBFMT has been associated with the labels for the job classification variable. We note again that the name given to the format *cannot* be a name that has been previously used for a variable.

PROCþMEANS;

VARþvariable name$_1$þvariable name$_2$ etc.;

The PROC MEANS step provides for the computation of descriptive statistics for quantitative variables. Referring to Figure 2.5, we have requested descriptive statistics for the variables AGE, EMPLOY, and SALARY.

PROCþFREQ;

$$\text{FORMATþ} \begin{Bmatrix} \text{variable} \\ \text{name} \end{Bmatrix} \text{þ} \begin{Bmatrix} \text{format} \\ \text{name} \end{Bmatrix} .;$$

TABLESþvariable$_1$þvariable$_2$ etc.;

The PROC FREQ step provides for a frequency distribution for each variable included in the TABLES statement. If a PROC FORMAT has previously been defined, the placing of a FORMAT statement with the variable name and its format name (followed by a period) after the **PROC FREQ** step will provide value labels for each category of the qualitative variable. Thus the placing of the **FORMAT SEX SEXFMT**.; statement after the **PROC FREQ** step will provide the labels "female" and "male" for the frequency distribution referring to "sex."

PROCꞵCHART;
$$\left. \begin{matrix} \text{VBAR} \\ \text{or} \\ \text{HBAR} \\ \text{or} \\ \text{PIE} \end{matrix} \right\} \text{ꞵvariable name}_1\text{ꞵvariable name}_2 \text{ etc.}/\{\text{options}\};$$

The **PROC CHART** step provides graphical representations for either qualitative or quantitative variables. Among the statements relating to **PROC CHART** are **VBAR** for a vertical bar chart (i.e., histogram), **HBAR** for a horizontal bar chart, and **PIE** for a pie chart. Any format statement which provides value labels should immediately follow the **PROC CHART** step. When obtaining charts for coded qualitative variables, the **DISCRETE** option should be used to inform SAS that only integer codes are provided. If we refer to Figure 2.5, we note that a horizontal bar chart has been requested for the variable JOBCLASS.

Now that we have examined the SAS program, we can evaluate its resulting output (see Figure 2.6). First, we observe that descriptive statistics have been computed (using **PROC MEANS**), including the mean, standard deviation, standard error of the mean, and the minimum and maximum values for each of the quantitative variables (AGE, EMPLOY, and SALARY). Moreover, we note that the FREQ procedure has provided the frequency and cumulative percentages, while the CHART procedure has plotted a horizontal bar chart for the selected qualitative variable.

2.5 INTRODUCTION TO BIOMEDICAL PROGRAMS (BMDP)

The series of programs called Biomedical Programs (BMDP) were originally developed in the early 1960s at the Health Sciences Computer Facility of the University of California at Los Angeles. This package was called BMD, but substantial improvements and simplifications were provided when the BMDP series became available in 1977 [see Dixon et. al. (1981)]. The BMDP series of programs provides an extremely large variety of elementary and sophisticated statistical procedures. Moreover, the syntax for BMDP is quite flexible since the columns to be used for control information are not rigidly specified.

One way in which BMDP differs from SPSS and SAS (see Sections 2.3 and 2.4) is that the particular BMDP program to be accessed must be indicated as part of the system (i.e., job control) statements. Thus only one program can be utilized for any given computer job run—although that program may contain

S T A T I S T I C A L   A N A L Y S I S   S Y S T E M

VARIABLE	LABEL	N	MEAN	STANDARD DEVIATION	MINIMUM VALUE	MAXIMUM VALUE	STD ERROR OF MEAN	SUM	VARIANCE
EMPLOY	MONTHS EMPLOYED AS OF JAN 1980	46	136.782609	136.546600	4.000000	475.000000	20.1326978	6292.0000	18644.9739
SALARY	WEEKLY SALARY AS OF JAN 1980	46	373.826087	110.976335	228.000000	798.000000	16.3625679	17196.0000	12315.7469
AGE	AGE IN YEARS AS OF JAN 1980	46	41.695652	13.521784	20.000000	68.000000	1.9936783	1918.0000	182.8386

SEX OF EMPLOYEE

SEX	FREQUENCY	CUM FREQ	PERCENT	CUM PERCENT
FEMALE	8	8	17.391	17.391
MALE	38	46	82.609	100.000

FREQUENCY BAR CHART

JOBCLASS	JOB CLASSIFICATION	FREQ	CUM. FREQ	PERCENT	CUM. PERCENT
CLERICAL	*******************	7	7	15.22	15.22
TECHNICAL	***	31	38	67.39	82.61
MANAGERIAL	********************	8	46	17.39	100.00

```
     ----+----+----+----+----+----+----+----+----+----+----+----+----+----+----
         2    4    6    8   10   12   14   16   18   20   22   24   26   28   30
                                   FREQUENCY
```

Figure 2.6 Partial output obtained from the SAS PROC MEANS, PROC FREQ, and PROC CHART procedures for the employee data.

41

numerous statistical methods as options. This feature contrasts with SPSS and SAS wherein the various statistical procedures are all incorporated as sub-programs (in SPSS) or procedure (PROC) steps (in SAS).

Figure 2.7 illustrates the BMDP program that has been written in order to compute various descriptive statistics by using program 1D of BMDP. As

```
{SYSTEM CARDS}
/PROBLEM  TITLE IS 'EMPLOYEE DATA'.
/INPUT    VARIABLES ARE 7.
          FORMAT IS '(F2.0,1X,F1.0,1X,F2.0,
          1X,F1.0,1X,F3.0,1X,F1.0,1X,F3.0)'.
          CASES ARE 46.
/VARIABLES NAMES ARE RESPNO,SEX,AGE,EDUC,
          EMPLOY,JOBCLASS,SALARY.
          LABEL IS RESPNO.
/GROUP    CODES(2) ARE 0,1.
          NAMES(2) ARE FEMALE,MALE.
          CODES(4) ARE 1,2,3,4,5.
          NAMES(4) ARE 'NON HS GRAD','HS GRAD',
          'POST HS EDUCATION','2 YR COLLEGE
          DEGREE','4 YR COLLEGE DEGREE'.
          CODES(6) ARE 1,2,3.
          NAMES(6) ARE CLERICAL,TECHNICAL,
          MANAGERIAL.
/END
01  1  40  3  207  2  458
02  0  21  2  025  1  235
:   :  :   :   :   :   :
46  1  59  5  293  3  450
{SYSTEM CARDS}
```

Figure 2.7 Using BMDP program 1D for employee data.

described in Section 2.1, we may observe from Figure 2.7 that there are system statements, control and procedure statements, and data. The control statements for BMDP are written in *paragraphs* that can be subdivided into *sentences*. Each sentence ends with a period (.), while paragraphs are separated by a slash (/). The names of variables are limited to *eight* characters. Any names, labels, or titles must be enclosed in apostrophes (') if they do not begin with a letter or if they contain a character that is not a letter or number (such as a blank). Formats that indicate the setup of the data cards must always be enclosed in apostrophes. In addition, the names of the paragraphs and sentences can often be abbreviated.

The program illustrated in Figure 2.7 can now be examined to obtain an understanding of the basic principles of BMDP. The general format for each sentence or paragraph will be provided and then followed by an explanation of how it was utilized in Figure 2.7.

S T A T I S T I C A L   A N A L Y S I S   S Y S T E M

VARIABLE	LABEL	N	MEAN	STANDARD DEVIATION	MINIMUM VALUE	MAXIMUM VALUE	STD ERROR OF MEAN	SUM	VARIANCE
EMPLOY	MONTHS EMPLOYED AS OF JAN 1980	46	136.782609	136.546600	4.000000	475.000000	20.1326978	6292.0000	18644.9739
SALARY	WEEKLY SALARY AS OF JAN 1980	46	373.826087	110.97335	228.000000	798.000000	16.3625679	17196.0000	12315.7469
AGE	AGE IN YEARS AS OF JAN 1980	46	41.695652	13.521784	20.000000	68.000000	1.9936783	1918.0000	182.8380

SEX OF EMPLOYEE

SEX	FREQUENCY	CUM FREQ	PERCENT	CUM PERCENT
FEMALE	8	8	17.391	17.391
MALE	38	46	82.659	100.000

FREQUENCY BAR CHART

JOBCLASS	JOB CLASSIFICATION	FREQ	CUM. FREQ	PERCENT	CUM. PERCENT
CLERICAL	|*************	7	7	15.22	15.22
TECHNICAL	|***	31	38	67.39	82.61
MANAGERIAL	|**************	8	46	17.39	100.00

```
          +--+--+--+--+--+--+--+--+--+--+--+--+--+--+--+
          2  4  6  8  10 12 14 16 18 20 22 24 26 28 30
                            FREQUENCY
```

Figure 2.6 Partial output obtained from the SAS PROC MEANS, PROC FREQ, and PROC CHART procedures for the employee data.

numerous statistical methods as options. This feature contrasts with SPSS and SAS wherein the various statistical procedures are all incorporated as sub-programs (in SPSS) or procedure (PROC) steps (in SAS).

Figure 2.7 illustrates the BMDP program that has been written in order to compute various descriptive statistics by using program 1D of BMDP. As

Figure 2.7 Using BMDP program 1D for employee data.

described in Section 2.1, we may observe from Figure 2.7 that there are system statements, control and procedure statements, and data. The control statements for BMDP are written in *paragraphs* that can be subdivided into *sentences*. Each sentence ends with a period (.), while paragraphs are separated by a slash (/). The names of variables are limited to *eight* characters. Any names, labels, or titles must be enclosed in apostrophes (') if they do not begin with a letter or if they contain a character that is not a letter or number (such as a blank). Formats that indicate the setup of the data cards must always be enclosed in apostrophes. In addition, the names of the paragraphs and sentences can often be abbreviated.

The program illustrated in Figure 2.7 can now be examined to obtain an understanding of the basic principles of BMDP. The general format for each sentence or paragraph will be provided and then followed by an explanation of how it was utilized in Figure 2.7.

$$/\text{PROBLEM\ss TITLE\ss IS'\{title\}'}.$$

The /PROBLEM paragraph provides a title (up to 160 characters) for the output of the program. The name PROBLEM can be abbreviated as PROB. Usually the title needs to be enclosed in apostrophes since blanks are often present between words. In Figure 2.7 the title 'EMPLOYEE DATA' has been utilized.

$$/\text{INPUT}\quad \text{VARIABLES\ss} \left\{ \begin{array}{c} = \\ \text{or} \\ \text{ARE} \end{array} \right\} \text{\ss}\{i\}.$$

$$\text{FORMAT\ss IS\ss'(format)'}.$$

$$\text{CASES\ss} \left\{ \begin{array}{c} = \\ \text{or} \\ \text{ARE} \end{array} \right\} \text{\ss} \left\{ \begin{array}{c} \text{sample} \\ \text{size} \end{array} \right\}.$$

The /INPUT paragraph follows the /PROBLEM paragraph. The VARI-ABLES sentence informs BMDP of the number of variables (i) contained by the input data. Since the employee data contain seven variables, the sentence VARIABLES ARE 7. is given in Figure 2.7. Note that VARIABLES can be abbreviated as VAR and that either the word ARE or an equal sign ($=$) can be utilized in this sentence.

The FORMAT sentence provides BMDP with information concerning the columns in which the variables are located. The format utilized in BMDP is usually F format [see Awad (1980), Dixon et al. (1981) and Sass (1974)]. An F format of the form $Fw.d$ can be interpreted to mean that the responses to the variable take up a width of w columns in which d columns follow the decimal point. For example, $F3.1$ format would mean that the responses to a variable require a total of three columns, including one column which follows the decimal point. When spaces are to be provided to separate responses of different variables, the symbol sX is utilized [where s represents the number of spaces (columns) to be left blank]. Thus, $2X$ would mean that two spaces are to be skipped. The format statement for each of the variables is presented in a sequential manner—beginning with the number of columns to be allocated to responses from the first variable and ending with the number of columns to be allocated to responses from the last variable (with commas separating the formats) and set off with parentheses and apostrophes. If the responses for each employee were contained on more than one card or line, the format of each successive card or line would be separated by a slash (/). The format indicated in Figure 2.7 can be interpreted as follows:

1. The responses to the first variable occupy two columns (columns 1 and 2), and the decimal point is only implicit.* This is followed by a space (column 3).

---

*Since none of the variables in the employee data problem are expressed in decimals, each variable has format $Fw.0$ implicit.

2. The responses to the second variable occupy one column (column 4). This is followed by a space (column 5).

3. The responses to the third variable occupy two columns (columns 6–7). This is followed by a space (column 8).

4. The responses to the fourth variable occupy one column (column 9). This is followed by a space (column 10).

5. The responses to the fifth variable occupy three columns (columns 11–13). This is followed by a space (column 14).

6. The responses to the sixth variable occupy one column (column 15). This is followed by a space (column 16).

7. The responses to the seventh variable occupy three columns (columns 17–19).

The CASES sentence indicates the number of observations or sample size obtained for the data. In our example, CASES ARE 46.

$$/\text{VARIABLESÞNAMESÞAREÞvar.name}_1,\text{var.name}_2, \ldots ,\text{var.name}_i.$$

$$\text{LABELÞISÞ} \left\{ \begin{array}{c} \text{name} \\ \text{of} \\ \text{labeling} \\ \text{variable} \end{array} \right\}.$$

The /VARIABLES paragraph describes the variables which have previously been defined in the /INPUT paragraph. The NAMES sentence provides names (up to eight characters) for each of the corresponding variables. Referring to Figure 2.7, we have RESPNO (respondent number), SEX, AGE (age in years as of January 1980), EDUC (education level), EMPLOY (length of employment in months), JOBCLASS (job classification), and SALARY (weekly salary). The LABEL sentence indicates which variables (if any) have been used to identify the observations in the data. In this employee study, RESPNO (respondent number) has been utilized to identify the various employees.

$$/\text{GROUPÞCODES } (i)\text{ÞAREÞvalue}_1,\text{value}_2, \text{ etc.}$$

$$\uparrow$$
$$\text{variable}$$
$$\text{number}$$
$$\downarrow$$
$$\text{NAMES}(i)\text{ÞAREÞ}\begin{array}{l}\text{value value} \\ \text{label}_1,\text{label}_2,\end{array} \text{ etc.}$$

The /GROUP paragraph is utilized to provide value labels for qualitative variables.* A CODES sentence can be established for each *qualitative* variable in order to indicate the number of the variable and its set of values. BMDP permits up to 10 different values for each variable. Referring to Figure 2.7, we

---

*The /GROUP paragraph can also be utilized to provide cutpoints (or class boundaries) for forming frequency distributions for quantitative variables.

observe that the variable SEX has a 2 in the CODES parentheses because it is the second variable on the list. Its possible values are 0 and 1. Note also that the variable EDUC is the fourth variable on the variable list and has code values of 1, 2, 3, 4, and 5. The NAMES sentence (for convenience) could directly follow the corresponding CODES sentence for a particular variable. Once again, the value in parentheses represents the variable number, but here the value labels are provided. Thus, as indicated, the names for the second variable (SEX) are presented as FEMALE, MALE. Moreover, we should note that the names for the fourth variable (EDUC) are enclosed with apostrophes since these value labels contain blank spaces.

/END

The /END paragraph signifies the end of the BMDP control statement information. The data immediately follow the /END paragraph and are then processed. No other sentences or paragraphs should follow the /END paragraph.

Now that we have examined the BMDP program in Figure 2.7, we can evaluate the generated output illustrated in Figure 2.8. Observe that the mean, standard deviation, standard error of the mean, coefficient of variation, smallest and largest value, Z score, range, and frequency (i.e., sample size) are printed for *all* variables, including qualitative variables (for which such interpretation is of course meaningless!).* In addition, the frequency count in each category of each qualitative variable is also displayed.

2.6 COMPARISON OF PACKAGES

2.6.1 Introduction

In Sections 2.3–2.5 we have introduced the basic features included in the computer packages SPSS, SAS, and BMDP, and we have written programs using each of these three packages to obtain descriptive statistics for the data obtained from the employee questionnaire displayed in Figure 2.1. In this section we shall briefly examine the syntax, features, and output of these three packages from a comparative perspective. We shall subdivide our discussion into three sections: comparison of control features, comparison of procedures, and comparison of the output obtained.

2.6.2 Comparison of Control Features

The three packages have both similarities and dissimilarities in the structure of their control features. These may be highlighted as follows: (1) In contrast to BMDP, which allows for the accessing of only one specific program,

---

*The descriptive summary measures are not presented for the "variable" RESPNO because it is merely used to identify employees. This was indicated to BMDP in the LABEL sentence.

EMPLOYEE DATA

VARIABLE NO. NAME	TOTAL FREQUENCY	MEAN	STANDARD DEVIATION	ST.ERR. OF MEAN	COEFF. OF VARIATION	SMALLEST VALUE Z-SCORE	LARGEST VALUE Z-SCORE	RANGE
2 SEX	40	0.820	0.383	0.0505	0.46390	0.0 -2.16	1.000 0.45	1.000
3 AGE	40	41.695	13.522	1.9937	0.32430	20.000 -1.60	68.000 1.95	48.000
4 EDUC	40	3.283	1.205	0.1776	0.36704	1.000 -1.89	5.000 1.43	4.000
5 EMPLOY	40	136.782	136.296	20.132	0.99627	9.000 -0.94	475.000 2.48	471.000
6 JOBCLASS	40	2.022	0.577	0.0851	0.28537	1.000 -1.77	3.000 1.70	2.300
7 SALARY	40	373.824	110.976	16.5625	0.29687	228.000 -1.31	798.000 3.82	570.000

VARIABLE NO. NAME	CATEGORY NAME	CATEGORY NUMBER	CATEGORY FREQUENCY	NO. OF VALUES MISSING OR OUTSIDE THE RANGE	TOTAL FREQUENCY
2 SEX	FEMALE	1	8		
	MALE	2	38	0	40
4 EDUC	NON HS	1	2		
	HS GRAD	2	13		
	POST HS	3	10		
	2 YR COL	4	12		
	4 YR COL	5	9	0	40
6 JOBCLASS	CLERICAL	1	1		
	TECHNICA	2	31		
	MANAGERI	3	8	0	40

Figure 2.8 Partial output obtained from the BMDP 1D program for the employee data.

an entire set of SPSS subprograms or SAS PROC steps can be included within a single job run. (2) BMDP and SAS are each highly flexible (being virtually in free format) in their specifications for locating control names and descriptions, while SPSS has a more rigid requirement in which the control or subprogram name must begin in column 1, while the description relating to the name must begin in column 16. (3) The naming of category labels in SAS is more cumbersome than in SPSS or BMDP. In SAS, a PROC FORMAT must be given for each categorical variable to be analyzed, and subsequently a FORMAT statement must be included as part of a procedure such as PROC FREQ or PROC CHART.

2.6.3 Comparison of Procedures

Each of the three packages contains procedures that relate to various descriptive summary measures (such as means and standard deviations), frequency distributions, and charts. Nevertheless, the statistics and charts currently available from SPSS are much more limited in comparison to those which can be obtained using SAS and BMDP. For example, only a histogram can be obtained from the SPSS FREQUENCIES subprogram, while either a horizontal or vertical bar chart or pie chart can be obtained as part of the SAS output. If the P-5D program of BMDP is accessed, more sophisticated types of plots (such as normal probability plots) can be obtained. In addition, a wider variety of descriptive summary measures is available from the UNIVARIATE procedure of SAS and the P-2D program of BMDP.

2.6.4 Comparison of Output

Figures 2.4, 2.6, and 2.8 illustrate partial output for the employee data obtained from SPSS, SAS, and BMDP. The computed descriptive summary measures are quite similar since the mean, standard deviation, standard error of the mean, and sample size can be obtained by accessing any of the packages. However, other measures such as skewness are available as options of the SPSS CONDESCRIPTIVE subprogram [see Nie et al. (1975)] while BMDP provides the minimum and maximum Z values observed for each variable. In addition, we should note that the P-1D program automatically prints these "descriptive statistics" for all variables (including each *qualitative variable*—for which interpretation is meaningless).

With respect to the frequency distributions and charts for qualitative variables, both SPSS (using the FREQUENCIES subprogram) and SAS (using PROC FREQ and PROC CHART) provide frequencies, percentages, cumulative percentages, and bar charts. In contrast, the P-1D program of BMDP provides only the frequency in each category for the qualitative variable of interest. If additional output such as percentages and bar charts are desired, the P-2D and P-5D programs must be accessed.

In the following chapters of this text, the use of these three packages will be explained and illustrated for a wide variety of statistical procedures.

PROBLEMS

2.1. The tax assessor for a large metropolitan county would like to conduct a survey of a sample of 200 single-family homes in the county. The following information is to be collected on the questionnaire:

Cols.	COUNTY TAX QUESTIONNAIRE
$\overline{1}\ \overline{2}\ \overline{3}$	Respondent Code Number
$\overline{}$ 4	1. Residential location: Urban 1 Surburban 2 Rural 3
$\overline{5}\ \overline{6}\ \overline{7}\ \overline{8}$	2. County taxes ($): _ _ _ _
$\overline{}$ 9	3. Type of house: Ranch 1 Split level 2 Colonial 3 Splanch 4 Other 5
$\overline{10}\ \overline{11}\ \overline{12}\ \overline{13}$	4. Number of rooms: _ _ . _
$\overline{14}$	5. Number of bathrooms: _
$\overline{15}$	6. Attached garage: Yes 1 No 2
$\overline{16}\ \overline{17}\ \overline{18}\ \overline{19}\ \overline{20}\ \overline{21}$	7. Lot size (square feet): _ _ _ _ _ _

If a sample of 200 homes is to be selected, write a computer program to compute descriptive statistics for the quantitative variables and frequency distributions for the qualitative variables using

(a) SPSS

(b) SAS

(c) BMDP

Note: Be sure to indicate where the system statements, control and procedure statements, and data are placed in the program.

2.2. You have been hired by the Dean of Graduate Studies to obtain some information pertaining to students who have applied for admission to the MBA program. With the dean's approval you have developed the following questionnaire:

Cols.	GRADUATE ADMISSIONS QUESTIONNAIRE
$\overline{1}\ \overline{2}\ \overline{3}$	Respondent Code Number
$\overline{5}$	1. Sex: Male 1 Female 2
$\overline{7}$	2. What is the applicant's current employment status?: Unemployed 1 Part-time employment 2 Full-time employment 3
$\overline{9}$	3. Did the applicant receive a baccalaureate degree more than 5 years ago?: Yes 1 No 2
$\overline{11}\ \overline{12}\ \overline{13}\ \overline{14}$	4. What was the applicant's college grade point index (or equivalent)?: _ . _ _
$\overline{16}\ \overline{17}\ \overline{18}$	5. What is the applicant's GMAT score?: _ _ _

If a sample of 50 applicants is selected, write a computer program to compute descriptive statistics for the quantitative variables and frequency distributions for the qualitative variable using

(a) SPSS

(b) SAS

(c) BMDP

Note: Be sure to indicate where the system statements, control and procedure statements, and data are placed in the program.

2.3. Referring to the set of real estate data presented in Appendix A, write a computer program to compute descriptive statistics for each quantitative variable and a frequency distribution for each qualitative variable using

(a) SPSS

(b) SAS

(c) BMDP

Note: Be sure to indicate where the systems statements, control and procedure statements, and data are placed in the program.   ■

Part II

LINEAR MODELS:
EXPERIMENTAL DESIGN
AND REGRESSION ANALYSIS

In Part II we shall deal with linear models. Chapters 3–7 focus primarily on methods of comparison among groups. Toward that end, experimental designs ranging from the simple one-factor completely randomized model (Chapter 3) to the more complex two-factor nonorthogonal model (Chapter 7) are considered in order to determine the existence of significant treatment effects. In this sense, we view such methods as extensions of the more fundamental one-sample and two-sample techniques reviewed in Chapter 1. In Chapters 8–14, however, emphasis shifts from such a *confirmatory* or *hypothesis testing* approach to one of a more *exploratory* nature, i.e., the development of regression models for prediction and evaluation purposes. This sequence of topics will provide us with a natural lead in to the subject of multivariate analysis (Part III) which, for the most part, we also view as a class of exploratory (as opposed to confirmatory) techniques whose purpose is to examine the underlying structure inherent in a set of data.

It is also important to note that the interrelationships between the two types of linear models considered in this part—experimental design models and regression models—are examined in depth in Chapter 12. This chapter expands on the interrelationships by focusing on certain types of designed experiments from a regression viewpoint. In particular, (1) a multiple regression model (Chapter 10) containing *dummy variables* is described; (2) the subject of *covariance analysis* as a link between ANOVA (Chapters 3 and 4) and regression (Chapters 8 and 10) is developed; and (3) a *generalized least-squares* approach to the study of nonorthogonal factorial experiments (Chapters 6–8) is discussed.

3

ONE-WAY ANOVA:
COMPLETELY RANDOMIZED DESIGN

3.1 INTRODUCTION

In Section 1.7.3 we tested for differences between the means of two independent groups. In particular, we used the t test to compare the mean earnings per share of commercial banking companies versus retailing companies. Suppose, however, there had been a third group of interest to the financial analyst such as utility companies. What hypothesis testing procedure would be appropriate?

In this chapter we shall be concerned with testing for differences between the means of several (c) independent groups (where $c \geq 3$). That is, we shall be testing the null hypothesis that there are no real differences among the c population group means

$$H_0: \quad \mu_{.1} = \mu_{.2} = \cdots = \mu_{.c}$$

against the alternative

$$H_1: \quad \text{not all } \mu_{.j} \text{ are equal} \quad (\text{where } j = 1, 2, \ldots, c)$$

If, based on statistical evidence, we conclude that there are significant differences between the means of at least two of the groups, we shall then be ready to use appropriate *a posteriori multiple comparison methods* to determine which of the c groups are significantly different as well as which of the c groups appear to differ from each other only by chance. Such procedures, however, will be described in Chapter 4.

3.2 TESTING FOR DIFFERENCES IN c MEANS: THE ONE-WAY F TEST

Methods developed for examining differences between the means of several groups are classified under the general title of **analysis of variance** or ANOVA.

Consider a situation in which the researcher wishes to compare the means of c independent groups. Let $Y_{11}, Y_{21}, Y_{31}, \ldots, Y_{i1}, \ldots, Y_{n_1 1}$ represent a random sample of n_1 observations taken from population 1. Similarly, let $Y_{12}, Y_{22}, Y_{32}, \ldots, Y_{i2}, \ldots, Y_{n_2 2}$ represent an independent random sample of n_2 observations drawn from population 2. Moreover, let $Y_{1j}, Y_{2j}, Y_{3j}, \ldots, Y_{ij}, \ldots, Y_{n_j j}$ and $Y_{1c}, Y_{2c}, Y_{3c}, \ldots, Y_{ic}, \ldots, Y_{n_c c}$, respectively, represent independent random samples from populations j and c. This is depicted in Table 3.1, where, using double-subscript notation, Y_{31} is the third observation in the first group, while Y_{2c} is the second observation in the cth (last) group, and so on.

TABLE 3.1 Data Matrix for ANOVA

Group 1	Group 2		Group j		Group c
Y_{11}	Y_{12}	$\cdots$	Y_{1j}	$\cdots$	Y_{1c}
Y_{21}	Y_{22}	$\cdots$	Y_{2j}	$\cdots$	Y_{2c}
Y_{31}	Y_{32}	$\cdots$	Y_{3j}	$\cdots$	Y_{3c}
.	.		.		.
.	.		.		.
.	.		.		.
Y_{i1}	Y_{i2}	$\cdots$	Y_{ij}	$\cdots$	Y_{ic}
.	.		.		.
.	.		.		.
.	.		.		.
$Y_{n_1 1}$	$Y_{n_2 2}$	$\cdots$	$Y_{n_j j}$	$\cdots$	$Y_{n_c c}$

If certain assumptions (see Section 3.4) can be met, we may test the null hypothesis

$$H_0: \quad \mu_{.1} = \mu_{.2} = \cdots = \mu_{.c}$$

against the alternative

$$H_1: \quad \text{not all } \mu_{.j} \text{ are equal} \qquad (\text{where } j = 1, 2, \ldots, c)$$

by computing the test statistic

$$F = \frac{\left\{\sum_{j=1}^{c}(Y_{.j}^2/n_j) - (Y_{..}^2/n)\right\}\Big/(c-1)}{\left[\sum_{j=1}^{c}\sum_{i=1}^{n_j}Y_{ij}^2 - \sum_{j=1}^{c}(Y_{.j}^2/n_j)\right]\Big/(n-c)} \tag{3.1}$$

where
$$\sum_{j=1}^{c} \sum_{i=1}^{n_j} Y_{ij}^2 = \text{sum of the squares of each observation in the data}$$

$$Y_{\cdot j} = \sum_{i=1}^{n_j} Y_{ij} = \text{total (sum) of the observations in group } j$$

$$Y_{\cdot j}^2 = \text{square of the total in group } j$$

$$Y_{\cdot\cdot} = \sum_{j=1}^{c} \sum_{i=1}^{n_j} Y_{ij} = \text{"grand" total (sum) over all observations in the data}$$

$$Y_{\cdot\cdot}^2 = \text{square of the grand total}$$

$$c = \text{number of groups being studied}$$

$$n_j = \text{sample size of group } j$$

$$n = \sum_{j=1}^{c} n_j = n_1 + n_2 + \cdots + n_c = \text{total number of observations in the data}$$

Recalling from Section 1.6.3 that the statistic F follows an F distribution (that is, $F \sim F_{c-1, n-c}$), we may reject the null hypothesis at the α level of significance if

$$F \geq F_{1-\alpha; c-1, n-c}$$

the upper-tail critical value of the F distribution having $c - 1$ and $n - c$ degrees of freedom (see Appendix B, Table B.4).* Since, under the null hypothesis, F had been defined as the ratio of two independent sample variances [Section 1.6.3 and Table 1.3 (formula G)], we see that although our primary interest is in comparing the c means, the terminology *analysis of variance* arises from the fact that this comparison is achieved by *analyzing variances*, as indicated in (3.1).

3.2.1 An Example

Let us now extend the case study of Section 1.7.3 so as to include a *third* business group: utility companies. Suppose that the financial analyst wished to compare the mean earnings per share of *three* types of nonindustrial business companies: commercial banks, retailers, and utilities. The sample data for the commercial banking companies and the retailing companies were displayed in Table 1.4 and summary computations were listed in Table 1.5. Recorded in Table 3.2 are the earnings per share (for the year 1979) for a random sample of 10 utility companies along with some summary computations. Combining this information, the data for the three nonindustrial business groupings are presented in Table 3.3 along with the appropriate summary computations.

The null hypothesis to be tested is that there are no differences between the three means

$$H_0: \quad \mu_{\cdot 1} = \mu_{\cdot 2} = \mu_{\cdot 3}$$

The alternative hypothesis is

$$H_1: \quad \text{not all } \mu_{\cdot j} \text{ are equal} \quad (\text{where } j = 1, 2, 3)$$

*The reason ANOVA methods yield one-tailed F tests is presented in Section 3.3.

TABLE 3.2 Earnings per Share for Ten Utility Companies in 1979

Company	Earnings per Share (Dollars)
Pacific Gas & Electric	3.55
Middle South Utilities	2.13
Consumers Power	3.24
American Natural Resources	6.47
Carolina Power & Light	3.06
Ohio Edison	1.80
Peoples Gas	5.29
Transco Companies	2.96
Arizona Public Service	2.90
Potomac Electric Power	1.73
Summed Observations	33.13
Mean	3.313
Sum of Squared Observations	130.2501
Square of the Summed Observations	1097.5969
Variance	2.2767
Standard Deviation	1.509

SOURCE: Peter D. Petre, "The Fortune Directory of the Largest Non-Industrial Companies," *Fortune* (July 14, 1980), pp. 146–159. © 1980 Time Inc. All rights reserved.

TABLE 3.3 Earnings per Share of Companies Under Three Nonindustrial Business Groupings

(1) Commercial Banking	(2) Retailing	(3) Utility	
6.42	3.52	3.55	
2.83	4.21	2.13	
8.94	4.36	3.24	
6.80	2.67	6.47	
5.70	3.49	3.06	
4.65	4.68	1.80	
6.20	3.30	5.29	
2.71	2.68	2.96	
8.34	7.25	2.90	
	.16	1.73	
n_j: 9	10	10	$n = 29$
$Y_{.j}$: 52.59	36.32	33.13	$Y_{..} = 122.04$
$\sum_{i=1}^{n_j} Y_{ij}^2$: 344.8411	160.9960	130.2501	$\sum_{j=1}^{c} \sum_{i=1}^{n_j} Y_{ij}^2 = 636.0872$

Using (3.1) and the summary computations in Table 3.3, we find that the numerator of the F statistic is

$$\frac{\sum\limits_{j=1}^{c} (Y_{.j}^2/n_j) - (Y_{..}^2/n)}{c - 1} = \left[\frac{(52.59)^2}{9} + \frac{(36.32)^2}{10} + \frac{(33.13)^2}{10} - \frac{(122.04)^2}{29}\right] \div (3 - 1)$$

$$= 35.3968 \div 2 = 17.6984$$

while the denominator is computed to be

$$\frac{\sum\limits_{j=1}^{c} \sum\limits_{i=1}^{n_j} Y_{ij}^2 - \sum\limits_{j=1}^{c} (Y_{.j}^2/n_j)}{n - c}$$

$$= \left\{636.0872 - \left[\frac{(52.59)^2}{9} + \frac{(36.32)^2}{10} + \frac{(33.13)^2}{10}\right]\right\} \div (29 - 3)$$

$$= 87.1124 \div 26 = 3.3505$$

Therefore the F statistic computed from (3.1) is

$$F = \frac{17.6984}{3.3505} = 5.28$$

Since F exceeds $F_{.95;\, 2,26} = 3.37$ (see Appendix B, Table B.4), the financial analyst may reject the null hypothesis at the .05 level and conclude that the means are not all equal. In fact, from Section 1.7.2, the P value—the probability of obtaining such a result as 5.28 or one even more extreme when H_0 is true— is between .01 and .025.

3.2.2 Interpreting the Results

Now that we have concluded that real differences in mean earnings per share exist among the three types of nonindustrial business groupings, the financial analyst would be ready to use a posteriori multiple comparison methods to determine whether each of the groups was significantly different or whether it was only one of the groups (say, commercial banking companies) that differed from the others. These and other procedures will be described in Chapter 4.

Prior to this, however, it is essential that the reader have a clear understanding of the underlying concepts behind the ANOVA procedures—not only so that they may be utilized more advantageously, but also to facilitate the development of more complex models in Chapters 5–7. Therefore, in Section 3.3 we shall deal with the conceptual development of ANOVA, while its assumptions are presented and examined in Section 3.4.

3.3 DEVELOPMENT

In designing an experiment, suppose the researcher has available a total of n *independent* subjects (i.e., *experimental units* to be measured)—be they persons, animals, plots of land, material, companies, etc. With respect to some *factor* of interest, the usual spread in the observed responses is an estimate of σ^2, while the sample mean is an estimate of μ. Let us now suppose that each of the n

subjects could be randomly assigned to one of c groups—with or without an equal number of subjects per group. For such situations the format of the experimental design is said to be *completely randomized*.

If the subjects in each of the c groups were "untreated," we would expect the groups to represent c random samples from the same population (or c identical populations) so that, except for chance differences, each provides similar estimates of σ^2 and μ. That is, we should expect $S^2_{.1}$, $S^2_{.2}$, ..., and $S^2_{.c}$ to differ only by chance, and, likewise, we should expect $\bar{Y}_{.1}$, $\bar{Y}_{.2}$, ..., and $\bar{Y}_{.c}$ to differ only by chance. No *treatment effect* could possibly exist since no treatment conditions were imposed on the c groups.

On the other hand, suppose that each of the c groups were to receive a particular *treatment* whereby the subjects in any one group are not affected or influenced by the imposed treatment conditions any more or less than the subjects of any other group experiencing their own particular treatment condition. We should again expect the c sample means $\bar{Y}_{.1}$, $\bar{Y}_{.2}$, ..., and $\bar{Y}_{.c}$ to differ only by chance because homogeneous subjects would tend to behave alike if treated alike. Again, in such instances, no treatment effect would exist.

Nevertheless, if we suppose that the particular treatment condition imposed on the subjects of one group influences that group significantly more or less than that of any other group, at least two of the c sample means would be expected to differ by an amount too great to be construed as due to chance. We would then conclude that at least two of the (population) means differ and that a significant treatment effect is present. Such a determination is the likely goal of the experiment.

The testing for treatment effects (when $c = 3$) is illustrated in Figure 3.1.

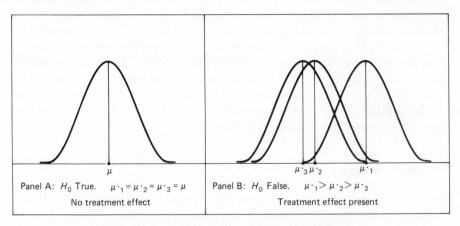

Figure 3.1   Testing the null hypothesis using ANOVA.

$$\frac{\sum_{j=1}^{c} (Y_{.j}^2/n_j) - (Y_{..}^2/n)}{c-1} = \left[\frac{(52.59)^2}{9} + \frac{(36.32)^2}{10} + \frac{(33.13)^2}{10} - \frac{(122.04)^2}{29} \right] \div (3-1)$$

$$= 35.3968 \div 2 = 17.6984$$

while the denominator is computed to be

$$\frac{\sum_{j=1}^{c} \sum_{i=1}^{n_j} Y_{ij}^2 - \sum_{j=1}^{c} (Y_{.j}^2/n_j)}{n-c}$$

$$= \left\{ 636.0872 - \left[\frac{(52.59)^2}{9} + \frac{(36.32)^2}{10} + \frac{(33.13)^2}{10} \right] \right\} \div (29-3)$$

$$= 87.1124 \div 26 = 3.3505$$

Therefore the F statistic computed from (3.1) is

$$F = \frac{17.6984}{3.3505} = 5.28$$

Since F exceeds $F_{.95; 2,26} = 3.37$ (see Appendix B, Table B.4), the financial analyst may reject the null hypothesis at the .05 level and conclude that the means are not all equal. In fact, from Section 1.7.2, the P value—the probability of obtaining such a result as 5.28 or one even more extreme when H_0 is true—is between .01 and .025.

3.2.2 Interpreting the Results

Now that we have concluded that real differences in mean earnings per share exist among the three types of nonindustrial business groupings, the financial analyst would be ready to use a posteriori multiple comparison methods to determine whether each of the groups was significantly different or whether it was only one of the groups (say, commercial banking companies) that differed from the others. These and other procedures will be described in Chapter 4.

Prior to this, however, it is essential that the reader have a clear understanding of the underlying concepts behind the ANOVA procedures—not only so that they may be utilized more advantageously, but also to facilitate the development of more complex models in Chapters 5–7. Therefore, in Section 3.3 we shall deal with the conceptual development of ANOVA, while its assumptions are presented and examined in Section 3.4.

3.3 DEVELOPMENT

In designing an experiment, suppose the researcher has available a total of n *independent* subjects (i.e., *experimental units* to be measured)—be they persons, animals, plots of land, material, companies, etc. With respect to some *factor* of interest, the usual spread in the observed responses is an estimate of σ^2, while the sample mean is an estimate of μ. Let us now suppose that each of the n

subjects could be randomly assigned to one of c groups—with or without an equal number of subjects per group. For such situations the format of the experimental design is said to be *completely randomized*.

If the subjects in each of the c groups were "untreated," we would expect the groups to represent c random samples from the same population (or c identical populations) so that, except for chance differences, each provides similar estimates of σ^2 and μ. That is, we should expect $S_{\cdot1}^2, S_{\cdot2}^2, \ldots,$ and $S_{\cdot c}^2$ to differ only by chance, and, likewise, we should expect $\bar{Y}_{\cdot1}, \bar{Y}_{\cdot2}, \ldots,$ and $\bar{Y}_{\cdot c}$ to differ only by chance. No *treatment effect* could possibly exist since no treatment conditions were imposed on the c groups.

On the other hand, suppose that each of the c groups were to receive a particular *treatment* whereby the subjects in any one group are not affected or influenced by the imposed treatment conditions any more or less than the subjects of any other group experiencing their own particular treatment condition. We should again expect the c sample means $\bar{Y}_{\cdot1}, \bar{Y}_{\cdot2}, \ldots,$ and $\bar{Y}_{\cdot c}$ to differ only by chance because homogeneous subjects would tend to behave alike if treated alike. Again, in such instances, no treatment effect would exist.

Nevertheless, if we suppose that the particular treatment condition imposed on the subjects of one group influences that group significantly more or less than that of any other group, at least two of the c sample means would be expected to differ by an amount too great to be construed as due to chance. We would then conclude that at least two of the (population) means differ and that a significant treatment effect is present. Such a determination is the likely goal of the experiment.

The testing for treatment effects (when $c = 3$) is illustrated in Figure 3.1.

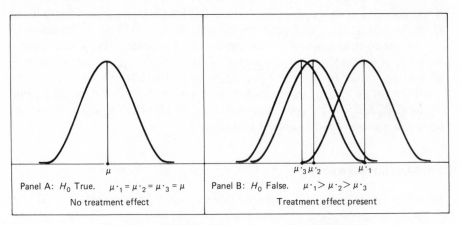

Panel A: H_0 True. $\mu_{\cdot1} = \mu_{\cdot2} = \mu_{\cdot3} = \mu$

No treatment effect

Panel B: H_0 False. $\mu_{\cdot1} > \mu_{\cdot2} > \mu_{\cdot3}$

Treatment effect present

Figure 3.1 Testing the null hypothesis using ANOVA.

3.3.1 Experimental Factor of Interest

The c groups may now be considered as c levels of some *experimental factor* of interest. An experimental factor may be defined as a controllable (i.e., independent) variable whose levels are of interest to the researcher for comparison purposes.* The factor may be *qualitative* or *quantitative*. As examples, "nonindustrial business classification" is a qualitative factor having such levels as commercial banking companies, retailing companies, and utility companies, while "drug dosage" is a quantitative factor having such levels as 5, 10, 15, and 20 mg.† Moreover, the levels of the factor may be considered as *fixed* or *random*. If the levels are fixed, the conclusions drawn pertain *only* to those specific levels of the factor. If the levels are random, inferences made by the researcher may be generalized to include *all* possible levels of that factor. In both the examples cited the levels of the factor are fixed. For instance, we may recall from Section 3.2.2 that the financial analyst concluded that there were significant differences in mean earnings per share among the three types of nonindustrial business groupings. Since the analyst was concerned only with the three selected levels—commercial banking companies, retailing companies, and utility companies—all inferences pertain strictly to these types of companies. Although it was not of interest here, had the financial analyst desired to draw conclusions

*A distinction can be made between an *experimental factor* and a *classification factor*. With designed (i.e., "planned") experiments the factor is a controllable variable whose c levels are established by the researcher so that the subjects randomly assigned to those c levels receive particular treatments. With survey research data, the "contrived" experiments are post hoc—the subjects assign themselves to particular levels of the (independent) classification factor. As an example of the latter, suppose in analyzing the responses to a large questionnaire a survey researcher wishes to determine the effects of attained education on income. Respondents (subjects) would be categorized according to the classification factor "attained education" (say, Below H. S. Degree, H. S. or Some College, College Degree, Some Graduate Work, or Graduate Degree) and then their incomes compared across all c educational levels.

†In a true experiment, the subjects (i.e., the *experimental units* of the study) randomly assigned to a particular level of the factor are then subjected to some treatment condition at that level. The objective of the researcher is to determine if a true treatment effect exists by comparing the results (i.e., subjects' responses) of the imposed treatment conditions at each level and evaluating whether or not such observed differences between those levels are significant. If, for example, the quantitative factor is "drug dosage," a pharmaceutical researcher might be interested in determining the effect (if any) of using different dosage levels (say 5, 10, 15, and 20 mg) of a newly developed drug on the diastolic blood pressure of hyperactive male rats. In such a situation the subjects—the hyperactive male rats—would randomly be assigned to one of the four groups (i.e., levels of the factor) receiving a particular treatment dosage. The results of the experiment—diastolic blood pressures being measured—constitute the response (i.e., *dependent*) variable of the study. If a treatment effect exists, the response variable is *dependent* on the level of the independent variable. This, then, would represent the relationship between ANOVA and regression analysis (to be studied in Chapters 8–14). See Problems 3.9, 4.10, and 8.7.

about mean earnings per share for *all* kinds of nonindustrial business companies (including such groupings as life insurance companies, diversified financial companies, transportation companies, etc.), the three selected levels would have been considered as a random sample out of all possible levels.

For the remainder of this chapter we shall deal only with experiments whose factor levels are fixed.

3.3.2 Developing the Model

To develop the classical ANOVA procedure for a completely randomized experimental design wherein the levels of the factor are fixed, Y_{ij}, the ith observation in group (level) j (where $i = 1, 2, \ldots, n_j$ and $j = 1, 2, \ldots, c$), can be represented by the model

$$Y_{ij} = \mu + \alpha_j + \epsilon_{ij} \tag{3.2}$$

where $\mu = $ *overall* effect or mean common to all the observations

$\alpha_j = \mu_{.j} - \mu$, a treatment effect peculiar to the jth level of the factor $(j = 1, 2, \ldots, c)$

$\epsilon_{ij} = Y_{ij} - \mu_{.j}$, *random variation* or *experimental error* associated with the ith observation in group j ($i = 1, 2, \ldots, n_j$ and $j = 1, 2, \ldots, c$)

$\mu_{.j} = $ true mean of the jth group (level)

From (3.2) we note that the experimental error ϵ_{ij} associated with the ith observation in the jth group represents the difference between that value and the mean of the jth group. Hence it is a measure of spread *within* a particular group.

Moreover, from (3.2) we also note that α_j, the treatment effect for the jth group, may be $+$, 0, or $-$ depending on whether $\mu_{.j}$ (the true mean of the jth level) exceeds, is equal to, or is exceeded by the overall mean μ. Now if the null hypothesis

$$H_0: \quad \mu_{.1} = \mu_{.2} = \cdots = \mu_{.c}$$

were true, each of the c population means would be equivalent to the overall mean μ; therefore there would be no treatment effects. That is, $\alpha_1 = \mu_{.1} - \mu$, $\alpha_2 = \mu_{.2} - \mu, \ldots$, and $\alpha_c = \mu_{.c} - \mu$ would all equal zero. In such instances, except for *chance differences*, we would expect each of the c sample means to be similar to $\bar{Y}_{..}$ (the mean based on all n observations). From our model (3.2), Y_{ij}, the ith observation in the jth sample (group), could be reexpressed as

$$Y_{ij} = \mu + \epsilon_{ij} \tag{3.3}$$

and the only differences from one observation to another would be due to chance, regardless of the population ($j = 1, 2, \ldots, c$) being sampled.

Using the layout of the data from Table 3.1, we can estimate the true parameters μ, α_j, and ϵ_{ij} by using the computed statistics $\hat{\mu}$, $\hat{\alpha}_j$, and $\hat{\epsilon}_{ij}$ where

$$\hat{\mu} = \bar{Y}..$$
$$\hat{\alpha}_j = \bar{Y}._j - \bar{Y}..$$
$$\hat{\epsilon}_{ij} = Y_{ij} - \bar{Y}._j$$

and

$$\bar{Y}.. = \frac{Y..}{n} = \frac{\sum\limits_{j=1}^{c}\sum\limits_{i=1}^{n_j} Y_{ij}}{n}$$

$$\bar{Y}._j = \frac{\sum\limits_{i=1}^{n_j} Y_{ij}}{n_j}$$

Thus, rewriting (3.2), we note that

$$Y_{ij} = \hat{\mu} + \hat{\alpha}_j + \hat{\epsilon}_{ij} \tag{3.4}$$

and

$$Y_{ij} = \bar{Y}.. + (\bar{Y}._j - \bar{Y}..) + (Y_{ij} - \bar{Y}._j) \tag{3.5}$$

If $\bar{Y}..$ is subtracted from both sides of (3.5), we get

$$Y_{ij} - \bar{Y}.. = (\bar{Y}._j - \bar{Y}..) + (Y_{ij} - \bar{Y}._j) \tag{3.6}$$

Furthermore, if each side of (3.6) is squared and then summed over both i and j, we obtain

$$\sum_{j=1}^{c}\sum_{i=1}^{n_j}(Y_{ij} - \bar{Y}..)^2 = \sum_{j=1}^{c}\sum_{i=1}^{n_j}(\bar{Y}._j - \bar{Y}..)^2 + \sum_{j=1}^{c}\sum_{i=1}^{n_j}(Y_{ij} - \bar{Y}._j)^2$$
$$+ 2\sum_{j=1}^{c}\sum_{i=1}^{n_j}(\bar{Y}._j - \bar{Y}..)(Y_{ij} - \bar{Y}._j) \tag{3.7}$$

However, from (3.7) it is not difficult algebraically to show that the last term, $2\sum_{j=1}^{c}\sum_{i=1}^{n_j}(\bar{Y}._j - \bar{Y}..)(Y_{ij} - \bar{Y}._j)$, equals zero so that the following important identity is obtained:

$$\sum_{j=1}^{c}\sum_{i=1}^{n_j}(Y_{ij} - \bar{Y}..)^2 \equiv \sum_{j=1}^{c}\sum_{i=1}^{n_j}(\bar{Y}._j - \bar{Y}..)^2 + \sum_{j=1}^{c}\sum_{i=1}^{n_j}(Y_{ij} - \bar{Y}._j)^2 \tag{3.8}$$

3.3.3 The Sums of Squares

The term on the left-hand side of (3.8) represents the summation of squared differences between every observation and the *overall* or "grand" mean common to all the observations. This is called the *total sum of squares* or *total variation* and is given by the symbol SST.

The first expression on the right-hand side of (3.8) can, with the use of summation notation, be restated as

$$\sum_{j=1}^{c}\sum_{i=1}^{n_j}(\bar{Y}._j - \bar{Y}..)^2 = \sum_{j=1}^{c} n_j(\bar{Y}._j - \bar{Y}..)^2 \tag{3.9}$$

When written in this manner, the term represents the summation of squared differences between each sample mean and the overall mean—*weighted* by the number of observations in each sample (group). This is called the *sum of squares among groups* or *among group variation* and is given by the symbol SSα.

The second expression on the right-hand side of (3.8) represents the summation of squared differences between each observation and its own sample mean—cumulated over all c levels (groups). This is called the *sum of squares within groups* or *within group variation* and is given by the symbol SSW.

Thus, from (3.8), the total sum of squares has been decomposed into two additive parts—the sum of squares among groups plus the sum of squares within groups:

$$SST = SS\alpha + SSW \qquad (3.10)$$

By using the data of interest to the financial analyst (see Table 3.3), the following statistics are computed:

$$\bar{Y}_{.1} = 5.843 \qquad \bar{Y}_{.2} = 3.632 \qquad \bar{Y}_{.3} = 3.313 \qquad \bar{Y}_{..} = 4.208$$

and the decomposition of the total sum of squares SST into its two parts, SSα and SSW, is depicted in Figure 3.2.

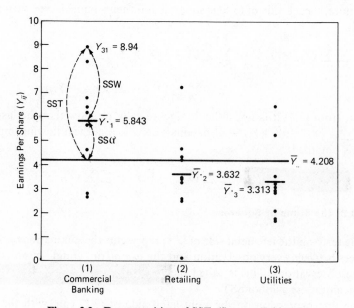

Figure 3.2   Decomposition of SST. (Source: Table 3.3.)

Unfortunately, in practical applications it is usually rather cumbersome to compute the sums of squares components from (3.8). Therefore, the following *computational* formulas for SSα, SSW, and SST are obtained through algebraic expansion:

$$SS\alpha = \sum_{j=1}^{c} \frac{Y_{\cdot j}^2}{n_j} - \frac{Y_{\cdot\cdot}^2}{n} \equiv \sum_{j=1}^{c} \sum_{i=1}^{n_j} (\bar{Y}_{\cdot j} - \bar{Y}_{\cdot\cdot})^2 \qquad (3.11)$$

$$SSW = \sum_{j=1}^{c} \sum_{i=1}^{n_j} Y_{ij}^2 - \sum_{j=1}^{c} \frac{Y_{\cdot j}^2}{n_j} \equiv \sum_{j=1}^{c} \sum_{i=1}^{n_j} (Y_{ij} - \bar{Y}_{\cdot j})^2 \qquad (3.12)$$

$$SST = \sum_{j=1}^{c} \sum_{i=1}^{n_j} Y_{ij}^2 - \frac{Y_{\cdot\cdot}^2}{n} \equiv \sum_{j=1}^{c} \sum_{i=1}^{n_j} (Y_{ij} - \bar{Y}_{\cdot\cdot})^2 \qquad (3.13)$$

so that (3.8) may be rewritten as

$$\sum_{j=1}^{c} \sum_{i=1}^{n_j} Y_{ij}^2 - \frac{Y_{\cdot\cdot}^2}{n} = \left(\sum_{j=1}^{c} \frac{Y_{\cdot j}^2}{n_j} - \frac{Y_{\cdot\cdot}^2}{n} \right) + \left(\sum_{j=1}^{c} \sum_{i=1}^{n_j} Y_{ij}^2 - \sum_{j=1}^{c} \frac{Y_{\cdot j}^2}{n_j} \right) \qquad (3.14)$$

Using (3.14) for the example of interest to the financial analyst, we observe from the summary computations in Table 3.3 that

$$122.5092 = 35.3968 + 87.1124$$

If each of these sums of squares is divided by its respective degrees of freedom, we obtain three *variances* or mean square terms—MSα, MSW, and MST:

$$S_{\hat{\alpha}}^2 = MS\alpha = \frac{SS\alpha}{c-1} \qquad (3.15)$$

$$S_W^2 = MSW = \frac{SSW}{n-c} \qquad (3.16)$$

$$S_T^2 = MST = \frac{SST}{n-1} \qquad (3.17)$$

Since there are c levels of the factor being compared, there are $c - 1$ degrees of freedom associated with the *mean squares among groups* term. On the other hand, there are $n - c$ degrees of freedom associated with the *mean squares within groups* term since each of the c levels contribute $n_j - 1$ degrees of freedom and

$$\sum_{j=1}^{c} (n_j - 1) = n - c \qquad (3.18)$$

Moreover, the *total mean squares* term contains $n - 1$ degrees of freedom because each observation Y_{ij} is being compared to the overall mean $\bar{Y}_{\cdot\cdot}$ based on all n observations.

We note that the degrees of freedom for the *total* is equal to the sum of the degrees of freedom associated with the *among* groups term plus that for the *within* groups term:

$$n - 1 = (c - 1) + (n - c) \qquad (3.19)$$

Hence we have observed that both degrees of freedom and sums of squares are additive; however, mean squares (i.e., variances) are not.

3.3.4 Obtaining the F Statistic

As previously stated, although the researcher is primarily interested in comparing the means to determine whether a treatment effect exists, the ANOVA procedure derives its name from the fact that this is achieved by analyzing variances. If the null hypothesis is true and there are no real differences in the c means, all three mean square terms—MSα, MSW, and MST—provide estimates of the true variance σ^2 inherent in the data.* Since it can be shown that MSα and MSW are independent, we may use the formulas from Section 1.6.3 [see (1.16) and (1.18)] as well as from Table 1.3 (formula G) to state that under the null hypothesis

$$F = \frac{MS\alpha}{MSW} \sim F_{c-1,n-c} \tag{3.20}$$

and the F statistic presented in (3.1) is obtained.

If H_0 is true, we expect the computed F statistic to be approximately equal to 1. On the other hand, if H_0 is false (and there are significant differences in the c means), we expect the computed F ratio to be significantly larger than 1 because the numerator, MSα, would be estimating the significant treatment effect in addition to the inherent variability in the data, whereas the denominator, MSW, would only be measuring the inherent variability. Hence the ANOVA yields a one-tailed F test in which the null hypothesis can be rejected only if the computed F statistic is sufficiently large enough to equal or exceed $F_{1-\alpha; c-1, n-c}$, the upper-tail critical value of the F distribution having $c - 1$ and $n - c$ degrees of freedom, as illustrated in Figure 3.3.

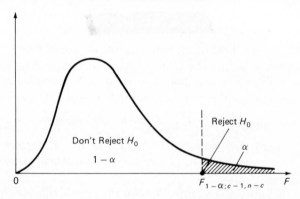

Figure 3.3   Regions of rejection and nonrejection when using ANOVA to test H_0.

3.3.5 The ANOVA Table

Since there are many steps involved in the determination of the F statistic, these computations should be summarized into an analysis of variance (ANOVA) table (see Table 3.4) which includes the sources of variation, the degrees of free-

---

*From Section 1.4 we recall that a variance is computed by dividing the sum of squared deviations (differences) by its appropriate degrees of freedom. Thus the mean square terms are all variances.

dom, the sums of squares, the mean squares (i.e., variances), and the calculated F value.

For the problem of interest to the financial analyst (see Section 3.2.1), the relevant computations can now be summarized as in Table 3.5.

TABLE 3.4   ANOVA Table

Source of Variation	Degrees of Freedom	Sum of Squares	Mean Square	F
Among Groups	$c - 1$	$SS\alpha = \sum_{j=1}^{c} \dfrac{Y_{\cdot j}^2}{n_j} - \dfrac{Y_{\cdot\cdot}^2}{n}$	$MS\alpha = \dfrac{SS\alpha}{c - 1}$	$F = \dfrac{MS\alpha}{MSW}$
Within Groups	$n - c$	$SSW = \sum_{j=1}^{c} \sum_{i=1}^{n_j} Y_{ij}^2 - \sum_{j=1}^{c} \dfrac{Y_{\cdot j}^2}{n_j}$	$MSW = \dfrac{SSW}{n - c}$	
Total	$n - 1$	$SST = \sum_{j=1}^{c} \sum_{i=1}^{n_j} Y_{ij}^2 - \dfrac{Y_{\cdot\cdot}^2}{n}$		

TABLE 3.5   ANOVA Table to Compare Earnings per Share Among Three Kinds of Companies

Source of Variation	Degrees of Freedom	Sum of Squares	Mean Square	F
Among Groups	2	35.3968	17.6984	5.28
Within Groups	26	87.1124	3.3505	
Total	28	122.5092		

SOURCE: Table 3.3.

3.4 ASSUMPTIONS

When the format for the planned experiment is a *completely randomized* design, the classical ANOVA procedure is based on the following four assumptions:

1. Independence
2. Normality
3. Homoscedasticity (i.e., homogeneity of variances)
4. Additivity

3.4.1 Independence Assumption

It must be assumed that the subjects or experimental units in each of the c independent sample groups (i.e., levels of the factor) are randomly and independently drawn so that an observed value in any one group has no effect or influence on any other observed value in that group or any of the other groups. This assumption of independence cannot be relaxed. If it is violated, the ANOVA procedure discussed here is invalid.

3.4.2 Normality and Homoscedasticity Assumptions

It is also assumed that the observed responses in each of the groups represent random samples drawn from c independent normal populations having equal variances (i.e., $\sigma_1^2 = \sigma_2^2 = \cdots = \sigma_c^2 = \sigma^2$). If the distributional assumption of normality (*shape*) and the property of homoscedasticity (*equal variability* from group to group) hold, any differences in the c populations must be attributed to central tendency, that is, differences in the means due to a significant treatment effect among the c levels of the factor. This was illustrated in panel B of Figure 3.1; however, as depicted in panel A of Figure 3.1, if the null hypothesis is true, each of the c samples can be considered as coming from the *same* normal population or from *identical* normal populations.

Normality. The assumption of normality implies that the observed responses Y_{ij} are *continuous* (see Sections 1.2 and 1.5.2) so that either an *interval* or *ratio* level of measurement (see Section 1.3) has been attained. It is interesting to note, however, that the assumption can be relaxed somewhat without affecting the results of the ANOVA. Thus, the one-way F test is considered robust to moderate departures from normality.*

Unless we have much experience with the kind of data being studied, it would be unlikely, in practice, that we would actually know whether our c samples have, in fact, been drawn from underlying normal populations. Although several procedures have been developed to test the assumption of normality [see, for example, Shapiro et al. (1968)], it is very difficult to make such a judgment when the c sample sizes are very small. If, however, we have good reason to believe this assumption has been violated, a good strategy is to seek appropriate transformations to normalize the data (see Section 1.8.1) or use appropriate *nonparametric* procedures [see Conover (1980)].

Homoscedasticity. When the c sample sizes are equal, the one-factor F test is also robust to moderate violations in the assumption of homoscedasticity [see Box and Andersen (1955)]. However, more than moderate departures from the assumption would seriously affect the α level and power, thus invalidating the usage of the test. Therefore, several procedures are available to test the assumption of homoscedasticity. Among the most well known is the F_{max} procedure developed by Hartley (1950). To test

$$H_0: \quad \sigma_1^2 = \sigma_2^2 = \cdots = \sigma_c^2 = \sigma^2$$

---

*Recent research [Berenson (1982)] has indicated that when the c populations are not normal, shape (symmetrical versus nonsymmetrical populations) has less of an influence on the ANOVA procedure than does density in the tails (heavy-tailed versus light-tailed populations).

against the alternative

$$H_1: \text{ not all } \sigma_j^2 \text{ are equal} \quad (j = 1, 2, \ldots, c)$$

each of the c sample variances $(S_{\cdot 1}^2, S_{\cdot 2}^2, \ldots, S_{\cdot c}^2)$ are computed, and the F_{max} statistic is simply obtained as the ratio of the largest of the c sample variances to that of the smallest. That is,

$$F_{max} = \frac{S_{largest}^2}{S_{smallest}^2} \sim F_{max(c, [\bar{n}.]-1)} \tag{3.21}$$

If, for a specified level of significance α, the computed F_{max} equals or exceeds $F_{max(1-\alpha);c,[\bar{n}.]-1}$, the upper-tail critical value of Hartley's F_{max} distribution based on c and $[\bar{n}.] - 1$ degrees of freedom (see Appendix B, Table B.5), the null hypothesis is rejected.

It is important to note that the Hartley test was devised for c equal-sized samples (so that $[\bar{n}.] \equiv n_j$, where $j = 1, 2, \ldots, c$). If, as in the example of interest to the financial analyst (Section 3.2.1), the sample sizes differ, an *approximate* F_{max} test can be attained by using $[\bar{n}.]$, the *integer* portion of the average sample size.

By using the data of Table 3.3,

$$S_{\cdot 1}^2 = 4.6925 \qquad S_{\cdot 2}^2 = 3.2313 \qquad S_{\cdot 3}^2 = 2.2767$$

and testing the null hypothesis

$$H_0: \quad \sigma_1^2 = \sigma_2^2 = \sigma_3^2$$

against the alternative

$$H_1: \text{ not all } \sigma_j^2 \text{ are equal} \quad (j = 1, 2, 3),$$

the F_{max} statistic is computed from (3.21) to be

$$F_{max} = \frac{4.6925}{2.2767} = 2.061$$

Here the sample sizes are unequal so that

$$[\bar{n}.] = \left[\frac{\sum_{j=1}^{c} n_j}{c} \right] = \left[\frac{9 + 10 + 10}{3} \right] = [9.67] = 9$$

and therefore $F_{max(.95); 3,8} = 6.00$ is the upper-tail critical value. Since 2.061 does not equal or exceed this upper-tail critical value, the null hypothesis cannot be rejected. Thus there is no evidence to conclude that the homoscedasticity assumption does not hold.

Unfortunately, however, the Hartley F_{max} procedure is not robust. It is extremely sensitive to departures from normality in the data and if there are such departures we should view the F_{max} test as inappropriate. Therefore, the F_{max} procedure should be used with caution. Unless the researcher is confident that the observed responses in each of the c groups are at least approximately normally distributed, the F_{max} test should not be used. In these situations the researcher should consider a modified Levene L test (which is robust, albeit more computationally complex).*

The L statistic is computed from

$$L = \frac{\left[\sum_{j=1}^{c} n_j(\bar{W}_{\cdot j} - \bar{W}_{\cdot\cdot})^2\right]\Big/(c-1)}{\left[\sum_{j=1}^{c}\sum_{i=1}^{n_j}(W_{ij} - \bar{W}_{\cdot j})^2\right]\Big/(n-c)} \sim F_{c-1,n-c} \qquad (3.22)$$

where $W_{ij} = |Y_{ij} - \mathfrak{M}_{\cdot j}|$ is the absolute difference between the ith observation in the jth group and the sample median of that jth group, $\bar{W}_{\cdot j} = \sum_{i=1}^{n_j} W_{ij}/n_j$ is the mean of the absolute differences in group j, and $\bar{W}_{\cdot\cdot} = \sum_{j=1}^{c}\sum_{i=1}^{n_j} W_{ij}/n$ is the *overall* mean common to all the absolute differences.

Comparing (3.22) to the expressions developed in Section 3.3, we see that the Levene L test is essentially an ANOVA procedure on the absolute differences W_{ij}. If, for a specified level of significance α, the computed L statistic equals or exceeds $F_{1-\alpha;\, c-1,n-c}$, the upper-tail critical value of the F distribution having $c-1$ and $n-c$ degrees of freedom (see Appendix B, Table B.4), the null hypothesis would be rejected.

By analogy to (3.1), for computational purposes we may re-express (3.22) as

$$L = \frac{\left\{\sum_{j=1}^{c}(W_{\cdot j}^2/n_j) - (W_{\cdot\cdot}^2/n)\right\}\Big/(c-1)}{\left[\sum_{j=1}^{c}\sum_{i=1}^{n_j} W_{ij}^2 - \sum_{j=1}^{c}(W_{\cdot j}^2/n_j)\right]\Big/(n-c)} \qquad (3.23)$$

where $\quad \sum_{j=1}^{c}\sum_{i=1}^{n_j} W_{ij}^2 =$ sum of the squares of each of the absolute differences

$$W_{\cdot j} = \sum_{i=1}^{n_j} W_{ij} = \text{total in group } j$$

$$W_{\cdot j}^2 = \text{square of the total in group } j$$

$$W_{\cdot\cdot} = \sum_{j=1}^{c}\sum_{i=1}^{n_j} W_{ij} = \text{``grand'' total of each of the absolute differences}$$

$$W_{\cdot\cdot}^2 = \text{square of the grand total}$$

For the financial analyst's data (Table 3.3), the three sample medians are

$$\mathfrak{M}_{\cdot 1} = 6.20 \qquad \mathfrak{M}_{\cdot 2} = 3.51 \qquad \mathfrak{M}_{\cdot 3} = 3.01$$

and the set of absolute deviations W_{ij} are displayed in Table 3.6.

From Table 3.6 the following summary computations are obtained:

$n_1 = 9$	$n_2 = 10$	$n_3 = 10$	$n = 29$
$W_{\cdot 1} = 14.61$	$W_{\cdot 2} = 11.72$	$W_{\cdot 3} = 10.09$	$W_{\cdot\cdot} = 36.42$
$W_{\cdot 1}^2 = 213.4521$	$W_{\cdot 2}^2 = 137.3584$	$W_{\cdot 3}^2 = 101.8081$	$W_{\cdot\cdot}^2 = 1326.4164$

$$\sum_{j=1}^{c}\sum_{i=1}^{n_j} W_{ij}^2 = 89.3242$$

*Conover et al. (1981) recently evaluated more than 50 procedures for testing the homogeneity of variance hypothesis and concluded that a Brown-Forsythe (1974) modification of the Levene test (1960) is among the most powerful and robust with respect to violations in the assumption of normality. Their modification involves the use of the sample median $\widehat{\mathfrak{M}}_{\cdot j}$ to obtain the absolute differences W_{ij} in lieu of the sample mean $\bar{Y}_{\cdot j}$ as initially described by Levene.

TABLE 3.6 Absolute Differences ($W_{ij} = |Y_{ij} - \hat{\mathfrak{M}}_{.j}|$) for Three Nonindustrial Business Groupings

(1) Commercial Banking	(2) Retailing	(3) Utility
.22	.01	.54
3.37	.70	.88
2.74	.85	.23
.60	.84	3.46
.50	.02	.05
1.55	1.17	1.21
.00	.21	2.28
3.49	.83	.05
2.14	3.74	.11
	3.35	1.28

SOURCE: Table 3.3.

By using (3.23), the null hypothesis

$$H_0: \quad \sigma_1^2 = \sigma_2^2 = \sigma_3^2$$

is tested against the alternative

$$H_1: \quad \text{not all } \sigma_j^2 \text{ are equal} \qquad (j = 1, 2, 3)$$

and the modified Levene L statistic is computed to be

$$L = \frac{1.89505 \div 2}{41.69065 \div 26} = .591$$

Since $L < F_{.95; \, 2, 26} = 3.37$, the null hypothesis would not be rejected. Hence there is insufficient evidence to conclude that the homoscedasticity assumption is violated. (Note that the same conclusion was obtained using the F_{max} procedure.)

As a consequence, the financial analyst would then have been ready to perform the ANOVA of Section 3.2.1 in order to compare the means of the various groups (levels).

Resolving the problem of unequal variances. If the researcher is unwilling to make the assumption of homogeneity of variances, there are three available options:

1. A variance-stabilizing data transformation may be used (see Section 1.8.1) with the ANOVA procedure performed on the transformed data.
2. An approximate solution to the c-sample Behrens-Fisher problem may be obtained directly from the observed data if it can still be assumed that the samples are drawn from c independent normal populations.

3. A *nonparametric* procedure [see Conover (1980)] may be employed to test the null hypothesis that each of the samples are drawn from the same population or from c independent and identically distributed populations against the alternative that not all the populations are the same—without requiring such stringent assumptions as normality and homoscedasticity.

Table 3.7 summarizes the most common types of transformations mentioned in Section 1.8.1. Such transformations are intended to normalize the data and stabilize the variances. The researcher must search for and then make the appropriate transformation prior to performing the ANOVA.

TABLE 3.7   Common Data Transformations

Situation	Transformation from Y_{ij} to $\mathcal{Y}_{ij}$
1. Y_{ij} is a proportion, percentage, or rate	$\mathcal{Y}_{ij} = \arcsin \sqrt{Y_{ij}}$; see Appendix B, Table B.7
2. Y_{ij} is a frequency count	$\mathcal{Y}_{ij} = \sqrt{Y_{ij}}$
3. $\dfrac{\bar{Y}_{\cdot 1}}{S_{\cdot 1}^2} \cong \dfrac{\bar{Y}_{\cdot 2}}{S_{\cdot 2}^2} \cong \cdots \cong \dfrac{\bar{Y}_{\cdot c}}{S_{\cdot c}^2}$ (where $Y_{ij} \geq 0$)	$\mathcal{Y}_{ij} = \sqrt{Y_{ij}}$
4. $\dfrac{\bar{Y}_{\cdot 1}}{S_{\cdot 1}} \cong \dfrac{\bar{Y}_{\cdot 2}}{S_{\cdot 2}} \cong \cdots \cong \dfrac{\bar{Y}_{\cdot c}}{S_{\cdot c}}$ (where $Y_{ij} \geq 0$)	$\mathcal{Y}_{ij} = \ln Y_{ij}$
5. $\dfrac{\bar{Y}_{\cdot 1}^2}{S_{\cdot 1}} \cong \dfrac{\bar{Y}_{\cdot 2}^2}{S_{\cdot 2}} \cong \cdots \cong \dfrac{\bar{Y}_{\cdot c}^2}{S_{\cdot c}}$ (where $Y_{ij} \geq 0$)	$\mathcal{Y}_{ij} = \dfrac{1}{Y_{ij}}$

On the other hand, we may recall that if the researcher believes that the c populations are normal but is unwilling to assume that their variances are equal, a transformation should not be used since one which stabilizes the variance would likely destroy the normality. In such circumstances, a c-sample Behrens-Fisher problem is said to exist, and the following approximation method suggested by Welch (1951) may be employed directly on the original data to compare the c means in lieu of the usual ANOVA procedure*:

$$F^* = \frac{\left[\sum_{j=1}^{c} \omega_j (\bar{Y}_{\cdot j} - \bar{Y}_{\cdot}^*)^2\right]\Big/(c-1)}{1 + \{[2(c-2)]/(c^2-1)\}\phi} \tag{3.24}$$

where

$$\omega_j = \frac{n_j}{S_{\cdot j}^2}$$

*An extensive simulation study by Kohr and Games (1974) demonstrated the value of the F^* test and led to the suggestion that it be selected over another useful approximation method devised from the work of Box (1954).

$$\bar{Y}^*_{\cdot\cdot} = \frac{\sum\limits_{j=1}^{c} \omega_j \bar{Y}_{\cdot j}}{\sum\limits_{j=1}^{c} \omega_j}$$

$$\phi = \sum_{j=1}^{c} \frac{1}{n_j - 1}\left(1 - \frac{\omega_j}{\sum\limits_{j=1}^{c}\omega_j}\right)^2 = \frac{1}{\left(\sum\limits_{j=1}^{c}\omega_j\right)^2}\sum_{j=1}^{c}\frac{\left(\sum\limits_{j=1}^{c}\omega_j - \omega_j\right)^2}{n_j - 1}$$

The null hypothesis of no real treatment effects would be rejected if F^* equals or exceeds $F_{1-\alpha;\,c-1,\,v^*}$, the upper-tail critical value of the F distribution having $c - 1$ and $v^* = (c^2 - 1)/3\phi$ degrees of freedom.

For illustrative purposes, had the financial analyst not wished to make the assumption of homoscedasticity, the Welch approximation method (3.24) could have been used directly on the data of Table 3.3 and the F^* statistic computed as follows:

$\bar{Y}_{\cdot j}$	$S^2_{\cdot j}$	n_j	$\omega_j = n_j/S^2_{\cdot j}$	$\omega_j \bar{Y}_{\cdot j}$
5.843	4.6925	9	1.918	11.207
3.632	3.2313	10	3.095	11.241
3.313	2.2767	10	4.392	14.551
			$\sum\limits_{j=1}^{c} \omega_j = 9.405$	$\sum\limits_{j=1}^{c} \omega_j \bar{Y}_{\cdot j} = 36.999$

$$\bar{Y}^*_{\cdot\cdot} = \frac{\sum\limits_{j=1}^{c} \omega_j \bar{Y}_{\cdot j}}{\sum\limits_{j=1}^{c} \omega_j} = \frac{36.999}{9.405} = 3.934$$

and

$$\phi = \frac{1}{\left(\sum\limits_{j=1}^{c}\omega_j\right)^2}\sum_{j=1}^{c}\frac{\left(\sum\limits_{j=1}^{c}\omega_j - \omega_j\right)^2}{n_j - 1}$$

$$= \frac{1}{(9.405)^2}\left[\frac{(9.405 - 1.918)^2}{9 - 1} + \frac{(9.405 - 3.095)^2}{10 - 1} + \frac{(9.405 - 4.392)^2}{10 - 1}\right]$$

$$= .160797$$

so that

$$F^* = \frac{\left[\sum\limits_{j=1}^{c}\omega_j(\bar{Y}_{\cdot j} - \bar{Y}^*_{\cdot\cdot})^2\right]\bigg/(c - 1)}{1 + \{[2(c - 2)]/(c^2 - 1)\}\phi}$$

$$= \frac{[(1.918)(5.843 - 3.934)^2 + (3.095)(3.632 - 3.934)^2 + (4.392)(3.313 - 3.934)^2]\bigg/(3 - 1)}{1 + \{[2(3 - 2)]/(3^2 - 1)\}(.160797)}$$

$$= 4.31$$

In this problem the critical value for F^* has $c - 1 = 2$ and $v^* = (c^2 - 1)/3\phi$ $= 16.7 \longrightarrow 17$ degrees of freedom. Since $F^* > F_{.95;\,2,17} = 3.59$, the null hypothesis would be rejected (the P value is between .025 and .05).

Such an approximation method depends on *separate variance estimates*—as did the method described in Table 1.3 (formula F) for the two-sample situation. It is important to note, however, that when the assumption of equal variances holds, each $S_{\cdot j}^2$ (where $j = 1, 2, \ldots, c$) is estimating the common population variance σ^2. Then, as in Table 1.3 (formula E) for the two-sample problem, the best estimate of σ^2 is the *pooled* or *combined* estimate S_P^2—a weighted average of the c sample variances (which utilizes the degrees of freedom as the weights). That is,

$$S_P^2 = \frac{(n_1 - 1)S_{\cdot 1}^2 + (n_2 - 1)S_{\cdot 2}^2 + \cdots + (n_c - 1)S_{\cdot c}^2}{(n_1 - 1) + (n_2 - 1) + \cdots + (n_c - 1)} \tag{3.25}$$

which may be re-expressed more simply as

$$S_P^2 = \sum_{j=1}^{c} \frac{(n_j - 1)S_{\cdot j}^2}{n_j - 1} = \frac{\sum_{j=1}^{c} (n_j - 1)S_{\cdot j}^2}{n - c} \tag{3.26}$$

The last equality in (3.26) may also be rewritten as

$$\frac{\sum_{j=1}^{c} (n_j - 1)S_{\cdot j}^2}{n - c} = \frac{\sum_{j=1}^{c} \sum_{i=1}^{n_j} (Y_{ij} - \bar{Y}_{\cdot j})^2}{n - c} = \frac{\sum_{j=1}^{c} \sum_{i=1}^{n_j} Y_{ij}^2 - \sum_{j=1}^{c} (Y_{\cdot j}^2/n_j)}{n - c} \tag{3.27}$$

which, of course, is the MSW term (3.16) for the F test in (3.1).

Finally, if the researcher is not willing to consider potentially appropriate data transformations merely for the purpose of employing the ANOVA procedure and, in addition, is also unwilling to assume the c populations are normal, nonparametric tests can be utilized. Such procedures, however, are outside the scope of this book [for example, see Conover (1980)].

3.4.3 Additivity Assumption

The final assumption necessary for performing the ANOVA procedure is that both the model and its effects are additive. In Section 3.3.2 we developed the F test for the additive, fixed-effects model

$$Y_{ij} = \mu + \alpha_j + \epsilon_{ij} \tag{3.2}$$

Of course, if this additive model does not hold, our F test is invalid.

We may now define the treatment effects in such a manner that $\sum_{j=1}^{c} \alpha_j = 0$ when the c sample sizes are equal and $\sum_{j=1}^{c} n_j \alpha_j = 0$ when the sample sizes are not all equal. If, however, the null hypothesis is true and no treatment effects are present (that is, $\alpha_1 = \alpha_2 = \cdots = \alpha_c = 0$), our model reduces to

$$Y_{ij} = \mu + \epsilon_{ij} \tag{3.3}$$

and, as previously mentioned in Section 3.3.2, the only differences from one observed response to another would be due to chance.

In summary, then, under a true null hypothesis each observation Y_{ij} would be randomly and independently drawn from the same (or identical) normal population(s) having overall mean μ and variance σ^2. Since μ is a constant, it is the random or experimental error term ϵ_{ij} which must be normally distributed with mean 0 and variance σ^2 [that is, $\epsilon_{ij} \sim \mathcal{N}(0,\sigma^2)$]. However, under a true alternative hypothesis we see from (3.2) that each observation Y_{ij} would be composed of μ (an overall constant), a treatment effect α_j (an additive constant for a particular level j), and experimental error ϵ_{ij}.* Hence the ϵ_{ij} must again be thought of as normally distributed with mean 0 and variance σ^2.

3.5 CONFIDENCE INTERVAL ESTIMATION

Now that we have examined the concepts of ANOVA and studied its assumptions, we are ready to return to the more practical aspects of data analysis. As part of the overall investigation of the results, the researcher may find it desirable to make confidence interval estimates of the mean response to particular levels of a factor. Thus a $100(1 - \alpha)\%$ confidence interval estimate of $\mu_{\cdot j}$, the true mean of the jth level, is given by

$$\bar{Y}_{\cdot j} - t_{1-\frac{\alpha}{2};\, n-c}\sqrt{\frac{\text{MSW}}{n_j}} \leq \mu_{\cdot j} \leq \bar{Y}_{\cdot j} + t_{1-\frac{\alpha}{2};\, n-c}\sqrt{\frac{\text{MSW}}{n_j}} \qquad (3.28)$$

where $j = 1, 2, \ldots, c$.

We note that since the homoscedasticity assumption would have already been tested (Section 3.4.2), it would be more appropriate to use the *pooled variance* estimate (i.e., the within groups mean square term MSW with its associated $n - c$ degrees of freedom) which "combines together" all c estimates of σ^2 than to use the particular (i.e., *separate*) sample variance estimate $S^2_{\cdot j}$ (based on only $n_j - 1$ degrees of freedom) as shown in Table 1.2, formula B.

As an example, using (3.28), the financial analyst would conclude that a 95% confidence interval estimate of the true mean earnings per share of all commercial banking companies is

$$\bar{Y}_{\cdot 1} - t_{.975;\, 26}\sqrt{\frac{\text{MSW}}{n_1}} \leq \mu_{\cdot 1} \leq \bar{Y}_{\cdot 1} + t_{.975;\, 26}\sqrt{\frac{\text{MSW}}{n_1}}$$

$$5.843 - 2.056\sqrt{\frac{3.3505}{9}} \leq \mu_{\cdot 1} \leq 5.843 + 2.056\sqrt{\frac{3.3505}{9}}$$

$$4.589 \leq \mu_{\cdot 1} \leq 7.097$$

Note that this interval differs somewhat from the interval using a separate sample variance estimate as presented in case 1 of Section 1.7.3.

---

*If the levels of the factor are fixed, α_j is a particular constant for each level j. If the levels of the factor are random, $\alpha_j \sim \mathcal{N}(0, \sigma^2_\alpha)$ for $j = 1, 2, \ldots, c$. Distinctions among fixed, random, and mixed models will be discussed in Chapter 6.

3.6 COMPUTER PACKAGES AND ONE-WAY ANOVA: USE OF SPSS, SAS, AND BMDP

3.6.1 Organizing the Data for Computer Analysis

In this section we shall explain how to use either SPSS, SAS, or BMDP to analyze a completely randomized experimental design model. For illustrative purposes we shall focus on the problem of interest to the financial analyst involving the earnings per share of companies under three nonindustrial groupings. For each company, two pieces of information are available: the earnings per share (in dollars) and the business grouping in which the company is classified. In addition, each company can be assigned an identification number ranging from 1 to 29 (the total number of observations). The data can then be organized in the format presented in Table 3.8. For the purpose of data entry the identification number can be assigned to columns 1–2, earnings per share to columns 4–7, and business group classification to column 9.

3.6.2 The SPSS Subprogram BREAKDOWN

The BREAKDOWN subprogram [see Nie et al. (1975)] can be utilized to perform a classical one-way analysis of variance since it "breaks down" a given quantitative variable into groups according to the values of a qualitative variable. The required BREAKDOWN procedure statements are

1   16

$$\text{BREAKDOWN} \quad \text{TABLES} = \left\{\begin{matrix}\text{quantitative (dependent)}\\ \text{variable } (Y) \text{ or list}\end{matrix}\right\} \emptyset\text{BY}\emptyset\left\{\begin{matrix}\text{qualitative}\\ \text{variable}\end{matrix}\right\}/$$

STATISTICS   1

Beginning in column 16, the quantitative (i.e., response) variable to be analyzed is named. This is followed (after the word BY) by the qualitative variable used to form groups. STATISTICS 1 must be utilized in order to obtain the analysis of variance table.

Figure 3.4 illustrates the complete SPSS program (including classical procedures as well as *multiple comparison* methods) that has been written for the problem of interest to the financial analyst.* Figure 3.5 represents annotated partial output.

3.6.3 The SAS GLM Procedure

The SAS GLM procedure [see SAS (1979)] can be utilized for classical analysis of variance models including cases in which there are unequal sample sizes in each group. The basic setup for using PROC GLM for a completely randomized experimental design model is

---

*Although multiple comparison methods are discussed in Chapter 4, they would be included along with the one-way ANOVA procedure when utilizing a computer package for a completely randomized model.

In summary, then, under a true null hypothesis each observation Y_{ij} would be randomly and independently drawn from the same (or identical) normal population(s) having overall mean μ and variance σ^2. Since μ is a constant, it is the random or experimental error term ϵ_{ij} which must be normally distributed with mean 0 and variance σ^2 [that is, $\epsilon_{ij} \sim \mathcal{N}(0, \sigma^2)$]. However, under a true alternative hypothesis we see from (3.2) that each observation Y_{ij} would be composed of μ (an overall constant), a treatment effect α_j (an additive constant for a particular level j), and experimental error ϵ_{ij}.* Hence the ϵ_{ij} must again be thought of as normally distributed with mean 0 and variance σ^2.

3.5 CONFIDENCE INTERVAL ESTIMATION

Now that we have examined the concepts of ANOVA and studied its assumptions, we are ready to return to the more practical aspects of data analysis. As part of the overall investigation of the results, the researcher may find it desirable to make confidence interval estimates of the mean response to particular levels of a factor. Thus a $100(1 - \alpha)\%$ confidence interval estimate of $\mu._j$, the true mean of the jth level, is given by

$$\bar{Y}._j - t_{1-\frac{\alpha}{2};\, n-c}\sqrt{\frac{\text{MSW}}{n_j}} \leq \mu._j \leq \bar{Y}._j + t_{1-\frac{\alpha}{2};\, n-c}\sqrt{\frac{\text{MSW}}{n_j}} \qquad (3.28)$$

where $j = 1, 2, \ldots, c$.

We note that since the homoscedasticity assumption would have already been tested (Section 3.4.2), it would be more appropriate to use the *pooled variance* estimate (i.e., the within groups mean square term MSW with its associated $n - c$ degrees of freedom) which "combines together" all c estimates of σ^2 than to use the particular (i.e., *separate*) sample variance estimate $S^2_{.j}$ (based on only $n_j - 1$ degrees of freedom) as shown in Table 1.2, formula B.

As an example, using (3.28), the financial analyst would conclude that a 95% confidence interval estimate of the true mean earnings per share of all commercial banking companies is

$$\bar{Y}._1 - t_{.975;\, 26}\sqrt{\frac{\text{MSW}}{n_1}} \leq \mu._1 \leq \bar{Y}._1 + t_{.975;\, 26}\sqrt{\frac{\text{MSW}}{n_1}}$$

$$5.843 - 2.056\sqrt{\frac{3.3505}{9}} \leq \mu._1 \leq 5.843 + 2.056\sqrt{\frac{3.3505}{9}}$$

$$4.589 \leq \mu._1 \leq 7.097$$

Note that this interval differs somewhat from the interval using a separate sample variance estimate as presented in case 1 of Section 1.7.3.

---

*If the levels of the factor are fixed, α_j is a particular constant for each level j. If the levels of the factor are random, $\alpha_j \sim \mathcal{N}(0, \sigma^2_\alpha)$ for $j = 1, 2, \ldots, c$. Distinctions among fixed, random, and mixed models will be discussed in Chapter 6.

3.6 COMPUTER PACKAGES AND ONE-WAY ANOVA: USE OF SPSS, SAS, AND BMDP

3.6.1 Organizing the Data for Computer Analysis

In this section we shall explain how to use either SPSS, SAS, or BMDP to analyze a completely randomized experimental design model. For illustrative purposes we shall focus on the problem of interest to the financial analyst involving the earnings per share of companies under three nonindustrial groupings. For each company, two pieces of information are available: the earnings per share (in dollars) and the business grouping in which the company is classified. In addition, each company can be assigned an identification number ranging from 1 to 29 (the total number of observations). The data can then be organized in the format presented in Table 3.8. For the purpose of data entry the identification number can be assigned to columns 1–2, earnings per share to columns 4–7, and business group classification to column 9.

3.6.2 The SPSS Subprogram BREAKDOWN

The BREAKDOWN subprogram [see Nie et al. (1975)] can be utilized to perform a classical one-way analysis of variance since it "breaks down" a given quantitative variable into groups according to the values of a qualitative variable. The required BREAKDOWN procedure statements are

1   16

$$\text{BREAKDOWN} \quad \text{TABLES} = \left\{ \begin{array}{c} \text{quantitative (dependent)} \\ \text{variable } (Y) \text{ or list} \end{array} \right\} \not{b} \text{BY} \not{b} \left\{ \begin{array}{c} \text{qualitative} \\ \text{variable} \end{array} \right\} /$$

STATISTICS   1

Beginning in column 16, the quantitative (i.e., response) variable to be analyzed is named. This is followed (after the word BY) by the qualitative variable used to form groups. STATISTICS 1 must be utilized in order to obtain the analysis of variance table.

Figure 3.4 illustrates the complete SPSS program (including classical procedures as well as *multiple comparison* methods) that has been written for the problem of interest to the financial analyst.* Figure 3.5 represents annotated partial output.

3.6.3 The SAS GLM Procedure

The SAS GLM procedure [see SAS (1979)] can be utilized for classical analysis of variance models including cases in which there are unequal sample sizes in each group. The basic setup for using PROC GLM for a completely randomized experimental design model is

*Although multiple comparison methods are discussed in Chapter 4, they would be included along with the one-way ANOVA procedure when utilizing a computer package for a completely randomized model.

TABLE 3.8 Organization of Earnings-Per-Share Data
for Computer Analysis

ID No.	Earnings Per Share ($)	Business Group
1	6.42	1
2	2.83	1
3	8.94	1
4	6.80	1
5	5.70	1
6	4.65	1
7	6.20	1
8	2.71	1
9	8.34	1
10	3.52	2
11	4.21	2
12	4.36	2
13	2.67	2
14	3.49	2
15	4.68	2
16	3.30	2
17	2.68	2
18	7.25	2
19	.16	2
20	3.55	3
21	2.13	3
22	3.24	3
23	6.47	3
24	3.06	3
25	1.80	3
26	5.29	3
27	2.96	3
28	2.90	3
29	1.73	3

PROCƀGLM;
    CLASSESƀ{name(s) of factor(s)};
    $$\text{MODEL}ƀ \left\{ \begin{array}{c} \text{dependent} \\ \text{(quantitative)} \\ \text{variable} \end{array} \right\} = \left\{ \begin{array}{c} \text{name of factor } a \\ \text{(qualitative} \\ \text{variable)} \end{array} \right\};$$
    $$\text{MEANS}ƀ \left\{ \begin{array}{c} \text{name of} \\ \text{factor } a \end{array} \right\};$$

The name of the factor is provided in the CLASSES statement, which *must* precede the MODEL statement. The MODEL statement indicates the dependent (quantitative) variable to be analyzed and the factor (or qualitative variable) to be used to form groups. The MEANS statement provides for the computation of the average value for each group specified.

Figure 3.6 illustrates the complete SAS program (including classical procedures as well as *multiple comparison* methods) written for the financial analyst's problem (see the preceding footnote). Figure 3.7 presents annotated partial output.

```
{SYSTEM CARDS}
RUN NAME       EARNINGS PER SHARE ANALYSIS
DATA LIST      FIXED(1)/1 IDNO 1-2,EARNINGS 4-7,GROUP 9
VAR LABELS     IDNO,IDENTIFICATION NUMBER/
               EARNINGS,EARNINGS PER SHARE ¤/
               GROUP,BUSINESS GROUP CLASSIFICATION/
VALUE LABELS   GROUP(1)BANKING(2)RETAILING(3)UTILITY/
READ INPUT DATA
01 6.42 1
02 2.83 1
...
29 1.73 3
END INPUT DATA
BREAKDOWN      TABLES=EARNINGS BY GROUP/
STATISTICS     1
ONEWAY         EARNINGS BY GROUP(1,3)/
               CONTRAST=0 1-1/
               CONTRAST=1 -.5 -.5/
               RANGES=TUKEY(.05)/
               RANGES=SCHEFFE(.05)/
FINISH
{SYSTEM CARDS}
```

Figure 3.4 SPSS program for earnings-per-share example.

```
- - - - - - - - - - - - - - - - -  D E S C R I P T I O N   O F   S U B P O P U L A T I O N S  - - - - - - - - - -
CRITERION VARIABLE   EARNINGS    EARNINGS PER SHARE $
   BROKEN DOWN BY    GROUP       BUSINESS GROUP CLASSIFICATION
- - - - - - - - - - - - - - - - - - - - - - - - - - - - - - - - - - - - - - - - - - - - - - - - - - - - - - - - - -

VARIABLE                   CODE    VALUE LABEL          SUM           MEAN        STD DEV      VARIANCE         N

FOR ENTIRE POPULATION                                 122.0400       4.2083        2.0917       4.3753   (     29)

GROUP                       1.     BANKING             52.5900       5.8433        2.1662       4.6925   (      9)
GROUP                       2.     RETAILING           36.3200       3.6320        1.7976       3.2313   (     10)
GROUP                       3.     UTILITY             33.1300       3.3130        1.5089       2.2767   (     10)

  TOTAL CASES =        29


              VARIABLE   EARNINGS      EARNINGS PER SHARE $
           BY VARIABLE   GROUP         BUSINESS GROUP CLASSIFICATION

                                              ANALYSIS OF VARIANCE                              (P VALUE)

                        SOURCE             D.F.    SUM OF SQUARES      MEAN SQUARES      F RATIO     F PROB.

                 BETWEEN GROUPS              2         35.3969            17.6984         5.282       0.0119

                 WITHIN GROUPS              26         87.1122             3.3505

                 TOTAL                      28        122.5091
```

Figure 3.5 Partial SPSS output for earnings-per-share data.

```
{SYSTEM CARDS}
DATA EARNINGS;
    INPUT IDNO 1-2 EARNINGS 4-7 GROUP 9;
LABEL IDNO=IDENTIFICATION NUMBER
        EARNINGS=EARNINGS PER SHARE $
        GROUP=BUSINESS GROUP CLASSIFICATION;
CARDS;
01 6.42 1
:   :   :
29 1.73 3
PROC FORMAT;
    VALUE GROUPF 1=BANKING
    2=RETAILING 3=UTILITY;
PROC GLM;
    CLASSES GROUP;
    MODEL EARNINGS=GROUP;
    MEANS GROUP; FORMAT GROUP GROUPF.;
    CONTRAST 'BANK VS OTHERS'
    GROUP 1 -.5 -.5;
    CONTRAST 'RETAIL VS UTIL'
    GROUP 0 1 -1;
{SYSTEM CARDS}
```

Figure 3.6 SAS program for earnings-per-share example.

3.6.4 The BMDP Program 7D

The BMDP program 7D [see Dixon et al. (1981)] can be used to analyze the results obtained from the classical one-way ANOVA model. Not only is the ANOVA table provided, but a histogram is plotted for each group, and the

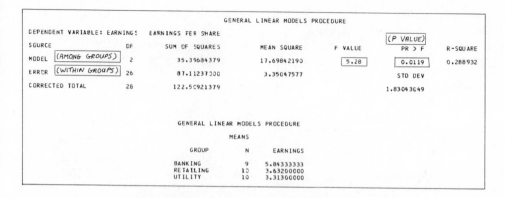

Figure 3.7 Partial SAS output for earnings-per-share data.

(unmodified) Levene test is used to test for the equality of variances. The paragraph of BMDP unique to program 7D is the /HISTOGRAM paragraph. For the one-way ANOVA, the format of this paragraph is

/HISTOGRAMƀGROUPINGƀISƀ{name of factor α}.

Thus, for the financial analyst's problem we would have

/HISTOGRAMƀGROUPINGƀISƀBUSGROUP.

Figure 3.8 illustrates the program using BMDP-7D that was written for the financial analyst's problem. We may observe that a GROUP sentence has been included in the /VARIABLE paragraph to indicate the basis for subdividing the data into separate groups. Figure 3.9 represents annotated partial output.

```
{SYSTEM CARDS}
/PROBLEM    TITLE IS 'ANALYSIS OF EARNINGS PER SHARE'.
/INPUT      VARIABLES ARE 3.
            FORMAT IS '(F2.0,1X,F4.0,1X,F1.0)'.
/VARIABLE   NAMES ARE IDNO, EARNINGS, BUSGROUP.
            LABEL IS IDNO.
            GROUPING IS BUSGROUP.
/GROUP      CODES(3) ARE 1,2,3.
            NAMES(3) ARE BANKING, RETAILING, UTILITY.
/HISTOGRAM  GROUPING IS BUSGROUP.
/END
01  6.42  1
02  2.83  1
 :   :    :
29  1.73  3
{SYSTEM CARDS}
```

Figure 3.8 Using BMDP program 7D for earnings-per-share example.

```
ESTIMATES OF MEANS
-----------------

              BANKING   RETAILIN   UTILITY   TOTAL
                1          2          3         4

EARNINGS  2   5.8433    3.6320     3.3130    4.2033


ONE WAY ANALYSIS OF VARIANCE FOR VARIABLE EARNINGS
**************************************************************************


ANALYSIS OF VARIANCE

SOURCE OF VARIANCE          D.F.    SUM OF SQ.   MEAN SQ.    F-VALUE    TAIL AREA PROBABILITY
EQUALITY OF CELL MEANS        2      35.3968     17.6984     5.2824          0.0119
ERROR                        26      87.1122      3.3505

                                                           (P VALUE)
```

Figure 3.9 Partial BMDP program 7D output for earnings-per-share data.

3.7 SUMMARY

In this chapter the procedures developed for testing for significant treatment effects pertain to a *one-way* analysis of variance since only one *factor* (with several levels) was considered. If the researcher wished to draw conclusions about a study in which two or more factors were simultaneously considered, the experimental design would necessarily be more complex. Such ANOVA procedures are presented in Chapters 5–7. Prior to this, however, it is necessary for us to expand our current investigation of the completely randomized experimental design so as to include methods which enable us to determine which means (or combinations of means) are significantly different from which others.

PROBLEMS

3.1. Hundal (1969) was interested in measuring the effects that knowledge of job performance had on the productivity of industrial workers. An experiment was designed whereby 18 male workers (of similar ability) were randomly assigned to three groups and then given the repetitive task of grinding a metallic piece to a specified size and shape. The data in Table P3.1 constitute the output of each subject during the experimental period.

TABLE P3.1   Output During Experimental Period

Level of Feedback Received		
(1) None	(2) Some	(3) Much
40	38	48
35	40	40
38	47	45
43	44	43
44	40	46
41	42	44

SOURCE: Extracted from Table 1 on p. 225 from P. S. Hundal, "Knowledge of Performance as an Incentive in Repetitive Industrial Work," *Journal of Applied Psychology*, Vol. 53 (1969), pp. 224–226.

(a) Assuming that the data are approximately normally distributed, analyze the results of the experiment. ($\alpha = .05$).

(b) As a research consultant, how would you explain to a client why you used (or didn't use) any data transformations during your analysis?

(c) As a research consultant, how would you explain the following to your client:
(1) Why the experiment is a completely randomized design.

(2) Whether the levels of the factor "feedback" should be considered quantitative or qualitative.

(3) Whether the levels of the factor "feedback" should be considered fixed or random (and how this affects the conclusions drawn).

(d) What can be said about the true mean productivity of such industrial workers if they have received "much feedback" on their job performance?

*3.2. In a test of the effectiveness of advertising on product perception, 30 subjects were randomly assigned to five treatment groups. Each group then received a particular advertisement regarding a ball-point pen. Advertising copy A and B tended to undersell the pen's characteristics, while C and D tended to overstate the pen's characteristics; E attempted to correctly state the pen's characteristics. After reading the advertisement and developing a sense of *product expectancy*, the subjects all received the *same* type of pen to evaluate. The subjects were permitted to test their pen and the plausibility of the advertising copy. The subjects were then asked to "value" the pen by stating the price they would expect to find if they shopped for such a pen at a national retail establishment. The results are given in Table P3.2. Using classical methods, analyze the data to determine whether product perception differs for various levels of expectations (i.e., advertising copy). ($\alpha = .05$).

TABLE P3.2   Price Expectations for $1.19 Ball-Point Pens

Advertising Copy Level				
A	B	C	D	E
1.39	1.39	.95	.85	1.10
1.60	1.50	.88	.90	1.29
1.50	1.89	.99	1.10	.99
1.75	1.45	1.25	1.29	1.19
1.72	1.68	1.00	.98	1.35
2.00	1.47	1.15	1.00	1.37

3.3. The data in Table P3.3 represent the percentage fat content for samples of seven

TABLE P3.3   Percentage Fat Content for Five Brands of Frankfurters

Brand of Frankfurters				
A	B	C	D	E
26	25	31	33	21
19	23	20	36	29
31	28	25	30	20
28	18	31	35	30
22	27	23	32	28
20	26	30	39	26
27	20	22	34	25

1-pound packages of frankfurters for each of five brands tested by a consumer's interest group. Analyze the data to determine whether there are any differences in the percentage fat content among the five brands. ($\alpha = .05$).

3.4. During the registration for the spring semester at a large university the Dean of Students conducted a survey to determine the aspirations of the student body. The data in Table P3.4 represent the anticipated starting salaries (in thousands of dollars) for samples of 10 freshmen, sophomores, juniors, and seniors majoring in business administration.

TABLE P3.4　Anticipated Starting Salaries of Business Administration Majors

	Class Designation		
Freshmen	Sophomores	Juniors	Seniors
14.9	15.8	15.3	15.5
16.4	16.7	16.5	16.2
19.3	14.7	16.0	16.6
18.5	16.3	18.6	15.5
24.5	18.1	16.3	17.0
21.1	22.4	15.2	17.7
17.3	17.5	17.2	16.8
28.0	17.3	17.7	15.9
18.5	16.5	16.9	16.4
16.7	16.6	18.1	16.3

(a) Analyze the data to determine the effects of class designation on anticipated starting salaries. ($\alpha = .05$).

(b) What can be said about the true mean anticipated starting salaries of seniors majoring in business administration?

3.5. An industrial psychologist wished to study the effects of alcoholic consumption on typing ability. Fifteen secretaries of similar experience and ability were randomly assigned to one of three alcoholic consumption levels (0 ounces—placebo, 1 ounce, and 2 ounces). Each secretary was then instructed to type the same standard page. The number of errors made by the secretary was then recorded as displayed in Table P3.5.

TABLE P3.5　Number of Typing Errors

	Alcoholic Consumption (ounces)	
0	1	2
2	7	10
5	5	6
3	6	10
6	3	12
4	9	12

(a) Analyze the data. ($\alpha = .05$).

(b) Reanalyze the data after applying an appropriate data transformation. Compare the results to those obtained in part (a).

3.6. A quality control engineer wished to study the effects of temperature on the life of a particular brand of automobile battery. A batch of 20 batteries of this type were randomly assigned to four groups (so that there were five batteries per group). Each group was then subjected to a particular temperature level: low, normal, high, and very high. All the batteries were simultaneously tested under these temperatures, and the times to failure (in hours) are recorded in Table P3.6.

TABLE P3.6   Effects of Temperature on Battery Life

	Temperature Level		
Low	Normal	High	Very High
8.0	7.6	6.0	5.1
8.1	8.2	6.3	5.6
9.2	9.8	7.1	5.9
9.4	10.9	7.7	6.7
11.7	12.3	8.9	7.8

(a) Analyze the data. ($\alpha = .05$).

(b) What can be said about the true mean battery life under normal temperature levels?

3.7. In an effort to determine the effects of employment on performance of full-time college students, the Dean of Students at a large university asked each member of a class of 33 seniors majoring in accountancy to state his/her cumulative grade point index and to classify himself/herself as having worked full time, part time, or not at all during the previous three academic years. The responses are displayed in Table P3.7 on the following page.

(a) Analyze the data. ($\alpha = .05$).

(b) What can be said about the true mean grade point index of senior accountancy majors who have not been employed?

(c) Discuss some of the problems inherent in obtaining the responses in the manner described and how these might invalidate the conclusions drawn from your analysis in part (a)?

3.8. A behavioral researcher wishes to examine the effects of noise on driving performance to see whether increases in the noise levels have an increasingly detrimental effect on performance. Twenty subjects of similar driving ability are randomly assigned to four treatment groups [i.e., constant noise levels of 95, 100, 105, and 110 decibels (dB)] of five subjects each, and a driving instructor administers the tests with the results as presented in Table P3.8 on page 84.

(a) Using classical methods, analyze the data to determine whether mean driving performance is different under various levels of constant noise. ($\alpha = .05$).

(b) What can be said about the true performance score of drivers subjected to a (constant) noise level of 100 dB?

TABLE P3.7 Measuring the Effects of Employment on Student Performance

	Employment Level	
Full Time	Part Time	Unemployed
2.22	3.13	2.94
2.39	2.44	3.09
2.07	2.90	2.52
2.82	2.93	3.00
2.71	2.64	3.12
$(n_1 = 5)$	3.49	2.79
	2.73	3.39
	3.03	2.85
	2.87	2.49
	2.83	2.88
	3.29	$(n_3 = 10)$
	2.80	
	2.34	
	2.56	
	2.70	
	2.37	
	3.10	
	3.00	
	$(n_2 = 18)$	

TABLE P3.8 Measuring Driving Performance Under Increasing Noise Levels

Treatments (Constant Noise Levels) (dB)			
110	105	100	95
67	64	79	71
65	73	75	82
63	76	66	78
75	68	70	80
72	74	77	72

3.9. A pharmaceutical statistician wished to investigate the effect on the diastolic blood pressure of hyperactive male albino rats under different dosage levels of a newly developed sedative. A group of 24 such rats were randomly assigned (6 each) to four dosage levels, and the diastolic blood pressures were recorded 15 minutes after treatment. The readings are displayed in Table P3.9.

(a) Analyze the data. ($\alpha = .05$).

(b) What can be said about the true mean diastolic blood pressure of such rats under a dosage of 10 milligrams?

TABLE P3.9 Diastolic Blood Pressures
(in Millimeters of Mercury)

Dosage Levels (milligrams)			
5	10	15	20
97	91	86	85
91	89	90	88
90	88	89	83
93	90	91	83
91	95	94	86
96	93	84	85

3.10. Using the relationship between t and F [see (1.20)], show that when the factor of interest has but two levels, the F test given by (3.1) is algebraically equivalent to the square of the two-tailed t test for two independent samples given in Table 1.3 (formula E).

3.11. Using the real estate data base (see Appendix A), select a random sample of 30 houses and determine at the .05 level of significance whether there is any difference in the selling price based on the geographic location.   ■

3.12. Using the real estate data base (see Appendix A), select a random sample of 30 houses and determine at the .01 level of significance whether there is any difference in the assessed value based on the type of garage or parking facility [categorized as none (N), carport (P), single (S), and double (D)].   ■

3.13. Using the real estate data base (see Appendix A), select a random sample of 30 houses and determine at the .05 level of significance whether there is any difference in the selling price based on the rating of a set of built-in features [categorized as poor (P), average (A), and good (G)].   ■

3.14. Reviewing concepts:
  (a) What is a completely randomized experimental design?
  (b) Discuss the differences between the following:
    (1) experimental versus classification factors;
    (2) quantitative versus qualitative factor levels;
    (3) fixed versus random factor levels.
  (c) Describe the ANOVA model.

MULTIPLE COMPARISONS
AND CONTRASTS

4.1 INTRODUCTION

As mentioned in Section 3.2.2, once the null hypothesis of no real differences among the c means has been rejected, it is imperative that the researcher employ *a posteriori* methods to determine specifically what kind of treatment effects exists.

On the other hand, there are situations in which the researcher initially plans an experiment with particular interest in certain comparisons among the c groups. In these circumstances, rather than simply performing an overall analysis by utilizing the classical methods of Chapter 3 (i.e., the one-way F test), the researcher may select an *a priori* method of analysis.

In this chapter both a posteriori and a priori methods will be considered. These types of procedures are invaluable tools for data analysis—not only when the researcher is concerned with completely randomized designs (Chapter 3) but with more sophisticated types of designs (Chapters 5–7) as well.

4.2 A POSTERIORI ANALYSIS

In an effort to determine which of the c means are significantly different from the others, it is improper for the researcher to use all possible two-sample t tests (see Table 1.3, formula E) to examine all *pairwise comparisons* between the means; all such combinations would not be independent and, if c was large enough, it is likely that the difference between the largest and smallest of the $\bar{Y}_{.j}$

would be declared significant even if the null hypothesis were true. That is, the greater the number of groups (i.e., levels of a factor) c, the greater the number of pairwise comparisons [i.e., $c(c-1)/2$] between means, and the more likely it would become to erroneously reject one or more of them—even if H_0 were true. Thus, if several pairwise comparisons were made, each at the α level, the probability of incorrectly rejecting H_0 at least once would increase with c and would exceed α.

In an effort to control α, several a posteriori multiple comparison procedures have been devised for investigating significant treatment effects once they have been determined through ANOVA [see Miller (1981) and Carmer and Swanson (1973)]. Among the most widely used classical a posteriori procedures are the Tukey T method (1953) and the Scheffé S method (1953). The former may be better for all pairwise comparisons, while the latter is better for more complex combinations of comparisons.

4.2.1 Tukey T Method

Once the researcher has determined through ANOVA that a significant treatment effect is present among the c groups, there are $\binom{c}{2} = \dfrac{c!}{2!\,(c-2)!}$ $= c(c-1)/2$ possible combinations of pairwise comparisons that may be analyzed.

The first step is to compute the differences $\bar{Y}_{.j} - \bar{Y}_{.j'}$ (where $j \neq j'$) among all $c(c-1)/2$ pairs of means. The *critical range* for the T method is then obtained from the quantity

$$\text{critical range} = Q_{1-\alpha;\,c,\,n-c}\sqrt{\frac{\text{MSW}}{\bar{n}.}} \tag{4.1}$$

where $Q_{1-\alpha;\,c,\,n-c}$ is the upper-tail critical value obtained from the Studentized range distribution having c and $n-c$ degrees of freedom (see Appendix B, Table B.6).* Moreover, MSW is the *mean square within groups* term (3.16) from the ANOVA, and $\bar{n}.$ is given by†

$$\bar{n}. = \frac{c}{\displaystyle\sum_{j=1}^{c}(1/n_j)} = \frac{c}{(1/n_1) + (1/n_2) + \cdots + (1/n_c)} \tag{4.2}$$

The final step is to compare each of the $c(c-1)/2$ pairs of means against the critical range and to declare any pair significant if the absolute difference in the sample means $|\bar{Y}_{.j} - \bar{Y}_{.j'}|$ equals or exceeds $Q_{1-\alpha;\,c,\,n-c}\sqrt{\text{MSW}/\bar{n}.}$

*Given two independent samples taken from the same (or identical) normal population(s), the Studentized range distribution was devised as the ratio of the *range* in a sample of data containing η observations to the standard deviation S_Y computed from a sample containing $v + 1$ observations. Thus, range/$S_Y \sim Q_{\eta,v}$, and the upper-tail critical values (95 and 99 percentiles) are presented in Appendix B, Table B.6.

†The expression for $\bar{n}.$ in (4.2) is called the *harmonic mean*. It is defined as the reciprocal of the average of the c reciprocals.

It is important to note here that the T method was originally devised for experiments containing equal-sized samples. Of course, $\bar{n}. \equiv n_j$ (where $j = 1, 2, \ldots, c$) if the sample sizes are all equal.

On the other hand, statisticians have maintained that the T method is *robust* to moderate departures from the assumption of equal sample sizes and that decent approximations can be made if the harmonic mean $\bar{n}.$ is utilized. These approximations are somewhat *conservative* because the value for the (harmonic) mean will always be less than the common n_j in the case of equal-sized samples, so that the critical range in (4.1) will be larger—making it more difficult to declare a pair of means as significantly different.

Example.   To apply the T method, we return to the example of Section 3.2.1. Using the ANOVA procedure, we recall that the financial analyst had concluded that significant differences exist in the mean earnings per share among the three nonindustrial business groupings.

To determine which differences among the means are in fact significant, the financial analyst would employ the T method. Since there are three groups, there are three possible pairwise comparisons to be made. From Table 3.3 the absolute mean differences are

1. $|\bar{Y}_{.1} - \bar{Y}_{.2}| = |5.843 - 3.632| = 2.211$
2. $|\bar{Y}_{.1} - \bar{Y}_{.3}| = |5.843 - 3.313| = 2.530$
3. $|\bar{Y}_{.2} - \bar{Y}_{.3}| = |3.632 - 3.313| = .319$

To determine the critical range, the financial analyst uses MSW = 3.3505 and, from (4.2), computes $\bar{n}. = 3/(\frac{1}{9} + \frac{1}{10} + \frac{1}{10}) = 9.64$. However, from Table B.6 in Appendix B the desired upper-tail critical value $Q_{.95; 3, 26}$ is not available and may be approximated through *linear interpolation* as 3.517. From (4.1), the critical range is then

$$\text{critical range} = 3.517\sqrt{\frac{3.3505}{9.64}} = 2.073$$

The financial analyst would therefore conclude that comparisons 1 and 2 are significant but 3 is not. That is, there are significant differences in mean earnings per share among (1) commercial banking companies and retailing companies as well as among (2) commercial banking companies and utility companies, but no real differences exist in the mean earnings per share between (3) retailing and utility companies. Therefore, we can conclude that only the commercial banking companies differ from the others.

Obtaining confidence interval estimates.   Using the T method, we may also establish a set of *simultaneous* confidence interval estimates for the true differences between each pair of means. This is achieved by adding and subtracting the critical range to the differences in each pair of sample means so that

$$(\bar{Y}_{.j} - \bar{Y}_{.j'}) - Q_{1-\alpha;\,c,\,n-c}\sqrt{\frac{\text{MSW}}{\bar{n}.}} \le (\mu_{.j} - \mu_{.j'})$$

$$\le (\bar{Y}_{.j} - \bar{Y}_{.j'}) + Q_{1-\alpha;\,c,\,n-c}\sqrt{\frac{\text{MSW}}{\bar{n}.}} \qquad (4.3)$$

where $j = 1, 2, \ldots, c$ and $j \ne j'$.

Using (4.3), the financial analyst would obtain the following set of 95% confidence interval estimates:

1. $2.211 - 2.073 \le \mu_{.1} - \mu_{.2} \le 2.211 + 2.073$
   $.138 \le \mu_{.1} - \mu_{.2} \le 4.284$
2. $2.530 - 2.073 \le \mu_{.1} - \mu_{.3} \le 2.530 + 2.073$
   $.457 \le \mu_{.1} - \mu_{.3} \le 4.603$
3. $.319 - 2.073 \le \mu_{.2} - \mu_{.3} \le .319 + 2.073$
   $-1.754 \le \mu_{.2} - \mu_{.3} \le 2.392$

The set of confidence interval estimates provides the researcher with the *same* conclusions that would be obtained from the previously described set of tests. As in comparisons (1) and (2), if a confidence interval does not include zero, the population means being compared are declared significantly different. On the other hand, if a confidence interval includes zero [comparison (3)], the population means being compared are said to differ only by chance.

Advantages of estimation.   There are two major advantages to the confidence interval approach, and both concern *data interpretation*.

First, the reseacher may estimate the *magnitude* of the differences between $\mu_{.j}$ and $\mu_{.j'}$. Here, for example, the financial analyst would conclude with 95% confidence that mean earnings per share for commercial banking companies is (1) between $.14 and $4.28 higher than that of retailing companies and (2) between $.46 and $4.60 higher than that of utility companies; however, (3) there is no real difference in mean earnings per share between retailing and utility companies.

Second, the researcher may more easily express the *meaning* behind the *experimentwise error rate* used to control for α when making a set of simultaneous statements once an initial analysis (ANOVA) has been performed. Thus, the confidence coefficient of 95% used by the financial analyst does not pertain to any single statement but rather to the set of simultaneous statements for all (three) possible pairs of means. This is called a *family confidence coefficient* $(1 - \alpha)$ and is the complement of the experimentwise error rate (α).

Definitions.   The following defines the family confidence coefficient and experimentwise error rate. If an experiment having c levels of a factor were repeated hundreds of times, in $100(1 - \alpha)\%$ of these replications we would expect *all* $c(c - 1)/2$ of the simultaneous interval statements to contain correctly the true mean differences $(\mu_{.j} - \mu_{.j'})$ within their limits [i.e., a family confidence coefficient of $100(1 - \alpha)\%$], while in $100\alpha\%$ of the replications *at least one* of the $c(c - 1)/2$ possible confidence intervals will fail to include the true mean differences within its limits (i.e., an experimentwise error rate of $100\alpha\%$).

4.2.2 Scheffé $\mathfrak{S}$ Method

Tukey's T method was used to evaluate all $c(c - 1)/2$ possible pairwise comparisons between means once an initial investigation (ANOVA) determined the existence of significant treatment effects. When the c sample sizes are vastly different, however, the T method may not be reliable (robust). In such circumstances a *multiple contrast* procedure developed by Scheffé (1953) can be used. This procedure, known as the Scheffé $\mathfrak{S}$ method, may be used not only for pairwise comparisons but for complex combinations of comparisons as well (i.e., for *contrasts*).

Comparisons and contrasts.   We may define a (linear) contrast $\mathfrak{L}_l$ among c means $(\mu_{.1}, \mu_{.2}, \ldots, \mu_{.c})$ as a linear combination

$$\mathfrak{L}_l = \sum_{j=1}^{c} C_{lj}\mu_{.j} = C_{l1}\mu_{.1} + C_{l2}\mu_{.2} + \cdots + C_{lc}\mu_{.c} \qquad (4.4)$$

such that $\sum_{j=1}^{c} C_{lj} = C_{l1} + C_{l2} + \cdots + C_{lc} = 0$. That is, the coefficients of the linear contrast are a set of numbers which sum to zero.

Thus, a pairwise comparison between means $\mu_{.j}$ and $\mu_{.j'}$ is a special case of a contrast. Among the c means, $\mu_{.j}$ would have a coefficient $C_{lj} = 1$, and $\mu_{.j'}$ would have a coefficient $C_{lj'} = -1$. All remaining means would have a coefficient of zero.

To evaluate a contrast, we merely replace $\mu_{.j}$ by its estimate $\bar{Y}_{.j}$ (where $j = 1, 2, \ldots, c$). That is,

$$\hat{\mathfrak{L}}_l = \sum_{j=1}^{c} C_{lj}\bar{Y}_{.j} \quad \text{where} \quad \sum_{j=1}^{c} C_{lj} = 0 \qquad (4.5)$$

Thus, with but three levels of the factor nonindustrial business groupings, the financial analyst may be interested in evaluating the following six contrasts:

$$\hat{\mathfrak{L}}_1 = \bar{Y}_{.1} - \bar{Y}_{.2}$$

$$\hat{\mathfrak{L}}_2 = \bar{Y}_{.1} - \bar{Y}_{.3}$$

$$\hat{\mathfrak{L}}_3 = \bar{Y}_{.2} - \bar{Y}_{.3}$$

$$\hat{\mathfrak{L}}_4 = \bar{Y}_{.1} - \left(\frac{\bar{Y}_{.2} + \bar{Y}_{.3}}{2}\right)$$

$$\hat{\mathfrak{L}}_5 = \bar{Y}_{.2} - \left(\frac{\bar{Y}_{.1} + \bar{Y}_{.3}}{2}\right)$$

$$\hat{\mathfrak{L}}_6 = \bar{Y}_{.3} - \left(\frac{\bar{Y}_{.1} + \bar{Y}_{.2}}{2}\right)$$

We may note that *contrasts* $\hat{\mathfrak{L}}_1$, $\hat{\mathfrak{L}}_2$, and $\hat{\mathfrak{L}}_3$ are simply the *pairwise comparisons* 1, 2, and 3 obtained using the Tukey T method. However, the Scheffé

S method also permits an evaluation of all possible contrasts—not just pairwise comparisons. Therefore $\hat{\mathcal{L}}_4$, $\hat{\mathcal{L}}_5$, and $\hat{\mathcal{L}}_6$, respectively, compare each mean against the average of the remaining two means.

The respective coefficients C_{lj} (where $l = 1, 2, \ldots, 6$ and $j = 1, 2, 3$) for such contrasts would be

	C_{l1}	C_{l2}	C_{l3}
$\hat{\mathcal{L}}_1$	1	-1	0
$\hat{\mathcal{L}}_2$	1	0	-1
$\hat{\mathcal{L}}_3$	0	1	-1
$\hat{\mathcal{L}}_4$	1	$-\frac{1}{2}$	$-\frac{1}{2}$
$\hat{\mathcal{L}}_5$	$-\frac{1}{2}$	1	$-\frac{1}{2}$
$\hat{\mathcal{L}}_6$	$-\frac{1}{2}$	$-\frac{1}{2}$	1

Note that the set of coefficients for each and every contrast sums to zero.

To obtain the coefficients for the more complex contrasts, we take, as an example, the estimated contrast $\hat{\mathcal{L}}_4$:

$$\hat{\mathcal{L}}_4 = \bar{Y}_{\cdot 1} - \left(\frac{\bar{Y}_{\cdot 2} + \bar{Y}_{\cdot 3}}{2}\right)$$

Observe that the coefficient for the first mean is unity, while the coefficients for the second and third means are each $-\frac{1}{2}$. Patterns of coefficients for $\hat{\mathcal{L}}_5$ and $\hat{\mathcal{L}}_6$ follow in a similar manner.

Once the contrasts have been listed, the standard deviation of each contrast may be estimated from

$$S_{\hat{\mathcal{L}}_l} = \sqrt{\text{MSW}\left(\sum_{j=1}^{c} \frac{C_{lj}^2}{n_j}\right)} = \sqrt{\text{MSW}\left(\frac{C_{l1}^2}{n_1} + \frac{C_{l2}^2}{n_2} + \cdots + \frac{C_{lc}^2}{n_c}\right)} \qquad (4.6)$$

From (4.6) we note that $S_{\hat{\mathcal{L}}_l}$ may change for each of the L contrasts ($l = 1, 2, \ldots, L$)—depending on the sample sizes and coefficients used. As examples, using $\hat{\mathcal{L}}_4$, the financial analyst estimates the standard deviation to be

$$S_{\hat{\mathcal{L}}_4} = \sqrt{3.3505\left(\frac{1}{9} + \frac{.25}{10} + \frac{.25}{10}\right)} = .735$$

whereas using $\hat{\mathcal{L}}_5$, the standard deviation is

$$S_{\hat{\mathcal{L}}_5} = \sqrt{3.3505\left(\frac{.25}{9} + \frac{1}{10} + \frac{.25}{10}\right)} = .715$$

The critical range for each contrast is obtained by multiplying its estimated standard deviation $S_{\hat{\mathcal{L}}_l}$ by a constant $\sqrt{(c-1)F_{1-\alpha;\, c-1, n-c}}$, the square root of the product of $c - 1$ with the upper-tail critical value of the F distribution having $c - 1$ and $n - c$ degrees of freedom. Thus, depending on $S_{\hat{\mathcal{L}}_l}$, the critical range may vary from contrast to contrast.

The final step in the S method is to declare significant any contrast wherein the absolute value of its estimate $|\hat{\mathcal{L}}_l|$ equals or exceeds the critical range $S_{\hat{\mathcal{L}}_l}\sqrt{(c-1)F_{1-\alpha(c-1,n-c)}}$.

Example. For the six contrasts of interest to the financial analyst the following results are obtained:

$$\bar{Y}_{.1} = 5.843, \qquad \bar{Y}_{.2} = 3.632, \qquad \bar{Y}_{.3} = 3.313, \qquad \text{MSW} = 3.3505,$$

$$\sqrt{(c-1)F_{1-\alpha;\,c-1,n-c}} = \sqrt{2(3.37)} = 2.596$$

Therefore, as indicated on page 93, contrasts $\mathcal{L}_1$, $\mathcal{L}_2$, and $\mathcal{L}_4$ are declared significant, but $\mathcal{L}_3$, $\mathcal{L}_5$, and $\mathcal{L}_6$ are not.

Obtaining confidence interval estimates. Using the S method, we may also establish a set of simultaneous confidence interval estimates for the true contrasts $\mathcal{L}_l$ (where $l = 1, 2, \ldots, L$) by adding and subtracting the critical range from the estimated contrasts. That is,

$$\hat{\mathcal{L}}_l - S_{\hat{\mathcal{L}}_l}\sqrt{(c-1)F_{1-\alpha;\,c-1,n-c}} \leq \mathcal{L}_l \leq \hat{\mathcal{L}}_l + S_{\hat{\mathcal{L}}_l}\sqrt{(c-1)F_{1-\alpha;\,c-1,n-c}} \qquad (4.7)$$

where $l = 1, 2, \ldots, L$.

From (4.7) the financial analyst would obtain the following set of 95% confidence interval estimates for the true contrasts:

$$\mathcal{L}_1: \qquad 2.211 - 2.183 \leq \mathcal{L}_1 \leq 2.211 + 2.183$$
$$.028 \leq \mathcal{L}_1 \leq 4.394$$

$$\mathcal{L}_2: \qquad 2.530 - 2.183 \leq \mathcal{L}_2 \leq 2.530 + 2.183$$
$$.347 \leq \mathcal{L}_2 \leq 4.713$$

$$\mathcal{L}_3: \qquad .319 - 2.126 \leq \mathcal{L}_3 \leq .319 + 2.126$$
$$-1.807 \leq \mathcal{L}_3 \leq 2.445$$

$$\mathcal{L}_4: \qquad 2.370 - 1.908 \leq \mathcal{L}_4 \leq 2.370 + 1.908$$
$$.462 \leq \mathcal{L}_4 \leq 4.278$$

$$\mathcal{L}_5: \qquad -.946 - 1.856 \leq \mathcal{L}_5 \leq -.946 + 1.856$$
$$-2.802 \leq \mathcal{L}_5 \leq .910$$

$$\mathcal{L}_6: \qquad -1.425 - 1.856 \leq \mathcal{L}_6 \leq -1.425 + 1.856$$
$$-3.281 \leq \mathcal{L}_6 \leq .431$$

From this set of confidence intervals the same conclusions regarding the true contrasts would be reached as were obtained using the hypothesis testing approach.

Using the family confidence coefficient of 95%, the financial analyst would conclude that mean earnings per share for commercial banking companies is between \$.03 and \$4.39 higher than that of retailing companies ($\mathcal{L}_1$) and is between \$.35 and \$4.71 higher than that of utility companies ($\mathcal{L}_2$). In addition,

Contrast	Estimated Contrast	$S_{\hat{L}_i}$	Critical Range	Conclusion
$\hat{L}_1 = \mu_{.1} - \mu_{.2}$	$\hat{L}_1 = 5.843 - 3.632 = 2.211$	$\sqrt{3.3505\left(\frac{1}{9} + \frac{1}{10} + 0\right)} = .841$	2.183	Significant
$\hat{L}_2 = \mu_{.1} - \mu_{.3}$	$\hat{L}_2 = 5.843 - 3.313 = 2.530$	$\sqrt{3.3505\left(\frac{1}{9} + 0 + \frac{1}{10}\right)} = .841$	2.183	Significant
$\hat{L}_3 = \mu_{.2} - \mu_{.3}$	$\hat{L}_3 = 3.632 - 3.313 = .319$	$\sqrt{3.3505\left(0 + \frac{1}{10} + \frac{1}{10}\right)} = .819$	2.126	Not significant
$\hat{L}_4 = \mu_{.1} - \left(\dfrac{\mu_{.2} + \mu_{.3}}{2}\right)$	$\hat{L}_4 = 5.843 - 3.473 = 2.370$	$\sqrt{3.3505\left(\frac{1}{9} + \frac{.25}{10} + \frac{.25}{10}\right)} = .735$	1.908	Significant
$\hat{L}_5 = \mu_{.2} - \left(\dfrac{\mu_{.1} + \mu_{.3}}{2}\right)$	$\hat{L}_5 = 3.632 - 4.578 = .946$	$\sqrt{3.3505\left(\frac{.25}{9} + \frac{1}{10} + \frac{.25}{10}\right)} = .715$	1.856	Not significant
$\hat{L}_6 = \mu_{.3} - \left(\dfrac{\mu_{.1} + \mu_{.2}}{2}\right)$	$\hat{L}_6 = 3.313 - 4.738 = 1.425$	$\sqrt{3.3505\left(\frac{.25}{9} + \frac{.25}{10} + \frac{1}{10}\right)} = .715$	1.856	Not significant

the mean earnings per share for commercial banking companies is between \$.46 and \$4.28 higher than that of retailing and utility companies taken together ($\mathcal{L}_4$). On the other hand, there is no real difference in mean earnings per share between retailing and utility companies ($\mathcal{L}_3$) or when each of these two types of non-industrial business groupings is compared against the other two taken together ($\mathcal{L}_5$ and $\mathcal{L}_6$).

Again, the 95% confidence coefficient pertains to the set of simultaneous statements for the contrasts considered and not to any particular contrast. If the experiment were to be replicated hundreds of times, we should expect that in 95% of these replications all the true contrasts would be included within their respective intervals (family confidence coefficient), while in only 5% of the replications (experimentwise error rate) one or more of all the confidence intervals would be in error.

4.2.3 Comparing the *T* Method and the *S* Method

The S method is desirable because of its *generality*. That is, the S method permits statements to be made about *all* contrasts (not just pairwise comparisons) and does not require equal-sized samples for each of the levels of the factor. Moreover, the S method would determine among *all* contrasts at least one wherein real (significant) effects are present (and which led to the initial rejection of the null hypothesis using ANOVA procedures). The *T* method, however, may not uncover any significant pairwise comparisons even when the ANOVA procedure had previously rejected the null hypothesis; that is, the means themselves may not differ significantly, but rather it may be complex combinations of the means which are representing the true treatment effects.

On the other hand, the *T* method is very simple to use because the computed critical range does not vary from one comparison to another. Moreover, the *T* method is robust to slight differences in the sample sizes (even though the procedure was developed specifically for experiments containing equal-sized samples). Most importantly, the *T* method yields powerful results [i.e., its confidence interval statements are narrower than the corresponding ones obtained by the S method—as indicated from the multiple comparisons 1, 2, 3 versus $\mathcal{L}_1$, $\mathcal{L}_2$, $\mathcal{L}_3$, as previously reported].

In summary, the *T* method is recommended for situations in which the researcher is interested only in pairwise comparisons of means—provided that the *c*-sample sizes are not very different. The more general S method is recommended under all other situations.

4.3 A PRIORI ANALYSIS:

There are occasions when a researcher is interested in making specific types of analyses among *c* levels of a factor without necessarily being interested in an overall ANOVA procedure. The most widely used method to achieve this is the *method of orthogonal contrasts*.

In advance of collecting the data, the researcher states up to $c - 1$ independent hypotheses (or contrasts) in lieu of the overall null hypothesis (i.e., $H_0: \mu_{.1} = \mu_{.2} = \cdots = \mu_{.c}$) particular to the ANOVA procedure. For each of the contrasts stated in the hypotheses, the researcher then *decomposes* the SSＱ (sums of squares among groups) term into a set of *independent* components — each having one degree of freedom. Once the data are collected, the researcher performs tests for each of these hypotheses and makes a set of (at most $c - 1$) independent decisions regarding significance.

Such procedures are called *a priori methods of analysis* because the hypotheses to be tested are established based on the interests of the researcher *prior to* collecting the data. No "data snooping" is involved. On the other hand, the Tukey T method and the Scheffé S method described in Section 4.2 are known as *a posteriori procedures* because they are used only *after* the researcher has initially performed ANOVA and determined that significant treatment effects are present. The various contrasts are established *after* the researcher has seen the data (data snooping).

4.3.1 Method of Orthogonal Contrasts

In Section 4.2 a contrast $\mathcal{L}_l$ was defined as a linear combination among the c means such that the coefficients C_{lj} sum to zero. An *orthogonal contrast* $\mathcal{L}_l^*$ is such a contrast; in addition, however, the product of its coefficients with those of any other orthogonal contrast (in the set of $c - 1$) must sum up to zero— provided that the c samples are of equal size. That is, $\mathcal{L}_l^*$ is an orthogonal contrast if

$$\mathcal{L}_l^* = \sum_{j=1}^{c} C_{lj}\mu_{.j} = C_{l1}\mu_{.1} + C_{l2}\mu_{.2} + \cdots + C_{lc}\mu_{.c}$$

where

$$\sum_{j=1}^{c} C_{lj} = C_{l1} + C_{l2} + \cdots + C_{lc} = 0$$

and

$$\sum_{j=1}^{c} C_{lj}C_{l'j} = C_{l1}C_{l'1} + C_{l2}C_{l'2} + \cdots + C_{lc}C_{l'c} = 0$$

where $n_1 = n_2 = \cdots = n_c$; $j = 1, 2, \ldots, c$; and $l \neq l'$ but $l = 1, 2, \ldots, c - 1$ and $l' = 1, 2, \ldots, c - 1$. On the other hand, if the c sample sizes are not all equal, the contrast $\mathcal{L}_l^* = \sum_{j=1}^{c} C_{lj}\mu_{.j}$ is an orthogonal contrast if $\sum_{j=1}^{c} C_{lj} = 0$ and, most importantly, $\sum_{j=1}^{c} (C_{lj}C_{l'j}/n_j) = 0$.

A set of contrasts may be described as *mutually orthogonal* if every contrast in the set is orthogonal to every other member of that set. If, for example, we return to the list of six contrasts established by the financial analyst in performing the Scheffé S method (Section 4.2.2), we note that only a properly chosen set of $c - 1 = 2$ of them are mutually orthogonal contrasts. Furthermore, we observe from the list on page 96 that several such sets (in this example three) would exist had the sample sizes all been equal. Prior to obtaining the data the

Contrast	(Set): Orthogonal Contrast[a]	Estimated Contrast	Coefficients		
			C_{l1}	C_{l2}	C_{l3}
$\mathcal{L}_1$	(1): $\mathcal{L}_1^* = \mu_{\cdot 1} - \mu_{\cdot 2}$	$\hat{\mathcal{L}}_1^* = \bar{Y}_{\cdot 1} - \bar{Y}_{\cdot 2}$	1	-1	0
$\mathcal{L}_6$	(1): $\mathcal{L}_2^* = \mu_{\cdot 3} - [(\mu_{\cdot 1} + \mu_{\cdot 2})/2]$	$\hat{\mathcal{L}}_2^* = \bar{Y}_{\cdot 3} - [(\bar{Y}_{\cdot 1} + \bar{Y}_{\cdot 2})/2]$	$-\frac{1}{2}$	$-\frac{1}{2}$	1
$\mathcal{L}_2$	(2): $\mathcal{L}_1^* = \mu_{\cdot 1} - \mu_{\cdot 3}$	$\hat{\mathcal{L}}_1^* = \bar{Y}_{\cdot 1} - \bar{Y}_{\cdot 3}$	1	0	-1
$\mathcal{L}_5$	(2): $\mathcal{L}_2^* = \mu_{\cdot 2} - [(\mu_{\cdot 1} + \mu_{\cdot 3})/2]$	$\hat{\mathcal{L}}_2^* = \bar{Y}_{\cdot 2} - [(\bar{Y}_{\cdot 1} + \bar{Y}_{\cdot 3})/2]$	$-\frac{1}{2}$	1	$-\frac{1}{2}$
$\mathcal{L}_3$	(3): $\mathcal{L}_1^* = \mu_{\cdot 2} - \mu_{\cdot 3}$	$\hat{\mathcal{L}}_1^* = \bar{Y}_{\cdot 2} - \bar{Y}_{\cdot 3}$	0	1	-1
$\mathcal{L}_4$	(3): $\mathcal{L}_2^* = \mu_{\cdot 1} - [(\mu_{\cdot 2} + \mu_{\cdot 3})/2]$	$\hat{\mathcal{L}}_2^* = \bar{Y}_{\cdot 1} - [(\bar{Y}_{\cdot 2} + \bar{Y}_{\cdot 3})/2]$	1	$-\frac{1}{2}$	$-\frac{1}{2}$

[a]Note that sets (1), (2), and (3) each contain a pair of mutually orthogonal contrasts provided that the sample sizes are all equal; however, for the financial analyst's data only set (3) contains a pair of mutually orthogonal contrasts ($\mathcal{L}_3$ and $\mathcal{L}_4$).

financial analyst would have had to select the *particular* set of $c - 1$ mutually orthogonal contrasts in which there was interest and establish the appropriate hypotheses.

Finding mutually orthogonal contrasts.   Suppose, for example, the financial analyst had been particularly interested in the third set of $c - 1$ orthogonal contrasts in lieu of the overall F test. That third set contains contrasts $\mathcal{L}_3$ and $\mathcal{L}_4$. They are orthogonal contrasts because their coefficients sum to zero and the product of the coefficients divided by the respective sample sizes also sum to zero. That is,

$$\mathcal{L}_3: \quad 0 + 1 + (-1) = 0$$

$$\mathcal{L}_4: \quad 1 + (-\tfrac{1}{2}) + (-\tfrac{1}{2}) = 0$$

and with respective sample sizes of 9, 10, and 10 we have

$$\frac{0(1)}{9} + \frac{1(-\tfrac{1}{2})}{10} + \frac{-1(-\tfrac{1}{2})}{10} = 0$$

However, the first set containing contrasts $\mathcal{L}_1$ and $\mathcal{L}_6$ is not orthogonal because the products of the coefficients divided by the respective sample sizes do not sum to zero:

$$\frac{1(-\tfrac{1}{2})}{9} + \frac{-1(-\tfrac{1}{2})}{10} + \frac{0(1)}{10} = -\frac{1}{180} = -.0056 \neq 0$$

On the other hand, the set of contrasts $\mathcal{L}_1$ and $\mathcal{L}_4$ would not be orthogonal regardless of whether the sample sizes were equal,

$$1(1) + -1(-\tfrac{1}{2}) + 0(-\tfrac{1}{2}) = 1.5000 \neq 0$$

or not equal,

$$\frac{1(1)}{9} + \frac{-1(-\tfrac{1}{2})}{10} + \frac{0(-\tfrac{1}{2})}{10} = .1611 \neq 0$$

The selection of a set of $c - 1$ mutually orthogonal contrasts is not always easy—especially if the c sample sizes differ and/or the number of levels of a factor (c) increases.* Frequently, in such circumstances, the researcher specifies a subset of h (where $h < c - 1$) orthogonal contrasts of interest and then merely "lumps together" the other $c - 1 - h$ "possible orthogonal contrasts" as an untested and unanalyzed remainder. Thus the choice is with the researcher as to whether to establish an entire set of $c - 1$ mutually orthogonal contrasts or only a particular subset of h such contrasts.

Performing the tests.   Nonetheless, for the set of orthogonal contrasts selected by the financial analyst, the following null hypotheses would be stated:

$$H_{0_1}: \quad \mathcal{L}_1^* = \mu_{.2} - \mu_{.3} = 0 \quad \text{or} \quad \mu_{.2} = \mu_{.3}$$

$$H_{0_2}: \quad \mathcal{L}_2^* = \mu_{.1} - \left(\frac{\mu_{.2} + \mu_{.3}}{2}\right) = 0 \quad \text{or} \quad \mu_{.1} = \frac{\mu_{.2} + \mu_{.3}}{2}$$

*For further information see the *hint* to Problem 4.8(b).

The respective alternatives are

$$H_{1_l}: \quad \mathcal{L}_l^* \neq 0 \qquad (\text{where } l = 1, 2, \ldots, c - 1)$$

To perform the $c - 1$ *separate* tests, each at the $\alpha = .05$ level, we compute the *sum of squares* for each orthogonal contrast $\mathcal{L}_l^*$ (where $l = 1, 2, \ldots, c - 1$) from

$$SS\mathcal{L}_l^* = \frac{(\hat{\mathcal{L}}_l^*)^2}{\sum\limits_{j=1}^{c} (C_{lj}^2/n_j)} \qquad (4.8)$$

where $\hat{\mathcal{L}}_l^* = \sum_{j=1}^{c} C_{lj}\bar{Y}_{.j}$. Since each of the $c - 1$ orthogonal contrasts in the set have but one degree of freedom, the *mean square* term $MS\mathcal{L}_l^*$ is given by

$$MS\mathcal{L}_l^* = SS\mathcal{L}_l^* \div 1 \equiv SS\mathcal{L}_l^* \qquad (4.9)$$

and to test the null hypothesis H_{0_l} for the orthogonal contrast $\mathcal{L}_l^*$, we compute the F ratio:

$$F_l = \frac{MS\mathcal{L}_l^*}{MSW} \sim F_{1, n-c} \qquad (4.10)$$

Thus, a set of $c - 1$ F ratios is computed, and each null hypothesis is separately tested—resulting in $c - 1$ specific decisions. A null hypothesis is rejected if F_l equals or exceeds $F_{1-\alpha; 1, n-c}$, the upper-tail critical value for the F distribution having one and $c - 1$ degrees of freedom.

Unlike the a posteriori multiple comparison procedures of Section 4.2 (the T and S methods) in which the decisions rendered were based on an overall family confidence coefficient or experimentwise error rate, the conclusions made here are specific to the individual a priori comparisons considered. Hence the researcher must interpret the results cautiously since, unfortunately, as c increases, the risk of reaching a false conclusion also increases.*

4.3.2 Obtaining Results

It is both important and interesting to point out that the method of orthogonal contrasts results in the decomposition of the sum of squares among groups (SSα) into $c - 1$ components (i.e., contrasts), each possessing but one degree of freedom. That is,

$$SS\alpha = \sum_{l=1}^{c-1} SS\mathcal{L}_l^* = \sum_{l=1}^{c-1} \frac{(\hat{\mathcal{L}}_l^*)^2}{\sum\limits_{j=1}^{c} (C_{lj}^2/n_j)} \qquad (4.11)$$

and the layout for the ANOVA table (Table 3.4) may be re-expressed as shown in Table 4.1.

---

*To resolve this dilemma, one could test each of the contrasts using a smaller level of significance [say $\alpha/(c - 1)$ instead of α] and thereby make it more difficult to declare a particular contrast as significant.

TABLE 4.1 ANOVA Table Reflecting the Decomposition of SSα for Orthogonal Contrasts

Source of Variation	Degrees of Freedom	Sum of Squares	Mean Square	F
Among Groups	$c-1$	$SS\alpha = \sum_{j=1}^{c} \dfrac{Y^2_{.j}}{n_j} - \dfrac{Y^2_{..}}{n}$	$MS\alpha = \dfrac{SS\alpha}{c-1}$	
$\mathcal{L}^*_1$	1	$SS\mathcal{L}^*_1 = \dfrac{(\hat{\mathcal{L}}^*_1)^2}{\sum_{j=1}^{c}(C^2_{1j}/n_j)}$	$MS\mathcal{L}^*_1 = SS\mathcal{L}^*_1$	$F_1 = \dfrac{MS\mathcal{L}^*_1}{MSW}$
$\mathcal{L}^*_2$	1	$SS\mathcal{L}^*_2 = \dfrac{(\hat{\mathcal{L}}^*_2)^2}{\sum_{j=1}^{c}(C^2_{2j}/n_j)}$	$MS\mathcal{L}^*_2 = SS\mathcal{L}^*_2$	$F_2 = \dfrac{MS\mathcal{L}^*_2}{MSW}$
$\cdots$	$\cdots$	$\cdots$	$\cdots$	$\cdots$
$\mathcal{L}^*_{c-1}$	1	$SS\mathcal{L}^*_{c-1} = \dfrac{(\hat{\mathcal{L}}^*_{c-1})^2}{\sum_{j=1}^{c}(C^2_{c-1,j}/n_j)}$	$MS\mathcal{L}^*_{c-1} = SS\mathcal{L}^*_{c-1}$	$F_{c-1} = \dfrac{MS\mathcal{L}^*_{c-1}}{MSW}$
Within groups	$n-c$	$SSW = \sum_{j=1}^{c}\sum_{i=1}^{n_j} Y^2_{ij} - \sum_{j=1}^{c} \dfrac{Y^2_{.j}}{n_j}$	$MSW = \dfrac{SSW}{n-c}$	
Total	$n-1$	$SST = \sum_{j=1}^{c}\sum_{i=1}^{n_j} Y^2_{ij} - \dfrac{Y^2_{..}}{n}$		

Example. For the problem of interest to the financial analyst,

$$\hat{\mathcal{L}}_1^* = \bar{Y}_{.2} - \bar{Y}_{.3} = 3.632 - 3.313 = .319$$

and, from (4.8),

$$SS\mathcal{L}_1^* = \frac{(\hat{\mathcal{L}}_1^*)^2}{\sum_{j=1}^{c} (C_{1j}^2/n_j)} = \frac{(.319)^2}{[(0)^2/9] + [(1)^2/10] + [(1)^2/10]} = .5088$$

whereas

$$\hat{\mathcal{L}}_2^* = \bar{Y}_{.1} - \left(\frac{\bar{Y}_{.2} + \bar{Y}_{.3}}{2}\right) = 5.8433 - 3.4725 = 2.3708$$

so that (except for rounding errors)

$$SS\mathcal{L}_2^* = \frac{(\hat{\mathcal{L}}_2^*)^2}{\sum_{j=1}^{c} (C_{2j}^2/n_j)} = \frac{(2.3708)^2}{[(1)^2/9] + [(.5)^2/10] + [(.5)^2/10]} = 34.8880$$

From (4.11),

$$SS\mathcal{Q} = .5088 + 34.8880 = 35.3968$$

as displayed in Tables 3.5 and 4.2. From Table B.4 in Appendix B the upper-tail

TABLE 4.2   Use of Orthogonal Contrasts to Study Mean Earnings per Share

Source of Variation	Degrees of Freedom	Sum of Squares	Mean Square	F
Among Groups	2	35.3968		
$\mathcal{L}_1^* = \mu_{.2} - \mu_{.3}$	1	.5088	.5088	.152
$\mathcal{L}_2^* = \mu_{.1} - [(\mu_{.2} + \mu_{.3})/2]$	1	34.8880	34.8880	10.41
Within Groups	26	87.1124	3.3505	
Total	28	122.5092		

critical value is $F_{.95;1,26} = 4.23$. Since $F_1 = .5088/3.3505 = .152$ is less than this critical value, the financial analyst could not reject the first null hypothesis and, at the .05 level of significance, would conclude there is no difference in mean earnings per share between retailing and utility companies. (The P value —the probability of obtaining such a result or one even more extreme when the null hypothesis is true—is almost as high as .75.) However, $F_2 = 34.8880/3.3505 = 10.41$ exceeds the critical value so that the financial analyst would reject the second null hypothesis and, at the .05 level, would conclude that the mean earnings per share of commercial banking companies significantly differs from those of the retailing and utility companies taken together. (The P value is smaller than .005.)

Using F versus t.   We may recall from (1.20) of Section 1.6.4 that whenever F has but one degree of freedom in the numerator there is an equivalence between $F_{1-\alpha;\,1,\,v}$ and the square of $t_{1-\frac{\alpha}{2};\,v}$. Therefore, even though we have presented the method of orthogonal contrasts by the decomposition of $SS\alpha$ and the utilization of the F test, we could have achieved the same results for each of the $c - 1$ possible contrasts by using the following two-tailed t tests for the null hypotheses (i.e., $H_{0_l}: \mathcal{L}_l^* = 0$),

$$t_l = \frac{\hat{\mathcal{L}}_l^*}{S_{\hat{\mathcal{L}}_l^*}} \sim t_{n-c} \tag{4.12}$$

where $l = 1, 2, \ldots, c - 1$ and where the standard deviation of the orthogonal contrast is given by the quantity

$$S_{\hat{\mathcal{L}}_l^*} = \sqrt{\text{MSW} \sum_{j=1}^{c} \frac{C_{lj}^2}{n_j}} \tag{4.13}$$

and by rejecting H_{0_l} if $t_l \geq t_{1-\frac{\alpha}{2};\,n-c}$ or if $t_l \leq t_{\frac{\alpha}{2};\,n-c}$.*

Confidence interval estimates of $\mathcal{L}_i^*$.   On the other hand, when the researcher is more interested in making confidence interval estimates of true contrasts, there is no choice between t or F. Interpretation of results is greatly facilitated by use of t since we are then dealing with the data as measured (rather than with squared units as necessary for F).

Instead of tests of hypotheses, a set of $c - 1$ *separate* confidence interval estimates of the true orthogonal contrasts $\mathcal{L}_i^*$ can be made as follows:

$$\hat{\mathcal{L}}_l^* - t_{1-\frac{\alpha}{2};\,n-c}S_{\hat{\mathcal{L}}_l^*} \leq \mathcal{L}_l^* \leq \hat{\mathcal{L}}_l^* + t_{1-\frac{\alpha}{2};\,n-c}S_{\hat{\mathcal{L}}_l^*} \tag{4.14}$$

where $l = 1, 2, \ldots, c - 1$ and where $S_{\hat{\mathcal{L}}_l^*}$ is obtained from (4.13).

Therefore, the financial analyst could have made the following two independent statements (each with 95% confidence):

$$\mathcal{L}_1^*: \quad .319 - 2.056\sqrt{3.3505(.2)} \leq \mu_{.2} - \mu_{.3} \leq .319 + 2.056\sqrt{3.3505(.2)}$$
$$-1.364 \leq \mu_{.2} - \mu_{.3} \leq 2.002$$

and

$$\mathcal{L}_2^*: \quad 2.3708 - 2.056\sqrt{3.3505(.1611)} \leq \mu_{.1} - \left(\frac{\mu_{.2} + \mu_{.3}}{2}\right)$$
$$\leq 2.3708 + 2.056\sqrt{3.3505(.1611)}$$
$$.8603 \leq \mu_{.1} - \left(\frac{\mu_{.2} + \mu_{.3}}{2}\right) \leq 3.8813$$

Since each statement is separate, the error rate is uncontrolled (see the footnote on p. 98). Thus the probability that both statements are correct is $(1 - \alpha)^2$

---

*Interestingly, for pairwise comparisons we note the similarity between the above t test and the t test for two independent samples (Table 1.3, formula E). The estimated orthogonal contrast $\hat{\mathcal{L}}_l^*$ is simply the difference in the two independent sample means $\bar{Y}_{.j} - \bar{Y}_{.j'}$ (where $j \neq j'$), while $S_{\hat{\mathcal{L}}_l^*}$ defined in (4.13) is identical to $\sqrt{S_p^2[(1/n_j) + (1/n_{j'})]}$. The only difference in the two types of t tests is that S_p^2 is the pooled variance estimate over two groups and is based on $n - 2$ degrees of freedom, whereas MSW is the pooled variance estimate over c groups (see Section 3.4.2) and is based on $n - c$ degrees of freedom.

$= .9025$, while the chance that at least one of them is in error is $1 - (1 - \alpha)^2$
$= .0975$.

4.4 COMPUTER PACKAGES AND MULTIPLE COMPARISONS AND CONTRASTS: USE OF SPSS, SAS, AND BMDP

4.4.1 The SPSS Subprogram ONEWAY

The SPSS subprogram ONEWAY not only performs a classical one-way analysis of variance, but in addition is capable of providing both multiple comparisons (using the Tukey and Scheffé methods) and orthogonal contrasts. The required ONEWAY procedure statement is

$$
\underset{\text{ONEWAY}}{\overset{1}{}} \quad \overset{16}{\begin{Bmatrix} \text{quantitative} \\ \text{(dependent)} \\ \text{variable or list} \end{Bmatrix}} \text{\textcent BY\textcent} \begin{Bmatrix} \text{independent} \\ \text{variable} \end{Bmatrix} \begin{pmatrix} \text{min.,} & \text{max.} \\ \text{value} & \text{value} \end{pmatrix} /
$$

Beginning in column 16, the quantitative (or response) variable to be analyzed is named, followed (after the word BY) by the qualitative variable used to form groups. Multiple comparisons using the Tukey T method or the Scheffé S method can be obtained by including the following statements after the ONEWAY procedure statement:

$$
\overset{16}{}
$$
RANGES = TUKEY (level of significance)/
RANGES = SCHEFFE (level of significance)/

For the data of interest to the financial analyst, if we refer to Figure 3.4 we can observe that the .05 level of significance has been chosen in obtaining multiple comparisons using the Tukey and Scheffé methods.

Orthogonal contrasts can also be obtained by using the ONEWAY subprogram as follows:

$$
\overset{16}{\text{CONTRAST}} = C_1 \text{\textcent} C_2 \text{\textcent} \ldots \text{\textcent} C_j/
$$

where $C_1, C_2, \ldots, C_j$ are the coefficients pertaining to the contrast of interest. Referring to Figure 3.4, note that two contrasts have been specified. The first contrast (CONTRAST = 0 1 −1) compares the second and third groups, while the second contrast (CONTRAST = 1 −.5 −.5) compares the first group to the average of the second and third groups.

Figure 4.1 represents annotated partial output from the ONEWAY subprogram for multiple comparisons and contrasts.

4.4.2 Using the SAS GLM Procedure for Multiple Contrasts

The SAS GLM procedure not only performs a classical analysis of variance for a one-way model but in addition is capable of providing both multiple comparisons (using the Tukey and Scheffé methods) and orthogonal contrasts.

```
        VARIABLE   EARNINGS     EARNINGS PER SHARE $
     BY VARIABLE   GROUP        BUSINESS GROUP CLASSIFICATION


CONTRAST COEFFICIENT MATRIX

            GRPO1      GRPC3
                 GRPO2

CONTRAST  1   0.0    1.0   -1.0

CONTRAST  2   1.0   -0.5   -0.5




                                   POOLED VARIANCE ESTIMATE
                    VALUE    S. ERROR    T VALUE   D.F.     T PROB.

CONTRAST  1         0.3190    0.8186       0.390    26.0      0.700

CONTRAST  2         2.3708    0.7347       3.227    26.0      0.003


TESTS FOR HOMOGENEITY OF VARIANCES

        COCHRANS C = MAX. VARIANCE/SUM(VARIANCES) = 0.4600, P = 0.482 (APPROX.)
        BARTLETT-BOX F =                             0.527, P = 0.591
        MAXIMUM VARIANCE / MINIMUM VARIANCE =        2.061


MULTIPLE RANGE TEST


TUKEY-HSD PROCEDURE
RANGES FOR THE 0.050 LEVEL -

          3.51   3.51

THE RANGES ABOVE ARE TABLE RANGES.  THE VALUE ACTUALLY COMPARED WITH MEAN(J)-MEAN(I) IS..
        1.2943 * RANGE * SQRT(1/N(I) + 1/N(J))


  (*) DENOTES PAIRS OF GROUPS SIGNIFICANTLY DIFFERENT AT THE 0.050 LEVEL


                              G  G  G
                              R  R  R
                              P  P  P
                              O  O  O
        MEAN      GROUP       3  2  1

        3.3130    GRPO3
        3.6320    GRPO2
        5.8433    GRPO1       *  *



SCHEFFE PROCEDURE
RANGES FOR THE 0.050 LEVEL -

          3.67   3.67

THE RANGES ABOVE ARE TABLE RANGES.  THE VALUE ACTUALLY COMPARED WITH MEAN(J)-MEAN(I) IS..
        1.2943 * RANGE * SQRT(1/N(I) + 1/N(J))


  (*) DENOTES PAIRS OF GROUPS SIGNIFICANTLY DIFFERENT AT THE 0.050 LEVEL


                              G  G  G
                              R  R  R
                              P  P  P
                              O  O  O
        MEAN      GROUP       3  2  1

        3.3130    GRPC3
        3.6320    GRPO2
        5.8433    GRPC1       *  *
```

Figure 4.1 Partial SPSS output for earnings-per-share data.

Multiple comparisons using the Tukey T method or the Scheffé S method can be obtained by using the following options as part of the MEANS statement:

MEANS ♭ {name of factor α}/TUKEY♭ SCHEFFE♭ ALPHA
$$= \{\text{level of significance}\};$$

The basic setup for obtaining orthogonal contrasts using the GLM procedure is

CONTRAST ♭'name or title'
$$\begin{Bmatrix} \text{name of} \\ \text{contrast} \\ \text{variable} \end{Bmatrix} ♭C_1 ♭ C_2 ♭ \ldots ♭ C_j;$$

Referring to Figure 3.6, note that two contrasts have been specified. The CONTRAST statement may be given a title up to 20 characters. The first contrast is entitled 'BANK VS OTHERS'. This contrast (GROUP $1 -.5 -.5$) involves a comparison of group 1 with the average of groups 2 and 3. The second contrast, entitled 'RETAILING VS UTILITIES', examines the difference between the second and third groups (GROUP $0\ 1\ -1$). We should note that all CONTRAST statements must follow the MODEL statement in the GLM procedure.

Figure 4.2 represents annotated partial output from the GLM procedure for orthogonal contrasts.

SOURCE	DF	TYPE I SS	F VALUE	PR > F
GROUP	2	35.39684379	5.28	0.0119

CONTRAST	DF	SS	F VALUE	PR > F
BANK VS OTHERS	1	34.88803879	10.41	0.0034
RETAIL VS UTIL	1	0.50880500	0.15	0.6999

GENERAL LINEAR MODELS PROCEDURE

MEANS

GROUP	N	EARNINGS
BANKING	9	5.84333333
RETAILING	10	3.63200000
UTILITY	10	3.31300000

Figure 4.2 Partial SAS output for earnings-per-share data.

4.4.3 The BMDP Program 1V

The BMDP program 1V not only performs a classical one-way analysis of variance but also provides orthogonal contrasts. The paragraph of BMDP unique to program 1V is the /DESIGN paragraph. For the one-way ANOVA (including orthogonal contrasts) the format of this paragraph is

/DESIGNþDEPENDENTþISþ{name of dependent variable}.
      CONTRASTþISþ$C_1, C_2, \ldots, C_J$.

The DEPENDENT sentence of the /DESIGN paragraph names the dependent variable to be analyzed, while the CONTRAST sentence provides the orthogonal contrast to be evaluated. For the problem of interest to the financial analyst, this paragraph would be

/DESIGNþDEPENDENTþISþEARNINGS.
      CONTRASTþISþ0, 1, −1.
      CONTRASTþISþ1, −.5, −.5.

The first contrast compares differences between group 2 and group 3, while the second contrast compares group 1 to the average of groups 2 and 3. We should note that as compared to the BMDP program 7D (see Section 3.6.4), the program using BMDP-1V needed to obtain orthogonal contrasts differs only with respect to the inclusion of the /DESIGN paragraph (in P-1V) instead of the /HISTOGRAM paragraph (in P-7D).

Figure 4.3 represents annotated partial output from program 1V for orthogonal contrasts.

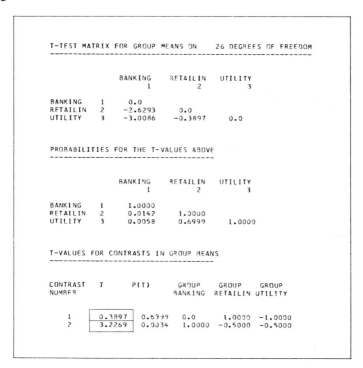

Figure 4.3 Partial BMDP program 1V output for earnings-per-share data.

4.5 DATA ANALYSIS: A PERSPECTIVE

The methods of Chapters 3 and 4 together provide the framework for analyzing (fixed-factor) experiments of the completely randomized type. However, the methods have been separated because the a priori and a posteriori multiple comparison techniques (Chapter 4) are not limited to situations in which the experimental design format is completely randomized (Chapter 3). When appropriately modified, these techniques can readily be utilized for more complex designs as well (see Chapters 5–7).

In summary, what we should have learned in these two chapters is a *methodological process* for analyzing data in designed experiments. Frequently the researcher must plan the entire experiment—selecting the appropriate experimental design model, choosing the levels of the factor(s) of interest, obtaining and randomly assigning the subjects to treatment conditions, and then completely analyzing the resulting data [see, for example, Cox (1958)]. It is indeed rare that the researcher is merely presented with a set of data and *told* to perform an F test! The researcher must *know* what procedures to use for specific situations.

For (fixed-factor) completely randomized designs, the researcher must first test the assumptions and, depending on the findings, perhaps make necessary tranformations or adjustments in the analysis. If specific hypotheses were not of interest prior to collecting the data, the researcher would use the classical F test to test for overall treatment effects. If H_0 is not rejected, the analysis ends —except, perhaps, for confidence interval estimates of particular treatment means. On the other hand, if H_0 is rejected, appropriate a posteriori techniques are used—depending on whether the researcher is interested only in pairwise comparisons or in more complex comparisons as well. However, for those situations in which the researcher had been interested in specific hypotheses prior to examining the data, appropriate a priori techniques would be used instead. This, then, is what is meant by a methodological process to analyze the data completely. In Chapters 5–7 we shall turn our attention to more sophisticated experimental design formats.

PROBLEMS

4.1a. Demonstrate your understanding of *contrasts* by listing all possible *unweighted and noncomplex* $\mathcal{L}_i$ terms and their corresponding set of coefficients C_{ij} for experiments involving $c = 4$ levels of a factor. *Hint:* There will be a total of 25 such contrasts: 6 simple pairwise comparisons ($\mathcal{L}_1, \mathcal{L}_2, \ldots, \mathcal{L}_6$), 4 contrasts involving one mean against the average of the remaining three ($\mathcal{L}_7, \mathcal{L}_8, \ldots, \mathcal{L}_{10}$), 12 contrasts involving one mean against the average of two others with the fourth excluded ($\mathcal{L}_{11}, \mathcal{L}_{12}, \ldots, \mathcal{L}_{22}$), and 3 contrasts involving the average of two means against two other means ($\mathcal{L}_{23}, \mathcal{L}_{24}, \mathcal{L}_{25}$).

4.1b. Demonstrate your understanding of *mutually orthogonal contrasts* by establishing a set of $c - 1 = 3$ mutually orthogonal contrasts for experiments containing four levels of a factor with equal sized samples. *Hint:* For one set, make two nonoverlapping pairwise comparisons plus a comparison of the averages of each of the above pairs. To verify your work, the coefficients for such a set of mutually orthogonal contrasts are shown below:

	C_{l1}	C_{l2}	C_{l3}	C_{l4}
$\mathcal{L}_1^*$	1	-1	0	0
$\mathcal{L}_2^*$	0	0	1	-1
$\mathcal{L}_3^*$	$\frac{1}{2}$	$\frac{1}{2}$	$-\frac{1}{2}$	$-\frac{1}{2}$

For another type of set, first make a pairwise comparison; then take the two just compared against a third mean; finally, take the average of these three means against the remaining fourth mean. The coefficients for such a set are as follows:

	C_{l1}	C_{l2}	C_{l3}	C_{l4}
$\mathcal{L}_1^*$	0	0	1	-1
$\mathcal{L}_2^*$	0	1	$-\frac{1}{2}$	$-\frac{1}{2}$
$\mathcal{L}_3^*$	1	$-\frac{1}{3}$	$-\frac{1}{3}$	$-\frac{1}{3}$

(Note again that these sets contain mutually orthogonal contrasts only if the c sample sizes are equal.)

4.2. From Problem 3.1, suppose, prior to having run the experiment, Hundal had specifically been interested in comparing (1) the mean productivity of workers receiving no feedback versus that of workers receiving some feedback and (2) the mean productivity of workers receiving much feedback against the average of the other two groups taken together.

(a) Assuming that the data are approximately normally distributed, analyze the results of the experiment.

(b) What might account for possible differences in your conclusions based on part (a) above versus those from part (a) of Problem 3.1?

(c) Are a posteriori methods of analysis appropriate here? Explain.

*4.3. (a) From Problem 3.2, complete your analysis of the data using multiple comparisons.

(b) Had the researcher been specifically interested in making certain comparisons among the levels of advertising copy before running the experiment, a priori methods of analysis could have been utilized. Based on what is known about the five advertising copy levels, set up an appropriate set of four mutually orthogonal contrasts and analyze the data.

4.4. From Problem 3.3, complete the analysis of the data if the researcher for the consumer's interest group were only interested in pairwise comparisons.

4.5. Had the Dean in Problem 3.4 been specifically interested in making certain comparisons among the four classes before conducting the survey, a priori methods of analysis could have been utilized. Set up an appropriate set of mutually orthogonal contrasts. If appropriate, analyze the data.

4.6. Had the industrial psychologist in Problem 3.5 been specific
making certain comparisons among the three levels of alc
before performing the experiment, a priori methods of ana
utilized. Set up an appropriate set of mutually orthogona
the data.

4.7. (a) From Problem 3.6, complete the analysis of the d
(b) Had the quality control engineer been specifically
comparisons among the four temperature levels
experiment, a priori methods of analysis could have be
appropriate set of mutually orthogonal contrasts, and ana

4.8. (a) From Problem 3.7, complete the analysis of the data, and
results.
(b) Had the Dean of Students been specifically interested in making ce
comparisons among these three levels of employment before conducting th
survey, a priori methods of analysis could have been utilized. Set up an
appropriate set of mutually orthogonal contrasts, and analyze the data.
Hint: In general, for an experiment with c groups you may construct a set
of $c - 1$ mutually orthogonal contrasts. These would then be $c(c - 1)/2$
possible constraining equations—$c - 1$ of the form $\sum_{j=1}^{c} C_{lj} = 0$ and

$(c - 1)(c - 2)/2$ of the form $\sum_{j=1}^{c} C_{lj}C_{l'j}/n_j = 0$. These $c(c - 1)/2$ possible
constraining equations would contain $c(c - 1)$ unknown contrast coefficients
C_{lj} (where $l = 1, 2, \ldots, c - 1$ and $j = 1, 2, \ldots, c$). Thus the selection of
appropriate values for coefficients is not unique. You must first specify a
total of $c(c - 1)/2$ coefficients in such a manner that not all of them are
chosen for any one contrast. The remaining unknown $c(c - 1)/2$ contrast
coefficients appearing in the $c(c - 1)/2$ constraining equations are then
obtained as the solutions to a set of simultaneous equations.

4.9. (a) From Problem 3.8, restate your conclusions based on your ANOVA. What
are the implications of those conclusions?
(b) Since the behavioral researcher was really interested in investigating whether
increases in the noise levels had an *increasingly detrimental* effect on driving
performance, a priori methods of analysis could have been used instead of
the overall F test (see Chapter 3) or linear regression methods (see Chapter 8
and Problem 8.6). Analyze the data to determine whether *increasing* noise
levels have a significantly *decreasing* effect on driving performance. *Hint:*
Set up a *linear contrast* with respective coefficients of $-3, -1, +1,$ and $+3$
for the four noise levels. Decompose SSα into this linear contrast plus an
unanalyzed and untested remainder.
(c) Discuss any differences in your conclusions to parts (a) and (b) and their
implications on appropriate data analysis.

4.10. (a) From Problem 3.9, complete the analysis of the data, and discuss your
results.
(b) Had the pharmaceutical statistician been really interested in studying
whether or not there was a significant *negative linear relationship* between
dosage level and diastolic pressure before performing the experiment, a
priori methods of analysis could have been used instead of linear regression

---

5

TWO-WAY ANOVA: RANDOMIZED COMPLETE BLOCK DESIGN

5.1 INTRODUCTION

In Chapter 3 we studied the properties of the completely ra
experiments in which a set of n *homogeneous* subjects were ra
the c levels (i.e., treatment conditions) of a factor of interes
test was performed to determine the existence of any si
effects. In this chapter, however, we turn our attention to
plete block design—experiments in which n *heterogen*
into r *homogeneous groups called blocks so that th*
then be randomly assigned, one each, to the c le
to the performance of a two-way F test to d
treatment effects. (Note that $n = rc$.)

5.1.1 Need for Blocking

The need for *blocking*
(treatment) levels become
Suppose a profess
for teaching element
(3) lecture/labor
In a c
randomly
period
me

valid indicator of the value of a particular instructional method, the completely randomized experiment just described is appropriate provided that the n subjects are truly homogeneous in their abilities. Since this is unlikely, i.e., students taking an introductory statistics course may differ in age, mathematical ability and background, intelligence, etc., a better analysis of potential treatment effects could be made if such *concomitant* variables as these were used to first *match* or block heterogeneous subjects into homogeneous strata. For example, if some measure of mathematical preparation (such as SAT score) were available, the top three students could form block 1, the next best three students could form block 2, and so on, until the three poorest students are grouped into block r. Within each of the r homogeneous blocks the $c = 3$ subjects could then be randomly assigned to one of the three methods of instruction. Since like subjects would, except for chance, tend to behave alike if treated alike, the null hypothesis of no treatment effects should not be rejected if there were no real differences in the three instructional methods. However, if the instructional methods were really different, it should be expected that this would be reflected through higher test scores being obtained by subjects associated with the "best" instructional method.

5.1.2 Purpose

Without such blocking, had a completely randomized design been utilized, it may have been possible, purely by chance, that the majority of the "better" students were assigned to one of the instructional methods, while a majority of the "poorer" students were assigned to another of the methods—thereby confusing the ability of the students with the methods themselves! That is, with such *confounding*, we would not be able to distinguish between students and instructional methods in the analysis. The danger, of course, is that faulty conclusions may be drawn. Based on the data, we may conclude that significant treatment effects (i.e., differences in instructional methods) exist when, in fact, the only "real" differences were in the abilities of the subjects.

Hence, whenever the subjects are heterogeneous with respect to certain concomitant variables that affect the responses (but are not of primary interest), randomized complete block experiments should be considered.

Since the primary objective for selecting a particular experimental design is to reduce experimental error (i.e., *variability within* the data), a better design could be obtained if subject variability is separated from the experimental error. Thus the purpose of selecting the randomized complete block design over the completely randomized design (Chapter 3) is to provide a more efficient analysis by reducing the experimental error and thereby obtaining more precise results.*
As will be described in Section 5.2, this will be achieved by decomposing the SSW term (3.12) into two parts: sums of squares due to blocks (i.e., "subject

---

*A comparison between the completely randomized design and the randomized complete block design will be presented in Section 5.5.

ability" in our example) and sums of squares due to random error. Thus, in our example, by choosing the randomized complete block design, we can, in effect, filter out and remove the variability among the subjects—enabling us to make more precise statements about the treatment effects (i.e., differences in instructional methods).

5.2 DEVELOPING THE MODEL

To develop the ANOVA procedure for a randomized complete block design note that Y_{ij}, the observation in the ith block ($i = 1, 2, \ldots, r$) under the jth factor level ($j = 1, 2, \ldots, c$), can be represented by the model

$$Y_{ij} = \mu + \alpha_j + B_i + \epsilon'_{ij} \tag{5.1}$$

where $\mu = overall$ effect or mean common to all the observations

$\alpha_j = \mu_{.j} - \mu$, a treatment effect peculiar to the jth level of the factor ($j = 1, 2, \ldots, c$)

$B_i = \mu_{i.} - \mu$, a block effect peculiar to the ith block ($i = 1, 2, \ldots, r$)

$\epsilon'_{ij} = random\ variation$ or $experimental\ error$ associated with the observation in the ith block under the jth factor level

$\mu_{.j} =$ true mean for the jth factor level

$\mu_{i.} =$ true mean for the ith block

Substituting for the treatment and block effects, we may write (5.1) as

$$Y_{ij} = \mu + (\mu_{.j} - \mu) + (\mu_{i.} - \mu) + \epsilon'_{ij} \tag{5.2}$$

and, by subtraction,

$$\epsilon'_{ij} = (Y_{ij} - \mu) - (\mu_{.j} - \mu) - (\mu_{i.} - \mu) \tag{5.3}$$

That is, experimental error represents the difference between the overall deviation and the treatment and block deviations—what is left over after taking into account both factor levels and blocks.

However, (5.3) may be simplified:

$$\epsilon'_{ij} = Y_{ij} - \mu_{.j} - \mu_{i.} + \mu \tag{5.4}$$

and thus from (5.2) and (5.4) we have

$$Y_{ij} = \mu + (\mu_{.j} - \mu) + (\mu_{i.} - \mu) + (Y_{ij} - \mu_{.j} - \mu_{i.} + \mu) \tag{5.5}$$

Using the sample data from the experiment, we can estimate the true parameters $\mu, \alpha_j, B_i,$ and ϵ'_{ij} by using the computed statistics $\hat{\mu}, \hat{\alpha}_j, \hat{B}_i,$ and $e'_{ij},$ where

$$\hat{\mu} = \bar{Y}_{..} = \frac{Y_{..}}{n} = \sum_{j=1}^{c} \sum_{i=1}^{r} \frac{Y_{ij}}{n}$$

$$\hat{\alpha}_j = \bar{Y}_{.j} - \bar{Y}_{..} \qquad \text{where } \bar{Y}_{.j} = \sum_{i=1}^{r} \frac{Y_{ij}}{r}$$

$$\hat{B}_i = \bar{Y}_{i.} - \bar{Y}_{..} \qquad \text{where } \bar{Y}_{i.} = \sum_{j=1}^{c} \frac{Y_{ij}}{c}$$
$$e'_{ij} = Y_{ij} - \bar{Y}_{.j} - \bar{Y}_{i.} + \bar{Y}_{..}$$

and where $n = rc$.

Rewriting (5.5), we obtain

$$Y_{ij} = \bar{Y}_{..} + (\bar{Y}_{.j} - \bar{Y}_{..}) + (\bar{Y}_{i.} - \bar{Y}_{..}) + (Y_{ij} - \bar{Y}_{.j} - \bar{Y}_{i.} + \bar{Y}_{..}) \qquad (5.6)$$

If $\bar{Y}_{..}$ is subtracted from both sides of (5.6), we get

$$Y_{ij} - \bar{Y}_{..} = (\bar{Y}_{.j} - \bar{Y}_{..}) + (\bar{Y}_{i.} - \bar{Y}_{..}) + (Y_{ij} - \bar{Y}_{.j} - \bar{Y}_{.i} + \bar{Y}_{..}) \qquad (5.7)$$

Now if each side of (5.7) is squared and then summed over both i and j, we obtain the following important identity:

$$\sum_{j=1}^{c} \sum_{i=1}^{r} (Y_{ij} - \bar{Y}_{..})^2 = \sum_{j=1}^{c} \sum_{i=1}^{r} (\bar{Y}_{.j} - \bar{Y}_{..})^2 + \sum_{j=1}^{c} \sum_{i=1}^{r} (\bar{Y}_{i.} - \bar{Y}_{..})^2$$
$$+ \sum_{j=1}^{c} \sum_{i=1}^{r} (Y_{ij} - \bar{Y}_{.j} - \bar{Y}_{i.} + \bar{Y}_{..})^2 \qquad (5.8)$$

5.2.1 The Sums of Squares

The term on the left-hand side of (5.8) is the *total variation* or *total sums of squares*, SST. As in Section 3.3.3, it represents the summation of squared differences between each observation and the overall mean. This is decomposed into three additive components. The first term on the right-hand side of (5.8) represents the summation of squared differences between the means at each of the factor levels and the overall mean, whereas the middle expression represents the summation of squared differences between the means of each block and the overall mean. The last term on the right-hand side of (5.8) represents *experimental error*—the variability in the data not accounted for by the blocks and factor levels (i.e., the treatments). Thus we see that the total variation is comprised of three additive parts: the among group variation SSα, the among block variation SSB, and the experimental error variation SSE:

$$\text{SST} = \text{SS}\alpha + \text{SSB} + \text{SSE} \qquad (5.9)$$

In practical applications, however, it is somewhat cumbersome to compute the sums of squares components from (5.8). Therefore, by using some algebra and summation notation, the following *computational* formulas for SSα, SSB, SSE, and SST are obtained through expansion:

$$\text{SS}\alpha = \sum_{j=1}^{c} \frac{Y_{.j}^2}{r} - \frac{Y_{...}^2}{n} \equiv \sum_{j=1}^{c} \sum_{i=1}^{r} (\bar{Y}_{.j} - \bar{Y}_{..})^2 \qquad (5.10)$$

$$\text{SSB} = \sum_{i=1}^{r} \frac{Y_{i.}^2}{c} - \frac{Y_{..}^2}{n} \equiv \sum_{j=1}^{c} \sum_{i=1}^{r} (\bar{Y}_{i.} - \bar{Y}_{..})^2 \qquad (5.11)$$

$$\text{SSE} = \sum_{j=1}^{c} \sum_{i=1}^{r} Y_{ij}^2 - \sum_{j=1}^{c} \frac{Y_{.j}^2}{r} - \sum_{i=1}^{r} \frac{Y_{i.}^2}{c} + \frac{Y_{..}^2}{n}$$

$$\equiv \sum_{j=1}^{c} \sum_{i=1}^{r} (Y_{ij} - \bar{Y}_{.j} - \bar{Y}_{i.} + \bar{Y}_{..})^2 \qquad (5.12)$$

$$\text{SST} = \sum_{j=1}^{c} \sum_{i=1}^{r} Y_{ij}^2 - \frac{Y_{..}^2}{n} \equiv \sum_{j=1}^{c} \sum_{i=1}^{r} (Y_{ij} - \bar{Y}_{..})^2 \qquad (5.13)$$

5.2.2 The Mean Squares

If each of these sums of squares are divided by their respective degrees of freedom, we obtain four mean square terms: MSα, MSB, MSE, and MST:

$$\text{MS}\alpha = \frac{\text{SS}\dot{\alpha}}{c-1} \qquad (5.14)$$

$$\text{MSB} = \frac{\text{SSB}}{r-1} \qquad (5.15)$$

$$\text{MSE} = \frac{\text{SSE}}{(r-1)(c-1)} \qquad (5.16)$$

$$\text{MST} = \frac{\text{SST}}{n-1} \qquad (5.17)$$

Since there are c levels of the factor being compared, there are $c - 1$ degrees of freedom associated with the *mean squares among groups* term. Similarly, with r blocks there are $r - 1$ degrees of freedom associated with the *mean squares among blocks* term. Furthermore, with a total of $n = rc$ observations, there are $n - 1$ degrees of freedom associated with the *total mean squares* term. Because the degrees of freedom for each component must add to the total degrees of freedom, we may obtain the remaining degrees of freedom for the *mean squares error* term by subtraction:

$$(n-1) - (c-1) - (r-1) = n - c - r + 1$$
$$= rc - c - r + 1$$
$$= (r-1)(c-1) \qquad (5.18)$$

Again, it is noted that both degrees of freedom and sums of squares are *additive*, but mean squares are not.

5.2.3 Forming the F Statistic

To test the null hypothesis of no treatment effects

$$H_0: \quad \mu_{.1} = \mu_{.2} = \cdots = \mu_{.c} \quad \text{(or } H_0: \quad \alpha_1 = \alpha_2 = \cdots = \alpha_c = 0\text{)}$$

against the alternative

$$H_1: \quad \text{not all } \mu_{.j} \text{ are equal} \quad \text{(or } H_1: \quad \text{not all } \alpha_j \text{ equal zero)}$$

we form the F statistic

$$F_\alpha = \frac{\text{MS}\alpha}{\text{MSE}} \sim F_{c-1,(r-1)(c-1)} \qquad (5.19)$$

and the null hypothesis would be rejected at the $\alpha\%$ level of significance if the computed F_α value equals or exceeds $F_{1-\alpha;\, c-1,(r-1)(c-1)}$, the upper-tail critical value from the F distribution having $c - 1$ and $(r - 1)(c - 1)$ degrees of freedom.

To examine whether it was advantageous to block, some researchers might want to test a null hypothesis of no block effects

$$H_0: \quad \mu_1. = \mu_2. = \cdots = \mu_r. \qquad (\text{or } H_0: \quad B_1 = B_2 = \cdots = B_r = 0)$$

against the alternative

$$H_1: \quad \text{not all } \mu_i. \text{ are equal} \qquad (\text{or } H_1: \quad \text{not all } B_i \text{ equal zero})$$

by forming the F statistic

$$F_B = \frac{\text{MSB}}{\text{MSE}} \sim F_{r-1,(r-1)(c-1)} \qquad (5.20)$$

and rejecting H_0 if $F_B \geq F_{1-\alpha;\, r-1,(r-1)(c-1)}$. However, it may be argued that this is unnecessary—that the sole purpose of establishing the blocks was to provide a more efficient means of testing for treatment effects by reducing the experimental error (through the control of heterogeneous subject-to-subject variation).*

5.2.4 The ANOVA Table

To summarize, the layout for the randomized complete block design is depicted in Table 5.1, while the ANOVA table is presented in Table 5.2.

TABLE 5.1   Layout for Randomized Complete Block Design

Block (B)	Factor α						Totals
	Level 1	Level 2	$\cdots$	Level j	$\cdots$	Level c	
1	Y_{11}	Y_{12}	$\cdots$	Y_{1j}	$\cdots$	Y_{1c}	$Y_1.$
2	Y_{21}	Y_{22}	$\cdots$	Y_{2j}	$\cdots$	Y_{2c}	$Y_2.$
.	.	.		.		.	.
.	.	.		.		.	.
.	.	.		.		.	.
i	Y_{i1}	Y_{i2}	$\cdots$	Y_{ij}	$\cdots$	Y_{ic}	$Y_i.$
.	.	.		.		.	.
.	.	.		.		.	.
.	.	.		.		.	.
r	Y_{r1}	Y_{r2}	$\cdots$	Y_{rj}	$\cdots$	Y_{rc}	$Y_r.$
Totals	$Y._1$	$Y._2$	$\cdots$	$Y._j$	$\cdots$	$Y._c$	$Y..$

*In essence, in a randomized complete block design the blocks are not given the status of a factor. In Chapters 6 and 7 we shall see that when the "blocks" are considered important enough to be a second factor in a particular study, the design format is appropriately called a two-factor experiment, and the tests for each factor effect would potentially be important.

TABLE 5.2 ANOVA Table

Source of Variation	Degrees of Freedom	Sum of Squares	Mean Square	F
Among Groups (α)	$c - 1$	$SS\alpha = \sum_{j=1}^{c} \dfrac{Y^2_{.j}}{r} - \dfrac{Y^2_{..}}{n}$	$MS\alpha = \dfrac{SS\alpha}{c-1}$	$F\alpha = \dfrac{MS\alpha}{MSE}$
Among Blocks (B)	$r - 1$	$SSB = \sum_{i=1}^{r} \dfrac{Y^2_{i.}}{c} - \dfrac{Y^2_{..}}{n}$	$MSB = \dfrac{SSB}{r-1}$	$F_B = \dfrac{MSB}{MSE}$
Error	$(r-1)(c-1)$	$SSE = \sum_{j=1}^{c} \sum_{i=1}^{r} Y^2_{ij} - \sum_{j=1}^{c} \dfrac{Y^2_{.j}}{r} - \sum_{i=1}^{r} \dfrac{Y^2_{i.}}{c} + \dfrac{Y^2_{..}}{n}$	$MSE = \dfrac{SSE}{(r-1)(c-1)}$	
Total	$n - 1$	$SST = \sum_{j=1}^{c} \sum_{i=1}^{r} Y^2_{ij} - \dfrac{Y^2_{..}}{n}$		

5.2.5 Assumptions

The assumptions of the ANOVA are basically the same for the randomized complete block design as they were for the completely randomized design (see Section 3.4): independence, normality, homoscedasticity, and additivity. The last two, however, merit special attention here.

Homoscedasticity. Since there is only a single observation Y_{ij} for each block (i) and each factor level (j) "combination," we may think of these as $n = rc$ different samples of size 1 coming from $n = rc$ normal populations each having the same underlying variance σ^2.

Additivity. When performing the ANOVA, we must assume that our model and its effects are additive. In Section 5.2.3 we developed the F test for the (assumed) additive, fixed-effects model,

$$Y_{ij} = \mu + \alpha_j + B_i + \epsilon'_{ij} \tag{5.1}$$

where, as in Section 3.4.3, without any loss in generality, we may also define the treatment effects and block effects in a manner such that $\sum_{j=1}^{c} \alpha_j = 0$ and $\sum_{i=1}^{r} B_i = 0$. If, however, the null hypothesis is true and no real treatment effects are present (that is, $\alpha_1 = \alpha_2 = \cdots = \alpha_c = 0$), our model reduces to

$$Y_{ij} = \mu + B_i + \epsilon'_{ij} \tag{5.21}$$

Furthermore, if the blocking is not advantageous (that is, $B_1 = B_2 = \cdots = B_r = 0$), the model reduces to

$$Y_{ij} = \mu + \epsilon'_{ij} \tag{5.22}$$

and the only differences from one observed response to another would be due to chance.*

Of course, if the additivity model does not hold, the previously described F tests are invalid. But to say that the effects are additive implies that there is no possible *interaction* between the levels of the factor and the established blocks. Using our hypothetical example, if a significant treatment effect is present so that some methods of instruction are deemed better than others, this finding should hold over all blocks. On the other hand, if the effects are not additive, one instructional method may be superior for blocks containing "brighter" students, while another instructional method might be superior for blocks containing "weaker" students, and we would then conclude that there is an *interaction* between the methods and the blocks.

The additivity assumption may be tested using a procedure developed by Tukey (1949), which will be described in Section 6.7.2. If the Tukey test for additivity results in a conclusion that an interaction is present, an appropriate

---

*In such a situation, all four mean square terms—MSα, MSB, MSE, and MST—would each be measuring the inherent variability in the data (σ^2).

data-stabilizing transformation (see Section 1.8.1) would be used and the ANOVA performed once the assumption appears to hold.

5.3 APPLICATION: VISUAL PATTERN RECOGNITION

An interesting application using the two-way ANOVA procedure is based in part on a behavioral investigation conducted by Levine et al. (1973) on visual pattern recognition. The practical ramifications of this physiological research effort are in the field of advertising and in communications theory.

As part of this study, eight subjects were each presented with a single numerical digit (0–9) for a variable on-time (12, 16, or 20 milliseconds) immediately followed (no delay time) by the presentation of a randomly selected masking noise field which remained on for a constant 500 milliseconds prior to extinction. A computer program randomly generated the order in which the numerical digits appeared as well as the length of on-time until 10 trials of each "digit-on-time" combination were observed by each subject.

If we wish to investigate the effects of different on-time levels on the percentage of correct responses, a randomized complete block design with one observation per cell (i.e., the percentage of correct responses out of 100 trials per on-time level) may be used. The null hypothesis of no treatment effects (i.e., no differences in mean percent correct recall over the three on-time levels)

$$H_0: \quad \mu_{.1} = \mu_{.2} = \mu_{.3}$$

can be tested against the alternative

$$H_1: \quad \text{not all } \mu_{.j} \text{ are equal} \quad (j = 1, 2, 3)$$

The data for the eight subjects (blocks) are given in Table 5.3.

TABLE 5.3   Percent Correct Response (for All Digits) by Subject

Subjects (Blocks)	On-Time Levels (msec)		
	12	16	20
i	55	68	67
ii	78	83	84
iii	34	53	54
iv	56	67	65
v	79	78	85
vi	20	29	30
vii	68	88	92
viii	59	58	72

SOURCE: Extracted from D. M. Levine, S. Wachspress, P. McGuire, and M. S. Mayzner, "Visual Information Processing of Numerical Inputs," *Bulletin of Psychonomic Society*, Vol. 1 (6A), (1973), pp. 404–406.

Since the observed data are percent correct recall scores (Y_{ij}), an arcsin root transformation (see Section 3.4.2 and Appendix B, Table B.7) is made before classical ANOVA methods can be used. The transformed data $(\mathcal{Y}_{ij})$ appear in Table 5.4 along with the relevant summary computations.

TABLE 5.4   Data After Arcsin Root Transformation

Subjects (Blocks)	On-Time Levels (msec)			Totals
	12	16	20	
i	47.9	55.6	54.9	158.4
ii	62.0	65.6	66.4	194.0
iii	35.7	46.7	47.3	129.7
iv	48.4	54.9	53.7	157.0
v	62.7	62.0	67.2	191.9
vi	26.6	32.6	33.2	92.4
vii	55.6	69.7	73.6	198.9
viii	50.2	49.6	58.0	157.8
Totals	389.1	436.7	454.3	1280.1

$$n = 24 \quad r = 8 \quad c = 3 \quad \sum_{j=1}^{c}\sum_{i=1}^{r} \mathcal{Y}_{ij}^2 = 71{,}763.33$$

SOURCE: Table 5.3 and Appendix B, Table B.7.

By using (5.10)–(5.13), the various sums of squares can now be computed on the transformed data:

$$SS\alpha = \sum_{j=1}^{c} \frac{\mathcal{Y}_{.j}^2}{r} - \frac{\mathcal{Y}_{..}^2}{n} = \left[\frac{(389.1)^2 + (436.7)^2 + (454.3)^2}{8}\right] - \frac{(1280.1)^2}{24}$$

$$= 284.44$$

$$SSB = \sum_{i=1}^{r} \frac{\mathcal{Y}_{i.}^2}{c} - \frac{\mathcal{Y}_{..}^2}{n} = \left[\frac{(158.4)^2 + (194.0)^2 + \cdots + (157.8)^2}{3}\right] - \frac{(1280.1)^2}{24}$$

$$= 3063.69$$

$$SSE = \sum_{j=1}^{c}\sum_{i=1}^{r} \mathcal{Y}_{ij}^2 - \sum_{j=1}^{c}\frac{\mathcal{Y}_{.j}^2}{r} - \sum_{i=1}^{r}\frac{\mathcal{Y}_{i.}^2}{c} + \frac{\mathcal{Y}_{..}^2}{n}$$

$$= 71{,}763.33 - \left[\frac{(389.1)^2 + (436.7)^2 + (454.3)^2}{8}\right]$$

$$- \left[\frac{(158.4)^2 + (194.0)^2 + \cdots + (157.8)^2}{3}\right] + \frac{(1280.1)^2}{24}$$

$$= 137.87$$

$$SST = \sum_{j=1}^{c}\sum_{i=1}^{r} \mathcal{Y}_{ij}^2 - \frac{\mathcal{Y}_{..}^2}{n} = 71{,}763.33 - \frac{(1280.1)^2}{24} = 3486.00$$

As in Table 5.2, the results are then summarized; see Table 5.5.

TABLE 5.5   ANOVA Table for Comparison of Transformed Percent Correct Recall Scores among Three On-Time Levels

Source of Variation	Degrees of Freedom	Sum of Squares	Mean Square	F
Among On-Times (Treatments)	2	284.44	142.22	14.44
Among Subjects (Blocks)	7	3063.69	437.67	44.43
Error	14	137.87	9.85	
Total	23	3486.00		

Using a 5% level of significance, from (5.19) the computed F_a statistic, 14.44, well exceeds the tabulated upper-tail critical value $F_{.95; 2, 14} = 3.74$, and the null hypothesis is rejected (the P value is .00043). It may be concluded that there is a significant treatment effect present. The transformed mean percent correct recall scores are not the same for all three on-time levels. A posteriori methods are needed to determine which treatments are different from each other.

Moreover, as a check on the effectiveness of blocking, from (5.20) the computed F_B statistic, 44.43, greatly exceeds the tabulated $F_{.95; 7, 14} = 2.76$, and, as expected, the blocking has been advantageous—there are highly significant differences from subject to subject.

5.4 MULTIPLE COMPARISON METHODS

5.4.1 A Posteriori Analysis: T Method and S Method

Once the null hypothesis has been rejected, to determine whether the three means are significantly different from each other or whether only one of them differs from the other two, we may again use the classical a posteriori procedures developed by Tukey (the T method) or by Scheffé (the S method)—depending on whether our interest is solely in pairwise comparisons or in all possible comparisons, respectively (see Section 4.2).

Tukey T method.   The *critical range* is computed from

$$\text{critical range} = Q_{1-\alpha; c, (r-1)(c-1)}\sqrt{\frac{\text{MSE}}{r}} \tag{5.23}$$

where $Q_{1-\alpha; c, (r-1)(c-1)}$ is the tabulated upper-tail critical value from the Studentized range distribution having c and $(r-1)(c-1)$ degrees of freedom (see Appendix B, Table B.6) and MSE is the mean square error term (5.16) from the ANOVA. As in Section 4.2.1, each of the $c(c-1)/2$ pairs of means is compared against this critical range, and any pair may be declared significant if

$$|\bar{Y}_{.j} - \bar{Y}_{.j'}| \geq Q_{1-\alpha;\, c,\, (r-1)(c-1)}\sqrt{\frac{\text{MSE}}{r}}$$

where $j \neq j'$.

For the visual pattern recognition study the three sample means from the transformed data are

$$\bar{\mathcal{Y}}_{.1} = 48.64 \qquad \bar{\mathcal{Y}}_{.2} = 54.59 \qquad \bar{\mathcal{Y}}_{.3} = 56.79$$

The critical range is computed to be

$$\text{critical range} = Q_{.95;\, 3,\, 14}\sqrt{\frac{9.85}{8}}$$

$$= (3.70)\sqrt{\frac{9.85}{8}} = 4.11$$

Using an experimentwise error rate of .05 for the three possible pairwise comparisons, it may be concluded that (1) the difference in mean percent correct recall at 12- versus 16-millisecond on-time levels is significant, as is (2) the difference in mean percent correct recall at 12- versus 20-millisecond on-time levels. However, (3) there are no real differences in mean percent correct recall at 16- versus 20-millisecond on-time levels. Therefore, it is the mean percent correct response to the shortest (12-millisecond) on-time level which differs from the others.

Obtaining confidence interval estimates.   By using the T method, a set of *simultaneous* confidence interval estimates for the difference in each pair of means could be established from

$$(\bar{Y}_{.j} - \bar{Y}_{.j'}) - Q_{1-\alpha;\, c,\, (r-1)(c-1)}\sqrt{\frac{\text{MSE}}{r}}$$

$$\leq (\mu_{.j} - \mu_{.j'}) \leq (\bar{Y}_{.j} - \bar{Y}_{.j'}) + Q_{1-\alpha;\, c,\, (r-1)(c-1)}\sqrt{\frac{\text{MSE}}{r}} \qquad (5.24)$$

where $j = 1, 2, \ldots, c$ and $j \neq j'$.

By using (5.24) with a *family* confidence coefficient of .95, the following set of confidence interval estimates are obtained from the transformed data:

1. $5.95 - 4.11 \leq (\mu'_{.2} - \mu'_{.1}) \leq 5.95 + 4.11$
    $1.84 \leq (\mu'_{.2} - \mu'_{.1}) \leq 10.06$
2. $8.15 - 4.11 \leq (\mu'_{.3} - \mu'_{.1}) \leq 8.15 + 4.11$
    $4.04 \leq (\mu'_{.3} - \mu'_{.1}) \leq 12.26$
3. $2.20 - 4.11 \leq (\mu'_{.3} - \mu'_{.2}) \leq 2.20 + 4.11$
    $-1.91 \leq (\mu'_{.3} - \mu'_{.2}) \leq 6.31$

Since these estimates are in *transformed* $\mathcal{Y}_{ij}$ units, they must be reconverted to percentages (Y_{ij}) to have proper meaning. Therefore, by using Table B.7 in Appendix B,

1. $.1\% \leq (\mu_{.2} - \mu_{.1}) \leq 3.1\%$
2. $.5\% \leq (\mu_{.3} - \mu_{.1}) \leq 4.5\%$
3. $-.1\% \leq (\mu_{.3} - \mu_{.2}) \leq 1.2\%$

Scheffé S method.   Similar to the method used in Section 4.2.2, the critical range for each possible contrast $\hat{\mathcal{L}}_l$ is obtained by multiplying the estimated standard deviation of the contrast $S_{\hat{\mathcal{L}}_l}$ by a constant $\sqrt{(c-1)F_{1-\alpha;\,c-1,(r-1)(c-1)}}$, the square root of the product of $c-1$ with the tabulated upper-tail critical value of the F distribution having $c-1$ and $(r-1)$ $(c-1)$ degrees of freedom. That is,

$$\text{critical range} = S_{\hat{\mathcal{L}}_l}\sqrt{(c-1)F_{1-\alpha;\,c-1,(r-1)(c-1)}} \qquad (5.25)$$

and

$$S_{\hat{\mathcal{L}}_l} = \sqrt{\frac{\text{MSE}}{r} \sum_{j=1}^{c} C_{lj}^2} \qquad (5.26)$$

where $l = 1, 2, \ldots, L$; $j = 1, 2, \ldots, c$; and C_{lj} is the coefficient for the jth term in the lth contrast (such that $\sum_{j=1}^{c} C_{lj} = 0$).

By using an experimentwise error rate α, any contrast is declared significant if the absolute value of its estimate $|\hat{\mathcal{L}}_l| = |\sum_{j=1}^{c} C_{lj}\bar{Y}_{.j}|$ equals or exceeds the critical range.

Obtaining confidence interval estimates.   By using the S method, with a *family* confidence coefficient of $1 - \alpha$, a set of *simultaneous* confidence interval estimates for the true contrasts $\mathcal{L}_l$ (where $l = 1, 2, \ldots, L$) may be obtained from

$$\hat{\mathcal{L}}_l - S_{\hat{\mathcal{L}}_l}\sqrt{(c-1)F_{1-\alpha;\,c-1,(r-1)(c-1)}} \leq \mathcal{L}_l \leq \hat{\mathcal{L}}_l + S_{\hat{\mathcal{L}}_l}\sqrt{(c-1)F_{1-\alpha;\,c-1,(r-1)(c-1)}}$$
$$(5.27)$$

As in the visual pattern recognition study, if the researcher is using transformed $\mathcal{Y}_{ij}$ units, the estimates obtained in (5.27) must be reconverted to the *original* data units (Y_{ij}) to have proper meaning.

5.4.2 A Priori Analysis: Orthogonal Contrasts

As in Section 4.3, there are occasions when, prior to obtaining the data, the researcher is *mainly* interested in only certain comparisons. In such situations a set of (up to) $c - 1$ hypotheses, each pertaining to a mutually orthogonal linear contrast, may be established and a priori methods of analysis employed.

For each of these orthogonal contrasts the null hypothesis ($H_0: \mathcal{L}_l^* = 0$) may be rejected at the $\alpha\%$ level of significance if

$$F_l = \frac{\text{MS}\mathcal{L}_l^*}{\text{MSE}} \geq F_{1-\alpha;\,1,(r-1)(c-1)} \qquad (5.28)$$

where $l = 1, 2, \ldots, c - 1$ and where

$$\text{MS}\mathcal{L}_i^* = \text{SS}\mathcal{L}_i^* = \frac{(\hat{\mathcal{L}}_i^*)^2}{\displaystyle\sum_{j=1}^{c} C_{ij}^2/r} \tag{5.29}$$

where $\hat{\mathcal{L}}_i^* = \sum_{j=1}^{c} C_{lj}\bar{Y}_{\cdot j}$.

On the other hand, for confidence interval estimation of the true contrast $\mathcal{L}_i^*$, interpretation is facilitated by using the relationship between t and F [see (1.20)] so that the estimates will not be in squared units of measurement. Hence, as in Section 4.3.2,

$$\hat{\mathcal{L}}_i^* - t_{1-\frac{\alpha}{2};\,(r-1)(c-1)} S_{\mathcal{L}_i^*} \leq \mathcal{L}_i^* \leq \hat{\mathcal{L}}_i^* + t_{1-\frac{\alpha}{2};\,(r-1)(c-1)} S_{\mathcal{L}_i^*} \tag{5.30}$$

where $l = 1, 2, \ldots, c - 1$ and where the estimated standard deviation of an orthogonal contrast is given by the quantity

$$S_{\mathcal{L}_i^*} = \sqrt{\frac{\text{MSE}}{r} \sum_{j=1}^{c} C_{lj}^2} \tag{5.31}$$

If, of course, the estimates in (5.30) are based on transformed $\mathcal{Y}_{ij}$ units, they must be converted back to the original Y_{ij} units of measurement to have proper meaning.

5.4.3 Multiple Comparison Methods: A Summary

In this section we have observed that under randomized complete block designs the testing and estimation formulas for the T and S methods and for the mutually orthogonal linear contrasts are similar to those developed for completely randomized designs in Chapter 4. The modifications here include the replacement of MSW [see (3.16)] and n_j with MSE and r, respectively, and a corresponding adjustment for degrees of freedom. In more sophisticated experimental design models (see Chapters 6 and 7) these methods may also be utilized with similar adjustments for the particular *mean square error* term, the number of observations comprising the means being compared, and degrees of freedom.

5.5 A COMPARISON OF THE RANDOMIZED COMPLETE BLOCK DESIGN TO THE COMPLETELY RANDOMIZED DESIGN

Now that we have developed the randomized complete block model and have used it in a visual pattern recognition study, the question arises as to what effect the blocking had on the analysis. That is, did the blocking result in an increase in precision for comparisons of treatment means in this specific experiment? Moreover, in general, how can the effects of blocking be evaluated?

The estimated relative efficiency (RE) of the randomized complete block design as compared to the completely randomized design may be computed from

$$\text{RE} = \frac{(r-1)\text{MSB} + r(c-1)\text{MSE}}{(n-1)\text{MSE}} \tag{5.32}$$

so that, from Table 5.5, for the visual pattern recognition experiment we have

$$RE = \frac{7(437.67) + 8(2)(9.85)}{23(9.85)} = 14.2$$

Thus roughly 14 times as many observations at each factor level would be needed in a completely randomized design to obtain the same precision for comparisons of treatment means as would be needed for our randomized complete block design.

5.6 COMPUTER PACKAGES AND THE RANDOMIZED BLOCK MODEL: USE OF SPSS, SAS, AND BMDP

5.6.1 Organizing the Data for Computer Analysis

In this section we shall explain how to use either SPSS, SAS, or BMDP to analyze a randomized block model. For illustrative purposes, we shall focus upon the visual pattern recognition problem (see Table 5.4) involving the percent correct response of subjects for various on-time levels of the set of digits. For each subject, two pieces of information are available: the transformed percent correct response and the on-time level. In addition, each subject is assigned an identification number from 1 up to 8 (the total number of subjects). The data can then be organized in the format presented in Table 5.6.

For the purposes of data entry, the subject identification number can be assigned to columns 1 and 2, the on-time levels to column 4, and the transformed percent correct response to columns 6–9.

5.6.2 The SPSS Subprogram ANOVA

The ANOVA subprogram [see Nie et al. (1975)] can be utilized for a randomized block design. For the ANOVA subprogram the required procedure statements are

1   16

ANOVA $\left\{\begin{matrix}\text{dependent}\\\text{variable } Y\end{matrix}\right\}$ ♭BY♭ $\left\{\begin{matrix}\text{block}\\ B\end{matrix}\right\}\left(\begin{matrix}\text{min. max.}\\\text{value, value}\end{matrix}\right)\left\{\text{factor } \alpha\right\}\left(\begin{matrix}\text{min. max.}\\\text{value, value}\end{matrix}\right)\Big/$.

OPTIONS   3

Beginning in column 16, the quantitative (or response) variable to be analyzed is followed (after the word BY) by the classifying variables used to form the block and factor. OPTIONS 3 is utilized here to form the proper sources of variation for the randomized block model (B, α, and error).

Figure 5.1 illustrates the complete SPSS program that has been written to develop a randomized block model for the visual pattern recognition problem. Figure 5.2 represents annotated partial output (see page 126).

TABLE 5.6 Organization of Percent Correct Response Data for Computer Analysis

Subject	On-Time Levels	Transformed Percent Correct
1	1	47.9
1	2	55.6
1	3	54.9
2	1	62.0
2	2	65.6
2	3	66.4
3	1	35.7
3	2	46.7
3	3	47.3
4	1	48.4
4	2	54.9
4	3	53.7
5	1	62.7
5	2	62.0
5	3	67.2
6	1	26.6
6	2	32.6
6	3	33.2
7	1	55.6
7	2	69.7
7	3	73.6
8	1	50.2
8	2	49.6
8	3	58.0

SOURCE: Table 5.4.

5.6.3 The SAS ANOVA Procedure

The SAS ANOVA procedure can be utilized for the randomized block model. This procedure is appropriate for analysis of variance models in which there are an equal number of observations per cell (such as in the randomized block model). The basic setup for using PROC ANOVA for a randomized block model is

PROCƀANOVA;

$$\text{CLASSESƀ} \begin{Bmatrix} \text{name of} \\ \text{block} \\ \text{and factor} \end{Bmatrix};$$

$$\text{MODELƀ} \begin{Bmatrix} \text{dependent} \\ \text{(quantitative)} \\ \text{variable} \end{Bmatrix} = \begin{Bmatrix} \text{name} \\ \text{of} \\ \text{block } B \end{Bmatrix} ƀ \begin{Bmatrix} \text{name} \\ \text{of} \\ \text{factor } \alpha \end{Bmatrix};$$

The names of the block and factor included in the randomized block model are provided in the CLASSES statement. This must precede the MODEL

```
{SYSTEM CARDS}
RUN NAME              VISUAL PATTERN RECOGNITION
DATA LIST             FIXED(1)/1 SUBJECT 1-2,ONTIME 4,
                      TPERCENT 6-9
VAR LABELS            SUBJECT,SUBJECT NUMBER/
                      ONTIME,ONTIME IN MILLISECONDS/
                      TPERCENT,TRANSFORMED PERCENT
                      CORRECT/
VALUE LABELS          ONTIME(1)12 MSEC(2)16 MSECS(3)20
                      MSECS/
READ INPUT DATA
01 1 47.9
 :  :  :
08 3 58.0
END INPUT DATA
ANOVA                 TPERCENT BY SUBJECT(1,8)ONTIME(1,3)/
OPTIONS               3
FINISH
{SYSTEM CARDS}
```

Figure 5.1   SPSS program for pattern recognition example.

```
* * * * * * * * * ANALYSIS  OF  VARIANCE * * * * * * * * * *
          TPERCENT TRANSFORMED PERCENT CORRECT
    BY SUBJECT   SUBJECT NUMBER
       ONTIME    ONTIME IN MILLISECONDS
* * * * * * * * * * * * * * * * * * * * * * * * * * * * * * *
```

SOURCE OF VARIATION		SUM OF SQUARES	DF	MEAN SQUARE	F	SIGNIF OF F
MAIN EFFECTS		3348.117	9	372.013	37.774	0.000
SUBJECT	(BLOCKS)	3063.690	7	437.670	44.441	0.000
ONTIME	(TREATMENT)	284.427	2	142.214	14.440	0.000
EXPLAINED		3348.117	9	372.013	37.774	0.000
RESIDUAL	(ERROR)	137.877	14	9.848		
TOTAL		3495.994	23	151.565		

Figure 5.2   Partial SPSS output for pattern recognition data.

statement. The MODEL statement indicates the dependent (quantitative) variable to be analyzed as well as the block and factor that are to be used in the model.

We should note that if contrasts are desired, PROC GLM should be used in lieu of PROC ANOVA.

Figure 5.3 illustrates the SAS program that has been written for the visual pattern recognition problem.* Figure 5.4 represents annotated partial output.

*We should note that the original percent correct response data could have been used as the input data, and SAS could then have been instructed to transform the data using an arcsin transformation [see SAS (1979)].

```
{SYSTEM CARDS}
DATA PATTERN;
    INPUT SUBJECT 1-2 ONTIME 4 TPERCENT 6-9;
LABEL SUBJECT=SUBJECT NUMBER
      ONTIME=ONTIME IN MILLISECONDS
      TPERCENT=TRANSFORMED PERCENT CORRECT;
CARDS;
01  1  47.9
 :  :   :
08  3  58.0
PROC FORMAT;
    VALUE ONTIMEF 1=12 MSECS
                  2=16 MSECS
                  3=20 MSECS;
PROC ANOVA;
    CLASSES SUBJECT ONTIME;
    MODEL TPERCENT=SUBJECT ONTIME;
    FORMAT ONTIME ONTIMEF.;
{SYSTEM CARDS}
```

Figure 5.3 SAS program for pattern recognition example.

ANALYSIS OF VARIANCE PROCEDURE

DEPENDENT VARIABLE TPERCENT   TRANSFORMED PERCENT CORRECT

SOURCE	DF	SUM OF SQUARES	MEAN SQUARE	F VALUE	PR > F
MODEL	9	3348.12958333	372.01439815	37.78	0.0001
ERROR	14	137.86666667	9.84761905		STD DEV
CORRECTED TOTAL	23	3485.99625000			3.13809163

SOURCE		DF	ANOVA SS	F VALUE	PR > F
SUBJECT	(BLOCKS)	7	3063.68958333	44.44	0.0001
ONTIME	(TREATMENT)	2	284.44000000	14.44	0.0004

Figure 5.4 Partial SAS output for pattern recognition data.

5.6.4 The BMDP Program 2V

The BMDP program 2V can be used for a randomized block model. The paragraph of BMDP unique to program 2V is the /DESIGN paragraph. For the randomized block model, the format of this paragraph is

$$/\text{DESIGN}b\text{DEPENDENT}b\text{IS}b\begin{Bmatrix} \text{name of} \\ \text{dependent} \\ \text{variable} \end{Bmatrix}.$$

$$\text{GROUPING}b\text{ARE}b\begin{Bmatrix} \text{name} \\ \text{of} \\ \text{block } B \end{Bmatrix}, \begin{Bmatrix} \text{name} \\ \text{of} \\ \text{factor } \alpha \end{Bmatrix}.$$

$$\text{INCLUDE} = 1, 2.$$

The DEPENDENT sentence of the /DESIGN paragraph names the dependent variable to be analyzed (transformed percent correct response for the pattern recognition problem). The GROUPING sentence names the variables used to classify the data into blocks and treatments (subjects and on-time for the pattern recognition problem). The INCLUDE sentence is needed for the randomized block design to provide a model in which the sources of variation consist of blocks, treatments, and error.

Figure 5.5 illustrates the program that has been written using BMDP-2V to develop a randomized block model for the visual pattern recognition problem. Figure 5.6 represents annotated partial output.

```
{SYSTEM CARDS}
/PROBLEM    TITLE IS 'VISUAL PATTERN RECOGNITION'.
/INPUT      VARIABLES ARE 3.
            FORMAT IS '(F2.0,1X,F1.0,1X,F4.1)'.
/VARIABLE   NAMES ARE SUBJECT,ONTIME,TPERCENT.
/DESIGN     DEPENDENT IS TPERCENT.
            GROUPING ARE SUBJECT,ONTIME.
            INCLUDE=1,2.
/GROUP      CODES(2) ARE 1,2,3.
            NAMES(2) ARE '12MSECS','16NSECS',
            '20MSECS'.
/END
01 1 47.9
 :  :  :
08 3 58.0
{SYSTEM CARDS}
```

Figure 5.5 Using BMDP program 2V for pattern recognition example.

VISUAL PATTERN RECOGNITION ANALYSIS

ANALYSIS OF VARIANCE FOR 1-ST
DEPENDENT VARIABLE - TPERCENT

SOURCE		SUM OF SQUARES	DEGREES OF FREEDOM	MEAN SQUARE	F	TAIL PROBABILITY
MEAN		68277.31975	1	68277.31975	6933.38	0.0000
SUBJECT	(BLOCKS)	3063.68996	7	437.66999	44.44	0.0000
ONTIME	(TREATMENT)	284.43996	2	142.21998	14.44	0.0004
ERROR		137.86672	14	9.84762		

Figure 5.6 Partial BMDP program 2V output for pattern recognition data.

5.7 SPECIAL PROBLEMS WITH RANDOMIZED BLOCK DESIGNS

5.7.1 Missing Values

In the completely randomized design of Chapter 3, experiments were not restricted to treatment groups containing equal sample sizes. On the other hand, in randomized *complete* block designs the c homogeneous subjects in every single block are randomly assigned, one each, to the c treatment groups (i.e., levels of the factor) so that the sample sizes are equal. Unfortunately, it sometimes happens in randomized complete block experiments that one or more observations cannot be obtained—either due to the nonresponse of a subject or the failure of equipment, etc. Since such a loss of one or more observations destroys the orthogonality of the randomized complete block design, one suggested approach is to substitute for these observations those values which minimize the SSE term (5.12).

As an example, using the transformed data of Table 5.4, suppose that an equipment failure occurs only during the 12-millisecond on-time level for subject iii so that an accurate response cannot be obtained. Rather than remove the whole block (subject) from the visual pattern recognition study, we may replace the missing value by the quantity

$$Y'_{ij} = \frac{rY'_{i.} + cY'_{.j} - Y'_{..}}{(r-1)(c-1)} \tag{5.33}$$

where $Y'_{i.}$ = total in the ith block excluding the missing value
$Y'_{.j}$ = total in the jth treatment group excluding the missing value
$Y'_{..}$ = grand total excluding the missing value

Thus, from Table 5.4, if the response for subject iii under the 12-millisecond on-time level (i.e., $y_{31} = 35.7$) had been missing, then $Y'_3. = 94.0$, $Y'_{.1} = 353.4$, and $Y'_{..} = 1244.4$. By using (5.33), the estimate for this missing value would be

$$Y'_{31} = \frac{8(94.0) + 3(353.4) - 1244.4}{7(2)} = 40.6$$

If this value were used in Table 5.4, the corresponding block, treatment, and grand totals would become

$$y_3. = 134.6 \qquad y_{.1} = 394.0 \qquad y_{..} = 1285.0$$

The resulting ANOVA table (which should now be verified by the reader) is displayed in Table 5.7.

TABLE 5.7 Visual Pattern Recognition Example Adjusted for a Missing Value

Source of Variation	Degrees of Freedom	Sum of Squares	Mean Square	F
Among On-times (Treatments)	2	240.38	120.19	12.59
Among Subjects (Blocks)	7	2971.67	424.52	44.45
Error	13	124.11	9.55	
Total	22	3336.16		

SOURCE: Tables 5.4 and 5.5 (with observation y_{31} missing).

Comparing Table 5.7 to the ANOVA table (Table 5.5) previously presented, we see that the resulting F tests are very similar.* However, we note that one degree of freedom has now been lost from the *error* term since only 23 of the 24 responses would actually have been observed.

For those occasions in which more than one observation is missing, (5.33) cannot be used. One approach, however, is to replace the missing values with "estimates" which make the SSE term (5.12) a minimum. This can easily be accomplished using the methods of differential calculus.†

5.7.2 Balanced Incomplete Randomized Blocks

Not all randomized block experiments with missing values are unplanned or "accidental." Occasionally, in randomized block experiments the desired number of treatment levels exceeds the number of subjects available for assignment within each of the blocks. In such situations a *balanced incomplete randomized block design*—one in which every combination of treatments occurs an equal number of times—should be used. Further discussion concerning balanced incomplete block experiments (including multiple comparisons) is presented by Hicks (1973) and Ott (1977).

5.8 RANDOMIZED BLOCK DESIGNS: A SUMMARY

In this chapter we considered the two-way analysis of variance problem with but one observation per cell. As mentioned in the footnote on p. 115, in randomized complete block experiments the blocks are not given the status of a factor—their purpose was to provide a more efficient means of testing for treatment effects by reducing experimental error attributable to variation among subjects. When two (or more) distinct factors are present in a particular study, the design format is called a factorial experiment. This will be the subject of Chapters 6 and 7.

---

*It should be noted, however, that this method of estimating the missing value usually results in a slight upward bias for the SSα term. Therefore, the researcher must be cautioned that when performing the ANOVA the computed F statistic will likely be slightly inflated. To adjust for this bias see Snedecor and Cochran (1980).

†From (5.12) we have

$$\text{SSE} = \sum_{j=1}^{c} \sum_{i=1}^{r} Y_{ij}^2 - \sum_{j=1}^{c} \frac{Y_{\cdot j}^2}{r} - \sum_{i=1}^{r} \frac{Y_{i\cdot}^2}{c} + \frac{Y_{\cdot\cdot}^2}{n}$$

As demonstrated by Hicks (1973), to obtain estimates Y'_{ij} of missing values Y_{ij} which will minimize (5.12), we take the partial derivatives of that expression with respect to each missing value and set each result equal to zero. That is, we obtain

$$\frac{\partial(\text{SSE})}{\partial Y'_{ij}} = 0$$

for each missing Y_{ij}. We then solve these equations for each missing Y_{ij}. Once the estimates Y'_{ij} are computed, we perform the ANOVA—removing from the error term one degree of freedom for each such Y_{ij} that had been missing.

PROBLEMS

5.1. In a taste testing experiment four brands of California Chablis are rated by nine subjects. To avoid any *carry-over* effects, the tasting sequence for the four wines is randomly determined for each of the nine subjects until a rating on a 10-point scale (1 = extremely unpleasing, 10 = extremely pleasing) is given for each of the following three characteristics: taste, aroma, and dryness. Table P5.1 displays the *summated* ratings—accumulated over all three characteristics.

TABLE P5.1 Summated Ratings of Four Brands of California Chablis

Subject	A	B	C	D
		Brand		
1	24	26	25	22
2	28	28	26	24
3	19	22	20	16
4	24	27	25	23
5	22	23	22	21
6	26	28	24	24
7	27	26	22	23
8	25	27	24	21
9	22	23	20	19

(a) Using $\alpha = .05$, completely analyze the results of the experiment.

(b) Discuss some of the problems inherent in obtaining the responses in the manner described and how these might invalidate the conclusions drawn from your analysis in part (a).

*5.2. An industrial psychologist wished to investigate the effects of drug usage on the reaction times of teenage drivers. Twenty-one teenagers who had just passed their driving tests were grouped, 3 each, according to aggressive driving performance so that the 3 "most aggressive drivers" comprised the first group, while the 3 "least aggressive drivers" comprised the seventh group. The members of each group were then randomly assigned to the 3 dosage levels of a certain barbiturate. Thirty minutes after administration of the drug, reaction times were recorded based on a test involving simulated driving performance. The data are presented in Table P5.2.

(a) Using $\alpha = .05$, completely analyze the results of the experiment.

(b) What can be said about the true mean reaction time (in seconds) of teenage drivers under the influence of 25 milligrams of the barbiturate?

5.3. An agronomist wishes to study the yield (in pounds per plot) of four different strains of Gallipoli wheat. A field was divided into six blocks. Within each block the four strains are randomly assigned, one each, to four plots for planting. The results of the experiment are displayed in Table P5.3.

(a) Using $\alpha = .05$, completely analyze the data to determine what differences in yield exist, if any, among the four strains of Gallipoli wheat.

TABLE P5.2 Reaction Times to Drug Dosage Levels
(in Seconds)

| | Dosage Level (milligrams) | | |
Groups	5	10	25
I	2.6	2.7	3.4
II	2.5	2.5	2.9
III	3.0	2.9	3.3
IV	2.8	3.1	3.7
V	3.2	3.4	3.9
VI	3.3	3.4	4.0
VII	3.5	3.5	4.6

TABLE P5.3 Wheat Yield (in Pounds per Planted Plot)

| | Gallipoli Wheat Strain | | | |
Blocks	A	B	C	D
I	30.4	28.8	33.0	31.8
II	33.9	25.5	32.7	33.5
III	32.7	27.3	34.5	34.5
IV	34.9	29.3	36.0	33.8
V	31.9	27.5	36.5	34.5
VI	35.4	28.3	34.2	36.0

(b) If, prior to running the experiment, the agronomist was specifically concerned with a comparison of strain B against the others taken together, set up the appropriate orthogonal contrast (and two other mutually orthogonal contrasts). What can be concluded?

(c) What may be said about the true mean yield (in pounds per plot) for strain B wheat?

5.4. The data in Table P5.4 represent the milliequivalents of sodium excreted by six subjects 2 hours after treatment with one of six diuretics assigned at random by a clinician over a 6-day period. Using $\alpha = .05$, completely analyze the data to determine whether there are any differences in the effectiveness of the diuretics. What would the clinician conclude?

5.5. Woodward (1970) sought to determine the fastest approach for a base runner to reach second base. He devised an experiment in which the time trials of various Cincinnati Reds baseball players would be evaluated over three different methods of rounding first base: round out, narrow angle, and wide angle. Suppose, in an effort to replicate Woodward's experiment, the data in Table P5.5 were obtained. Using $\alpha = .05$, completely analyze the data to determine whether or not there are any differences in the three base running methods.

TABLE P5.4 Amount of Excreted Sodium

Subjects	Treatments (Diuretics)					
	A	B	C	D	E	F
I	3.9	30.6	25.2	4.4	29.4	38.9
II	5.6	30.1	33.5	7.9	30.7	33.1
III	5.8	16.9	25.5	4.0	32.9	39.2
IV	4.3	23.2	18.9	4.4	28.2	28.1
V	5.9	26.7	20.5	4.2	23.4	38.2
VI	4.3	10.9	26.7	4.4	12.0	26.7

SOURCE: Extracted from F. Wilcoxon, and R. A. Wilcox, *Some Rapid Approximate Statistical Procedures*, p. 11, Copyright 1949, 1964, Lederle Laboratories Division of American Cyanamid Company, All Rights Reserved, and Reprinted with Permission.

TABLE P5.5 Comparison of Methods for Rounding First Base

Player	Methods		
	Round Out	Narrow Angle	Wide Angle
A	5.65	5.60	5.40
B	5.90	5.85	5.70
C	5.25	5.15	5.00
D	5.85	5.80	5.70
E	5.55	5.55	5.35
F	5.40	5.50	5.55

5.6. The data in Table P5.6 represent the earnings per share for the seven largest diversified financial companies (based on 1979 assets) over three time periods.

TABLE P5.6 Earnings Per Share ($)

Company	Time Periods		
	1969	1978	1979
Aetna Life & Casualty	.89	6.19	7.25
Travelers Corp.	1.58	8.43	9.20
American Express	1.27	4.31	4.83
H. F. Ahmanson	1.73	5.13	5.11
Merrill Lynch & Co.	.94	2.00	3.26
First Charter Financial	.95	3.55	3.05
Great Western Financial	.97	4.01	4.15

SOURCE: Peter D. Petre, "The Fortune Directory of the Largest Non-industrial Companies," *Fortune* (July 14, 1980), pp. 146–159. © 1980 Time Inc. All rights reserved.

(a) Using $\alpha = .05$, completely analyze the data to determine whether or not there are any differences in earnings per share over the three time periods.

(b) If, in advance of obtaining the data, the researcher was primarily interested in (1) assessing differences in earnings per share from 1978 to 1979 and (2) determining whether there are any differences from 1969 to 1978–1979 taken together, perform the appropriate a priori analysis.

(c) If, in advance of obtaining the data, the researcher was primarily interested in studying whether or not earnings per share has been *monotonically increasing* over (*increasing*) time periods, perform the appropriate a priori analysis. *Hint:* Set up a *linear contrast* with respective coefficients of -1, 0, and $+1$ for the three time periods. Decompose SSα into this linear contrast plus an unanalyzed and untested remainder.

5.7. Brady (1969) was interested in evaluating the effects of the rhythmicity of a metronome on the speech of stutterers. Twelve subjects classified as severe stutterers were asked to speak extemporaneously for 3 minutes under three conditions: N (no metronome aid), A (irregular metronome aid), and R (regular metronome aid). Table P5.7 indicates the number of dysfluencies under each treatment condition.

TABLE P5.7   Number of Dysfluencies Under Three Treatment Conditions

	Condition		
Subject	N	A	R
1	15	5	3
2	11	3	3
3	18	3	1
4	21	4	5
5	6	2	2
6	17	2	0
7	10	2	0
8	8	3	0
9	13	2	0
10	4	0	1
11	11	4	2
12	17	1	2

SOURCE: J. P. Brady, "Studies on the Metronome Effect on Stuttering," *Behaviour Research and Therapy*, Vol. 7 (1969), pp. 197–204. Copyright 1969, Pergamon Press, Ltd. Reprinted with permission.

(a) Using $\alpha = .05$, completely analyze the data.

(b) Discuss why a data transformation should (or should not) be used prior to any classical analysis of this experiment.

5.8. Parker and Holford (1968) describe a study on the application of medical research to crime. Suppose that Table P5.8 represents a replication of their study. Table P5.8 presents the logarithms of the concentration of three halogens (in parts per million) in the hair of 10 human subjects. Using $\alpha = .05$, completely analyze the data.

TABLE P5.8   **Logarithm of Concentration of Halogens in Hair**

	Trace Element		
Subject	Cl	I	Br
1	7.6	2.3	3.5
2	7.5	3.6	3.1
3	8.4	1.6	3.0
4	8.8	2.1	3.5
5	7.7	4.5	2.9
6	5.9	2.6	5.7
7	6.0	2.4	2.4
8	8.1	2.6	3.8
9	7.6	2.2	2.8
10	8.3	2.9	3.9

5.9. In an experiment concerning physiological effects under hypnosis, Damaser et al. (1963) measured skin potential (adjusted for initial level) in millivolts for eight subjects under the emotions of fear, happiness, depression, and calmness—which had been requested in a random manner. The data are presented in Table P5.9.

TABLE P5.9   **Skin Potential Under Four Hypnotically Requested Emotions**

	Emotion			
Subject	Fear	Happiness	Depression	Calmness
1	23.1	22.7	22.5	22.6
2	57.6	53.2	53.7	53.1
3	10.5	9.7	10.8	8.3
4	23.6	19.6	21.1	21.6
5	11.9	13.8	13.7	13.3
6	54.6	47.1	39.2	37.0
7	21.0	13.6	13.7	14.8
8	20.3	23.6	16.3	14.8

SOURCE: E. C. Damaser, R. E. Shor, and M. T. Orne, "Physiological Effects During Hypnotically Requested Emotions," *Psychosomatic Medicine*, Vol. 25 (1963), pp. 334–343. Reprinted with permission of the publisher. Copyright 1963 by The American Psychosomatic Society, Inc.

(a) Using $\alpha = .05$, completely analyze the data.

(b) Suppose, prior to obtaining the data, the researchers had really been interested in studying whether skin potential under calmness differs from that under the other three emotions taken together. Set up the appropriate orthogonal contrast (plus the unanalyzed and untested remainder), and analyze the data.

5.10. Kline (1950) investigated a problem involving intelligence and hypnosis. In an effort to replicate this study, suppose the data in Table P5.10 indicate the IQ scores (adjusted to requested age under hypnosis) based on an Otis test of mental ability for 10 subjects who were told under hypnosis to regress to ages 15, 10, and 8. Using $\alpha = .05$, completely analyze the data.

TABLE P5.10   IQ Scores Under Hypnosis

Subject	Treatment			
	Waking	Regress to 15	Regress to 10	Regress to 8
A	126	127	127	125
B	118	119	122	120
C	119	121	123	119
D	110	109	107	107
E	113	112	110	112
F	111	113	112	111
G	110	114	112	112
H	118	118	118	119
I	123	126	127	125
J	124	123	126	125

5.11. Reviewing concepts:
  (a) What is a randomized complete block design?
  (b) What are the fundamental differences between the randomized complete block experiment and the completely randomized experiment?
  (c) Describe the ANOVA model for a randomized complete block design.

6

TWO-WAY ANOVA
WITH EQUAL OBSERVATIONS
IN EACH CELL

6.1 INTRODUCTION

In our discussions of analysis of variance procedures in Chapters 3 and 5 we have been concerned with experiments that involved but one factor of interest. In Chapter 3 we explored the most fundamental analysis of variance procedure, the completely randomized model wherein only one factor varied over several levels of interest. This model was modified in Chapter 5 to consider the effect of blocking in the development of a more efficient experimental design. In this chapter we shall extend our discussion to a model in which two factors are being simultaneously considered. In addition to studying how each of these two factors vary over different levels, we shall also examine the way in which the two factors interact together.

6.2 THE TWO-WAY ANALYSIS OF VARIANCE MODEL WITH EQUAL REPLICATION

6.2.1 Model Development

The two-way analysis of variance model which we shall consider in this chapter assumes that there are equal sample sizes (replications) for each combination of the levels of factor α with those of factor $\mathcal{B}$.* The model can be

---

*See Chapters 7 and 12 for a discussion of ANOVA models with unequal sample sizes.

expressed as

$$Y_{ijk} = \mu + \alpha_i + \mathcal{B}_j + \alpha\mathcal{B}_{ij} + \epsilon_{ijk} \qquad (6.1)$$

where $i = 1, \ldots, r$; $j = 1, \ldots, c$; $k = 1, \ldots, n$; Y_{ijk} is the kth observation for level i of factor α and level j of factor $\mathcal{B}$; μ represents the overall mean effect; α_i represents the effect of the ith level of factor α; $\mathcal{B}_j$ represents the effect of the jth level of factor $\mathcal{B}$; $\alpha\mathcal{B}_{ij}$ represents the interaction effect at level i of factor α and level j of factor $\mathcal{B}$; and ϵ_{ijk} represents the random error present in the kth observation in cell ij.

The random error term ϵ_{ijk} is assumed to be normally distributed with mean zero and variance σ^2. The general format for a two-way analysis of variance problem is illustrated in Table 6.1.

TABLE 6.1   Tabular Format for a Two-Way ANOVA with Replication

| Factor α | Factor $\mathcal{B}$ | | | | Total |
	1	2	$\cdots$	c	
1	Y_{111} Y_{112} $\vdots$ Y_{11n} $\quad Y_{11.}$	Y_{121} Y_{122} $\vdots$ Y_{12n} $\quad Y_{12.}$	$\cdots$	Y_{1c1} Y_{1c2} $\vdots$ Y_{1cn} $\quad Y_{1c.}$	$Y_{1..}$
2	Y_{211} Y_{212} $\vdots$ Y_{21n} $\quad Y_{21.}$	Y_{221} Y_{222} $\vdots$ Y_{22n} $\quad Y_{22.}$	$\cdots$	Y_{2c1} Y_{2c2} $\vdots$ Y_{2cn} $\quad Y_{2c.}$	$Y_{2..}$
$\vdots$	$\vdots$	$\vdots$	$\cdots$	$\vdots$	$\vdots$
r	Y_{r11} Y_{r12} $\vdots$ Y_{r1n} $\quad Y_{r1.}$	Y_{r21} Y_{r22} $\vdots$ Y_{r2n} $\quad Y_{r2.}$	$\cdots$	Y_{rc1} Y_{rc2} $\vdots$ Y_{rcn} $\quad Y_{rc.}$	$Y_{r..}$
Total	$Y_{.1.}$	$Y_{.2.}$	$\cdots$	$Y_{.c.}$	$Y_{...}$

The information expressed in Table 6.1 consists of the following:

Y_{ijk} = value of the kth observation for level i of factor α and level j of factor $\mathcal{B}$

$$H_{0_\alpha}: \quad \mu_{1..} = \mu_{2..} = \cdots = \mu_{r..} \ (\text{or } \alpha_i = 0 \text{ for all } i)$$

$$H_{1_\alpha}: \quad \text{not all } \mu_{i..} \text{ are equal} \quad (\text{or not all } \alpha_i = 0, \ i = 1, 2, \ldots, r)$$

cision rule is to reject H_{0_α} if $F_\alpha = MS\alpha/MSE \geq F_{1-\alpha;\ r-1, rc(n-1)}$.

econd, we wish to determine whether there is any difference in the average

se for different levels of factor $\mathcal{B}$:

$$H_{0_\mathcal{B}}: \quad \mu_{.1.} = \mu_{.2.} = \cdots = \mu_{.c.} \ (\text{or } \mathcal{B}_j = 0 \text{ for all } j)$$

$$H_{1_\mathcal{B}}: \quad \text{not all } \mu_{.j.} \text{ are equal} \quad (\text{or not all } \mathcal{B}_j = 0, \ j = 1, 2, \ldots, c)$$

ecision rule is to reject $H_{0_\mathcal{B}}$ if $F_\mathcal{B} = MS\mathcal{B}/MSE \geq F_{1-\alpha;\ c-1, rc(n-1)}$.

Third, we wish to determine whether there is an interacting effect between

s α and $\mathcal{B}$:

$$H_{0_{\alpha\mathcal{B}}}: \quad \alpha\mathcal{B}_{ij} = 0 \ (\text{for all } i \text{ and } j)$$

$$H_{0_{\alpha\mathcal{B}}}: \quad \alpha\mathcal{B}_{ij} \neq 0$$

decision rule is to reject $H_{0_{\alpha\mathcal{B}}}$ if $F_{\alpha\mathcal{B}} = MS\alpha\mathcal{B}/MSE \geq F_{1-\alpha;\ (r-1)(c-1), rc(n-1)}$.

6.2.3 Computational Formulas

As we previously noted in Chapters 3 and 5, computational formulas are

able for obtaining the sums of squares for each source of variation. Such

ulas are useful because they reduce the amount of computational drudgery

would otherwise be required when solving problems without using a com-

r package.

These computational formulas are as follows:

$$SS\alpha = \frac{\sum_{i=1}^{r} Y_{i..}^2}{cn} - \frac{Y_{...}^2}{rcn} \tag{6.8}$$

$$SS\mathcal{B} = \frac{\sum_{j=1}^{c} Y_{.j.}^2}{rn} - \frac{Y_{...}^2}{rcn} \tag{6.9}$$

$$SS\alpha\mathcal{B} = \frac{\sum_{i=1}^{r}\sum_{j=1}^{c} Y_{ij.}^2}{n} - \frac{\sum_{i=1}^{r} Y_{i..}^2}{cn} - \frac{\sum_{j=1}^{c} Y_{.j.}^2}{rn} + \frac{Y_{...}^2}{rcn} \tag{6.10}$$

$$SSE = \sum_{i=1}^{r}\sum_{j=1}^{c}\sum_{k=1}^{n} Y_{ijk}^2 - \frac{\sum_{i=1}^{r}\sum_{j=1}^{c} Y_{ij.}^2}{n} \tag{6.11}$$

$$SST = \sum_{i=1}^{r}\sum_{j=1}^{c}\sum_{k=1}^{n} Y_{ijk}^2 - \frac{Y_{...}^2}{rcn} \tag{6.12}$$

able 6.3 presents the ANOVA table for the two-way model with equal repli-

ation using these computational formulas.

$Y_{ij.}$ = sum of the values in cell ij

$Y_{i..}$ = sum of the values in row i

$Y_{.j.}$ = sum of the values in column j

$Y_{...}$ = grand total of all values over all rows and columns

Understanding the notation for two-way ANOVA. The individual values Y_{ijk} represent the items within each cell combination of factor α and factor $\mathcal{B}$. For example, Y_{121} represents the first observation for level 1 of factor α and level 2 of factor $\mathcal{B}$. The cell totals $Y_{ij.}$ represent the total (or sum) for all observations collected for a particular combination (cell) of factor α and factor $\mathcal{B}$. For example, $Y_{22.}$ represents the sum of all observations at level 2 of factor α and level 2 of factor $\mathcal{B}$. The row totals $Y_{i..}$ represent the total (or sum) of all observations for a particular level of factor α. Thus, $Y_{1..}$ is the sum of all values for row 1, the first level of factor α. The column totals $Y_{.j.}$ represent the total of all observations for a particular level of factor $\mathcal{B}$. For example, $Y_{.2.}$ represents the total of all observations in column 2, the second level of factor $\mathcal{B}$. Finally, the grand total $Y_{...}$ represents the sum of all levels of factor α and all levels of factor $\mathcal{B}$.

Partitioning the sum of squares. Now that we have defined the notation to be utilized in this two-way analysis of variance model, we can examine the way in which the total variation or sum of squares can be subdivided into the appropriate set of independent components. We may recall (see Section 3.3) that for the one-way ANOVA model we had

$$SST = SS\alpha + SSW \tag{3.10}$$

while for the randomized blocks model (see Section 5.2) we had

$$SST = SS\alpha + SSB + SSE \tag{5.9}$$

Since the two-way ANOVA model (6.1) contains effects due to factor α, due to factor $\mathcal{B}$, due to the interaction of factors α and $\mathcal{B}$, and due to random error, the total sum of squares (SST) can be subdivided in the following manner:

$$SST = SS\alpha + SS\mathcal{B} + SS\alpha\mathcal{B} + SSE \tag{6.2}$$

These sums of squares are defined as follows for the two-way ANOVA model with equal observations in each cell:

$$SST = \sum_{i=1}^{r}\sum_{j=1}^{c}\sum_{k=1}^{n}(Y_{ijk} - \bar{Y}...)^2 \tag{6.3}$$

$$SS\alpha = cn\sum_{i=1}^{r}(\bar{Y}_{i..} - \bar{Y}...)^2 \tag{6.4}$$

$$SS\mathcal{B} = rn\sum_{j=1}^{c}(\bar{Y}_{.j.} - \bar{Y}...)^2 \tag{6.5}$$

$$SS\alpha\mathcal{B} = n\sum_{i=1}^{r}\sum_{j=1}^{c}(\bar{Y}_{ij.} - \bar{Y}_{i..} - \bar{Y}_{.j.} + \bar{Y}...)^2 \tag{6.6}$$

$$\text{SSE} = \sum_{i=1}^{r} \sum_{j=1}^{c} \sum_{k=1}^{n} (Y_{ijk} - \bar{Y}_{ij\cdot})^2 \tag{6.7}$$

where

$$\bar{Y}_{\cdots} = \frac{Y_{\cdots}}{rcn} \qquad \bar{Y}_{i\cdot\cdot} = \frac{Y_{i\cdot\cdot}}{cn} \qquad \bar{Y}_{\cdot j\cdot} = \frac{Y_{\cdot j\cdot}}{rn} \qquad \bar{Y}_{ij\cdot} = \frac{Y_{ij\cdot}}{n}$$

The total sum of squares measures the total variation in the response variable (Y) and is based on squared differences between the individual values (Y_{ijk}) and the overall mean of all values ($\bar{Y}_{\cdots}$). The sum of squares due to factor α (SSα) represents differences among the various levels of factor α; it is based on the squared differences between the mean for each level of factor α ($\bar{Y}_{i\cdot\cdot}$) and the overall mean ($\bar{Y}_{\cdots}$). Similarly, the sum of squares due to factor $\mathcal{B}$ (SS$\mathcal{B}$) represents differences among the various levels of factor $\mathcal{B}$; it is based on the squared differences between the mean for each level of factor $\mathcal{B}$ ($\bar{Y}_{\cdot j\cdot}$) and the overall mean ($\bar{Y}_{\cdots}$). The sum of squares due to the interaction of factors α and $\mathcal{B}$ (SS$\alpha\mathcal{B}$) represents the effect of various combinations of levels for factors α and $\mathcal{B}$; it is based on the squared difference between the cell mean ($\bar{Y}_{ij\cdot}$), the mean for level i of factor α ($\bar{Y}_{i\cdot\cdot}$), the mean for level j of factor $\mathcal{B}$ ($\bar{Y}_{\cdot j\cdot}$), and the overall mean ($\bar{Y}_{\cdots}$). The concept of *interaction* will be discussed further in Section 6.4. Finally, the error sum of squares (SSE) measures the random error in the response variable that is not affected by the level of factor α or factor $\mathcal{B}$. SSE is based on the squared difference between the individual values (Y_{ijk}) and the cell means ($\bar{Y}_{ij\cdot}$).

6.2.2 The Analysis of Variance Table

Now that the total sum of squares has been subdivided into its additive components, we need to determine the number of degrees of freedom associated with each source of variation. As in the case of the one-way ANOVA model, the degrees of freedom for the total sum of squares are the sample size minus 1. However, in this two-factor model with replication the total sample size is rcn, so the total degrees of freedom are $rcn - 1$. The degrees of freedom for factor α are $r - 1$, the number of different levels for factor α minus 1. In the same manner, the degrees of freedom for factor $\mathcal{B}$ are $c - 1$, the number of different levels for factor $\mathcal{B}$ minus 1. The degrees of freedom for the interaction of factors α and $\mathcal{B}$ are the product $(r - 1)(c - 1)$. Finally, the degrees of freedom for the error are $rc(n - 1)$.

The sum of squares and degrees of freedom for all sources of variation are summarized in the analysis of variance table (assuming a fixed-effects model*) presented as Table 6.2. From Table 6.2 we observe that there are three distinct tests of hypothesis in a two-way analysis of variance model with equal observations. First we would like to determine whether there is any difference in the average response for different levels of factor α:

*A discussion of fixed-, random-, and mixed-effects models is presented in Section 6.5.

TABLE 6.2   Analysis of Variance Table for a Two-Factor Model with Equal Replication (Fixed Effects)

Source	Degrees of Freedom	Sum of Squares	Mean Square (MS) or Variance	F
α	$r - 1$	$cn\sum_{i=1}^{r}(\bar{Y}_{i\cdot\cdot} - \bar{Y}_{\cdots})^2$	$\text{MS}\alpha = \dfrac{\text{SS}\alpha}{r-1}$	$\dfrac{\text{MS}\alpha}{\text{MSE}}$
$\mathcal{B}$	$c - 1$	$rn\sum_{j=1}^{c}(\bar{Y}_{\cdot j\cdot} - \bar{Y}_{\cdots})^2$	$\text{MS}\mathcal{B} = \dfrac{\text{SS}\mathcal{B}}{c-1}$	$\dfrac{\text{MS}\mathcal{B}}{\text{MSE}}$
$\alpha\mathcal{B}$	$(r-1)(c-1)$	$n\sum_{i=1}^{r}\sum_{j=1}^{c}(\bar{Y}_{ij\cdot} - \bar{Y}_{i\cdot\cdot} - \bar{Y}_{\cdot j\cdot} + \bar{Y}_{\cdots})^2$	$\text{MS}\alpha\mathcal{B} = \dfrac{\text{SS}\alpha\mathcal{B}}{(r-1)(c-1)}$	$\dfrac{\text{MS}\alpha\mathcal{B}}{\text{MSE}}$
Error	$rc(n - 1)$	$\sum_{i=1}^{r}\sum_{j=1}^{c}\sum_{k=1}^{n}(Y_{ijk} - \bar{Y}_{ij\cdot})^2$	$\text{MSE} = \dfrac{\text{SSE}}{rc(n-1)}$	
Total	$rcn - 1$	$\sum_{i=1}^{r}\sum_{j=1}^{c}\sum_{k=1}^{n}(Y_{ijk} - \bar{Y}_{\cdots})^2$		

TABLE 6.3 Analysis of Variance Table for a Two-Factor Fixed Effects Model with Equal Replication Using Computational Formulas

Source	Degrees of Freedom	Sum of Squares	Mean Square	F
α	$r-1$	$\sum_{i=1}^{r} \dfrac{Y_{i..}^2}{cn} - \dfrac{Y_{...}^2}{rcn}$	$MS\alpha = \dfrac{SS\alpha}{r-1}$	$\dfrac{MS\alpha}{MSE}$
β	$c-1$	$\sum_{j=1}^{c} \dfrac{Y_{.j.}^2}{rn} - \dfrac{Y_{...}^2}{rcn}$	$MS\beta = \dfrac{SS\beta}{c-1}$	$\dfrac{MS\beta}{MSE}$
$\alpha\beta$	$(r-1)(c-1)$	$\sum_{i=1}^{r}\sum_{j=1}^{c} \dfrac{Y_{ij.}^2}{n} - \sum_{i=1}^{r}\dfrac{Y_{i..}^2}{cn} - \sum_{j=1}^{c}\dfrac{Y_{.j.}^2}{rn} + \dfrac{Y_{...}^2}{rcn}$	$MS\alpha\beta = \dfrac{SS\alpha\beta}{(r-1)(c-1)}$	$\dfrac{MS\alpha\beta}{MSE}$
Error	$rc(n-1)$	$\sum_{i=1}^{r}\sum_{j=1}^{c}\sum_{k=1}^{n} Y_{ijk}^2 - \dfrac{\sum_{i=1}^{r}\sum_{j=1}^{c} Y_{ij.}^2}{n}$	$MSE = \dfrac{SSE}{rc(n-1)}$	
Total	$rcn-1$	$\sum_{i=1}^{r}\sum_{j=1}^{c}\sum_{k=1}^{n} Y_{ijk}^2 - \dfrac{Y_{...}^2}{rcn}$		

6.2.4 Application

Now that the basic principles of the two-way ANOVA model with equal replication have been developed, we may turn to the following example: Suppose an auditor for the Internal Revenue Service (IRS) wanted to study the effect on the processing time for individual tax returns of two factors, the complexity of the tax return and the period in which the tax form was submitted. The first factor, complexity of tax return, was categorized into three levels: simple, somewhat complex, and highly complex. The second factor, time of submission of the tax form, was categorized into two levels: early (submitted more than 1 month prior to the deadline) and on time (submitted within 1 month of the deadline). A random sample of three tax returns for each of the six combinations of complexity and submission time was selected from a regional IRS office. The results are presented in Table 6.4.

From Table 6.3, we should note that the formulas require the computation of the following quantities:

$$\sum_{i=1}^{r}\sum_{j=1}^{c}\sum_{k=1}^{n} Y_{ijk}^2 \qquad \frac{\sum_{i=1}^{r} Y_{i..}^2}{cn} \qquad \frac{\sum_{j=1}^{c} Y_{.j.}^2}{rn} \qquad \frac{\sum_{i=1}^{r}\sum_{j=1}^{c} Y_{ij.}^2}{n} \qquad Y_{...}$$

TABLE 6.4 Processing Time for a Random Sample of 18 Tax Returns Classified by Complexity and Submission Time

Complexity	Submission Time				Total
	Early		On Time		
Simple	24 34 32	$Y_{11.} = 90$	49 56 39	$Y_{12.} = 144$	$Y_{1..} = 234$
Somewhat Complex	54 40 56	$Y_{21.} = 150$	82 70 73	$Y_{22.} = 225$	$Y_{2..} = 375$
Highly Complex	56 66 73	$Y_{31.} = 195$	93 82 80	$Y_{32.} = 255$	$Y_{3..} = 450$
Total		$Y_{.1.} = 435$		$Y_{.2.} = 624$	$Y_{...} = 1059$

From the data in Table 6.4 we have $r = 3$, $c = 2$, $n = 3$, and

$$\sum_{i=1}^{3}\sum_{j=1}^{2}\sum_{k=1}^{3} Y_{ijk}^2 = (24)^2 + (34)^2 + \cdots + (80)^2 = 69{,}013.0$$

$$\sum_{i=1}^{3} \sum_{j=1}^{2} \frac{Y_{ij.}^2}{n} = \frac{(90)^2 + (144)^2 + (150)^2 + (225)^2 + (195)^2 + (255)^2}{3}$$

$$= \frac{205,011}{3} = 68,337.0$$

$$\sum_{i=1}^{3} \frac{Y_{i..}^2}{cn} = \frac{(234)^2 + (375)^2 + (450)^2}{2(3)} = \frac{397,881}{6} = 66,313.5$$

$$\sum_{j=1}^{2} \frac{Y_{.j.}^2}{rn} = \frac{(435)^2 + (624)^2}{3(3)} = \frac{578,601}{9} = 64,289.0$$

$$\frac{Y_{...}^2}{rcn} = \frac{(1059)^2}{3(2)(3)} = \frac{1,121,481}{18} = 62,304.5$$

By using (6.8)–(6.12), the various sums of squares are now computed.

$$SS\alpha = \frac{\sum_{i=1}^{r} Y_{i..}^2}{cn} - \frac{Y_{...}^2}{rcn} = 66,313.5 - 62,304.5 = 4009.0$$

$$SS\mathfrak{B} = \frac{\sum_{j=1}^{c} Y_{.j.}^2}{rn} - \frac{Y_{...}^2}{rcn} = 64,289.0 - 62,304.5 = 1984.5$$

$$SS\alpha\mathfrak{B} = \frac{\sum_{i=1}^{r}\sum_{j=1}^{c} Y_{ij.}^2}{n} - \frac{\sum_{i=1}^{r} Y_{i..}^2}{cn} - \frac{\sum_{j=1}^{c} Y_{.j.}^2}{rn} + \frac{Y_{...}^2}{rcn}$$

$$= 68,337.0 - 66,313.5 - 64,289.0 + 62,304.5 = 39.0$$

$$SSE = \sum_{i=1}^{r}\sum_{j=1}^{c}\sum_{k=1}^{n} Y_{ijk}^2 - \frac{\sum_{i=1}^{r}\sum_{j=1}^{c} Y_{ij.}^2}{n} = 69,013.0 - 68,337.0 = 676.0$$

$$SST = \sum_{i=1}^{r}\sum_{j=1}^{c}\sum_{k=1}^{n} Y_{ijk}^2 - \frac{Y_{...}^2}{rcn} = 69,013.0 - 62,304.5 = 6708.5$$

Using the format of Table 6.3, the ANOVA table for the tax return problem can be displayed as Table 6.5.

TABLE 6.5   Analysis of Variance Table for the Tax Return Problem Using Computational Formulas

Source	Degrees of Freedom	Sum of Squares	Mean Square	F
α (Complexity)	$r - 1 = 2$	4009.0	2004.5	35.583
$\mathfrak{B}$ (Submission Time)	$c - 1 = 1$	1984.5	1984.5	35.228
$\alpha\mathfrak{B}$ (Complexity × Submission Time)	$(r - 1)(c - 1) = 2$	39.0	19.5	.346
Error	$rc(n - 1) = 12$	676.0	56.33	
Total	$rcn - 1 = 17$	6708.5		

From the results presented in Table 6.5, we can test hypotheses relating to factor α (complexity), factor $\mathcal{B}$ (submission time), and the interaction of complexity and submission time. In testing the hypothesis concerning factor α,

$H_{0\alpha}$: $\mu_{1..} = \mu_{2..} = \mu_{3..}$ (or $\alpha_i = 0$ for all i)

$H_{1\alpha}$: not all $\mu_{i..}$ are equal ($i = 1, 2, 3$) That is, the levels of complexity do not have equal mean processing times.

We have (at the .05 level of significance) $F_\alpha = 35.583 > F_{.95; \, 2,12} = 3.89$. Thus we reject the null hypothesis and conclude that there is a significant difference in the mean processing times based on the complexity level of the tax returns.

In testing the hypothesis concerning factor $\mathcal{B}$,

$H_{0\mathcal{B}}$: $\mu_{.1.} = \mu_{.2.}$ (or $\mathcal{B}_j = 0$ for all j)

$H_{1\mathcal{B}}$: $\mu_{.1.} \neq \mu_{.2.}$. That is, the different submission times do not have equal mean processing times.

We have (at the .05 level of significance) $F_\mathcal{B} = 35.228 > F_{.95; \, 1,12} = 4.75$. Therefore, we reject the null hypothesis and conclude that there is a difference in the mean processing time based on differences in the submission times of the tax returns.

Finally, we can test whether there is an interacting effect between factor α complexity) and factor $\mathcal{B}$ (submission time):

$$H_{0\alpha\mathcal{B}}: \alpha\mathcal{B}_{ij} = 0 \text{ (for all } i \text{ and } j)$$

$$H_{1\alpha\mathcal{B}}: \alpha\mathcal{B}_{ij} \neq 0$$

We have (at the .05 level of significance) $F_{\alpha\mathcal{B}} = .346 < F_{.95; \, 2,12} = 3.89$. Therefore, we do not reject the null hypothesis. We can conclude that there is no evidence of an interacting effect between the complexity and submission time of tax returns.

6.2.5 The Concept of Interaction

Now that we have performed these three statistical tests, we can obtain a better understanding of the interpretation of the concept of interaction by plotting the cell means on a two-dimensional graph (see Figure 6.1). Since $\bar{Y}_{ij.} = Y_{ij.}/n$ we have

$$\bar{Y}_{11.} = \frac{Y_{11.}}{3} = \frac{90}{3} = 30 \qquad \bar{Y}_{12.} = \frac{Y_{12.}}{3} = \frac{144}{3} = 48$$

$$\bar{Y}_{21.} = \frac{Y_{21.}}{3} = \frac{150}{3} = 50 \qquad \bar{Y}_{22.} = \frac{Y_{22.}}{3} = \frac{225}{3} = 75$$

$$\bar{Y}_{31.} = \frac{Y_{31.}}{3} = \frac{195}{3} = 65 \qquad \bar{Y}_{32.} = \frac{Y_{32.}}{3} = \frac{255}{3} = 85$$

In Figure 6.1, we have plotted the average processing time for each level of complexity and submission times. For our data, the two lines (representing early and on-time submissions) appear roughly parallel. This phenomenon can be interpreted to mean that the *difference* in mean processing time between early and on-time submissions is virtually the same for the various levels of complexity. In other words, there is no *interaction* between these two factors. This is clearly substantiated from the *F* test that indicated $(F_{\alpha\beta} = .346 < F_{.95; 2,12} = 3.89)$ no evidence of an interacting effect between complexity and submission time on the response variable (processing time). Thus the effect of each factor in the model (6.1) can be considered *additive*, since the level of one factor does not have any influence on the effect of the other factor. Therefore, from Table 6.5 and Figure 6.1, we can conclude that mean processing time appears to increase with the complexity of the tax form and the "lateness" of the submission. These conclusions will be discussed in further detail in Section 6.3 when multiple comparison procedures will be extended to the two-way model to determine which levels of a factor are significantly different.

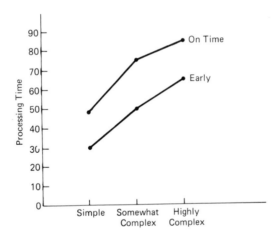

Figure 6.1   Processing time based on complexity for different submission times.

6.3 MULTIPLE COMPARISONS FOR TWO-WAY ANOVA MODELS

In Section 4.2 we explored the use of multiple comparison procedures to deter-mine (a posteriori) which levels of a factor differed in their average response. Tukey's *T* procedure was utilized to make pairwise comparisons between two groups, while Scheffé's § method was used for more complex comparisons involving sets of groups and/or unequal sample sizes. Although both the Tukey and Scheffé multiple comparison procedures can be easily extended to the two-factor model, we shall examine only the Tukey *T* method here (see Section 7.5 for the Scheffé § method).

Tukey's confidence intervals can be used for either rows (factor α) or columns (factor $\mathscr{B}$).

For factor α we have

$$(\bar{Y}_{i..} - \bar{Y}_{i'..}) \pm Q_{1-\alpha;\, r, rc(n-1)}\sqrt{\frac{\text{MSE}}{cn}} \tag{6.13}$$

where $\bar{Y}_{i..}$ is the average value for row i of factor α and $\bar{Y}_{i'..}$ is the average value for row i' ($i' \neq i$) of factor α.

For factor $\mathscr{B}$ we have

$$(\bar{Y}_{.j.} - \bar{Y}_{.j'.}) \pm Q_{1-\alpha;\, c, rc(n-1)}\sqrt{\frac{\text{MSE}}{rn}} \tag{6.14}$$

where $\bar{Y}_{.j.}$ is the average value for column j of factor $\mathscr{B}$ and $\bar{Y}_{.j'.}$ is the average value for column j' ($j' \neq j$) of factor $\mathscr{B}$.

In either case, the Q value represents the $1 - \alpha$ percentile on the Studentized range distribution (see Appendix B, Table B.6). Moreover, for either set there is $100(1 - \alpha)\%$ confidence that for *all* pairs of comparisons the interval developed correctly estimates the true difference between the two groups.

Thus if we refer to the tax return example of Section 6.2, we can apply the Tukey T method to determine which levels of complexity differ in terms of mean processing time. From Tables 6.4 and 6.5 we have

$$\bar{Y}_{1..} = 39 \qquad \bar{Y}_{2..} = 62.5 \qquad \bar{Y}_{3..} = 75$$
$$\text{MSE} = 56.33 \qquad r = 3 \qquad c = 2 \qquad n = 3$$

Using (6.13), with 95% confidence we obtain the following paired comparisons:

1. *Simple vs. somewhat complex:*

$$(\bar{Y}_{1..} - \bar{Y}_{2..}) \pm 3.77\sqrt{\frac{56.33}{6}}$$
$$(39 - 62.5) \pm 11.551$$
$$-35.051 \leq \mu_{1..} - \mu_{2..} \leq -11.949$$

2. *Simple vs. highly complex:*

$$(\bar{Y}_{1..} - \bar{Y}_{3..}) \pm 3.77\sqrt{\frac{56.33}{6}}$$
$$(39 - 75) \pm 11.551$$
$$-47.551 \leq \mu_{1..} - \mu_{3..} \leq -24.449$$

3. *Somewhat complex vs. highly complex:*

$$(\bar{Y}_{2..} - \bar{Y}_{3..}) \pm 3.77\sqrt{\frac{56.33}{6}}$$
$$(62.5 - 75) \pm 11.551$$
$$-24.051 \leq \mu_{2..} - \mu_{3..} \leq -.949$$

Therefore, since each of the confidence interval estimates does not include zero, we can conclude that all pairwise comparisons are significant with 95% confidence. Thus the mean processing time differs significantly for each level of complexity.

6.4 INTERPRETING INTERACTING EFFECTS

6.4.1 Introduction

In our previous example concerning the processing of tax returns, we determined that the factors of complexity and submission time appeared to be independent, i.e., not to interact. However, in many instances factors under investigation clearly interact with each other. In some applications the two interacting factors can have either a *synergistic* effect or an *interfering* effect on each other. A synergistic effect occurs when the two factors interact in a joint, multiplicative manner that is greater than the mere presence of the two effects taken together. On the other hand, an interfering effect occurs when the level of one factor interferes with, or reduces, the effect of the particular level of the second factor. In such cases the effect of the two factors on the response variable is much less than what could be expected from the presence (in an additive manner) of the two factors taken together.

6.4.2 Application

To further investigate the interpretation of interaction, we shall examine the following: Suppose a large industrial corporation was interested in studying the effect of alternative workweek scheduling arrangements on absenteeism. Four different scheduling arrangements were available for a 40-hour workweek:

1. Standard "9–5" workweek.
2. Flexible daily hours: Employees can choose starting times within a fixed range of hours.
3. Biweekly day off: Hours are arranged during a 2-week period to produce eight 9-hour days, one 8-hour day, and one day off.
4. Four-day workweek: Hours are arranged into four 10-hour work days so that workers have one day off per week.

The experiment was to run for a period of 1 year at four different company locations that were matched approximately according to both size and qualitative characteristics. Each plant location was randomly assigned to a work scheduling arrangement. In addition to studying different work scheduling arrangements, three groups of employees were to be examined: supervisory, clerical, and production workers. A random sample of two workers in each group was selected in each of the four locations. The results, which indicate the yearly number of days absent (adjusted to reflect an 8-hour work day), are presented in Table 6.6. From Table 6.6, we have $r = 4$, $c = 3$, $n = 2$, and

TABLE 6.6 Days Absent for a Random Sample of 24 Workers Classified by Type of Work Scheduling and Type of Job

	Type of Job			
Work Scheduling	Production	Clerical	Supervisory	Total
Standard	13.0 16.0 $Y_{11\cdot} = 29$	10.0 8.0 $Y_{12\cdot} = 18$	7.0 5.0 $Y_{13\cdot} = 12$	$Y_{1\cdot\cdot} = 59$
Flexible	10.0 12.0 $Y_{21\cdot} = 22$	9.0 8.0 $Y_{22\cdot} = 17$	5.0 6.0 $Y_{23\cdot} = 11$	$Y_{2\cdot\cdot} = 50$
Biweekly Day Off	9.3 10.1 $Y_{31\cdot} = 19.4$	8.5 8.3 $Y_{32\cdot} = 16.8$	4.5 4.3 $Y_{33\cdot} = 8.8$	$Y_{3\cdot\cdot} = 45$
Four-Day Workweek	7.5 7.0 $Y_{41\cdot} = 14.5$	7.0 7.8 $Y_{42\cdot} = 14.8$	5.0 5.5 $Y_{43\cdot} = 10.5$	$Y_{4\cdot\cdot} = 39.8$
Total	$Y_{\cdot1\cdot} = 84.9$	$Y_{\cdot2\cdot} = 66.6$	$Y_{\cdot3\cdot} = 42.3$	$Y_{\cdot\cdot\cdot} = 193.8$

$$\sum_{i=1}^{4} \sum_{j=1}^{3} \sum_{k=1}^{2} Y_{ijk}^2 = (13)^2 + (16)^2 + \cdots + (5.5)^2 = 1751.72$$

$$\frac{\sum_{i=1}^{4} \sum_{j=1}^{3} Y_{ij\cdot}^2}{n} = \frac{(29)^2 + (18)^2 + \cdots + (10.5)^2}{2} = \frac{3478.58}{2} = 1739.29$$

$$\frac{\sum_{i=1}^{4} Y_{i\cdot\cdot}^2}{cn} = \frac{(59)^2 + (50)^2 + (45)^2 + (39.8)^2}{3(2)} = \frac{9590.04}{6} = 1598.34$$

$$\frac{\sum_{j=1}^{3} Y_{\cdot j\cdot}^2}{rn} = \frac{(84.9)^2 + (66.6)^2 + (42.3)^2}{4(2)} = \frac{13,432.86}{8} = 1679.1075$$

$$\frac{Y_{\cdot\cdot\cdot}^2}{rcn} = \frac{(193.8)^2}{4(3)(2)} = \frac{37,558.44}{24} = 1564.935$$

By using (6.8) – (6.12), the various sums of squares are computed and the appropriate ANOVA table is then presented as Table 6.7.

$$\text{SS}\alpha = \frac{\sum_{i=1}^{r} Y_{i\cdot\cdot}^2}{cn} - \frac{Y_{\cdot\cdot\cdot}^2}{rcn} = 1598.34 - 1564.935 = 33.405$$

$$\text{SS}\beta = \frac{\sum_{j=1}^{c} Y_{\cdot j\cdot}^2}{rn} - \frac{Y_{\cdot\cdot\cdot}^2}{rcn} = 1679.1075 - 1564.935 = 114.1725$$

$$\text{SS}\alpha\beta = \frac{\sum_{i=1}^{r} \sum_{j=1}^{c} Y_{ij\cdot}^2}{n} - \frac{\sum_{i=1}^{r} Y_{i\cdot\cdot}^2}{cn} - \frac{\sum_{j=1}^{c} Y_{\cdot j\cdot}^2}{rn} + \frac{Y_{\cdot\cdot\cdot}^2}{rcn}$$

$$= 1739.29 - 1598.34 - 1679.1075 + 1564.935 = 26.775$$

$$\text{SSE} = \sum_{i=1}^{r} \sum_{j=1}^{c} \sum_{k=1}^{n} Y_{ijk}^2 - \frac{\sum_{i=1}^{r} \sum_{j=1}^{c} Y_{ij\cdot}^2}{n} = 1751.72 - 1739.29 = 12.43$$

$$\text{SST} = \sum_{i=1}^{r} \sum_{j=1}^{c} \sum_{k=1}^{n} Y_{ijk}^2 - \frac{Y_{\cdot\cdot\cdot}^2}{rcn} = 1751.72 - 1564.935 = 186.785$$

From the results shown in Table 6.7 we can determine whether there is a significant effect on absenteeism due to work schedules and due to type of job. In addition, we can determine whether there is an interacting effect between these two factors. The .05 level of significance is utilized for each test. For factor α,

$H_{0\alpha}$: $\mu_{1\cdot\cdot} = \mu_{2\cdot\cdot} = \mu_{3\cdot\cdot} = \mu_{4\cdot\cdot}$ (or $\alpha_i = 0$ for all i)

$H_{1\alpha}$: not all $\mu_{i\cdot\cdot}$ are equal (The different work scheduling arrangements do not have the same mean absenteeism.)

From Table 6.7, $F_\alpha = 10.75 > F_{.95; \, 3, 12} = 3.49$. Therefore, we reject the null hypothesis and conclude that there is a difference in mean absenteeism among the four work scheduling arrangements.

For factor $\mathcal{B}$,

$H_{0_{\mathcal{B}}}$: $\mu_{.1.} = \mu_{.2.} = \mu_{.3.}$ (or $\mathcal{B}_j = 0$ for all j)

$H_{1_{\mathcal{B}}}$: not all $\mu_{.j.}$ are equal (The different job types do not have the same mean absenteeism.)

From Table 6.7, $F_{\mathcal{B}} = 55.113 > F_{.95; 2,12} = 3.89$. Once again we reject the null hypothesis and conclude that there is a difference in mean absenteeism among the various job types.

TABLE 6.7 **Analysis of Variance Table for the Absenteeism Problem**

Source	Degrees of Freedom	Sum of Squares	Mean Square	F
$\mathcal{A}$ (Work Schedule)	$r - 1 = 3$	33.405	11.135	10.75
$\mathcal{B}$ (Type of Job)	$c - 1 = 2$	114.1725	57.0863	55.113
$\mathcal{A}\mathcal{B}$ (Work Schedule × Type of Job)	$(r - 1)(c - 1) = 6$	26.775	4.4629	4.309
Error	$rc(n - 1) = 12$	12.43	1.0358	
Total	$rcn - 1 = 23$	186.785		

In testing the interaction of factors $\mathcal{A}$ and $\mathcal{B}$, we form the hypothesis

$$H_{0_{\mathcal{A}\mathcal{B}}}: \quad \mathcal{A}\mathcal{B}_{ij} = 0 \text{ (for all } i \text{ and } j)$$
$$H_{1_{\mathcal{A}\mathcal{B}}}: \quad \mathcal{A}\mathcal{B}_{ij} \neq 0$$

and from Table 6.7, $F_{\mathcal{A}\mathcal{B}} = 4.309 > F_{.95; 6,12} = 3.00$. Thus in this example we reject the null hypothesis and conclude that there is an interacting effect on absenteeism based on work schedule and type of job.

We can obtain a more thorough understanding of the meaning and consequences of this interaction effect by computing the average value for each cell as follows:

$$\bar{Y}_{11.} = \frac{29}{2} = 14.5 \qquad \bar{Y}_{12.} = \frac{18}{2} = 9.0 \qquad \bar{Y}_{13.} = \frac{12}{2} = 6.0$$

$$\bar{Y}_{21.} = \frac{22}{2} = 11.0 \qquad \bar{Y}_{22.} = \frac{17}{2} = 8.5 \qquad \bar{Y}_{23.} = \frac{11}{2} = 5.5$$

$$\bar{Y}_{31.} = \frac{19.4}{2} = 9.7 \qquad \bar{Y}_{32.} = \frac{16.8}{2} = 8.4 \qquad \bar{Y}_{33.} = \frac{8.8}{2} = 4.4$$

$$\bar{Y}_{41.} = \frac{14.5}{2} = 7.25 \qquad \bar{Y}_{42.} = \frac{14.8}{2} = 7.4 \qquad \bar{Y}_{43.} = \frac{10.5}{2} = 5.25$$

These cell means are plotted (see Figure 6.2) so that each point represents the average absenteeism for each work schedule and type of job. If we examine Figure 6.2, it is clearly evident why there is a significant interacting effect between work schedule and type of job. Certainly the lines which represent absenteeism for the different work schedules are not parallel for the various job types. Thus the differences in mean absenteeism among various job types are not constant for all work scheduling arrangements. For example, there appears to be a large difference in absenteeism between production and clerical workers for a "standard" work scheduling arrangement. However, this difference narrows when other work scheduling arrangements are involved, and, in fact, when a 4-day week is considered, production workers have a slightly lower rate of mean absenteeism. Thus the existence of an interacting effect *complicates the interpretation* of the significance of the main effects (work schedule and job type). Hence we cannot directly conclude that there is a difference in mean absenteeism among the various job types because this difference *is not the same* for all work schedules. Likewise, we cannot directly conclude that there is a difference in mean absenteeism among the work schedules because this difference *is not the same* for all job types. Thus the existence of a significant interaction effect has created a more complex set of conclusions for our data.

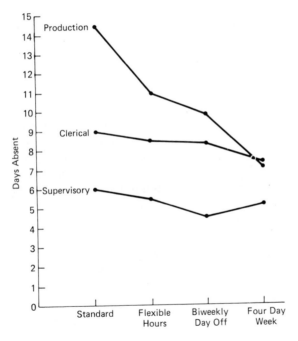

Figure 6.2 Absenteeism based on work schedules for different types of jobs.

6.5 FIXED, RANDOM, AND MIXED MODELS

In our discussion of analysis of variance models we have not focused upon the manner in which the various levels of a factor have been selected. From this perspective, there are three alternative models:

1. Fixed-effects model (Model I)
2. Random-effects model (Model II)
3. Mixed-effects model (Model III)

The first model, the fixed-effects model (Model I) described in (6.1), assumes that the levels of a factor have been *specifically* selected for analysis. This means that the levels of the factor have *not* been randomly selected from a population and that no inferences can be drawn about any other levels except the ones used in the study.

In contrast to the fixed-effects model, Model II, the random-effects model, contains factors in which the levels are *randomly selected* from a population. The objective for a random-effects model is not necessarily to examine differences among levels but, more importantly, to estimate the variability due to each factor [see Hicks (1973)]. For example, if we wished to study the effect on productivity of different workers and different machines, we might randomly select a sample of machines and assign a random sample of workers to each machine for a given number of days. Not only would we be able to measure whether workers and machines have significant effects on productivity, but we would also be able to estimate the variability due to different machines and the variability due to different workers. The random effects model can be expressed as in (6.1), but here α_i, $\mathcal{B}_j$, $\alpha\mathcal{B}_{ij}$, and ϵ_{ijk} are each assumed to be normally distributed.

The third model, the mixed-effects model (Model III), contains a mixture of fixed and random effects. This model, with factor α assumed to be a fixed effect and factor $\mathcal{B}$ assumed to be a random effect, can also be expressed as in (6.1), where $\mathcal{B}_j$, $\alpha\mathcal{B}_{ij}$, and ϵ_{ijk} are each assumed to be normally distributed. For example, referring to the absenteeism problem of Section 6.4, if the types of jobs had been randomly selected from a larger set of jobs, we would have had a mixed-effects model in which factor α (work schedules) was a fixed effect and factor $\mathcal{B}$ (types of jobs) was a random effect.

Although the random- and mixed-effects models can be discussed in much greater depth [see Hicks (1973) and Winer (1971)], our focus involves the consequences of the various models on the F test. Since the components of the models differ in their assumptions, they also lead to different F tests in evaluating the significance of the main effects (factors α and $\mathcal{B}$). Therefore, the appropriate F tests for each of the three models are summarized in Table 6.8.

As we observe in Table 6.8, the tests for the main effects differ depending on the type of model selected. For the fixed-effects model, the F tests involve the ratio of MSα or MS$\mathcal{B}$ to the error variance (MSE). For the random-effects model, the F tests (for the main effects) involve the ratio of MSα or MS$\mathcal{B}$ to the interaction variance (MS$\alpha\mathcal{B}$). For the mixed model with factor α fixed and factor $\mathcal{B}$ random, the F test for factor α involves the ratio of MSα to MS$\alpha\mathcal{B}$, while the test for factor $\mathcal{B}$ involves the ratio of MS$\mathcal{B}$ to MSE.

TABLE 6.8   *F* Tests for Two-Factor ANOVA Models with Replication

Null Hypothesis	Fixed Effects (α & β Fixed)	Random Effects (α & β Random)	Mixed Effects (α Fixed, β Random)
$\alpha_i = 0$	$F_\alpha = \dfrac{MS\alpha}{MSE}$	$F_\alpha = \dfrac{MS\alpha}{MS\alpha\beta}$	$F_\alpha = \dfrac{MS\alpha}{MS\alpha\beta}$
$\beta_j = 0$	$F_\beta = \dfrac{MS\beta}{MSE}$	$F_\beta = \dfrac{MS\beta}{MS\alpha\beta}$	$F_\beta = \dfrac{MS\beta}{MSE}$
$\alpha\beta_{ij} = 0$	$F_{\alpha\beta} = \dfrac{MS\alpha\beta}{MSE}$	$F_{\alpha\beta} = \dfrac{MS\alpha\beta}{MSE}$	$F_{\alpha\beta} = \dfrac{MS\alpha\beta}{MSE}$

6.6 TWO-FACTOR ANOVA MODEL WITHOUT REPLICATION

6.6.1 Development

In our discussion of the two-factor model in Sections 6.1–6.5, we have assumed that there were at least two observations in each cell. For such a model we were able to measure directly the interaction effect since the error variance was based on differences within each cell. However, in many circumstances only one observation can be obtained for each combination of factor α and factor β. When there is only one observation per cell, the interaction and error sources of variation cannot be separately estimated. However, if the assumption is made that there is no interaction between the two variables, we can formulate the following two-factor model without replication:

$$Y_{ij} = \mu + \alpha_i + \beta_j + \epsilon_{ij} \tag{6.15}$$

where $i = 1, \ldots, r; j = i, \ldots, c$; and the ϵ_{ij} are normally distributed.

The general format for the two-factor ANOVA model without replication is presented in Table 6.9, where Y_{ij} is the value at level i of factor α and level j of factor β, $Y_i.$ is the total (sum) of the values for level i of factor α, $Y_{.j}$ is the total (sum) of the values for level j of factor β, and $Y_{..}$ is the grand total of all values over all rows and columns.

In this two-factor model without replication, the total sum of squares (SST) can be subdivided as follows:

$$SST = SS\alpha + SS\beta + SSE \tag{6.16}$$

These sums of squares can be defined as

$$SST = \sum_{i=1}^{r} \sum_{j=1}^{c} (Y_{ij} - \bar{Y}..)^2 \tag{6.17}$$

$$SS\alpha = c \sum_{i=1}^{r} (\bar{Y}_i. - \bar{Y}..)^2 \tag{6.18}$$

TABLE 6.9 **Tabular Format for a Two-Way ANOVA Model Without Replication**

Factor α	Factor β				
	1	2	$\cdots$	c	Total
1	Y_{11}	Y_{12}	$\cdots$	Y_{1c}	$Y_{1.}$
2	Y_{21}	Y_{22}	$\cdots$	Y_{2c}	$Y_{2.}$
$\cdot$	$\cdot$	$\cdot$	$\cdots$	$\cdot$	$\cdot$
$\cdot$	$\cdot$	$\cdot$		$\cdot$	$\cdot$
r	Y_{r1}	Y_{r2}	$\cdots$	Y_{rc}	$Y_{r.}$
Total	$Y_{.1}$	$Y_{.2}$	$\cdots$	$Y_{.c}$	$Y_{..}$

$$\text{SS}\beta = r \sum_{j=1}^{c} (\bar{Y}_{.j} - \bar{Y}_{..})^2 \tag{6.19}$$

$$\text{SSE} = \sum_{i=1}^{r} \sum_{j=1}^{c} (Y_{ij} - \bar{Y}_{i.} - \bar{Y}_{.j} + \bar{Y}_{..})^2 \tag{6.20}$$

where

$$\bar{Y}_{..} = \frac{Y_{..}}{rc} \qquad \bar{Y}_{i.} = \frac{Y_{i.}}{c} \qquad \bar{Y}_{.j} = \frac{Y_{.j}}{r}$$

As in the case of other ANOVA models, computational formulas are available for obtaining the various sums of squares:

$$\text{SS}\alpha = \frac{\sum_{i=1}^{r} Y_{i.}^2}{c} - \frac{Y_{..}^2}{rc} \tag{6.21}$$

$$\text{SS}\beta = \frac{\sum_{j=1}^{c} Y_{.j}^2}{r} - \frac{Y_{..}^2}{rc} \tag{6.22}$$

$$\text{SSE} = \sum_{i=1}^{r} \sum_{j=1}^{c} Y_{ij}^2 - \frac{\sum_{i=1}^{r} Y_{i.}^2}{c} - \frac{\sum_{j=1}^{c} Y_{.j}^2}{r} + \frac{Y_{..}^2}{rc} \tag{6.23}$$

$$\text{SST} = \sum_{i=1}^{r} \sum_{j=1}^{c} Y_{ij}^2 - \frac{Y_{..}^2}{rc} \tag{6.24}$$

The ANOVA table is presented as Table 6.10.

6.6.2 Application

A large consumer testing organization is interested in evaluating the durability of different brands of transistor batteries used in electronic calculators. Three different brands of batteries are to be tested: a "famous" American brand,

TABLE 6.10 Analysis of Variance Table for a Two-Factor Model without Replication

Source	Degrees of Freedom	Sum of Squares	Mean Square	F
α	$r-1$	$\sum_{i=1}^{r}\dfrac{Y_{i\cdot}^2}{c} - \dfrac{Y_{\cdot\cdot}^2}{rc}$	$MS\alpha = \dfrac{SS\alpha}{r-1}$	$\dfrac{MS\alpha}{MSE}$
β	$c-1$	$\sum_{j=1}^{c}\dfrac{Y_{\cdot j}^2}{r} - \dfrac{Y_{\cdot\cdot}^2}{rc}$	$MS\beta = \dfrac{SS\beta}{c-1}$	$\dfrac{MS\beta}{MSE}$
Error	$(r-1)(c-1)$	$\sum_{i=1}^{r}\sum_{j=1}^{c} Y_{ij}^2 - \dfrac{\sum_{i=1}^{r} Y_{i\cdot}^2}{c} - \dfrac{\sum_{j=1}^{c} Y_{\cdot j}^2}{r} + \dfrac{Y_{\cdot\cdot}^2}{rc}$	$MSE = \dfrac{SSE}{(r-1)(c-1)}$	
Total	$rc-1$	$\sum_{i=1}^{r}\sum_{j=1}^{c} Y_{ij}^2 - \dfrac{Y_{\cdot\cdot}^2}{rc}$		

an "unknown" American brand, and a Japanese brand. In addition to studying differences among the brands, the testing agency wishes to determine whether the durability of the various brands of batteries differs for various calculator models from a particular manufacturer. Three models were selected for the experiment: a basic function calculator, a business calculator, and a scientific calculator. Since only one battery of each brand is available for each type of calculator, it is assumed that there is no interaction between brand of battery and type of calculator on battery durability.

The results of the study in which durability has been measured according to the life of the battery in hours (under continuous usage) are presented in Table 6.11.

TABLE 6.11 Battery Life (in Hours) Classified by Brand and Type of Calculator

Battery Brand	Type of Calculator			Total
	Basic Function	Business	Scientific	
Famous American	17.6	16.1	13.7	$Y_1. = 47.4$
Unknown American	11.8	10.0	9.1	$Y_2. = 30.9$
Japanese	15.3	13.5	12.0	$Y_3. = 40.8$
Total	$Y._1 = 44.7$	$Y._2 = 39.6$	$Y._3 = 34.8$	$Y.. = 119.1$

For the data in Table 6.11, the following summary computations are obtained: $r = 3$, $c = 3$, and

$$\sum_{i=1}^{r} \sum_{j=1}^{c} Y_{ij}^2 = (17.6)^2 + (16.1)^2 + \cdots + (12.0)^2 = 1639.05$$

$$\frac{\sum_{i=1}^{r} Y_{i.}^2}{c} = \frac{(47.4)^2 + (30.9)^2 + (40.8)^2}{3} = \frac{4866.21}{3} = 1622.07$$

$$\frac{\sum_{j=1}^{c} Y_{.j}^2}{r} = \frac{(44.7)^2 + (39.6)^2 + (34.8)^2}{3} = \frac{4777.29}{3} = 1592.43$$

$$\frac{Y..^2}{rc} = \frac{(119.1)^2}{3(3)} = 1576.09$$

By using (6.21) − (6.24), the various sums of squares are computed and the ANOVA table is established as Table 6.12.

$$SS\alpha = \frac{\sum\limits_{i=1}^{r} Y_{i.}^2}{c} - \frac{Y_{..}^2}{rc} = 1622.07 - 1576.09 = 45.98$$

$$SS\mathcal{B} = \frac{\sum\limits_{j=1}^{c} Y_{.j}^2}{r} - \frac{Y_{..}^2}{rc} = 1592.43 - 1576.09 = 16.34$$

$$SSE = \sum\limits_{i=1}^{r}\sum\limits_{j=1}^{c} Y_{ij}^2 - \frac{\sum\limits_{i=1}^{r} Y_{i.}^2}{c} - \frac{\sum\limits_{j=1}^{c} Y_{.j}^2}{r} + \frac{Y_{..}^2}{rc}$$

$$= 1639.05 - 1622.07 - 1592.43 + 1576.09 = .64$$

$$SST = \sum\limits_{i=1}^{r}\sum\limits_{j=1}^{c} Y_{ij}^2 - \frac{Y_{..}^2}{rc} = 1639.05 - 1576.09 = 62.96$$

TABLE 6.12   Analysis of Variance Table for the Battery Problem

Source	Degrees of Freedom	Sum of Squares	Mean Square	F
α (Brands)	$r - 1 = 2$	45.98	22.99	143.688
$\mathcal{B}$ (Type of Calculators)	$c - 1 = 2$	16.34	8.17	51.063
Error	$(r - 1)(c - 1) = 4$	0.64	.16	
Total	$rc - 1 = 8$	62.96		

For the results obtained in Table 6.12, we can test hypotheses concerning the significance of factor α (brands) and factor $\mathcal{B}$ (type of calculators). For factor α, given the hypotheses

$H_{0\alpha}$: $\mu_1. = \mu_2. = \mu_3.$ (or $\alpha_i = 0$ for all i)

$H_{1\alpha}$:   not all $\mu_{i.}$ are equal   [the brands do not have the same durability (life)]

we have (at the .01 level of significance) $F_\alpha = 143.688 > F_{.99; 2,4} = 18.0$ so that the null hypothesis is rejected. We would conclude that there is a difference in durability for the various brands of batteries.

For factor $\mathcal{B}$, given the hypotheses

$H_{0\mathcal{B}}$: $\mu_{.1} = \mu_{.2} = \mu_{.3}$ (or $\mathcal{B}_j = 0$ for all j)

$H_{1\mathcal{B}}$:   not all $\mu_{.j}$ are equal   (the durability is not the same for all types of calculators)

we have (at the .01 level of significance) $F_\mathcal{B} = 51.063 > F_{.99; 2,4} = 18.0$, and again the null hypothesis is rejected. Here we would conclude that there is a difference in the durability of batteries when used in different types of calculators.

6.7 TESTING FOR ADDITIVITY

In our discussion of the two-factor model without replication, since the interaction effect could not be separated from the error effect the assumption was implicitly made that there was no interaction between the two factors. In this section we shall discuss ways in which the existence of any possible interaction effect could be determined—first by using a graphical approach and second by employing a well-known statistical procedure.

6.7.1 Graphical Approach to Additivity

In our previous graphical interpretation of interaction the cell means were plotted (see Figures 6.1 and 6.2). A similar graph which displays each Y_{ij} could also be plotted for a model without replication. Using the data from Table 6.11, Figure 6.3 indicates the battery life for each brand of battery for each type of calculator.

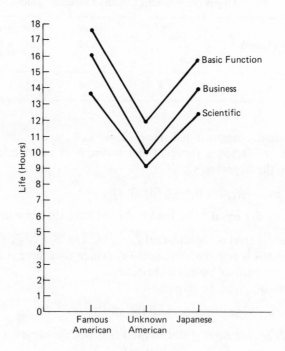

Figure 6.3 Battery life based on brands for different types of calculators.

We can observe from Figure 6.3 that the lines connecting the battery brands for each type of calculator seem to be roughly parallel. This would appear to provide some descriptive evidence to indicate that there is no interaction between brands and type of calculator.

6.7.2 Tukey Test for Additivity

However, a more quantitative approach toward the measurement of inter-action in the two-factor model was developed by Tukey (1949) and is called the *test for additivity*. This procedure assumes that any interaction γ_{ij} that may exist is a *multiplicative* function of the main effects of the two factors so that

$$\gamma_{ij} = \mathfrak{D}\alpha_i\mathfrak{B}_j \qquad (6.25)$$

Thus the null and alternative hypotheses can be stated as

$$H_0: \quad \mathfrak{D} = 0 \qquad \text{(no interaction is present)}$$
$$H_1: \quad \mathfrak{D} \neq 0 \qquad \text{(interaction is present)}$$

To perform this test of additivity, the error sum of squares given in (6.20) must be partitioned into two components, *nonadditivity* and *pure error*, so that

$$\text{SSE} = \text{SSNA} + \text{SSPE} \qquad (6.26)$$

where

$$\text{SSNA} = \frac{\left[\sum\limits_{i=1}^{r}\sum\limits_{j=1}^{c} Y_{ij}(\bar{Y}_{i.} - \bar{Y}_{..})(\bar{Y}_{.j} - \bar{Y}_{..})\right]^2}{[(\text{SS}\alpha)(\text{SS}\mathfrak{B})]/rc} \qquad (6.27)$$

and

$$\text{SSPE} = \text{SSE} - \text{SSNA} \qquad (6.28)$$

The computations for SSNA are illustrated in Table 6.13, where

$$\bar{Y}_{i.} = \frac{Y_{i.}}{c} \qquad \bar{Y}_{.j} = \frac{Y_{.j}}{r} \qquad \bar{Y}_{..} = \frac{Y_{..}}{rc}$$

$$\bar{Y}_{1.} = \frac{47.4}{3} = 15.8 \qquad \bar{Y}_{2.} = \frac{30.9}{3} = 10.3 \qquad \bar{Y}_{3.} = \frac{40.8}{3} = 13.6$$

$$\bar{Y}_{.1} = \frac{44.7}{3} = 14.9 \qquad \bar{Y}_{.2} = \frac{39.6}{3} = 13.2 \qquad \bar{Y}_{.3} = \frac{34.8}{3} = 11.6$$

$$\bar{Y}_{..} = \frac{119.1}{3(3)} = 13.233333$$

Using Table 6.13 and (6.27), we obtain

$$\text{SSNA} = \frac{(5.364334)^2}{[45.98(16.34)]/[3(3)]} = .3447$$

From (6.28),

$$\text{SSPE} = \text{SSE} - \text{SSNA} = .64 - .3447$$
$$= .2953$$

Since it can be shown that the nonadditivity component has one degree of freedom while the pure error component has $rc - r - c$ degrees of freedom, the appropriate test statistic is given by

TABLE 6.13   Computing SSNA

Y_{ij}	$\bar{Y}_{i.} - \bar{Y}_{..}$	$\bar{Y}_{.j} - \bar{Y}_{..}$	$Y_{ij}(\bar{Y}_{i.} - \bar{Y}_{..})(\bar{Y}_{.j} - \bar{Y}_{..})$
$Y_{11} = 17.6$	$15.8 - 13.233333$	$14.9 - 13.233333$	75.288889
$Y_{12} = 16.1$	$15.8 - 13.233333$	$13.2 - 13.233333$	-1.377444
$Y_{13} = 13.7$	$15.8 - 13.233333$	$11.6 - 13.233333$	-57.433444
$Y_{21} = 11.8$	$10.3 - 13.233333$	$14.9 - 13.233333$	-57.688889
$Y_{22} = 10.0$	$10.3 - 13.233333$	$13.2 - 13.233333$	$.977778$
$Y_{23} = \ 9.1$	$10.3 - 13.233333$	$11.6 - 13.233333$	43.599111
$Y_{31} = 15.3$	$13.6 - 13.233333$	$14.9 - 13.233333$	9.35
$Y_{32} = 13.5$	$13.6 - 13.233333$	$13.2 - 13.233333$	$-.165$
$Y_{33} = 12.0$	$13.6 - 13.233333$	$11.6 - 13.233333$	-7.186667
			5.364334

$$F = \frac{\text{SSNA}}{\text{SSPE}/(rc - r - c)} \sim F_{1,rc-r-c} \tag{6.29}$$

and the decision rule is to reject the null hypothesis of additivity (i.e., no interaction) if $F \geq F_{1-\alpha;\ 1,rc-r-c}$.

Therefore, for our battery example,

$$F = \frac{.3447}{.2953/3} = 3.502$$

At the .05 level of significance, since $F = 3.502 < F_{.95;\ 1,3} = 10.1$, we do not reject the null hypothesis. We would conclude that there is no evidence of an interacting effect between battery brand and type of calculator on battery durability. However, in other applications it is quite possible that a significant multiplicative interaction might be found. In such cases, a data transformation (possibly including logarithms) could be utilized to remove any multiplicative interaction effect (see Section 3.4.2).

6.8 HIGHER-WAY ANOVA MODELS

In this chapter we have extended our discussion of ANOVA models to consider those situations in which two factors (and their interaction) are involved. The basic concepts of the two-factor model can be extended to situations in which three or more factors are to be evaluated. Such models provide additional complexities that preclude their detailed development in this text. Interested readers are referred to Box et al. (1978), Cochran and Cox (1957), Hicks (1973), or Winer (1971).

However, we shall present in Table 6.14 tests of hypothesis for the three-way ANOVA model for various fixed, random, and mixed effects.

TABLE 6.14 *F* **Tests for Three-Way ANOVA Models**
with Equal Cell Sample Size

Null Hypothesis	$\alpha, \mathcal{B}, \mathcal{C}$ Fixed	$\alpha, \mathcal{B}$ Fixed, $\mathcal{C}$ Random	$\alpha, \mathcal{B}$ Random, $\mathcal{C}$ Fixed	$\alpha, \mathcal{B}, \mathcal{C}$ Random
$\alpha_i = 0$	MSα/MSE	MSα/MS$\alpha\mathcal{C}$	MSα/MS$\alpha\mathcal{B}$	No exact test[a]
$\mathcal{B}_j = 0$	MS$\mathcal{B}$/MSE	MS$\mathcal{B}$/MS$\mathcal{B}\mathcal{C}$	MS$\mathcal{B}$/MS$\alpha\mathcal{B}$	No exact test[a]
$\mathcal{C}_k = 0$	MS$\mathcal{C}$/MSE	MS$\mathcal{C}$/MSE	No exact test[a]	No exact test[a]
$\alpha\mathcal{B}_{ij} = 0$	MS$\alpha\mathcal{B}$/MSE	MS$\alpha\mathcal{B}$/MS$\alpha\mathcal{B}\mathcal{C}$	MS$\alpha\mathcal{B}$/MSE	MS$\alpha\mathcal{B}$/MS$\alpha\mathcal{B}\mathcal{C}$
$\alpha\mathcal{C}_{ik} = 0$	MS$\alpha\mathcal{C}$/MSE	MS$\alpha\mathcal{C}$/MSE	MS$\alpha\mathcal{C}$/MS$\alpha\mathcal{B}\mathcal{C}$	MS$\alpha\mathcal{C}$/MS$\alpha\mathcal{B}\mathcal{C}$
$\mathcal{B}\mathcal{C}_{jk} = 0$	MS$\mathcal{B}\mathcal{C}$/MSE	MS$\mathcal{B}\mathcal{C}$/MSE	MS$\mathcal{B}\mathcal{C}$/MS$\alpha\mathcal{B}\mathcal{C}$	MS$\mathcal{B}\mathcal{C}$/MS$\alpha\mathcal{B}\mathcal{C}$
$\alpha\mathcal{B}\mathcal{C}_{ijk} = 0$	MS$\alpha\mathcal{B}\mathcal{C}$/MSE	MS$\alpha\mathcal{B}\mathcal{C}$/MSE	MS$\alpha\mathcal{B}\mathcal{C}$/MSE	MS$\alpha\mathcal{B}\mathcal{C}$/MSE

[a]Approximate tests for these null hypotheses are available [see Hicks (1973) or Winer (1971)].

6.9 COMPUTER PACKAGES AND TWO-WAY ANOVA WITH EQUAL REPLICATION: USE OF SPSS, SAS, AND BMDP

6.9.1 Organizing the Data for Computer Analysis

In this section we shall explain how to use either SPSS, SAS, or BMDP to analyze a two-way analysis of variance model with equal replication. For illustrative purposes, we shall refer to the tax return problem of Section 6.2 (see Table 6.4). For each tax return three types of information are available: the processing time, the complexity, and the submission time. In addition to these three variables, each of the 18 tax returns selected can be provided with an identification number. Thus the data can be organized in the format presented in Table 6.15.

For the purposes of data entry, for each tax return the identification number is to be placed in columns 1 and 2, the processing time in columns 4 and 5, the complexity level in column 7, and the submission time in column 9.

6.9.2 Using the SPSS Subprogram ANOVA for the Two-Way Analysis of Variance Model

The ANOVA subprogram of SPSS can be used to analyze an ANOVA model that includes up to five factors.* The basic setup of the ANOVA procedure statement is

1   16

ANOVA $\left\{\begin{array}{l}\text{dependent}\\\text{variable } (Y)\\\text{or list}\end{array}\right\}$ ♭BY♭ $\left\{\begin{array}{l}\text{independent}\\\text{variable or}\\\text{factor } \alpha\end{array}\right\}$ $\left(\begin{array}{l}\text{min.}\quad\text{max.}\\\text{value, value}\end{array}\right)$ etc. /

*For the two-factor model without replication, set up your program as discussed in Section 5.6.2.

TABLE 6.15   Organization of Tax Return Data

ID No.	Processing Time	Complexity Level	Submission Time
1	24	1	1
2	34	1	1
3	32	1	1
4	49	1	2
5	56	1	2
6	39	1	2
7	54	2	1
8	40	2	1
9	56	2	1
10	82	2	2
11	70	2	2
12	73	2	2
13	56	3	1
14	66	3	1
15	73	3	1
16	93	3	2
17	82	3	2
18	80	3	2

Beginning in column 16, the dependent or response variable is named; this is followed (after the word BY) by the names of each factor (or independent variable). The lowest and highest coded value for each factor *must* be placed in parentheses after the factor name. Up to five dependent and five independent variables can be analyzed in a single ANOVA procedure statement.

Figure 6.4 illustrates the complete SPSS program that was written to analyze the tax return problem (see Tables 6.4 and 6.15). We note that, although various OPTIONS and STATISTICS are available when using the ANOVA subprogram, only STATISTICS 3 has been requested. This statistic prints the mean and total for each cell of the ANOVA table. Figure 6.5 displays annotated partial output of the SPSS subprogram ANOVA for the tax return problem.

6.9.3 Using the SAS PROC ANOVA for the Two-Way Analysis of Variance Model

As we indicated previously for the randomized blocks model (Section 5.6.3), the SAS ANOVA procedure can be used for an analysis of variance model in which there are an equal number of observations per cell.* The basic setup when using PROC ANOVA for the two-factor model is

*For the two-factor model without replication, set up your program as discussed in Section 5.6.3.

```
{SYSTEM CARDS}
RUN NAME          ANALYSIS OF TAX RETURNS
DATA LIST         FIXED(1)/1 IDNO 1-2,PROCESS 4-5,
                  COMPLEX 7,SUBMIT 9
VAR LABELS        IDNO,IDENTIFICATION NUMBER/
                  PROCESS,PROCESSING TIME IN DAYS/
                  COMPLEX,COMPLEXITY LEVEL/
                  SUBMIT,SUBMISSION TIME/
VALUE LABELS      COMPLEX(1)SIMPLE(2)SOMEWHAT COMPLEX
                  (3)HIGHLY COMPLEX/
                  SUBMIT(1)EARLY(2)ON TIME/
READ INPUT DATA
01 24  1  1
02 34  1  1
 .  .  .  .
 .  .  .  .
18 80  3  2
END INPUT DATA
ANOVA             PROCESS BY COMPLEX(1,3)SUBMIT(1,2)/
STATISTICS        3
FINISH
{SYSTEM CARDS}
```

Figure 6.4 SPSS program for tax return example.

```
* * * * * * * * * * * A N A L Y S I S   O F   V A R I A N C E * * * * * * * * * * * *
             PROCESS  PROCESSING TIME IN DAYS
        BY COMPLEX  COMPLEXITY LEVEL
           SUBMIT   SUBMISSION TIME
* * * * * * * * * * * * * * * * * * * * * * * * * * * * * * * * * * * * * * * * * *
```

SOURCE OF VARIATION		SUM OF SQUARES	DF	MEAN SQUARE	F	SIGNIF OF F
MAIN EFFECTS		5993.496	3	1997.832	35.465	0.000
COMPLEX	(α)	4009.000	2	2004.500	35.583	0.000
SUBMIT	(β)	1984.500	1	1984.500	35.228	0.000
2-WAY INTERACTIONS		39.000	2	19.500	0.346	0.714
COMPLEX SUBMIT	$(\alpha \times \beta)$	39.000	2	19.500	0.346	0.714
EXPLAINED		6032.496	5	1206.499	21.417	0.000
RESIDUAL	ERROR	675.996	12	56.333		
TOTAL		6708.492	17	394.617		

Figure 6.5 Partial SPSS output for tax return data.

PROCƀANOVA;

CLASSESƀ $\begin{Bmatrix} \text{names of factors or} \\ \text{independent variables} \end{Bmatrix}$;

MODELƀ $\begin{Bmatrix} \text{dependent} \\ \text{variable} \end{Bmatrix} = \begin{Bmatrix} \text{name of} \\ \text{factor } \alpha \end{Bmatrix} \; \Big| \Big| \; \begin{Bmatrix} \text{name of} \\ \text{factor } \beta \end{Bmatrix}$ etc.

165

The names of the factors are provided in the CLASSES statement. These are separated by a single space. The CLASSES statement must precede the MODEL statement. The MODEL statement indicates the dependent variable and the factors involved in the ANOVA model. A vertical line (|) is placed between the factors to indicate the presence of a *crossed design*—one in which each level of factor ⍺ appears with each level of factor ℬ.

Figure 6.6 illustrates the complete SAS program that has been written to analyze the tax return problem (see Tables 6.4 and 6.15). From Figure 6.6 we

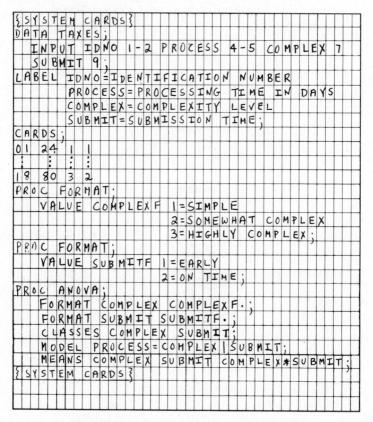

Figure 6.6   SAS program for tax return example.

can observe that a MEANS statement has been provided as part of the PROC ANOVA. This statement will enable SAS to compute means for each level of each factor and for each cell. We should also note that formats are indicated as part of the PROC ANOVA step so that the value labels for the factors will be printed. Figure 6.7 represents annotated partial output of the SAS ANOVA analysis for the tax return problem.

ANALYSIS OF VARIANCE PROCEDURE

DEPENDENT VARIABLE: PROCESS   PROCESSING TIME IN DAYS

SOURCE	DF	SUM OF SQUARES	MEAN SQUARE	F VALUE	PR > F
MODEL	5	6032.50000000	1206.50000000	21.42	0.0001
ERROR	12	676.00000000	56.33333333		STD DEV
CORRECTED TOTAL	17	6708.50000000			7.50555350

SOURCE	DF	ANOVA SS	F VALUE	PR > F
COMPLEX (a)	2	4009.00000000	35.58	0.0001
SUBMIT (β)	1	1984.50000000	35.23	0.0001
COMPLEX*SUBMIT $(a \times \beta)$	2	39.00000000	0.35	0.7142

Figure 6.7 Partial output from the SAS PROC ANOVA for tax return data.

6.9.4 Using BMDP Program 2V for the Two-Way Analysis of Variance Model

The BMDP program 2V can be used for two-way analysis of variance models as well as for the randomized block design (see Section 5.6.4).* The only paragraph of BMDP (see Section 2.7) unique to 2V is the /DESIGN paragraph. For the two-factor model the form of this paragraph is

/DESIGN   DEPENDENTþISþ$\begin{Bmatrix}\text{name of}\\\text{dependent variable}\end{Bmatrix}$.

GROUPINGþAREþ$\begin{Bmatrix}\text{names of factors or}\\\text{independent variables}\end{Bmatrix}$.

The DEPENDENT sentence names the dependent variable, while the GROUPING sentence names the factors to be analyzed, separated by commas.

Figure 6.8 illustrates the program (using P-2V) that has been written to

```
{SYSTEM CARDS}
/PROBLEM TITLE IS 'ANALYSIS OF TAX RETURNS'.
/INPUT VARIABLES ARE 4.
       FORMAT IS '(F2.0,1X,F2.0,1X,F1.0,1X,F1.0)'.
/VARIABLE NAMES ARE IDNO,PROCESS,COMPLEX,SUBMIT.
/DESIGN DEPENDENT IS PROCESS.
        GROUPING ARE COMPLEX,SUBMIT.
/GROUP CODES(3) ARE 1,2,3.
       NAMES(3) ARE SIMPLE,'SOMEWHAT COMPLEX',
       'HIGHLY COMPLEX'.
       CODES(4) ARE 1,2.
       NAMES(4) ARE EARLY,'ON TIME'.
/END
01 24 1 1
 :    :  :
18 80 3 2
{SYSTEM CARDS}
```

Figure 6.8 Using BMDP program 2V for tax return example.

*For the two-factor model without replication, set up your program as discussed in Section 5.6.4.

analyze the tax return problem (see Tables 6.4 and 6.15). The /GROUP paragraph is required so that the labels for each factor can be provided. Codes and names are given for both complexity (the third variable named) and submission time (the fourth variable named). Figure 6.9 represents annotated partial output obtained from the BMDP program 2V analysis of the tax return problem.

ANALYSIS OF TAX RETURNS
ANALYSIS OF VARIANCE FOR   1-ST
DEPENDENT VARIABLE - PROCESS

SOURCE		SUM OF SQUARES	DEGREES OF FREEDOM	MEAN SQUARE	F	TAIL PROBABILITY
MEAN		62304.50000	1	62304.50000	1106.00	0.0000
COMPLEX	(α)	4009.00000	2	2004.50000	35.58	0.0000
SUBMIT	(β)	1984.50000	1	1984.50000	35.23	0.0001
CS	$(\alpha \times \beta)$	39.00000	2	19.50000	0.35	0.7142
ERROR		676.00000	12	56.33333		

Figure 6.9   Partial BMDP program 2V output for tax return data.

PROBLEMS

*6.1. The retailing manager of a food chain wishes to determine whether product location will have any effect on the sale of pet toys. Two factors are to be studied: (1) height of shelf: top, middle, and bottom; and (2) location in aisle: front, middle, and rear. A random sample of 18 stores was selected, and 2 stores were randomly assigned to each combination of shelf height and aisle location. The size of the display area was constant for all 18 stores. At the end of a 1-week trial period, the sales of the product in each store were as follows:

	Height of Shelf		
Aisle Location	Top	Middle	Bottom
Front	86 72	62 54	50 40
Middle	32 24	20 14	18 16
Rear	60 46	40 28	28 22

At the .05 level of significance,
(a) Is there an effect due to aisle location?
(b) Is there an effect due to shelf height?
(c) Is there an interaction between shelf height and aisle location?
(d) Plot a graph of average sales for each aisle location for each shelf height.
(e) If appropriate, use Tukey's T method to determine which aisle locations and which shelf heights differ in sales.

(f) Based on the results of parts (a)–(e), what conclusions can you reach concerning sales?

6.2. An industrial psychologist would like to determine the effect of alcoholic consumption on the typing ability of a group of secretaries. Two factors are to be considered: the amount of consumption (0, 1, and 2 ounces) and the type of manuscript (technical, nontechnical). A group of 12 secretaries (of similar typing ability) were randomly assigned to an alcoholic consumption level and a manuscript type. Each secretary was instructed to type a standard page (either technical or nontechnical) following the appropriate alcoholic consumption. The number of errors made by each secretary was recorded with the following results:

	Type of Manuscript	
Alcoholic Consumption (ounces)	Technical	Nontechnical
0	5	0
	3	2
1	12	3
	14	6
2	18	10
	21	7

At the .01 level of significance,

(a) Is there an effect due to level of alcoholic consumption?

(b) Is there an effect due to type of manuscript?

(c) Is there an interaction between alcoholic consumption and manuscript type?

(d) Plot a graph of average errors for each level of alcoholic consumption for each type of manuscript.

(e) If appropriate, use Tukey's T method to determine which levels of alcoholic consumption differ in number of errors.

(f) Based on the results of parts (a)–(e), what conclusions can you reach concerning the number of errors?

(g) Using an appropriate transformation, reanalyze the data and compare the results to those obtained in parts (a) to (f).

6.3. A large home builders' association would like to study the effect of style of house and source of energy on home heating costs for single-family homes. Four styles of house (split level, ranch, colonial, and splanch) and three sources of heating energy (oil, gas, and electricity) were studied. A random sample of two houses of each style and source of energy was selected within a large housing development. The monthly heating costs in dollars (averaged over the year) for the houses are shown at the top of page 170:

At the .05 level of significance,

(a) Is there an effect due to type of house?

(b) Is there an effect due to energy source?

(c) Is there an interaction between type of house and energy source?

| | Energy Source | | |
Type of House	Natural Gas	Oil	Electricity
Split Level	65	110	145
	78	92	160
Ranch	55	95	120
	67	80	105
Colonial	60	92	134
	70	100	148
Splanch	72	105	165
	81	120	180

(d) Plot a graph of average monthly heating cost for each type of house for each energy source.

(e) If appropriate, use Tukey's T method to determine which types of houses and which sources of energy differ in home heating costs.

(f) Based on the results of parts (a)–(e), what conclusions can you draw about monthly heating costs?

6.4. A professor of statistics wanted to conduct an experiment concerning three computer packages, SPSS, SAS, and BMDP. This experiment was to be conducted in an advanced statistical methods class that consisted of 27 students, equally divided among computer, statistics, and "other" majors. Three students in each major were to analyze independently a set of data using a randomly assigned computer package. The number of minutes of work required to write the program needed for analyzing the data were recorded with the following results:

| | Computer Package | | |
Major	SPSS	SAS	BMDP
Computer	12	10	14
	8	13	9
	11	9	13
Statistics	12	8	15
	10	12	11
	15	14	16
Other	11	13	20
	17	18	17
	15	16	19

At the .05 level of significance,

(a) Is there an effect due to major?

(b) Is there an effect due to computer packages?

(c) Is there an interaction of major and computer package?

(d) Plot a graph of the minutes required for each major for each computer package.

(e) If appropriate, use Tukey's T method to determine which majors and which computer packages differ in the amount of time required.

(f) Based on the results in parts (a)–(e), what conclusions can you draw about the time required to complete the assigned task?

6.5. An investigation was to be carried out to determine the effect of different individuals (raters) evaluating the service at various branches of a fast-food chain. A *random sample* of three evaluators was selected from a pool of evaluators available, and a *random sample* of four restaurants of the chain was selected in a particular metropolitan area. Each rater evaluated each restaurant on two different occasions. The ratings were as follows:

	Restaurants			
Raters	A	B	C	D
I	70	61	82	74
	77	75	88	76
II	76	67	90	80
	80	63	96	76
III	84	66	92	84
	78	68	98	86

At the .01 level of significance,

(a) Is there an effect due to raters?

(b) Is there an effect due to restaurants?

(c) Is there an interaction between raters and restaurants?

(d) If the three raters were specifically (rather than randomly) selected, how would your analysis in parts (a)–(c) be affected?

6.6. The board of education of a large state wishes to study differences in class size between elementary, intermediate, and high schools in various cities. A random sample of three cities within the state was selected. Two schools at each level were chosen within each city, and the average class size for the school was recorded as indicated on page 172:

At the .05 level of significance,

(a) Is there an effect due to educational level?

(b) Is there an effect due to cities?

(c) Is there an interaction between educational level and city?

(d) Plot the average class size for each educational level for each city.

Educational Level	City		
	A	B	C
Elementary	32	26	20
	34	30	23
Intermediate	35	33	24
	39	30	27
High School	43	37	31
	38	34	28

(e) If appropriate, use Tukey's T method to determine which educational levels differ in class size.

(f) If the three cities had been specifically (not randomly) selected, how would your analysis in parts (a)–(c) be affected?

6.7. A large American car manufacturer wishes to study the mileage attained by four different models (subcompact, compact, intermediate, and full size) using four different brands of gasoline. One car of each model was tested with each brand of gasoline with the following results (in miles per gallon):

Model	Brand			
	A	B	C	D
Subcompact	26.8	27.4	24.3	25.9
Compact	23.6	25.2	22.7	24.0
Intermediate	21.3	23.4	19.8	22.4
Full Size	18.3	19.2	19.6	17.9

At the .05 level of significance,

(a) Using Tukey's test for additivity, is there evidence of a multiplicative interaction between car model and brand of gasoline?

(b) Is there a difference among models?

(c) Is there a difference among brands?

6.8. The marketing department of a supermarket chain wanted to study the effects of product placement and price on sales of a disposable razor. Three different product placements (checkout counter, front display, and health and beauty aisle) and four different prices for a package of two razors (69, 79, 89, and 99 cents) were to be utilized. Twelve equal-sized stores were randomly selected, and 1 store was assigned to each combination of product placement and price. The results, indicating the number of packages sold during a 1-week period, were as follows:

Product Placement	Price (cents)			
	69	79	89	99
Checkout Counter	164	139	127	122
Front Display	125	97	92	80
Health and Beauty Aisle	105	92	84	78

At the .01 level of significance,

(a) Using Tukey's test for additivity, is there a multiplicative interaction between product placement and price?

(b) Is there a difference in sales among the various product placements?

(c) Is there a difference in sales for different prices?

6.9. The Environmental Protection Agency of a large suburban county is studying coliform bacteria counts (in parts per thousand) at beaches within the county. Three types of beaches are to be considered—ocean, bay, and sound—in three geographic areas of the county—west, central, and east. One beach of each type is randomly selected in each region of the county. The coliform bacteria counts at each beach on a particular day were as follows:

Type of Beach	Geographical Area		
	West	Central	East
Ocean	22	9	4
Bay	35	20	10
Sound	29	15	6

At the .05 level of significance,

(a) Using Tukey's test for additivity, is there a multiplicative interaction between type of beach and geographical area within the county?

(b) Is there a difference in coliform count among types of beaches?

(c) Is there a difference in coliform count among geographical areas?

(d) Use an appropriate data transformation and reanalyze the problem. Compare and contrast the results.

6.10. Reviewing concepts:

(a) What is meant by a two-way (or two-factor) ANOVA model?

(b) What is meant by interaction?

(c) What is a mixed model?

(d) What is the purpose of Tukey's test for additivity?

7

ANALYSIS
OF NONORTHOGONAL
FACTORIAL DESIGNS

7.1 INTRODUCTION

In Chapter 6 we described the (balanced) two-factor ANOVA design wherein each cell contained an equal number of observations. In this chapter we shall focus on the more complex two-factor ANOVA problem where there are an unequal number of observations in each cell. Such situations as these are not uncommon to either business or to the social and medical sciences. Observational (i.e., *post hoc*) studies are often formulated in this manner—especially when the data are obtained from survey research. In establishing the design, the subjects are first categorized in accordance with the two classification factors of interest, and their responses are then recorded in the appropriate cell. In these kinds of situations there is of course no way to "control" for unequal cell allocations. Nevertheless, it sometimes happens that even when the researcher has properly designed a two-factor ANOVA experiment the resulting data do not appear in equal numbers in the various cells. Despite appropriate planning, these types of situations occur if subjects fail to complete the experiment. For example, laboratory animals may become ill or die, human subjects might not show up, or testing equipment may malfunction.

7.2 THE PROBLEM OF NONORTHOGONALITY

Unfortunately, the problem which arises when the factorial experiment does not contain equal-sized samples in the cells is that the design is *nonorthogonal*. In other words, the total sums of squares in the ANOVA table (see Table 6.2)

cannot be decomposed into a series of additive components which permit the analysis of the separate effects. That is, from (6.2) and Table 6.2,

$$SST \neq SS\alpha + SS\mathcal{B} + SS\alpha\mathcal{B} + SSE \qquad (7.1)$$

It must be noted, however, that there are but two possible situations when the cell sample sizes are unequal: the cell frequencies are either allocated *proportionately* or *disproportionately*.

If the number of observations is allocated *proportionately* among the cells, no serious problem exists; only minor "adjustments" to the $SS\alpha\mathcal{B}$ term in (7.1) need be made, and the researcher may then proceed with the usual two-factor ANOVA as described in Section 6.2. One such application of *proportional allocation* will be presented in Section 7.4.

On the other hand, the usual ANOVA procedures cannot be utilized if the number of observations is allocated *disproportionately* among the cells because this results in a *correlation* between the factors. It then becomes difficult to determine the magnitude of the separate effects that each factor has on the response variable. Thus for such situations there has been much debate among statisticians as to the proper method of analysis. Some statisticians [see Glass and Stanley (1970)] have suggested a random discarding of data to achieve either equal cell frequencies or proportional cell frequencies. Moreover, others [see Snedecor and Cochran (1980)] have advocated *approximate F tests* using the *method of unweighted means* provided that certain conditions were met. This approach will be described in Section 7.6. In addition, other statisticians [see Overall and Spiegel (1969), Winer (1971), and Appelbaum and Cramer (1974)] have argued for a *least-squares regression approach* to obtain *exact F* tests for the nonorthogonal ANOVA problem. This will be described in Section 12.9.

7.3 THE MODEL: LAYOUT AND NOTATION

The general layout for data obtained in a two-factor experiment with unequal numbers of observations in the cells is presented in Table 7.1. The corresponding summary of totals, sample sizes, and means in the cells, rows, and columns is displayed in Table 7.2.

The model for the nonorthogonal two-factor fixed-effects ANOVA is presented in (7.2); we note that except for the fact that the sample sizes in each of the rc cells are now permitted to vary, this model would otherwise be identical to that indicated in (6.1) for the balanced two-factor fixed-effects experimental design: Thus,

$$Y_{ijk} = \mu + \alpha_i + \mathcal{B}_j + \alpha\mathcal{B}_{ij} + \epsilon_{ijk} \qquad (7.2)$$

where $i = 1, 2, \ldots, r; j = 1, 2, \ldots, c;$ and $k = 1, 2, \ldots, n_{ij};$ and where

$Y_{ijk} = k$th observation for level i of factor α and level j of factor $\mathcal{B}$
$\mu =$ overall or grand mean effect
$\alpha_i =$ effect of the ith level of factor α

$\mathcal{B}_j$ = effect of the jth level of factor $\mathcal{B}$

$\mathcal{A}\mathcal{B}_{ij}$ = interaction effect at level i of factor $\mathcal{A}$ and level j of factor $\mathcal{B}$

ϵ_{ijk} = random error present in the kth observation in cell ij [where we assume $\epsilon_{ijk} \sim \mathcal{N}(0, \sigma^2)$]

TABLE 7.1   Data Layout for a Two-Factor Nonorthogonal Design

Factor $\mathcal{A}$	Factor $\mathcal{B}$					
	1	2	$\cdots$	j	$\cdots$	c
1	Y_{111} Y_{112} $\cdot$ $\cdot$ $\cdot$ $Y_{11n_{11}}$	Y_{121} Y_{122} $\cdot$ $\cdot$ $\cdot$ $Y_{12n_{12}}$	$\cdots$	Y_{1j1} Y_{1j2} $\cdot$ $\cdot$ $\cdot$ $Y_{1jn_{1j}}$	$\cdots$	Y_{1c1} Y_{1c2} $\cdot$ $\cdot$ $\cdot$ $Y_{1cn_{1c}}$
2	Y_{211} Y_{212} $\cdot$ $\cdot$ $\cdot$ $Y_{21n_{21}}$	Y_{221} Y_{222} $\cdot$ $\cdot$ $\cdot$ $Y_{22n_{22}}$	$\cdots$	Y_{2j1} Y_{2j2} $\cdot$ $\cdot$ $\cdot$ $Y_{2jn_{2j}}$	$\cdots$	Y_{2c1} Y_{2c2} $\cdot$ $\cdot$ $\cdot$ $Y_{2cn_{2c}}$
$\cdot$ $\cdot$ $\cdot$	$\cdot$	$\cdot$	$\cdots$	$\cdot$	$\cdots$	$\cdot$
i	Y_{i11} Y_{i12} $\cdot$ $\cdot$ $\cdot$ $Y_{i1n_{i1}}$	Y_{i21} Y_{i22} $\cdot$ $\cdot$ $\cdot$ $Y_{i2n_{i2}}$	$\cdots$	Y_{ij1} Y_{ij2} $\cdot$ $\cdot$ $\cdot$ $Y_{ijn_{ij}}$	$\cdots$	Y_{ic1} Y_{ic2} $\cdot$ $\cdot$ $\cdot$ $Y_{icn_{ic}}$
$\cdot$ $\cdot$ $\cdot$	$\cdot$	$\cdot$	$\cdots$	$\cdot$	$\cdots$	$\cdot$
r	Y_{r11} Y_{r12} $\cdot$ $\cdot$ $\cdot$ $Y_{r1n_{r1}}$	Y_{r21} Y_{r22} $\cdot$ $\cdot$ $\cdot$ $Y_{r2n_{r2}}$	$\cdots$	Y_{rj1} Y_{rj2} $\cdot$ $\cdot$ $\cdot$ $Y_{rjn_{rj}}$	$\cdots$	Y_{rc1} Y_{rc2} $\cdot$ $\cdot$ $\cdot$ $Y_{rcn_{rc}}$

For the information expressed in Tables 7.1 and 7.2, the *dot notation* ($\cdot$) permits the following descriptions:

Y_{ijk} = value corresponding to the kth observation in cell ij (where $i = 1, 2, \ldots, r; j = 1, 2, \ldots, c;$ and $k = 1, 2, \ldots, n_{ij}$)

TABLE 7.2 Summary Information for a Two-Factor Nonorthogonal Design

Factor α	Factor $\mathcal{B}$						Row (Factor α) Summaries
	1	2	$\cdots$	j	$\cdots$	c	
1	$Y_{11\cdot}$ n_{11} $\bar{Y}_{11\cdot}$	$Y_{12\cdot}$ n_{12} $\bar{Y}_{12\cdot}$	$\cdots$	$Y_{1j\cdot}$ n_{1j} $\bar{Y}_{1j\cdot}$	$\cdots$	$Y_{1c\cdot}$ n_{1c} $\bar{Y}_{1c\cdot}$	$Y_{1\cdot\cdot}$ $n_{1\cdot}$ $\bar{Y}_{1\cdot\cdot}$
2	$Y_{21\cdot}$ n_{21} $\bar{Y}_{21\cdot}$	$Y_{22\cdot}$ n_{22} $\bar{Y}_{22\cdot}$	$\cdots$	$Y_{2j\cdot}$ n_{2j} $\bar{Y}_{2j\cdot}$	$\cdots$	$Y_{2c\cdot}$ n_{2c} $\bar{Y}_{2c\cdot}$	$Y_{2\cdot\cdot}$ $n_{2\cdot}$ $\bar{Y}_{2\cdot\cdot}$
$\vdots$	$\vdots$	$\vdots$	$\cdots$	$\vdots$	$\cdots$	$\vdots$	$\vdots$
i	$Y_{i1\cdot}$ n_{i1} $\bar{Y}_{i1\cdot}$	$Y_{i2\cdot}$ n_{i2} $\bar{Y}_{i2\cdot}$	$\cdots$	$Y_{ij\cdot}$ n_{ij} $\bar{Y}_{ij\cdot}$	$\cdots$	$Y_{ic\cdot}$ n_{ic} $\bar{Y}_{ic\cdot}$	$Y_{i\cdot\cdot}$ $n_{i\cdot}$ $\bar{Y}_{i\cdot\cdot}$
$\vdots$	$\vdots$	$\vdots$	$\cdots$	$\vdots$	$\cdots$	$\vdots$	$\vdots$
r	$Y_{r1\cdot}$ n_{r1} $\bar{Y}_{r1\cdot}$	$Y_{r2\cdot}$ n_{r2} $\bar{Y}_{r2\cdot}$	$\cdots$	$Y_{rj\cdot}$ n_{rj} $\bar{Y}_{rj\cdot}$	$\cdots$	$Y_{rc\cdot}$ n_{rc} $\bar{Y}_{rc\cdot}$	$Y_{r\cdot\cdot}$ $n_{r\cdot}$ $\bar{Y}_{r\cdot\cdot}$
Column (Factor $\mathcal{B}$) Summaries	$Y_{\cdot1\cdot}$ $n_{\cdot1}$ $\bar{Y}_{\cdot1\cdot}$	$Y_{\cdot2\cdot}$ $n_{\cdot2}$ $\bar{Y}_{\cdot2\cdot}$	$\cdots$	$Y_{\cdot j\cdot}$ $n_{\cdot j}$ $\bar{Y}_{\cdot j\cdot}$	$\cdots$	$Y_{\cdot c\cdot}$ $n_{\cdot c}$ $\bar{Y}_{\cdot c\cdot}$	$Y_{\cdots}$ = grand total $n_{\cdot\cdot}$ = total sample size $\bar{Y}_{\cdots}$ = grand mean

$Y_{ij\cdot}$ = summed total of values in cell ij

n_{ij} = sample size in cell ij

$\bar{Y}_{ij\cdot}$ = mean of cell ij

$Y_{i\cdot\cdot}$ = total in row i (i.e., the sum of the values for the ith level of factor α)

$n_{i\cdot}$ = number of observations in row i

$\bar{Y}_{i\cdot\cdot}$ = mean in row i

$Y_{\cdot j\cdot}$ = total in column j (i.e., the sum of the values for the jth level of factor $\mathcal{B}$)

$n_{\cdot j}$ = number of observations in column j

$\bar{Y}_{\cdot j\cdot}$ = mean in column j

$Y_{\cdots}$ = grand total over all rows and columns

$n_{\cdot\cdot}$ = total sample size

$\bar{Y}_{\cdots}$ = overall or grand mean in the data

7.4 PROPORTIONATE CELL FREQUENCIES

7.4.1 Development

When the unequal sample sizes are allocated proportionately among the set of cells (as determined by the particular row and column levels), it will always be true that

$$n_{ij} = \frac{n_{i.} \cdot n_{.j}}{n_{..}} \tag{7.3}$$

for all the rc cells. That is, the number of observations in cell ij, the ith level of factor α with the jth level of factor $\mathcal{B}$, is the product of the sample size for the ith level of factor α and the sample size for the jth level of factor $\mathcal{B}$ divided by the total sample size in the experiment.

When dealing with two-factor experiments having unequal cell frequencies, the researcher should always check for proportionality prior to selecting an appropriate method of analysis. If the relation given by (7.3) holds in each one of the rc cells, the cell frequencies are said to follow a proportional pattern, and the usual two-factor ANOVA methods of Section 6.2 may be used—after taking into account, of course, that the number of observations may differ from cell to cell. The computational formulas for these *adjusted* sums of squares terms are

$$SS\alpha = \sum_{i=1}^{r} \frac{Y_{i..}^2}{n_{i.}} - \frac{Y_{...}^2}{n_{..}} \tag{7.4}$$

$$SS\mathcal{B} = \sum_{j=1}^{c} \frac{Y_{.j.}^2}{n_{.j}} - \frac{Y_{...}^2}{n_{..}} \tag{7.5}$$

$$SS\alpha\mathcal{B} = \sum_{i=1}^{r} \sum_{j=1}^{c} \frac{Y_{ij.}^2}{n_{ij}} - \sum_{i=1}^{r} \frac{Y_{i..}^2}{n_{i.}} - \sum_{j=1}^{c} \frac{Y_{.j.}^2}{n_{.j}} + \frac{Y_{...}^2}{n_{..}}$$

$$\equiv \sum_{i=1}^{r} \sum_{j=1}^{c} \frac{Y_{ij.}^2}{n_{ij}} - SS\alpha - SS\mathcal{B} - \frac{Y_{...}^2}{n_{..}} \tag{7.6}$$

$$SSE = \sum_{i=1}^{r} \sum_{j=1}^{c} \sum_{k=1}^{n_{ij}} Y_{ijk}^2 - \sum_{i=1}^{r} \sum_{j=1}^{c} \frac{Y_{ij.}^2}{n_{ij}} \tag{7.7}$$

$$SST = \sum_{i=1}^{r} \sum_{j=1}^{c} \sum_{k=1}^{n_{ij}} Y_{ijk}^2 - \frac{Y_{...}^2}{n_{..}} \tag{7.8}$$

We may note the similarities between (7.4)–(7.8) and the corresponding sums of squares terms [(6.8)–(6.12)] for the balanced two-factor experiment. Moreover, we may also observe that the degrees of freedom associated with each of the sources of variation are identical to those given in the balanced two-factor experiment.* As usual, the mean square terms are of course obtained by dividing each of these sums of squares by the corresponding degrees of freedom.

*In the balanced two-factor experiment wherein each of the rc cells are of equal sample size n, the degrees of freedom for the *experimental error* term was presented as $rc(n-1)$. However, for nonorthogonal two-factor experiments we must write the degrees of freedom for experimental error as $n_{..} - rc$. The results are equivalent because $rc(n-1) = rcn - rc$ and, from Chapter 6, rcn is the total number of observations in the experiment—defined here as $n_{..}$.

Table 7.3 summarizes the appropriate F tests—depending on whether the experimental design is a fixed-effects, random-effects, or mixed-effects model (see Section 6.5).

TABLE 7.3   Summary of Possible F Tests

Null Hypothesis	Model I, Fixed Effects	Model II, Random Effects	Model III, Mixed Effects (α Fixed, β Random)
$\alpha_i = 0$	$F_\alpha = \text{MS}\alpha/\text{MSE}$	$F_\alpha = \text{MS}\alpha/\text{MS}\alpha\beta$	$F_\alpha = \text{MS}\alpha/\text{MS}\alpha\beta$
$\beta_j = 0$	$F_\beta = \text{MS}\beta/\text{MSE}$	$F_\beta = \text{MS}\beta/\text{MS}\alpha\beta$	$F_\beta = \text{MS}\beta/\text{MSE}$
$\alpha\beta_{ij} = 0$	$F_{\alpha\beta} = \text{MS}\alpha\beta/\text{MSE}$	$F_{\alpha\beta} = \text{MS}\alpha\beta/\text{MSE}$	$F_{\alpha\beta} = \text{MS}\alpha\beta/\text{MSE}$

7.4.2 Application

An educational psychologist wanted to examine the effects of two factors, preschool training and parental education, on a female child's vocabulary development. Fifty second-grade girls who had been brought up in the same community were classified according to their preschool training as well as to their parents' attained level of education. The girls were then given the Peabody picture vocabulary test, and the results are recorded in Table 7.4. The summary computations for each cell, row, and column are displayed in Table 7.5. Since the sample sizes obviously differ in each cell of this nonorthogonal 2×3 factorial experiment, we check whether the unequal sample sizes are allocated proportionately among the six cells. Using (7.3), we obtain

TABLE 7.4   Scores on the Peabody Picture Vocabulary Test for 50 Second-Grade Girls

Nursery School Training	Parental College Education		
	Both	One	None
No	65 84	76 63 71 94 72 74 87 74 60 69	64 82 64 66 63 65 67 66
Yes	79 96 86	85 78 82 87 75 72 83 80 74 74 51 90 64 75 72	87 70 78 81 80 74 72 88 77 80 60 75

TABLE 7.5 Summary Computations

Nursery School Training	Parental College Education			Row (Factor α) Summaries
	Both	One	None	
No	$Y_{11\cdot} = 149$	$Y_{12\cdot} = 740$	$Y_{13\cdot} = 537$	$Y_{1\cdot\cdot} = 1426$
	$n_{11} = 2$	$n_{12} = 10$	$n_{13} = 8$	$n_{1\cdot} = 20$
	$\bar{Y}_{11\cdot} = 74.5$	$\bar{Y}_{12\cdot} = 74.0$	$\bar{Y}_{13\cdot} = 67.1$	$\bar{Y}_{1\cdot\cdot} = 71.3$
Yes	$Y_{21\cdot} = 261$	$Y_{22\cdot} = 1142$	$Y_{23\cdot} = 922$	$Y_{2\cdot\cdot} = 2325$
	$n_{21} = 3$	$n_{22} = 15$	$n_{23} = 12$	$n_{2\cdot} = 30$
	$\bar{Y}_{21\cdot} = 87.0$	$\bar{Y}_{22\cdot} = 76.1$	$\bar{Y}_{23\cdot} = 76.8$	$\bar{Y}_{2\cdot\cdot} = 77.5$
Column (Factor β) Summaries	$Y_{\cdot1\cdot} = 410$	$Y_{\cdot2\cdot} = 1882$	$Y_{\cdot3\cdot} = 1459$	$Y_{\cdot\cdot\cdot} = 3751$
	$n_{\cdot1} = 5$	$n_{\cdot2} = 25$	$n_{\cdot3} = 20$	$n_{\cdot\cdot} = 50$
	$\bar{Y}_{\cdot1\cdot} = 82.0$	$\bar{Y}_{\cdot2\cdot} = 75.3$	$\bar{Y}_{\cdot3\cdot} = 73.0$	$\bar{Y}_{\cdot\cdot\cdot} = 75.0$

$$\sum_{i=1}^{r} \sum_{j=1}^{c} \sum_{k=1}^{n_{ij}} Y_{ijk}^2 = 285,863$$

$$n_{11} \overset{?}{=} \frac{n_{1\cdot} \, n_{\cdot1}}{n_{\cdot\cdot}} \qquad n_{12} \overset{?}{=} \frac{n_{1\cdot} \, n_{\cdot2}}{n_{\cdot\cdot}} \qquad n_{13} \overset{?}{=} \frac{n_{1\cdot} \, n_{\cdot3}}{n_{\cdot\cdot}}$$

$$2 = \frac{20(5)}{50} \qquad 10 = \frac{20(25)}{50} \qquad 8 = \frac{20(20)}{50}$$

$$\checkmark \quad \checkmark \qquad \checkmark \quad \checkmark \qquad \checkmark \quad \checkmark$$

$$n_{21} \overset{?}{=} \frac{n_{2\cdot} \, n_{\cdot1}}{n_{\cdot\cdot}} \qquad n_{22} \overset{?}{=} \frac{n_{2\cdot} \, n_{\cdot2}}{n_{\cdot\cdot}} \qquad n_{23} \overset{?}{=} \frac{n_{2\cdot} \, n_{\cdot3}}{n_{\cdot\cdot}}$$

$$3 = \frac{30(5)}{50} \qquad 15 = \frac{30(25)}{50} \qquad 12 = \frac{30(20)}{50}$$

$$\checkmark \quad \checkmark \qquad \checkmark \quad \checkmark \qquad \checkmark \quad \checkmark$$

and we observe that proportional cell frequencies are present. Therefore the sums of squares may be computed from (7.4)–(7.8) as follows:

$$SS\alpha = \sum_{i=1}^{r} \frac{Y_{i\cdot\cdot}^2}{n_{i\cdot}} - \frac{Y_{\cdot\cdot\cdot}^2}{n_{\cdot\cdot}} = \left[\frac{(1426)^2}{20} + \frac{(2325)^2}{30} \right] - \frac{(3751)^2}{50} = 461.3$$

$$SS\beta = \sum_{j=1}^{c} \frac{Y_{\cdot j\cdot}^2}{n_{\cdot j}} - \frac{Y_{\cdot\cdot\cdot}^2}{n_{\cdot\cdot}} = \left[\frac{(410)^2}{5} + \frac{(1882)^2}{25} + \frac{(1459)^2}{20} \right] - \frac{(3751)^2}{50} = 331.0$$

$$SS\alpha\beta = \sum_{i=1}^{r} \sum_{j=1}^{c} \frac{Y_{ij\cdot}^2}{n_{ij}} - SS\alpha - SS\beta - \frac{Y_{\cdot\cdot\cdot}^2}{n_{\cdot\cdot}}$$

$$= \left[\frac{(149)^2}{2} + \frac{(740)^2}{10} + \cdots + \frac{(922)^2}{12} \right] - 461.3 - 331.0 - \frac{(3751)^2}{50}$$

$$= 205.9$$

$$SSE = \sum_{i=1}^{r} \sum_{j=1}^{c} \sum_{k=1}^{n_{ij}} Y_{ijk}^2 - \sum_{i=1}^{r} \sum_{j=1}^{c} \frac{Y_{ij.}^2}{n_{ij}}$$

$$= 285{,}863 - \left[\frac{(149)^2}{2} + \frac{(740)^2}{10} + \cdots + \frac{(922)^2}{12} \right] = 3464.8$$

$$SST = \sum_{i=1}^{r} \sum_{j=1}^{c} \sum_{k=1}^{n_{ij}} Y_{ijk}^2 - \frac{Y_{...}^2}{n_{..}} = 285{,}863 - \frac{(3751)^2}{50} = 4463.0$$

Since the levels of each factor are considered to be *fixed*, by using Table 7.3, we can display the appropriate ANOVA table as Table 7.6. The following hypotheses are tested:

$H_{0\alpha}: \quad \mu_{1.} = \mu_{2.}$ \qquad (or no preschool training effects)

$H_{1\alpha}: \quad \mu_{1.} \neq \mu_{2.}$

$H_{0\mathcal{B}}: \quad \mu_{.1} = \mu_{.2} = \mu_{.3}$ \qquad (or no parental education effects)

$H_{1\mathcal{B}}: \quad$ not all $\mu_{.j}$ are equal \qquad $(j = 1, 2, 3)$

$H_{0\alpha\mathcal{B}}: \quad \mu_{ij} \equiv$ for all rc cells \qquad (or no interaction effects)

$H_{1\alpha\mathcal{B}}: \quad$ not all μ_{ij} are equal \qquad $(i = 1, 2$ and $j = 1, 2, 3)$

TABLE 7.6   Summary ANOVA Table for Nonorthogonal Two-Factor Fixed-Effects Model with Proportional Frequencies

Source of Variation	Degrees of Freedom	Sum of Squares	Mean Square	F	P Values
Nursery School Training (α)	$r - 1 = 1$	461.3	461.30	5.86	$.01 < P < .025$ (significant)
Parental Education ($\mathcal{B}$)	$c - 1 = 2$	331.0	165.50	2.10	$P > .10$ (not significant)
Interaction ($\alpha\mathcal{B}$)	$(r - 1)(c - 1) = 2$	205.9	102.95	1.31	$P > .10$ (not significant)
Error	$n.. - rc = 44$	3464.8	78.75	—	—
Total	$n.. - 1 = 49$	4463.0	—	—	—

At the .05 level of significance we observe that there is no evidence of an interaction effect, nor is there any reason to suspect an effect due to achieved parental education on the child's vocabulary. However, since $F_\alpha = MS\alpha/MSE = 5.86 > F_{.95; 1, 44} \cong 4.07$, the (approximate) upper-tail critical value for the F distribution having 1 and 44 degrees of freedom (see Appendix B, Table B.4), we may reject $H_{0\alpha}$ and conclude that female children having nursery school training scored significantly higher on the vocabulary test than did children without such preschool training.

Since the significant factor had but two levels, our analysis is concluded. On the other hand, if a significant factor possesses more than two levels, a posteriori multiple comparison methods may be employed to determine which of the levels (or combinations thereof) are significantly different.

7.5 MULTIPLE COMPARISONS IN NONORTHOGONAL FACTORIAL DESIGNS: THE SCHEFFÉ S METHOD

When the sample sizes are unequal among the rc cells of a two-factor fixed-effects experiment, the Scheffé S method (see Section 4.2.2) may be employed to evaluate all possible contrasts—provided that the null hypothesis has been rejected.

If, for example, the effects of factor $\mathcal{A}$ are declared significant, we may evaluate the contrast $\mathcal{L}_l$ by its estimate

$$\hat{\mathcal{L}}_l = \sum_{i=1}^{r} C_{li}\hat{\mu}_{i\cdot}. \tag{7.9}$$

where

$$\sum_{i=1}^{r} C_{li} = 0$$

and where

$$\hat{\mu}_{i\cdot} = \frac{\sum_{j=1}^{c} \bar{Y}_{ij\cdot}}{c}$$

It should be noted that $\hat{\mu}_{i\cdot}$ is an unbiased estimator of the true (unweighted) factor level mean $\mu_{i\cdot}$ (where $i = 1, 2, \ldots, r$). Moreover, it should also be pointed out that the estimator $\hat{\mu}_{i\cdot}$ is not equal to the *weighted* sample mean $\bar{Y}_{i\cdot\cdot}$ for each level of factor $\mathcal{A}$ because the cell sample sizes are not equal [see Neter and Wasserman (1974)].

Since $\hat{\mu}_{i\cdot}$ is merely a simple average of cell means, we may re-express (7.9) as

$$\hat{\mathcal{L}}_l = \sum_{i=1}^{r} C_{li}\left(\frac{\sum_{j=1}^{c} \bar{Y}_{ij\cdot}}{c}\right) \tag{7.10}$$

and estimate the standard deviation of each contrast from

$$S_{\hat{\mathcal{L}}_l} = \sqrt{\frac{\text{MSE}}{c^2} \sum_{i=1}^{r} \sum_{j=1}^{c} \frac{C_{li}^2}{n_{ij}}} \tag{7.11}$$

From (7.11) we note that $S_{\hat{\mathcal{L}}_l}$ may change for each of the L contrasts ($l = 1, 2, \ldots, L$)—depending on the sample sizes and coefficients used.

The *critical range* for each contrast is obtained by multiplying $S_{\hat{\mathcal{L}}_l}$ by a constant $\sqrt{(r-1)F_{1-\alpha;\, r-1,\, n\cdot\cdot - rc}}$.

The final step in the S method is to declare significant any contrast wherein the absolute value of its estimate $|\hat{\mathcal{L}}_l|$ equals or exceeds the critical range. That is, for factor $\mathcal{A}$ we would declare the contrast $\mathcal{L}_l$ to be significant if

$$|\hat{\mathcal{L}}_l| \geq S_{\hat{\mathcal{L}}_l}\sqrt{(r-1)F_{1-\alpha;\, r-1,\, n\cdot\cdot - rc}} \tag{7.12}$$

In a similar manner, if the effects of factor $\mathcal{B}$ were declared significant, we could estimate the contrast $\mathcal{L}_{l'}$ by using the unweighted mean $\hat{\mu}_{\cdot j} = \sum_{i=1}^{r} \bar{Y}_{ij\cdot}/r$ as an estimator of $\mu_{\cdot j}$, the true mean of the jth level of factor $\mathcal{B}$ (where $j = 1, 2,$

$\ldots, c$). Here the estimates

$$\hat{\mathcal{L}}_{l'} = \sum_{j=1}^{c} C_{l'j}\hat{\mu}_{\cdot j} = \sum_{j=1}^{c} C_{l'j}\left(\frac{\sum_{i=1}^{r} \bar{Y}_{ij\cdot}}{r}\right) \tag{7.13}$$

would be obtained for each of the desired contrasts (where $l' = 1, 2, \ldots, L'$), and we would declare the contrast $\mathcal{L}_{l'}$ as significant if

$$|\hat{\mathcal{L}}_{l'}| \geq \sqrt{\frac{\text{MSE}}{r^2} \sum_{i=1}^{r} \sum_{j=1}^{c} \frac{C_{l'j}^2}{n_{ij}}} \sqrt{(c-1)F_{1-\alpha;\, c-1;\, n\cdots -rc}} \tag{7.14}$$

Finally, if the interaction effect had been declared significant, we could use the rc cell means $\bar{Y}_{ij\cdot}$ to evaluate a contrast $\mathcal{L}_{l''}$ by obtaining its estimate

$$\hat{\mathcal{L}}_{l''} = \sum_{i=1}^{r} \sum_{j=1}^{c} C_{l''ij}\bar{Y}_{ij\cdot}. \tag{7.15}$$

where $\sum_{i=1}^{r} \sum_{j=1}^{c} C_{l''ij} = 0$ and where $l'' = 1, 2, \ldots, L''$. Any contrast $\mathcal{L}_{l''}$ would then be declared significant if

$$|\hat{\mathcal{L}}_{l''}| \geq \sqrt{\text{MSE} \sum_{i=1}^{r} \sum_{j=1}^{c} \frac{C_{l''ij}^2}{n_{ij}}} \sqrt{(rc-1)F_{1-\alpha;\, rc-1,\, n\cdots -rc}} \tag{7.16}$$

7.6 DISPROPORTIONATE CELL FREQUENCIES

As previously stated in Section 7.2, if a proportional allocation of observations among the cells of a nonorthogonal two-factor experiment is not obtained, the usual two-factor ANOVA methods cannot be used. Therefore, in this section we shall describe an often used and fairly simple approximation procedure called the *method of unweighted means*. This procedure yields an approximate F test which may be employed in nonorthogonal two-factor experiments provided: (1) that no cell is void of data (i.e., *empty*) and (2) that the largest cell sample is not more than twice the size of the smallest cell sample.*

7.6.1 Method of Unweighted Means

To perform the unweighted means analysis, we first replace the n_{ij} observations in each of the rc cells of Table 7.1 by their respective cell means $\bar{Y}_{ij\cdot}$ (where $i = 1, 2, \ldots, r$ and $j = 1, 2, \ldots, c$). The general layout now appears as Table 7.7. We may observe that the format of Table 7.7 is identical to that of Table 6.9—the layout for the two-factor ANOVA model with one observation per cell—except that here the one "observation" per cell is a mean. Using the symbol $\mathcal{Y}_{ij\cdot}$ to represent the "observations" $\bar{Y}_{ij\cdot}$ in each of the rc cells, we may display the corresponding two-way table (including summary totals and means) as in Table 7.8. From Table 7.8 it should be pointed out that the *unweighted*

---

*A least-squares regression approach (which does not require such conditions) is presented in Section 12.9.

TABLE 7.7 Data Layout for Method of Unweighted Means

Factor α	Factor β					
	1	2	$\cdots$	j	$\cdots$	c
1	$\bar{Y}_{11\cdot}$	$\bar{Y}_{12\cdot}$	$\cdots$	$\bar{Y}_{1j\cdot}$	$\cdots$	$\bar{Y}_{1c\cdot}$
2	$\bar{Y}_{21\cdot}$	$\bar{Y}_{22\cdot}$	$\cdots$	$\bar{Y}_{2j\cdot}$	$\cdots$	$\bar{Y}_{2c\cdot}$
$\vdots$	$\vdots$	$\vdots$	$\cdots$	$\vdots$	$\cdots$	$\vdots$
i	$\bar{Y}_{i1\cdot}$	$\bar{Y}_{i2\cdot}$	$\cdots$	$\bar{Y}_{ij\cdot}$	$\cdots$	$\bar{Y}_{ic\cdot}$
$\vdots$	$\vdots$	$\vdots$	$\cdots$	$\vdots$	$\cdots$	$\vdots$
r	$\bar{Y}_{r1\cdot}$	$\bar{Y}_{r2\cdot}$	$\cdots$	$\bar{Y}_{rj\cdot}$	$\cdots$	$\bar{Y}_{rc\cdot}$

factor level means $\bar{\mathcal{Y}}_{i\cdot\cdot}$ and $\bar{\mathcal{Y}}_{\cdot j\cdot}$ are, respectively, equivalent to the estimators $\hat{\mu}_{i\cdot}$ and $\hat{\mu}_{\cdot j}$ used for a posteriori multiple comparisons as described in Section 7.5. That is,

$$\bar{\mathcal{Y}}_{i\cdot\cdot} = \frac{\sum_{j=1}^{c} \mathcal{Y}_{ij\cdot}}{c} = \frac{\mathcal{Y}_{i\cdot\cdot}}{c} \equiv \hat{\mu}_{i\cdot} \qquad (\text{where } i = 1, 2, \dots, r) \qquad (7.17)$$

and

$$\bar{\mathcal{Y}}_{\cdot j\cdot} = \frac{\sum_{i=1}^{r} \mathcal{Y}_{ij\cdot}}{r} = \frac{\mathcal{Y}_{\cdot j\cdot}}{r} \equiv \hat{\mu}_{\cdot j} \qquad (\text{where } j = 1, 2, \dots, c) \qquad (7.18)$$

The second step of the unweighted means analysis is to compute the sums of squares and mean squares as in a two-factor experiment with one observation per cell (see Table 6.10):

$$\text{SS}\alpha = \frac{\sum_{i=1}^{r} \mathcal{Y}_{i\cdot\cdot}^2}{c} - \frac{\mathcal{Y}_{\cdot\cdot\cdot}^2}{rc} \qquad (7.19)$$

$$\text{MS}\alpha = \frac{\text{SS}\alpha}{r-1} \qquad (7.20)$$

$$\text{SS}\beta = \frac{\sum_{j=1}^{c} \mathcal{Y}_{\cdot j\cdot}^2}{r} - \frac{\mathcal{Y}_{\cdot\cdot\cdot}^2}{rc} \qquad (7.21)$$

$$\text{MS}\beta = \frac{\text{SS}\beta}{c-1} \qquad (7.22)$$

TABLE 7.8 Summary Table for Method of Unweighted Means

Factor α	Factor β 1	2	$\cdots$	j	$\cdots$	c	Row (Factor α) Summaries Totals	Means
1	$y_{11\cdot}$	$y_{12\cdot}$	$\cdots$	$y_{1j\cdot}$	$\cdots$	$y_{1c\cdot}$	$y_{1\cdot\cdot}$	$\bar{y}_{1\cdot\cdot}$
2	$y_{21\cdot}$	$y_{22\cdot}$	$\cdots$	$y_{2j\cdot}$	$\cdots$	$y_{2c\cdot}$	$y_{2\cdot\cdot}$	$\bar{y}_{2\cdot\cdot}$
$\cdot$	$\cdot$	$\cdot$	$\cdot$	$\cdot$	$\cdot$	$\cdot$	$\cdot$	$\cdot$
$\cdot$	$\cdot$	$\cdot$	$\cdot$	$\cdot$	$\cdot$	$\cdot$	$\cdot$	$\cdot$
i	$y_{i1\cdot}$	$y_{i2\cdot}$	$\cdots$	$y_{ij\cdot}$	$\cdots$	$y_{ic\cdot}$	$y_{i\cdot\cdot}$	$\bar{y}_{i\cdot\cdot}$
$\cdot$	$\cdot$	$\cdot$	$\cdot$	$\cdot$	$\cdot$	$\cdot$	$\cdot$	$\cdot$
$\cdot$	$\cdot$	$\cdot$	$\cdot$	$\cdot$	$\cdot$	$\cdot$	$\cdot$	$\cdot$
r	$y_{r1\cdot}$	$y_{r2\cdot}$	$\cdots$	$y_{rj\cdot}$	$\cdots$	$y_{rc\cdot}$	$y_{r\cdot\cdot}$	$\bar{y}_{r\cdot\cdot}$
Column (Factor β) Summaries — Totals	$y_{\cdot1\cdot}$	$y_{\cdot2\cdot}$	$\cdots$	$y_{\cdot j\cdot}$	$\cdots$	$y_{\cdot c\cdot}$	$y_{\cdot\cdot\cdot}$ = Grand Total	
Means	$\bar{y}_{\cdot1\cdot}$	$\bar{y}_{\cdot2\cdot}$	$\cdots$	$\bar{y}_{\cdot j\cdot}$	$\cdots$	$\bar{y}_{\cdot c\cdot}$	$\bar{y}_{\cdot\cdot\cdot}$ = Grand Unweighted Mean	

$$SS\alpha\beta = \sum_{i=1}^{r} \sum_{j=1}^{c} \mathcal{Y}_{ij}^2 - \frac{\sum_{i=1}^{r} \mathcal{Y}_{i..}^2}{c} - \frac{\sum_{j=1}^{c} \mathcal{Y}_{.j.}^2}{r} + \frac{\mathcal{Y}_{...}^2}{rc}$$

$$\equiv \sum_{i=1}^{r} \sum_{j=1}^{c} \mathcal{Y}_{ij}^2 - SS\alpha - SS\beta - \frac{\mathcal{Y}_{...}^2}{rc} \tag{7.23}$$

$$MS\alpha\beta = \frac{SS\alpha\beta}{(r-1)(c-1)} \tag{7.24}$$

We may recall that in such experiments the interaction and experimental error sources of variation cannot be separated; as described in Section 6.6, the analysis then proceeds under the assumption that no interaction effect is present. Here, however, we do treat these *unseparated* sources of variation as interaction (7.23) rather than as experimental error [see (6.23)].

The third step of the unweighted means analysis is to estimate the experimental error term in the usual manner from the *original* data layout and summary (Tables 7.1 and 7.2). Using (7.7), we have

$$SSE = \sum_{i=1}^{r} \sum_{j=1}^{c} \sum_{k=1}^{n_{ij}} Y_{ijk}^2 - \sum_{i=1}^{r} \sum_{j=1}^{c} \frac{Y_{ij.}^2}{n_{ij}} \tag{7.7}$$

and

$$MSE = \frac{SSE}{n_{..} - rc} \tag{7.25}$$

The MSE is a pooled estimate of the variability within each cell of the original data. Under the assumption of homoscedasticity σ^2 is, of course, the true variance in each cell. From Table 7.8, however, we recall that the method of unweighted means deals with but one "observation" (i.e., the unweighted mean) per cell.

The fourth step of the unweighted means analysis then is to estimate the average variability among the observations $\mathcal{Y}_{ij.}$ over all rc cells. Since $\mathcal{Y}_{ij.} = \bar{Y}_{ij.}$, the variance of the (unweighted mean) observation $\mathcal{Y}_{ij.}$ is σ^2/n_{ij}. Hence the average of these rc variances is given by

$$\frac{\sum_{i=1}^{r} \sum_{j=1}^{c} (\sigma^2/n_{ij})}{rc} = \frac{\sigma^2}{rc} \sum_{i=1}^{r} \sum_{j=1}^{c} \frac{1}{n_{ij}} \tag{7.26}$$

MS_{avg}, the estimated average variance in the unweighted means, is obtained by replacing σ^2 in (7.26) with its estimate MSE. Thus

$$MS_{avg} = \frac{MSE}{rc} \sum_{i=1}^{r} \sum_{j=1}^{c} \frac{1}{n_{ij}} \tag{7.27}$$

and since (7.27) is the estimate of the experimental error in the method of unweighted means, we may proceed to analyze the data (see Table 7.9).*

*The MS_{avg} term in (7.27) may be equivalently defined as MSE divided by the *harmonic mean* of the rc cell sample sizes. The latter definition of MS_{avg} will be useful when utilizing a computer package to extract necessary information for an unweighted means analysis (see Section 7.7).

TABLE 7.9 ANOVA Table for the Method of Unweighted Means

Source of Variation	Degrees of Freedom	Sum of Squares	Mean Square
α	$r-1$	$SS\alpha = \dfrac{\sum_{i=1}^{r} y_{i..}^2}{c} - \dfrac{y_{...}^2}{rc}$	$MS\alpha = \dfrac{SS\alpha}{r-1}$
β	$c-1$	$SS\beta = \dfrac{\sum_{j=1}^{c} y_{.j.}^2}{r} - \dfrac{y_{...}^2}{rc}$	$MS\beta = \dfrac{SS\beta}{c-1}$
$\alpha\beta$	$(r-1)(c-1)$	$SS\alpha\beta = \sum_{i=1}^{r}\sum_{j=1}^{c} y_{ij}^2 - \dfrac{\sum_{i=1}^{r} y_{i..}^2}{c} - \dfrac{\sum_{j=1}^{c} y_{.j.}^2}{r} + \dfrac{y_{...}^2}{rc}$	$MS\alpha\beta = \dfrac{SS\alpha\beta}{(r-1)(c-1)}$
Error	$n_{..} - rc$		$MS_{avg} = \dfrac{MSE}{rc}\sum_{i=1}^{r}\sum_{j=1}^{c}\dfrac{1}{n_{ij}}$

where

$$MSE = \frac{\sum_{i=1}^{r}\sum_{j=1}^{c}\sum_{k=1}^{n_{ij}} Y_{ijk}^2 - \sum_{i=1}^{r}\sum_{j=1}^{c} \dfrac{Y_{ij.}^2}{n_{ij}}}{n_{..} - rc}$$

The fifth step in the unweighted means analysis is to test the hypotheses

$H_{0\alpha}$:   no factor α effects are present

$H_{0\mathcal{B}}$:   no factor $\mathcal{B}$ effects are present

$H_{0\alpha\mathcal{B}}$:   no interaction effects are present

by forming the appropriate approximate F ratios—depending on whether the experimental design is considered to be a fixed-effects, random-effects, or mixed-effects model—as summarized in Table 7.10. The F tests are all *approximate*

TABLE 7.10   Determining the Appropriate Approximate F Ratios

Hypothesis Test	Model I Fixed-Effects	Model II Random-Effects	Model III Mixed-Effects (α Fixed, $\mathcal{B}$ Random)
$H_{0\alpha}$: No Factor α Effects	$F_\alpha \cong \dfrac{\text{MS}\alpha}{\text{MS}_{\text{avg}}}$	$F_\alpha \cong \dfrac{\text{MS}\alpha}{\text{MS}\alpha\mathcal{B}}$	$F_\alpha \cong \dfrac{\text{MS}\alpha}{\text{MS}\alpha\mathcal{B}}$
$H_{0\mathcal{B}}$: No Factor $\mathcal{B}$ Effects	$F_\mathcal{B} \cong \dfrac{\text{MS}\mathcal{B}}{\text{MS}_{\text{avg}}}$	$F_\mathcal{B} \cong \dfrac{\text{MS}\mathcal{B}}{\text{MS}\alpha\mathcal{B}}$	$F_\mathcal{B} \cong \dfrac{\text{MS}\mathcal{B}}{\text{MS}_{\text{avg}}}$
$H_{0\alpha\mathcal{B}}$: No Interaction	$F_{\alpha\mathcal{B}} \cong \dfrac{\text{MS}\alpha\mathcal{B}}{\text{MS}_{\text{avg}}}$	$F_{\alpha\mathcal{B}} \cong \dfrac{\text{MS}\alpha\mathcal{B}}{\text{MS}_{\text{avg}}}$	$F_{\alpha\mathcal{B}} \cong \dfrac{\text{MS}\alpha\mathcal{B}}{\text{MS}_{\text{avg}}}$

because the assumption of homogeneity of variance within each cell has been violated in the unweighted means analysis. In other words, the variance (of the unweighted mean $\bar{Y}_{ij.}$) in each cell is σ^2/n_{ij}, and, of course, homoscedasticity could only have been achieved had the rc cell sample sizes been equal. However, provided that no $n_{ij} = 0$ or that any cell sample size is more than twice that of any other, the method of unweighted means offers useful approximate F tests by utilizing the estimated average variance in the cell means (7.27) to measure the experimental error in the data.

7.6.2 Application

The Dean of the College of Business Administration at a large university desired to examine the effects of sex and graduate school intention on the grade point index. Thus, 42 upperclassmen majoring in business were first classified according to sex and intention to attend graduate school, and then their grade point indexes were recorded as in Table 7.11. The required summary computations for each cell, row, and column are presented in Table 7.12. Using (7.3), we immediately determine that

$$n_{11} \neq \frac{n_1. \, n_{.1}}{n_{..}}$$

$$9 \neq \frac{22(16)}{42} = 8.38$$

TABLE 7.11 Grade Point Indexes of 42 Upperclassmen Majoring in Business

Sex	Yes	No	Not Sure
		Graduate School Intent	
Male	3.0 2.8 2.9 2.8 3.2 2.7 3.8 2.6 3.6	2.6 2.0 3.1 2.4 2.8 2.6	2.9 3.0 2.6 3.2 2.7 2.5 3.2
Female	2.8 3.3 3.8 3.0 2.9 3.9 2.8	2.3 2.8 2.5 3.3 2.9 2.2 2.4 2.9	2.5 2.7 3.0 3.5 2.8

TABLE 7.12 Required Summary Computations

Sex	Yes	No	Not Sure	Row (Factor α) Summaries
		Graduate School Intent		
Male	$Y_{11\cdot} = 27.4$ $n_{11} = 9$ $\bar{Y}_{11\cdot} = 3.04$	$Y_{12\cdot} = 15.5$ $n_{12} = 6$ $\bar{Y}_{12\cdot} = 2.58$	$Y_{13\cdot} = 20.1$ $n_{13} = 7$ $\bar{Y}_{13\cdot} = 2.87$	$n_{1\cdot} = 22$
Female	$Y_{21\cdot} = 22.5$ $n_{21} = 7$ $\bar{Y}_{21\cdot} = 3.21$	$Y_{22\cdot} = 21.3$ $n_{22} = 8$ $\bar{Y}_{22\cdot} = 2.66$	$Y_{23\cdot} = 14.5$ $n_{23} = 5$ $\bar{Y}_{23\cdot} = 2.90$	$n_{2\cdot} = 20$
Column (Factor $\mathcal{B}$) Summaries	$n_{\cdot 1} = 16$	$n_{\cdot 2} = 14$	$n_{\cdot 3} = 12$	$n_{\cdot\cdot} = 42$ $\sum\limits_{i=1}^{r}\sum\limits_{j=1}^{c}\sum\limits_{k=1}^{n_{ij}} Y_{ijk}^2 = 357.65$

and so we conclude that the unequal cell frequencies are disproportional. However, since no cell is void of data and since the largest cell sample ($n_{11} = 9$) is not more than twice the size of the smallest cell sample ($n_{23} = 5$), the method of unweighted means may be used to analyze the data. From Tables 7.11 and 7.12, the required data and summary information are displayed in Table 7.13. The appropriate sums of squares and mean squares are then computed:

$$SS\alpha = \frac{\sum\limits_{i=1}^{r} \mathcal{Y}_{i\cdot\cdot}^2}{c} - \frac{\mathcal{Y}_{\cdots}^2}{rc} = \left[\frac{(8.49)^2 + (8.77)^2}{3}\right] - \frac{(17.26)^2}{6} = .0130$$

$$MS\alpha = \frac{SS\alpha}{r-1} = \frac{.0130}{2-1} = .0130$$

$$SS\mathcal{B} = \frac{\sum\limits_{j=1}^{c} \mathcal{Y}_{\cdot j\cdot}^2}{r} - \frac{\mathcal{Y}_{\cdots}^2}{rc} = \left[\frac{(6.25)^2 + (5.24)^2 + (5.77)^2}{2}\right] - \frac{(17.26)^2}{6} = .2552$$

189

TABLE 7.13 Method of Unweighted Means

Sex	Graduate School Intent			Row (Factor α) Summaries	
	Yes	No	Not Sure	Totals	Means
Male	3.04	2.58	2.87	$y_{1..} = 8.49$	$\bar{y}_{1..} = 2.830$
Female	3.21	2.66	2.90	$y_{2..} = 8.77$	$\bar{y}_{2..} = 2.923$
Column (Factor β) Summaries					
Totals	$y_{.1.} = 6.25$	$y_{.2.} = 5.24$	$y_{.3.} = 5.77$	$y_{...} = 17.26$	
Means	$\bar{y}_{.1.} = 3.125$	$\bar{y}_{.2.} = 2.620$	$\bar{y}_{.3.} = 2.885$	$\bar{y}_{...} = 2.877$	

$$MS\mathfrak{B} = \frac{SS\mathfrak{B}}{c-1} = \frac{.2552}{3-1} = .1276$$

$$SS\mathfrak{A}\mathfrak{B} = \sum_{i=1}^{r} \sum_{j=1}^{c} \bar{y}_{ij.}^2 - SS\mathfrak{A} - SS\mathfrak{B} - \frac{\bar{y}_{...}^2}{rc} = [(3.04)^2 + (2.58)^2 + \cdots$$

$$+ (2.90)^2] - .0130 - .2552 - \frac{(17.26)^2}{6} = .0051$$

$$MS\mathfrak{A}\mathfrak{B} = \frac{SS\mathfrak{A}\mathfrak{B}}{(r-1)(c-1)} = \frac{.0051}{(2-1)(3-1)} = .00255$$

$$SSE = \sum_{i=1}^{r} \sum_{j=1}^{c} \sum_{k=1}^{n_{ij}} Y_{ijk}^2 - \sum_{i=1}^{r} \sum_{j=1}^{c} \frac{Y_{ij.}^2}{n_{ij}}$$

$$= 357.65 - \left[\frac{(27.4)^2}{9} + \frac{(15.5)^2}{6} + \cdots + \frac{(14.5)^2}{5} \right] = 5.39$$

$$MSE = \frac{SSE}{n_{..} - rc} = \frac{5.39}{42 - 6} = .1497$$

$$MS_{avg} = \frac{MSE}{rc} \sum_{i=1}^{r} \sum_{j=1}^{c} \frac{1}{n_{ij}} = \left(\frac{.1497}{6} \right) \left(\frac{1}{9} + \frac{1}{6} + \frac{1}{7} + \frac{1}{7} + \frac{1}{8} + \frac{1}{5} \right) = .02217$$

Since the levels of each factor are considered *fixed*, the appropriate ANOVA table is presented as Table 7.14.

TABLE 7.14   ANOVA Table for the Method of Unweighted Means for a Fixed-Effects Model with Disproportionate Cell Frequencies

Source of Variation	Degrees of Freedom	Sum of Squares	Mean Square	F	Approximate P Values
Sex ($\mathfrak{A}$)	1	.0130	.0130	.59	$P \cong .5$ (not significant)
School Intent ($\mathfrak{B}$)	2	.2552	.1276	5.76	$P < .01$ (significant)
Sex-Intent Interaction ($\mathfrak{A}\mathfrak{B}$)	2	.0051	.00255	.12	$P > .5$ (not significant)
Error	36	—	.02217	—	—

The following hypotheses are tested:

$$H_{0\mathfrak{A}}: \quad \mu_1. = \mu_2. \quad \text{(or no sex effect)}$$
$$H_{1\mathfrak{A}}: \quad \mu_1. \neq \mu_2.$$

$$H_{0\mathfrak{B}}: \quad \mu._1 = \mu._2 = \mu._3 \quad \text{(or no intent effect)}$$
$$H_{1\mathfrak{B}}: \quad \text{not all } \mu._j \text{ are equal} \quad (j = 1, 2, 3)$$

and

$$H_{0\mathfrak{A}\mathfrak{B}}: \quad \mu_{ij} \equiv \text{ for all } rc \text{ cells} \quad \text{(or no interaction effect)}$$
$$H_{1\mathfrak{A}\mathfrak{B}}: \quad \text{not all } \mu_{ij} \text{ are equal} \quad (i = 1, 2 \text{ and } j = 1, 2, 3)$$

At the .05 level of significance we observe that there is no evidence of an interaction effect; nor is there any reason to suspect the existence of an effect due to sex on the grade point index. However, since $F_\mathcal{B} \cong MS\mathcal{B}/MS_{avg} = 5.76 >$ $F_{.95;2,36} = 3.26$, we may reject $H_{0_\mathcal{B}}$ and conclude that there are significant differences in the grade point index based on graduate school intent.

7.6.3 A Posteriori Multiple Comparisons

To determine what these "differences" are, we employ the Scheffé $\mathcal{S}$ method. Using (7.13), (7.14), and (7.25) as well as the results from Tables 7.12 and 7.13, we obtain the six contrast estimates displayed on page 193.

Using an experimentwise error rate of .05 for the family of contrasts, we may conclude that significant differences exist in the grade point index between upperclassmen planning to attend graduate school versus those not planning to attend ($\mathcal{L}_1$). Moreover, there are significant differences in the grade point index for those students planning to attend graduate school and those who are not or are undecided taken together ($\mathcal{L}_4$). Furthermore, there are also significant differences in the grade point index between upperclassmen who are not planning to attend and those who are either undecided or planning to attend taken together ($\mathcal{L}_5$).

7.7 COMPUTER PACKAGES AND UNBALANCED FACTORIAL DESIGNS: USE OF SPSS, SAS, AND BMDP

7.7.1 Introduction

As mentioned in Section 7.2, there are several approaches for dealing with factorial experiments in which there are an unequal number of observations in each cell. Various *regression* approaches (see Section 12.10) are utilized by computer packages as part of their linear model procedures (ANOVA in SPSS, PROC GLM in SAS and P-2V in BMDP). However, in this chapter we have described the method of *unweighted means* as an approach to analyzing unbalanced factorial experiments. Thus in this section we shall explain how to use SPSS, SAS, or BMDP in order to obtain the results based on this approach.

7.7.2 Organizing the Data

A two-step process is required in order to obtain the appropriate MS terms needed for the unweighted means analysis. First, a two-way analysis of variance (see Section 6.9) may be applied to the raw data in order to extract the MSE term and each of the rc cell means. Using (7.27), manual computations would provide MS_{avg}. The cell means obtained from the first output are then used as the response or dependent variable input to the second run in order to compute the $MS\alpha$, $MS\mathcal{B}$, and $MS\alpha\mathcal{B}$ terms. The ANOVA table needed for the

Note: $\sqrt{(c-1)F_{1-\alpha;c-1,n\cdots-rc}} = \sqrt{2F_{.95;2,36}} = 2.5534$:

Estimated Contrast $\hat{\mathcal{L}}_{l'}$	$S_{\hat{\mathcal{L}}_{l'}}$	Critical Range	Conclusion
$\hat{\mathcal{L}}_1 = \hat\mu_{\cdot 1} - \hat\mu_{\cdot 2} = 3.125 - 2.620 = .505$	$\sqrt{\dfrac{.1497}{(2)^2}\left(\dfrac{1}{9} + \dfrac{1}{6} + 0 + \dfrac{1}{7} + \dfrac{1}{8} + 0\right)} = .1429$	.365	Significant
$\hat{\mathcal{L}}_2 = \hat\mu_{\cdot 1} - \hat\mu_{\cdot 3} = 3.125 - 2.885 = .240$	$\sqrt{\dfrac{.1497}{(2)^2}\left(\dfrac{1}{9} + 0 + \dfrac{1}{7} + \dfrac{1}{7} + 0 + \dfrac{1}{5}\right)} = .1495$	.382	Not significant
$\hat{\mathcal{L}}_3 = \hat\mu_{\cdot 2} - \hat\mu_{\cdot 3} = 2.620 - 2.885 = -.265$	$\sqrt{\dfrac{.1497}{(2)^2}\left(0 + \dfrac{1}{6} + \dfrac{1}{7} + 0 + \dfrac{1}{8} + \dfrac{1}{5}\right)} = .1541$	.393	Not significant
$\hat{\mathcal{L}}_4 = \hat\mu_{\cdot 1} - \left(\dfrac{\hat\mu_{\cdot 2}+\hat\mu_{\cdot 3}}{2}\right) = 3.125 - 2.752 = .373$	$\sqrt{\dfrac{.1497}{(2)^2}\left(\dfrac{1}{9} + \dfrac{.25}{6} + \dfrac{.25}{7} + \dfrac{1}{7} + \dfrac{.25}{8} + \dfrac{.25}{5}\right)} = .1243$	.317	Significant
$\hat{\mathcal{L}}_5 = \hat\mu_{\cdot 2} - \left(\dfrac{\hat\mu_{\cdot 1}+\hat\mu_{\cdot 3}}{2}\right) = 2.620 - 3.005 = -.385$	$\sqrt{\dfrac{.1497}{(2)^2}\left(\dfrac{.25}{9} + \dfrac{1}{6} + \dfrac{.25}{7} + \dfrac{.25}{7} + \dfrac{1}{8} + \dfrac{.25}{5}\right)} = .1285$	.328	Significant
$\hat{\mathcal{L}}_6 = \hat\mu_{\cdot 3} - \left(\dfrac{\hat\mu_{\cdot 1}+\hat\mu_{\cdot 2}}{2}\right) = 2.885 - 2.872 = .013$	$\sqrt{\dfrac{.1497}{(2)^2}\left(\dfrac{.25}{9} + \dfrac{.25}{6} + \dfrac{1}{7} + \dfrac{.25}{7} + \dfrac{.25}{8} + \dfrac{1}{5}\right)} = .1339$	.342	Not significant

unweighted means analysis could then be simply constructed manually and the appropriate hypotheses tests made.

The organization of the raw data needed for the first output is presented as Table 7.15, while the organization of the *cell means* data needed for the second output is presented as Table 7.16. We should note that for data entry purposes the identification number may be placed in columns 1 and 2, GPI in columns 4–7, SEX in column 9, and graduate school intention (INTENT) in column 11.

TABLE 7.15   Organizing the Data for Computer Analysis

ID No.	GPI	Sex	Graduate School Intention
1	3.00	1	1
2	2.80	1	1
3	3.80	1	1
4	2.80	1	1
5	3.20	1	1
.	.	.	.
.	.	.	.
.	.	.	.
38	2.50	2	3
39	3.00	2	3
40	2.80	2	3
41	2.70	2	3
42	3.50	2	3

SOURCE: Table 7.11.

TABLE 7.16   Organizing the Cell Means for Computer Analysis

ID No.	GPI	Sex	Graduate School Intention
1	3.04	1	1
2	2.58	1	2
3	2.87	1	3
4	3.21	2	1
5	2.66	2	2
6	2.90	2	3

SOURCE: Table 7.13.

7.7.3 Using SPSS for Unweighted Means ANOVA

Figure 7.1 illustrates the program that has been written using the ANOVA subprogram of SPSS (see Sections 5.6 and 6.9) to perform two-way ANOVA on the raw data in order to obtain the MSE term and the cell means. The second part of the analysis would involve substituting the cell means (see Tables 7.13

and 7.16) for the raw data and rerunning the program illustrated in Figure 7.1. In this case GPI would represent the average GPI for a particular combination of sex and graduate school intention. In addition, OPTIONS 3 would have to be substituted for STATISTICS 3 since there would be only one observation per cell. Figure 7.2 represents annotated partial output obtained from these SPSS programs.

```
{SYSTEM CARDS}
RUN NAME                   UNWEIGHTED MEANS ANALYSIS OF GPI
DATA LIST                  FIXED(1)/1 ID 1-2,GPI 4-7,SEX 9,
                           INTENT 11
VAR LABELS                 GPI,GRADE POINT INDEX/
                           SEX, SEX OF STUDENT/
                           INTENT,GRADUATE SCHOOL INTENTION/
VALUE LABELS               SEX(1)MALE(2)FEMALE/
                           INTENT(1)YES(2)NO(3)NOT SURE/
READ INPUT DATA
01 3.00 1 1
 :    :     :   :
42 3.50 2 3
END INPUT DATA
ANOVA                      GPI BY SEX(1,2) INTENT(1,3)/
STATISTICS                 3
FINISH
{SYSTEM CARDS}
```

Figure 7.1 SPSS program for graduate school example.

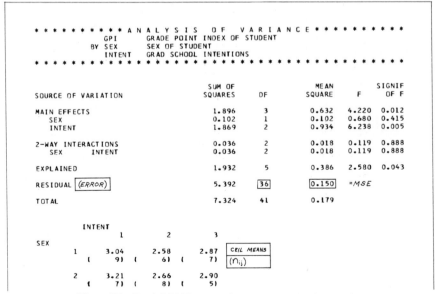

Figure 7.2 Partial SPSS output for graduate school example.

```
* * * * * * * * * * * A N A L Y S I S   O F   V A R I A N C E * * * * * * * * * *
                    GPI       GRADE POINT INDEX
             BY SEX           SEX OF STUDENT
                INTENT        GRADUATE SCHOOL INTENTION
* * * * * * * * * * * * * * * * * * * * * * * * * * * * * * * * * * * * * * * *
```

SOURCE OF VARIATION	SUM OF SQUARES	DF	MEAN SQUARE	F	SIGNIF OF F
MAIN EFFECTS	0.268	3	0.089	35.538	0.027
SEX (α)	0.013	1	0.013	5.192	0.150
INTENT (β)	0.255	2	0.128	50.711	0.019
EXPLAINED	0.268	3	0.089	35.538	0.027
RESIDUAL (α × β)	0.005	2	0.003		
TOTAL	0.273	5	0.055		

Figure 7.2   (continued)

7.7.4 Using SAS for Unweighted Means ANOVA

Figure 7.3 illustrates the program that has been written using PROC GLM of SAS (see Sections 5.6 and 6.9) to perform a two-way ANOVA on the raw data in order to obtain the MSE term and the cell means. The second part

```
{SYSTEM CARDS}
DATA UNWEIGHT;
     INPUT ID 1-2 GPI 4-7 SEX 9 INTENT 11;
LABEL ID=IDENTIFICATION NUMBER
      GPI=GRADE POINT INDEX
      SEX=SEX OF STUDENT
      INTENT=GRADUATE SCHOOL INTENTION;
CARDS;
01 3.00 1 1
 :   :  : :
42 3.50 2 3
PROC FORMAT;
     VALUE SEXF 1=MALE
                2=FEMALE;
PROC FORMAT;
     VALUE INTENTF 1=YES 2=NO
                   3=NOT SURE;
PROC GLM;
     FORMAT SEX SEXF.;
     FORMAT INTENT INTENTF.;
     CLASSES SEX INTENT;
     MODEL GPI=SEX|INTENT;
     MEANS SEX*INTENT;
{SYSTEM CARDS}
```

Figure 7.3 SAS program for graduate school example.

of the analysis involves the substitution of the cell means (see Tables 7.13 and 7.16) for the raw data and rerunning the program illustrated in Figure 7.3. In this case, GPI would represent the average GPI for a particular combination of sex and graduate school intention. In addition, for this second computer run the MODEL statement would be

MODELþGPI = SEXþINTENT;

since there is only one observation per cell. Figure 7.4 represents annotated partial output obtained from these SAS programs.

SEX	INTENT	N	GPI
1	1	9	3.04444444
1	2	6	2.58333333
1	3	7	2.87142857
2	1	7	3.21428571
2	2	8	2.66250000
2	3	5	2.90000000

DEPENDENT VARIABLE: GPI   GRADE POINT INDEX

SOURCE	DF	SUM OF SQUARES	MEAN SQUARE
MODEL	5	1.93188492	0.38637698
ERROR	36	5.39216270	0.14978230
CORRECTED TOTAL	41	7.32404762	

DEPENDENT VARIABLE: GPI   GRADE POINT INDEX

SOURCE	DF	SUM OF SQUARES	MEAN SQUARE
MODEL	3	0.26830000	0.08943333
ERROR $(\alpha \times \beta)$	2	0.00503333	0.00251667
CORRECTED TOTAL	5	0.27333333	

SOURCE	DF	ANOVA SS	F VALUE	PR > F
SEX (α)	1	0.01306667	5.19	0.1503
INTENT (β)	2	0.25523333	50.71	0.0193

Figure 7.4   Partial SAS output for graduate school example.

7.7.5 Using BMDP for Unweighted Means ANOVA

Figure 7.5 illustrates the program that has been written using BMDP program 2V (see Sections 5.6 and 6.9) to perform a two-way ANOVA on the raw data in order to obtain the MSE term and the cell means. The second part of the analysis involves the substitution of the cell means (see Tables 7.13 and 7.16) for the raw data and rerunning the program illustrated in Figure 7.5. In this case GPI would represent the average GPI for a particular combination of sex and graduate school intention. In addition, for this second computer run the

$$\text{INCLUDE} = 1, 2.$$

sentence must be part of the /DESIGN paragraph since there is only one observation in each cell. Figure 7.6 represents annotated partial output obtained from these BMDP programs.

```
{SYSTEM CARDS}
/PROBLEM    TITLE IS 'ANALYSIS OF INTENTIONS'.
/INPUT      VARIABLES ARE 4.
            FORMAT IS '(F2.0,1X,F4.2,1X,F1.0,1X,F1.0)'.
/VARIABLE   NAMES ARE ID,GPI,SEX,INTENT.
            LABEL IS ID.
/GROUP      CODES(3) ARE 1,2.
            NAMES(3) ARE MALE,FEMALE.
            CODES(4) ARE 1,2,3.
            NAMES(4) ARE YES,NO,NOTSURE.
/DESIGN     DEPENDENT IS GPI.
            GROUPING ARE SEX,INTENT.
/END
01  3.00  1  1
 :   :    :  :
42  3.50  2  3
{SYSTEM CARDS}
```

Figure 7.5   Using BMDP program 2V for graduate school example.

CELL MEANS FOR 1-ST DEPENDENT VARIABLE

							MARGINAL
SEX =	MALE	MALE	MALE	FEMALE	FEMALE	FEMALE	
INTENT =	YES	NO	NOTSURE	YES	NO	NOTSURE	
GPI	3.04444	2.58333	2.87143	3.21429	2.66250	2.90000	2.88809
COUNT	9	6	7	7	8	5	42

ANALYSIS OF VARIANCE FOR 1-ST
DEPENDENT VARIABLE - GPI

SOURCE	SUM OF SQUARES	DEGREES OF FREEDOM	MEAN SQUARE	F	TAIL PROBABILITY
MEAN	335.91726	1	335.91726	2242.70	0.0000
SEX	0.08672	1	0.08672	0.58	0.4517
INTENT	1.88083	2	0.94041	6.28	0.0046
SI	0.03557	2	0.01779	0.12	0.8884
ERROR	5.39216	36	0.14978		

SOURCE	SUM OF SQUARES	DEGREES OF FREEDOM	MEAN SQUARE
MEAN	49.65126	1	49.65126
SEX $(\alpha \times \beta)$	0.01307	1	0.01307
INTENT (α)	0.25523	2	0.12762
ERROR (β)	0.00503	2	0.00252

Figure 7.6   Partial BMDP program 2V output for graduate school example.

7.8 SUMMARY

When the cell frequencies are proportionate, we have observed that (with only a slight adjustment to the interaction sums of squares) the two-factor ANOVA procedures described in Chapter 6 can be readily applied. Moreover, we have also observed that when the cell frequencies are disproportionate, the method of unweighted means can provide an excellent approximation to the nonorthogonal two-factor experimental design problem.

However, there would be many occasions when this simple method of unweighted means would yield unsatisfactory results and should be avoided. If cell sample sizes are very different or if empty cells exist in the rc table, the more complex least-squares regression methods of Chapter 12 would be better suited. In fact, Winer (1971) has argued that a least-squares regression approach is *always* more appropriate if, as in survey research, the data are classified within their natural strata. In addition, for more complex nonorthogonal experiments concerning three or more classification factors, a least-squares regression approach is more appropriate—regardless of whether the cell frequencies are proportionate or disproportionate.

In Chapters 8–14 we shall focus on regression methods. Therefore, we shall again return to this nonorthogonal factorial experiment in Section 12.9 and re-examine the ANOVA problem using a regression approach.

PROBLEMS

7.1. For corporations reporting a profit in the year 1979, a financial analyst wanted to compare the earnings per share of the "top two" companies in each of five industries against the "next four" based on annual sales. The data are presented in Table P7.1. Using $\alpha = .05$, completely analyze the data and discuss your conclusions.

TABLE P7.1 Earnings per Share for 1979

Annual Sales Ranking	Industrial Grouping				
	Petro-leum	Motor Vehi-cles	Office Equip-ment	Elec-tronics	Metal Manu-facturing
Top 2	9.74	10.04	5.16	6.20	6.03
	9.46	9.75	11.89	2.65	6.31
Next 4	6.48	7.33	6.35	3.65	4.82
	10.44	6.07	8.78	2.81	14.29
	6.78	5.28	7.20	4.87	6.56
	10.23	7.10	4.10	6.30	7.49

7.2. During a physical examination for entrance into the armed forces a group of 40 males aged 18 were first classified according to ethnicity and residence, and then their diastolic blood pressures were recorded as in Table P7.2. Using $\alpha = .05$, completely analyze the data and discuss your conclusions.

TABLE P7.2 Diastolic Blood Pressure

	Residence		
Ethnicity	Urban	Suburb	Rural
Caucasian	81 80 81 75 80 79 83 88 73 68 84 81	83 80 74 82	86 88 72 81
Black	87 84 81 89 86 92	84 88	93 82
Other	84 74 72 82 83 77	87 82	86 80

7.3. In a survey of television viewing habits a sample of 50 cable TV subscribers were first classified according to employment status and age level. The data in Table P7.3 represent the reported number of weekly television viewing hours by the subscribers. Using $\alpha = .05$, completely analyze the data and discuss your conclusions.

TABLE P7.3 Number of Hours Watching Television

	Age Level		
Employment Status	Under 40	40–60	Over 60
Full Time	11 15 16 22 23 16 19 13	12 23 17 24 8 16 20 17	19 28 22 22
Part Time	18 19 24 25	19 18 17 23 20 26 25	17 26 24 30
Unemployed	25 21 29 31	18 26 21 20 17	27 20 25 31 33 32

7.4. A sample of 40 first-grade children were classified according to sex and aggressiveness. The children were then asked to respond to some anger-provoking stimuli, and their reaction times (in seconds) are recorded in Table P7.4. Using $\alpha = .05$, completely analyze the data and discuss your conclusions.

***7.5.** A hospital administrator wished to examine postsurgical hospitalization periods following knee surgery. A random sample of 40 patients was classified according to type of knee surgery and age group, and then the number of required postsurgical hospitalization days were recorded as shown in Table P7.5. Using $\alpha = .05$, completely analyze the data and discuss your conclusions.

TABLE P7.4   Reaction Times (in Seconds)

Sex	Aggressiveness Level								
	High			Normal			Low		
Boy	.98	.89	1.02	1.04	1.01	.95	1.05	1.06	.98
	1.00	.97	.71	.87	1.02		.84	1.14	1.06
	.73	.86	.95				1.04		
Girl	.89	1.00	1.04	.93	.85	1.16	1.04	1.07	1.10
	1.06	.95	.99	1.04	1.03		1.03	.94	1.06
							1.06	1.11	

TABLE P7.5   Number of Postsurgical Hospitalization Days

Type of Knee Surgery	Age Group of Patient													
	Under 30					30–50					Over 50				
Arthroscopy	1	3	2	2	3	4	3	2	3	2	3	5	2	3	3
	2	6	2			2	3								
Arthrotomy	3	3	6	4	8	4	5	11	5	6	4	8	12	10	3
	10	5	7	8							4				

7.6. A security analyst wished to examine the behavior pattern of price-to-earnings ratios of various stocks. A random sample of 30 issues were first classified according to listed exchange and annual dividend payment records, and then their price-to-earnings ratios were recorded as shown in Table P7.6. Using $\alpha = .05$, completely analyze the data and discuss your conclusions.

TABLE P7.6   Price-to-Earnings Ratios

Stock Exchange	Annual Dividend Payments											
	Yes						No					
New York	9	7	8	4	8	8	4	3	7	6	5	7
	5	5	10	6	7							
American	7	8	5	4	3	6	5	6	6	7	4	3
							5					

7.7. Using the real estate data base (see Appendix A), select a random sample of 30 houses and determine ($\alpha = .05$) whether there is any difference in the selling price based on the type of garage and the presence of a fireplace. If appropriate, use the method of unweighted means. ∎

7.8. Using the real estate data base (see Appendix A), select a random sample of 30 houses and determine ($\alpha = .01$) whether there is any difference in the assessed value based on the presence or absence of central air-conditioning and a fireplace. If appropriate, use the method of unweighted means. ∎

7.9. Using the real estate data base (see Appendix A), select a random sample of 30 houses and determine ($\alpha = .05$) whether there is any difference in the age of houses based on the presence or absence of a fireplace and central heating. If appropriate, use the method of unweighted means. ∎

7.10. Reviewing concepts:
   (a) What is meant by a nonorthogonal factorial design?
   (b) Describe the differences between balanced and unbalanced factorial designs.
   (c) What is the purpose of the method of unweighted means?

8

SIMPLE LINEAR REGRESSION

AND

CORRELATION ANALYSIS

8.1 INTRODUCTION AND PREVIEW

Up to this point we have focused primarily on methods of comparison among groups [i.e., levels of some factor(s) of interest]. Depending on the situation, experimental designs ranging in format from the simple one-factor completely randomized model (Chapter 3) to the more complex two-factor nonorthogonal model (Chapter 7) were considered in order to determine the existence of significant treatment effects. Starting with this chapter, however, emphasis shifts from such a *confirmatory* or hypothesis testing approach to one of a more *exploratory* nature—that is, the development of models for prediction and evaluation purposes. Specifically, in Chapter 8 we shall deal with simple linear regression and correlation analysis. **Regression** is concerned with *prediction*, that is, the ability to build a statistical model which uses information about a set of *independent* or *predictor* variables in order to estimate the expected value of some *dependent* or *response* variable, while **correlation** is concerned with measuring the *relationship* or *strength of association* among sets of variables.

It is assumed that in an introductory course the reader would have been exposed to the subject of simple linear regression and correlation analysis—at least from a descriptive if not inferential point of view. [For references, see Chapter 14 of Berenson and Levine (1983).] Hence in this chapter we shall highlight the essential concepts of simple linear regression and correlation and focus primarily on the fundamental aspects of model development and evaluation through methods of statistical inference.

8.2 SIMPLE LINEAR REGRESSION

In a simple linear regression analysis we attempt to develop a linear model from which the values of a dependent (i.e., response) variable can be predicted based on particular values of a single independent variable. To develop the model a sample of n independent pairs of observations $(X_1, Y_1), (X_2, Y_2), \ldots, (X_n, Y_n)$ are obtained, where X_i represents the ith value of the independent or predictor variable X and where Y_i represents the corresponding response—that is, the ith value of the dependent variable Y.

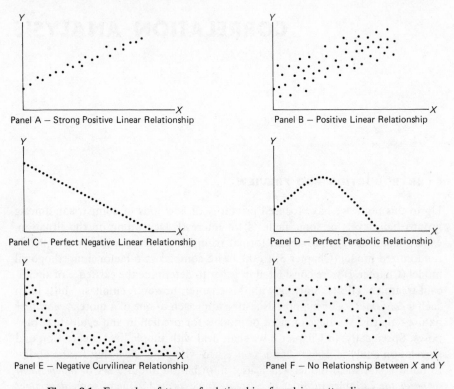

Panel A — Strong Positive Linear Relationship

Panel B — Positive Linear Relationship

Panel C — Perfect Negative Linear Relationship

Panel D — Perfect Parabolic Relationship

Panel E — Negative Curvilinear Relationship

Panel F — No Relationship Between X and Y

Figure 8.1 Examples of types of relationships found in scatter diagrams.

To study the possible underlying relationship between X and Y, the n individual pairs of observations can be plotted on a two-dimensional graph called a scatter diagram. The dependent variable Y is plotted on the vertical axis, while the independent variable X is plotted on the horizontal axis. Figure 8.1 depicts several kinds of possible underlying relationships between the X and Y variables. Simple linear relationships are displayed in panels A, B, and C. In panel A there is a very strong positive relationship between X and Y, whereas in panel B the positive linear relationship is much weaker—there is much (wider) *scatter* in the data. In each case, however, the values of Y are generally increasing as X

increases. On the other hand, a "perfect" negative linear relationship is presented in panel C. That is, we could fit (draw) a straight line that passes through all the plotted points and there would be no scatter above or below such a line—all the points would lie on it. Thus the values of Y are decreasing linearly as X increases. More "complex" mathematical relationships are displayed in panels D and E. The "upside-down" U-shaped curve in panel D represents a perfect parabolic relationship between X and Y. That is, a parabola (second-degree curve) could be fitted to the data in such a manner that there would be no scatter above or below the curve. This of course would not be the case in panel E— although a fairly strong exponential or negative curvilinear relationship between X and Y is indicated. In such situations Y is observed to decrease at a decreasing rate with increases in X. On the other hand, the scatter diagram in panel F represents a set of data for which there is no relationship between X and Y. High and low values of Y appear at each value of X. Any type of mathematical model developed for these data would fit badly—there would be so much scatter above and below the fitted regression equation as to render it useless for prediction purposes.

8.3 DEVELOPING THE REGRESSION EQUATION

The scatter diagram aids the researcher in selecting an appropriate regression model. By examining the plotted sample points, the researcher attempts to project the underlying mathematical relationship that may exist between X and Y.

In a simple regression model wherein there is but one predictor variable X, this functional relationship can be expressed as

$$Y_i = f(X_i) + \epsilon_i \qquad (8.1)$$

$$\text{Observed Data} = \text{Fitted Model} + \text{Residual}$$

where any observed value Y_i in the population would be a function of the true mathematical model $f(X_i)$ plus some *residual* ϵ_i. The ϵ_i term represents scatter above and below the regression equation as demonstrated in panels A, B, E, and F of Figure 8.1. On the other hand, if the model were to have fit the data perfectly, as in panels C and D of Figure 8.1, there would be no scatter about the regression equation; that is, $\epsilon_i = 0$ and $Y_i = f(X_i)$ for all i.

If the scatter diagram indicates a possible linear relationship between X and Y, the population regression model (8.1) can be re-expressed as

$$Y_i = \beta_0 + \beta_1 X_i + \epsilon_i \qquad (8.2)$$

where the two unknown parameters β_0 and β_1 are necessary for determining a straight line. β_0 is the true *intercept*, a constant factor in the regression model representing the expected or fitted value of Y when $X = 0$. β_1 is the true *slope*;

it represents the amount that Y changes (either positively or negatively) per *unit* change in X.

Since we do not have access to the entire population, we cannot compute the parameters β_0 and β_1 and obtain the population regression model. The objective then becomes one of obtaining estimates b_0 (for β_0) and b_1 (for β_1) from the sample. Usually this is accomplished by employing the **method of least squares**. With this method the statistics b_0 and b_1 are computed from the sample in such a manner that the best possible fit within the constraints of the least squares model is achieved. That is, we obtain the linear regression equation

$$\hat{Y}_i = b_0 + b_1 X_i \tag{8.3}$$

such that $\sum_{i=1}^{n}(Y_i - \hat{Y}_i)^2 = \sum_{i=1}^{n} e_i^2$ is *minimized*.

In using the least-squares method, the following two *normal equations* are developed:

$$\text{I.} \quad \sum_{i=1}^{n} Y_i = nb_0 + b_1 \sum_{i=1}^{n} X_i$$

$$\text{II.} \quad \sum_{i=1}^{n} X_i Y_i = b_0 \sum_{i=1}^{n} X_i + b_1 \sum_{i=1}^{n} X_i^2 \tag{8.4}$$

and, solving simultaneously for b_1 and b_0, we compute

$$b_1 = \frac{n \sum_{i=1}^{n} X_i Y_i - \left(\sum_{i=1}^{n} X_i\right)\left(\sum_{i=1}^{n} Y_i\right)}{n \sum_{i=1}^{n} X_i^2 - \left(\sum_{i=1}^{n} X_i\right)^2} \tag{8.5}$$

and

$$b_0 = \bar{Y} - b_1 \bar{X} = \left(\frac{\sum_{i=1}^{n} Y_i}{n}\right) - b_1 \left(\frac{\sum_{i=1}^{n} X_i}{n}\right) \tag{8.6}$$

so that the sample regression equation $\hat{Y}_i = b_0 + b_1 X_i$ is obtained.*

8.4 APPLICATION

As an example, suppose the Director of Career Development at a university wishes to investigate the relationship between the achieved college grade point index and the starting salary of recent graduates majoring in business so that when advising students she may build a model to predict the starting salary of business majors based on the grade point index. A random sample of 30 recent

---

*Analogous to (8.2) the sample regression model can be expressed as

$$Y_i = b_0 + b_1 X_i + e_i$$

which can also be written as $Y_i = \hat{Y}_i + e_i$. By subtraction, $Y_i - (b_0 + b_1 X_i) = Y_i - \hat{Y}_i = e_i$. By using the least-squares principle, $L = \sum_{i=1}^{n} e_i^2 = $ a *minimum total*. Thus, from the calculus, by taking the *partial derivatives* of L with respect to b_0 and b_1 and setting the results equal to zero (that is, $\partial L/\partial b_0 = 0$ and $\partial L/\partial b_1 = 0$) we obtain the two normal equations presented in (8.4) and solve for the sample regression equation (8.3) as indicated.

graduates from the College of Business is drawn, and the data pertaining to the grade point index (GPI), starting salary (in thousands of dollars), and business major are recorded for each individual as in Table 8.1.

TABLE 8.1   **Starting Salary and Grade Point Index for a Random Sample of 30 Recent Graduates Majoring in Business**

Individual No.	(X) GPI	(Y) Starting Salary ($000)	Business Major
1	2.7	17.0	Accountancy
2	3.1	17.7	Accountancy
3	3.0	18.6	Accountancy
4	3.3	20.5	Accountancy
5	3.1	19.1	Accountancy
6	2.4	16.4	Accountancy
7	2.9	19.3	Accountancy
8	2.1	14.5	Accountancy
9	2.6	15.7	Accountancy
10	3.2	18.6	Accountancy
11	3.0	19.5	Accountancy
12	2.2	15.0	Accountancy
13	2.8	18.0	Accountancy
14	3.2	20.0	Accountancy
15	2.9	19.0	Accountancy
16	3.0	17.4	Marketing/Management
17	2.6	17.3	Marketing/Management
18	3.3	18.1	Marketing/Management
19	2.9	18.0	Marketing/Management
20	2.4	16.2	Marketing/Management
21	2.8	17.5	Marketing/Management
22	3.7	21.3	Marketing/Management
23	3.1	17.2	Marketing/Management
24	2.8	17.0	Marketing/Management
25	3.5	19.6	Marketing/Management
26	2.7	16.6	Marketing/Management
27	2.6	15.0	Marketing/Management
28	3.2	18.4	Marketing/Management
29	2.9	17.3	Marketing/Management
30	3.0	18.5	Marketing/Management

The scatter diagram is displayed in Figure 8.2. A brief examination of Figure 8.2 clearly indicates a positive linear relationship between the grade point index X and the starting salary Y. As the grade point index increases, the starting salary also tends to increase.

From the data in Table 8.1 the summary computations recorded in Table 8.2 are needed not only for computing the intercept b_0 and slope b_1 of the regression equation (8.3), but also for studying other aspects of regression and correlation to be discussed throughout this chapter.

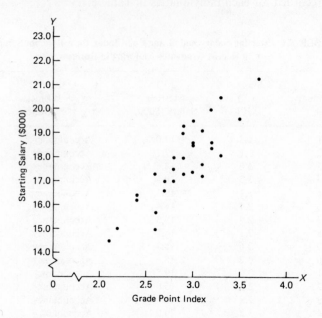

Figure 8.2 Scatter diagram of grade point index versus starting salary of 30 recent graduates majoring in business. (Source: Data are taken from Table 8.1.)

**TABLE 8.2 Summary Computations
Needed for Simple Regression
and Correlation Analysis**

$\sum\limits_{i=1}^{n} X_i = 87.0$	$\bar{X} = 2.90$
$\sum\limits_{i=1}^{n} Y_i = 534.3$	$\bar{Y} = 17.81$
$\sum\limits_{i=1}^{n} X_i Y_i = 1564.24$	$\sum\limits_{i=1}^{n} X_i^2 = 256.06$
$n = 30$	$\sum\limits_{i=1}^{n} Y_i^2 = 9593.41$

SOURCE: Table 8.1.

By using (8.5) and (8.6), the slope is

$$b_1 = \frac{n\sum\limits_{i=1}^{n} X_i Y_i - \left(\sum\limits_{i=1}^{n} X_i\right)\left(\sum\limits_{i=1}^{n} Y_i\right)}{n\sum\limits_{i=1}^{n} X_i^2 - \left(\sum\limits_{i=1}^{n} X_i\right)^2} = \frac{30(1564.24) - 87\,0(534.3)}{30(256.06) - (87.0)^2} = +3.92819$$

while the intercept is

$$b_0 = \bar{Y} - b_1\bar{X} = 17.81 - 3.92819(2.90) = +6.41825$$

Thus the equation for the *best straight line* for these data is

$$\hat{Y}_i = 6.41825 + 3.92819X_i$$

The slope, $b_1 = +3.92819$, indicates that for each one full unit (*point*) increase in the grade point index the starting salary increases by approximately 3.93 thousands of dollars (i.e., \$3930). The Y intercept, $b_0 = +6.41825$, represents a constant portion of the starting salary (i.e., approximately \$6420) that does not vary with the grade point index.

The sample regression equation will be used for prediction purposes in Section 8.7.

8.5 ASSUMPTIONS OF SIMPLE LINEAR REGRESSION

In our investigations into hypothesis testing and numerous ANOVA models we have noted that the appropriate application of a particular statistical procedure is dependent on how well a set of assumptions for that procedure is met. The assumptions necessary for regression analysis are analogous to those of the analysis of variance since they both fall under the general heading of *linear models* [see Box et al. (1978)].

1. Normality
2. Linearity
3. Independence
4. Homoscedasticity

8.5.1 Normality

The first assumption, *normality,* is necessary for the purpose of inference. In regression analysis we consider the independent variable X to be *fixed* at specific levels. Moreover, at each fixed X the dependent variable Y is considered a random variable having a certain probability density function denoted by $f(Y|X)$ with mean $\mu_{Y|X}$ and variance $\sigma^2_{Y|X}$. In other words, the entire population is decomposed into several (sub-) populations—one for each fixed X—wherein the random variable Y is distributed according to a particular density $f(Y|X)$. For descriptive purposes, it would only have been necessary to assume that at each fixed X the probability density function for Y is identical. More generally, however, for both descriptive and inferential purposes it is assumed that at each fixed X the (sub-) population of Y values follows a normal distribution (see Figure 8.3). That is,

$$Y \sim f(Y|X) = \mathcal{N}(\mu_{Y|X}, \sigma^2_{Y|X})$$

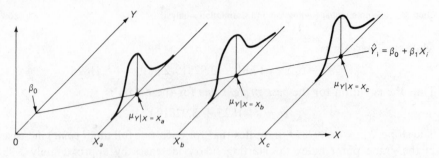

Figure 8.3 Assumptions of simple linear regression.

The assumption of normality will be investigated further in Sections 8.6 and 8.11.2. It should be noted, however, that like the t test and the analysis of variance F test, regression analysis is *robust* against moderate departures from the normality assumption. That is, as long as the distribution of Y values around each X is not very different from a normal distribution, inferences about the line of regression and the regression coefficients will not be seriously affected.

8.5.2 Linearity

The second assumption, *linearity*, specifies the functional relationship between X and Y. From Figure 8.3 we observe that at each fixed X the corresponding $\mu_{Y|X}$ is a straight-line function of X (that is, $\mu_{Y|X} = \beta_0 + \beta_1 X$). For example, as depicted in Figure 8.3, there is a positive linear relationship between X and Y. Note that when X is fixed at X_a the corresponding (sub-) population of Y values is normally distributed with mean $\mu_{Y|X=X_a}$. Similarly, when X is fixed at a larger value, say X_b, the corresponding (sub-) population of Y values is normally distributed with a larger mean $\mu_{Y|X=X_b}$, and so on.

The linearity assumption will be examined further in Sections 8.10 and 8.11.

8.5.3 Independence

The third assumption, *independence*, requires that the observed Y values be independent of one another for each value of X.* Referring to Figure 8.3, if X is fixed at X_a, the observed Y values from the corresponding (sub-) population are assumed to be independent of each other. Similarly, if X is fixed at X_b, the observed Y values from the corresponding (sub-) population are also assumed to be independent, and so on. In addition, the (sub-) population of possible

*More formally we may speak of this as the *independence of error* assumption. The independence of error assumption requires that the *error* e_i (that is, the *residual* difference between an observed and a predicted value of Y) should be independent for each value of X. If at each fixed X level the Y_i values are independent and normally distributed [that is, $Y \sim \mathcal{N}(\mu_{Y|X}, \sigma_{Y|X}^2)$], then the residual values $e_i = (Y_i - \hat{Y}_i)$ are also independent and normally distributed [that is, $e \sim \mathcal{N}(0, \sigma_{Y|X}^2)$].

210

Y values for one level of X is independent of the (sub-) population of possible Y values for any other level of X.

The independence assumption often concerns data that are collected over a period of time. For example, in economic data the observed Y values (say sales revenues or stock market prices) for one particular time period may be correlated with the previous and subsequent time periods. These types of models fall under the general heading of time series analysis [see, for example, Berenson and Levine (1983, Chapter 16) or Bowerman and O'Connell (1979)]. The independence assumption will be investigated further in Section 8.11.2 and Chapter 14.

8.5.4 Homoscedasticity

The fourth assumption, *homoscedasticity*, requires that the variation or scatter about the line of regression $\sigma^2_{Y|X}$ be constant for all values of X. As depicted in Figure 8.3, this means that Y varies the same amount when X is fixed at a low value as when X is fixed at a high value.

Uniformity in spread (i.e., homoscedasticity) about the regression line is quite an important assumption for using the least-squares method. If there are serious departures from this assumption, either data transformations or *weighted least-squares* methods can be applied. The homoscedasticity assumption will be examined further in Section 8.11.2 and Chapter 14.

8.5.5 Interpreting the Assumptions

To summarize for the model thus far developed, we note that for any fixed level of X the (sub-) population of *independent* Y values has a *normal distribution* whose particular *mean* $\mu_{Y|X}$ *changes linearly* with changes in X but whose *variance* $\sigma^2_{Y|X}$ *remains constant* under changes in X. In terms of the problem of interest to the Director of Career Development (see Section 8.4), it should be assumed that at each possible grade point index there exists an entire (sub-) population of *independent* starting salary values which are *normally distributed* with a mean $\mu_{Y|X}$ (which changes linearly with changes in the grade point index) and with a variance $\sigma^2_{Y|X}$ (which remains constant with changes in the grade point index). That is, starting salaries are a linear function of the grade point index. Moreover, for recent graduates with a particular grade point index the possible starting salary values are independent and normally distributed. Furthermore, variation in starting salaries obtained by recent graduates with low grade point indexes is the same as the variation in starting salaries obtained by recent graduates having high grade point indexes.

8.6 THE STANDARD ERROR OF ESTIMATE

In Section 8.3 the least-squares method was used to develop a linear regression equation to predict a starting salary based on a grade point index. Although the least-squares method results in a line that fits the data with the minimum amount of variation, the sample regression equation is not a perfect predictor—unless, of course, there is no variability in the population regression equation so that all the observed data points fall on the predicted regression line. If, from Section 8.5, we assume that for any fixed value of X, say X_i, the corresponding possible Y values are independent and normally distributed with mean $\mu_{Y|X=X_i}$ (i.e. the fitted value), then we should expect some observed Y values to be much larger than the mean, others should be much smaller, while most should be "clustering" around $\mu_{Y|X=X_i}$. Moreover, since we also assumed in Section 8.5 that this spread of possible Y values around the mean (i.e., the fitted value) is homogeneous— regardless of the level of X—we need to develop a statistic that estimates this population variance $\sigma_{Y|X}^2$ if, as would be typical in practice, we are only dealing with a sample. Such a measure, denoted by $S_{Y|X}^2$, would be defined as the variance around the regression line. Its square root, $S_{Y|X}$, would then be called the *standard deviation of the regression equation*. More commonly, however, statisticians refer to this measure of scatter around the regression line ($S_{Y|X}$) as the *standard error of estimate*. It is defined as

$$S_{Y|X} = \sqrt{\frac{\sum_{i=1}^{n}(Y_i - \hat{Y}_i)^2}{n - 2}} \tag{8.7}$$

where Y_i is the observed value of Y for a given X_i and $\hat{Y}_i$, the sample estimate of $\mu_{Y|X=X_i}$, is the predicted value of Y for a given X_i. The computation of $S_{Y|X}$ using (8.7) is cumbersome. This laborious computation can be simplified, however, because of the following identity:

$$\sum_{i=1}^{n}(Y_i - \hat{Y}_i)^2 \equiv \sum_{i=1}^{n} e_i^2 \equiv \sum_{i=1}^{n} Y_i^2 - b_0 \sum_{i=1}^{n} Y_i - b_1 \sum_{i=1}^{n} X_i Y_i \tag{8.8}$$

The standard error of estimate can then be obtained using the following computational formula:

$$S_{Y|X} = \sqrt{\frac{\sum_{i=1}^{n} Y_i^2 - b_0 \sum_{i=1}^{n} Y_i - b_1 \sum_{i=1}^{n} X_i Y_i}{n - 2}} \tag{8.9}$$

By using the summary calculations in Table 8.2 as well as the regression coefficients obtained in the problem of interest to the Director of Career Development, the standard error of estimate $S_{Y|X}$ can be computed from (8.9) as

$$S_{Y|X} = \sqrt{\frac{9593.41 - 6.41825(534.3) - 3.92819(1564.24)}{30 - 2}}$$

$$= \sqrt{\frac{19.5070994}{28}} = \sqrt{.69668} = .835$$

The standard error of estimate is measured in units of the dependent variable Y. Hence this standard error of estimate, equal to .835 (in thousands of dollars), represents the *average* variation around the fitted regression line.*

Because of the normality assumption in Section 8.5.1, if, for fixed values of X the $S_{Y|X}$ is added to and subtracted from the corresponding fitted Y values and parallel lines drawn, theoretically, for large samples, we should expect approximately 68.26% of the observed set of (sample) points plotted on the scatter diagram to be included within these boundaries.† Actually, from Figure 8.4 we observe that 20 of the 30 sample points (that is, 66.7%) are so enclosed. This, then, is one method of evaluating the appropriateness of the normality assumption (see also Section 8.11.2).

As we shall observe in the section which follows, the standard error of estimate will be used extensively in statistical inference. Intuitively, the closer $S_{Y|X}$ is to zero, the better the model fits the observed data and vice versa (see again Figure 8.1).

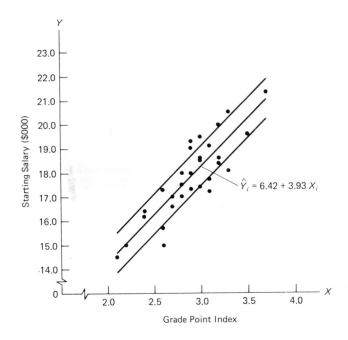

Figure 8.4   The fitted regression line $\pm$ standard error of estimate.

*Note that in computing the standard error of estimate we divide the sum of squared residuals by $n - 2$. Two degrees of freedom are lost because two other statistics—b_0 and b_1—are used in (8.9) in order to compute the statistic $S_{Y|X}$.

†Similarly, for large samples approximately 95.44% of the data points are expected to be encompassed between the fitted regression line $\pm 2S_{Y|X}$, and 99.73% of the data points should be included between the fitted regression line $\pm 3S_{Y|X}$.

8.7 INFERENTIAL METHODS IN SIMPLE LINEAR REGRESSION: AN OVERVIEW

Now that we have developed the least-squares linear regression equation and measured the scatter about the sample regression line, we are ready to determine whether all our endeavors have been fruitful. That is, does a significant linear relationship exist for prediction purposes? If the answer is affirmative, how appropriate is the *linear* fit? Might not a *curvilinear* model fit better?

Based on these answers, we would then be ready to develop confidence interval estimates of the true regression parameters. In addition, we would also be ready to utilize the regression equation for prediction purposes by forming both confidence interval estimates of the *mean response* $\mu_{Y|X}$ at selected values of X as well as *prediction* interval estimates for an *individual response* $\hat{Y}_I$ at selected values of X.

8.8 PARTITIONING THE TOTAL VARIATION

To determine how well the independent variable predicts the dependent variable in our regression equation, we need to develop several measures of variation. The first measure, the *total variation*, is a measure of variation of the observed Y values around their mean $\bar{Y}$. It measures the variability in the Y values without taking into consideration the X values at all.

The total variation or *total sums of squares* SST can be computed from

$$SST = \sum_{i=1}^{n}(Y_i - \bar{Y})^2 = \sum_{i=1}^{n}Y_i^2 - \frac{\left(\sum_{i=1}^{n}Y_i\right)^2}{n} \tag{8.10}$$

The total variation can now be partitioned into two components: the *explained variation* (that which is attributable to the relationship between X and Y postulated by the model) and the *unexplained variation* (that which is not accounted for by the model relating X and Y). That is,

$$\text{Total Variation} = \text{Explained Variation} + \text{Unexplained Variation} \tag{8.11}$$

These different measures of variation are depicted in Figure 8.5.

The explained variation or *sums of squares due to regression* SSR is obtained from

$$SSR = \sum_{i=1}^{n}(\hat{Y}_i - \bar{Y})^2 = b_0 \sum_{i=1}^{n}Y_i + b_1 \sum_{i=1}^{n}X_iY_i - \frac{\left(\sum_{i=1}^{n}Y_i\right)^2}{n} \tag{8.12}$$

From Figure 8.5 and the middle term of (8.12) we observe that SSR represents the differences between the $\hat{Y}_i$ (the values of Y that would be predicted from the regression relationship) and $\bar{Y}$ (the average value of Y that exists even without considering the regression relationship). Intuitively, if a good fitting regression

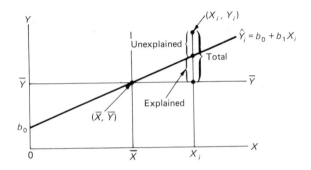

Figure 8.5 Measures of variation in regression and correlation.

equation is obtained, then SSR, the variability explained by regression, will represent a large portion of the total variation SST. Of course, a *perfect fit* occurs if $Y_i \equiv \hat{Y}_i$ at each value X_i. All the observed Y values would then lie on the computed regression line, and SSR $\equiv$ SST.

The unexplained variation or *error sums of squares* SSE is obtained from

$$\text{SSE} = \sum_{i=1}^{n} (Y_i - \hat{Y}_i)^2 = \sum_{i=1}^{n} Y_i^2 - b_0 \sum_{i=1}^{n} Y_i - b_1 \sum_{i=1}^{n} X_i Y_i \qquad (8.13)$$

From Figure 8.5 and the middle term of (8.13) we see that SSE represents the *residual* differences between the observed and predicted Y values. Hence the unexplained variation is a measure of scatter about the regression line. Moreover, from (8.7) and (8.9) we recall that SSE is the numerator under the square root in the computation of the standard error of estimate $S_{Y|X}$. Intuitively, if a good-fitting regression equation is obtained, then SSE, the variability unexplained by the regression, would represent a small portion of SST. If a perfect fit were obtained, there would be no scatter about the regression line and SSE would be zero (as would, of course, $S_{Y|X}$).

It is important to re-emphasize that when measuring scatter about the regression line (SSE and $S_{Y|X}$), we are examining that portion of the variability in Y that has not been accounted for by the fitted mathematical relationship between X and Y. There are two important sources of residual error here. The first source may be due to a *lack of fit*. That is, a simple linear regression model might not be the appropriate fit for the data—either because a curvilinear model would fit better (see Chapter 11) or because not all the appropriate predictor variables have been considered (see Chapter 10). If such model adjustments are made, SSE, the unexplained portion of variability in Y, can be reduced. The second source of residual variability may simply be thought of as the existence of *pure error*. Such "chance" or random variations in the data are uncontrollable and represent the imperfections in the fitted relationship. This will be examined further in our study of lack of fit (see Section 8.10).

8.9 TESTING THE SIGNIFICANCE OF THE LINEAR RELATIONSHIP

To test the significance of the linear relationship, we may use either an ANOVA approach or a t test.

8.9.1 An ANOVA Approach

Now that the total variation in Y has been partitioned into two component sums of squares, we may set up the ANOVA table as in Table 8.3. We recall from Chapters 3–7 that through orthogonal partitioning of sources of variation both the component sum of squares and degrees of freedom are additive. The former is obviously true from the identity in (8.11). The latter, however, requires some explanation. In the first footnote on p. 213 we noted that when computing $S^2_{Y|X}$ in (8.9) two degrees of freedom were lost from the total number of observations in the sample n. Similarly, when computing the variance of the Y values (only the statistic $\bar{Y}$ is involved), one degree of freedom would be lost. Hence, by subtraction, the degrees of freedom for the explained variation would be $(n-1) - (n-2) = 1$. It corresponds to the one degree of freedom needed for testing the existence of the true slope β_1.

Thus, to test the null hypothesis

$$H_0: \quad \beta_1 = 0 \qquad \text{(i.e., no simple linear regression is present)}$$

against the alternative

$$H_1: \quad \beta_1 \neq 0 \qquad \text{(i.e., there is a significant simple linear regression present)}$$

we obtain the F ratio

$$F = \frac{\text{MSR}}{\text{MSE}} \sim F_{1, n-2} \tag{8.14}$$

and, using an α level of significance, the null hypothesis is rejected if $F \geq F_{1-\alpha; 1, n-2}$.

By using (8.10)–(8.13) for the example of interest to the Director of Career Development, the three measures of variation are computed from the summary data in Table 8.2:

$$\text{SST} = \sum_{i=1}^{n} Y_i^2 - \frac{\left(\sum_{i=1}^{n} Y_i\right)^2}{n}$$

$$= 9593.41 - \frac{(534.3)^2}{30} = 77.5270$$

$$\text{SSR} = b_0 \sum_{i=1}^{n} Y_i + b_1 \sum_{i=1}^{n} X_i Y_i - \frac{\left(\sum_{i=1}^{n} Y_i\right)^2}{n}$$

$$= 6.41825(534.3) + 3.92819(1564\ 24) - \frac{(534.3)^2}{30} = 58.0199$$

$$\text{SSE} = \sum_{i=1}^{n} Y_i^2 - b_0 \sum_{i=1}^{n} Y_i - b_1 \sum_{i=1}^{n} X_i Y_i$$

$$= 9593.41 - 6.41825(534.3) - 3.92819(1564.24) = 19.5071$$

TABLE 8.3 Testing the Linear Relationship by ANOVA

Source of Variation	Degrees of Freedom	Sum of Squares	Mean Square	F
Due to Regression (i.e., Explained)	1	$\text{SSR} = \sum_{i=1}^{n} (\hat{Y}_i - \bar{Y})^2 = b_0 \sum_{i=1}^{n} Y_i + b_1 \sum_{i=1}^{n} X_i Y_i - \dfrac{\left(\sum_{i=1}^{n} Y_i\right)^2}{n}$	$\text{MSR} = \dfrac{\text{SSR}}{1}$	$F = \dfrac{\text{MSR}}{\text{MSE}}$
Error (i.e., Unexplained)	$n - 2$	$\text{SSE} = \sum_{i=1}^{n} (Y_i - \hat{Y}_i)^2 = \sum_{i=1}^{n} Y_i^2 - b_0 \sum_{i=1}^{n} Y_i - b_1 \sum_{i=1}^{n} X_i Y_i$	$\text{MSE} = \dfrac{\text{SSE}}{n-2}$ $= S_{\hat{Y}|x}^2$	
Total	$n - 1$	$\text{SST} = \sum_{i=1}^{n} (Y_i - \bar{Y})^2 = \sum_{i=1}^{n} Y_i^2 - \dfrac{\left(\sum_{i=1}^{n} Y_i\right)^2}{n}$	$\text{MST} = \dfrac{\text{SST}}{n-1}$ $= S_{\hat{Y}}^2$	

Hence the ANOVA table shown as Table 8.4 is obtained. Since, at the .05 level of significance,

$$F = \frac{\text{MSR}}{\text{MSE}} = 83.28 > F_{.95;\,1,\,28} = 4.20$$

the null hypothesis is rejected, and it may be concluded that a highly significant simple linear relationship exists (the P value is smaller than .005).

TABLE 8.4   Testing the Linear Relationship

Source of Variation	Degrees of Freedom	Sum of Squares	Mean Square	F	P Value
Due to Regression (i.e., Explained)	1	58.0199	58.0199	83.28	$P < .005$ (highly significant)
Error (i.e., Unexplained)	28	19.5071	.69668		
Total	29	77.5270			

The next step, however, is to study the appropriateness of the simple linear model using the methods to be considered in Sections 8.10 and 8.11. Perhaps a curvilinear model would fit the data significantly better. On the other hand, had the hypothesis test proved insignificant, there would be no point using the fitted regression model further. In such cases the researcher would still want to test the appropriateness of the linear model (see Section 8.10). The results of such a test would either assist the researcher in searching for an appropriate curvilinear relationship between X and Y (see Chapter 11) or suggest that it would be better to seek other predictor variables and/or perhaps even develop a multiple regression model (see Chapter 10). If none of these could be achieved, the researcher would than make inferential statements using only the observed set of responses (Y)—thereby not utilizing the insignificant relationship with X.

8.9.2 The t-Test Approach

Recall from (1.20) of Section 1.6.4 the identity

$$(t_{1-\frac{\alpha}{2};\,v})^2 \equiv F_{1-\alpha;\,1,\,v} \tag{1.20}$$

so that whenever F has but one degree of freedom in the numerator there is an equivalence between it and an appropriate t. Therefore, the t test may also be used to test

$$H_0: \quad \beta_1 = 0$$

against the alternative

$$H_1: \quad \beta_1 \neq 0$$

The test is given by

$$b_1$$

$$t = \frac{b_1 - \beta_1}{S_{b_1}} \sim t_{n-2} \tag{8.15}$$

where the standard deviation of the sample slope can be obtained from

$$S_{b_1} = \sqrt{\frac{S_{Y|X}^2}{\sum_{i=1}^{n} X_i^2 - \left[\left(\sum_{i=1}^{n} X_i\right)^2 / n\right]}} = \frac{S_{Y|X}}{\sqrt{n-1} \, S_X} \tag{8.16}$$

and where S_X is the standard deviation of the X values. By using a level of significance of α, the decision rule is to reject H_0 if $t \geq t_{1-\frac{\alpha}{2};n-2}$ or if $t \leq t_{\frac{\alpha}{2};n-2}$. For our example, $S_{Y|X}^2 = \text{MSE} = .69668$, and

$$(n-1)S_X^2 = \sum_{i=1}^{n} X_i^2 - \frac{\left(\sum_{i=1}^{n} X_i\right)^2}{n} = 256.06 - \frac{(87.0)^2}{30} = 3.76$$

so that

$$S_{b_1} = \sqrt{\frac{.69668}{3.76}} = .43045$$

and

$$t = \frac{3.92819}{.43045} = +9.126$$

By using an .05 level of significance, $t = 9.126 > t_{.975;28} = 2.048$, and H_0 is rejected.*

8.10 TESTING FOR THE APPROPRIATENESS
OF THE SIMPLE LINEAR REGRESSION MODEL

Consider the two scatter diagrams displayed in Figure 8.6. Panel A depicts a situation in which a significant simple linear relationship between X and Y is said to exist. Despite this fact, however, it is obvious that a simple curvilinear model would fit the data significantly better. On the other hand, panel B demonstrates a situation in which there is no evidence of a simple linear relationship. Here, too, however, a simple curvilinear model would provide a statistically significant relationship between X and Y. Thus, regardless of the result of the test for the significance of the simple linear relationship, it is still necessary to determine if such a model provides an appropriate fit. Perhaps a curvilinear model (see Chapter 11) would fit the data significantly better. The appropriateness of the *simple linear* model will be examined here; in Chapter 10 we shall consider the appropriateness of including other predictor variables into a multiple regression model.

*Except for rounding errors, note that the computed F, 83.28, is the square of the computed t. Similarly, from Tables B.4 and B.3 in Appendix B, $F_{.95;1,28} = 4.20 = (t_{.975;28})^2$.

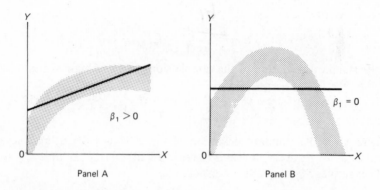

Figure 8.6 Studying the appropriateness of the simple linear regression model.

To test the appropriateness of the simple linear relationship, we shall partition the unexplained variation SSE into its two aforementioned sources: *lack of fit* and *pure error*. Recall from Section 8.8 that *lack of fit* is that portion of unexplained variation due to (1) the inappropriate choice of model (for example, a straight line instead of a curve) or (2) the omission of important predictor variables. On the other hand, *pure error* was described as that portion of unexplained variation which measures the random fluctuations or inherent scatter in the response variable. Thus

Unexplained Variation (SSE)

$$= \text{Lack of Fit (SSLF)} + \text{Pure Error (SSPE)} \qquad (8.17)$$

It is extremely important to note that pure error can only be estimated if the sample contains at least one level of X for which at least two independent measurements on Y have been obtained. If each of the n sample observations have differing X values, we cannot test the appropriateness of the regression model—we could, however, evaluate its aptness by examining the residuals (see Section 8.11).

The pure error sum of squares SSPE may be computed as follows: At *each* level of X for which at least two independent measurements on Y have been observed, first calculate the particular sample mean of the Y values $(\bar{Y}_j)$ and then compute the summation of squared differences of these observed Y values around their particular sample mean. If the various *summation of squared differences* are now added together, SSPE is obtained.*

More formally, suppose that X is fixed at X_j (where $j = 1, 2, \ldots, l$) where there exist n_j independent Y values $(Y_{j1}, Y_{j2}, \ldots, Y_{jn_j})$. First compute $\bar{Y}_j$. Then compute $\sum_{k=1}^{n_j} (Y_{jk} - \bar{Y}_j)^2$. Summing these results together over all l

---

*It is appropriate to *pool* the separate results because of the assumption of homoscedasticity in Section 8.5.4. If at each level of X a pure error estimate of the true variance in Y is obtained, the pooled or average estimate taken over all repeating levels of X is a better estimate of this true variance in Y.

levels of repeated X, we obtain

$$\text{SSPE} = \sum_{j=1}^{l} \sum_{k=1}^{n_j} (Y_{jk} - \bar{Y}_j)^2 \tag{8.18}$$

Note that at any particular fixed level of X, say X_j, the degrees of freedom are $n_j - 1$. Thus the degrees of freedom associated with pure error are $\sum_{j=1}^{l} (n_j - 1) = \sum_{j=1}^{l} n_j - l$, where $\sum_{j=1}^{l} n_j$ is the total number of measurements based on repeating X levels and l is the number of different X levels that repeat.

MSPE is computed by

$$\text{MSPE} = \frac{\text{SSPE}}{\sum_{j=1}^{l} n_j - l} \tag{8.19}$$

Because of the relationship between unexplained variation, lack of fit, and pure error, the lack of fit sum of squares can be obtained through subtraction. That is,

$$\text{SSLF} = \text{SSE} - \text{SSPE} \tag{8.20}$$

The degrees of freedom associated with the lack of fit can also be determined through subtraction:

$$\begin{pmatrix} \text{Lack of Fit} \\ \text{d.f.} \end{pmatrix} = \begin{pmatrix} \text{Error} \\ \text{d.f.} \end{pmatrix} - \begin{pmatrix} \text{Pure Error} \\ \text{d.f.} \end{pmatrix}$$

$$= (n-2) - \left(\sum_{j=1}^{l} n_j - l \right) = n - 2 - \sum_{j=1}^{l} n_j + l$$

MSLF, the mean square for lack of fit, is now obtained in the usual manner; that is,

$$\text{MSLF} = \frac{\text{SSLF}}{n - 2 - \sum_{j=1}^{l} n_j + l} \tag{8.21}$$

Table 8.5 represents the ANOVA table adjusted for the decomposition of the unexplained variation (error) term. Hence, to test the null hypothesis

$$H_0: \quad \text{the simple linear model is an appropriate fit}$$

against the alternative

$$H_1: \quad \text{the simple linear model is not an appropriate fit}$$

we have

$$F = \frac{\text{MSLF}}{\text{MSPE}} \sim F_{n-2-\sum_{j=1}^{l} n_j + l, \ \sum_{j=1}^{l} n_j - l} \tag{8.22}$$

and the null hypothesis would be rejected at the α level of significance if $F \geq F_{1-\alpha;n-2-\sum_{j=1}^{l} n_j + l, \sum_{j=1}^{l} n_j - l}$.

Note that when H_0 is true, both MSLF and MSPE are estimating the inherent variability in the Y values. Except for chance, the F ratio should equal 1. On the other hand, if H_0 is false, MSLF is estimating this inherent variability in addition to lack of fit, and the F ratio will significantly exceed 1.

TABLE 8.5 ANOVA Table Adjusted for the Decomposition of Unexplained Variation

Source of Variation	Degrees of Freedom	Sum of Squares	Mean Square	F
Due to Regression (i.e., Explained)	1	SSR	$\text{MSR} = \dfrac{\text{SSR}}{1}$	$F = \dfrac{\text{MSR}}{\text{MSE}}$
Error (i.e., Unexplained)	$n - 2$	SSE	$\text{MSE} = \dfrac{\text{SSE}}{n-2}$	
Lack of Fit	$n - 2 - \sum_{j=1}^{l} n_j + l$	SSLF	$\text{MSLF} = \dfrac{\text{SSLF}}{n - 2 - \sum_{j=1}^{l} n_j + l}$	$F = \dfrac{\text{MSLF}}{\text{MSPE}}$
Pure Error	$\sum_{j=1}^{l} n_j - l$	SSPE	$\text{MSPE} = \dfrac{\text{SSPE}}{\sum_{j=1}^{l} n_j - l}$	
Total	$n - 1$	SST	$\text{MST} = \dfrac{\text{SST}}{n-1}$	

For the problem of interest to the Director of Career Development we observe from Table 8.1 that 26 of the 30 sample observations had repeating X values (grade point indexes). Moreover, there are nine different grade point indexes that repeat. These are summarized in Table 8.6 along with the corresponding starting salaries, and SSPE is computed.

From Table 8.6, SSPE $= 14.4800$, and there are 17 degrees of freedom. Using (8.20), we obtain

$$\text{SSLF} = \text{SSE} - \text{SSPE} = 19.5071 - 14.4800 = 5.0271$$

with

$$n - 2 - \sum_{j=1}^{l} n_j + l = 30 - 2 - 26 + 9 = 11 \text{ degrees of freedom}$$

Hence, from (8.22),

$$F = \frac{\text{MSLF}}{\text{MSPE}} = \frac{5.0271/11}{14.4800/17} = \frac{.457009}{.851765} = .537$$

TABLE 8.6   Obtaining Pure Error Estimates from Repeated Observations on Grade Point Index

X_j	Y_{jk}	n_j	$\bar{Y}_j$	Contributing Sum of Squares, $\sum_{k=1}^{n_j} (Y_{jk} - \bar{Y}_j)^2$	Contributing Degrees of Freedom, $n_j - 1$
2.4	16.2, 16.4	2	16.3	.02	1
2.6	15.0, 15.7, 17.3	3	16.0	2.78	2
2.7	16.6, 17.0	2	16.8	.08	1
2.8	17.0, 17.5, 18.0	3	17.5	.50	2
2.9	17.3, 18.0, 19.0, 19.3	4	18.4	2.54	3
3.0	17.4, 18.5, 18.6, 19.5	4	18.5	2.22	3
3.1	17.2, 17.7, 19.1	3	18.0	1.94	2
3.2	18.4, 18.6, 20.0	3	19.0	1.52	2
3.3	18.1, 20.5	2	19.3	2.88	1
				SSPE $= 14.4800$	$\sum_{j=1}^{l} (n_j - 1) = 17$

SOURCE: Table 8.1.

Using an .05 level of significance, we have

$$F = .537 < F_{.95; 11, 17} \cong 2.41$$

(obtained through interpolation in Appendix B, Table B.4) so that the null hypothesis is not rejected. Therefore, there is no evidence to suspect that the simple linear regression model is an inappropriate fit to these data.

8.11 ANALYZING THE RESIDUALS

To test the appropriateness of the fitted regression model using the methods of Section 8.10, an estimate of pure error is required. However, such an estimate can only be made if the sample contains at least one level of X for which at least two independent measurements on Y have been obtained. If, as in survey research, the levels of the predictor variable are *uncontrolled* (see Section 8.13), it may well be the case that the n sample observations would each have differing X values, and thus a lack of fit test could not be made. In these situations, the primary means of evaluating the aptness of the fitted model is an analysis of the residuals.

The individual residual values e_i can be obtained as part of the computer output (see Section 10.12) or can be computed manually at each value of X using the sample regression equation developed from (8.3). Residual analysis provides an invaluable aid to the researcher in the model-building process. Not only are we able to evaluate the aptness of a fitted model, but also we can study potential violations in the assumptions.

8.11.1 Evaluating the Aptness of the Fitted Model

The aptness of the fitted regression model can best be evaluated by plotting the residuals on the vertical axis against the corresponding values of the independent variable X on the horizontal axis for all n observations. For the problem of interest to the Director of Career Development, this scatter diagram is presented as Figure 8.7. We observe that there is no apparent pattern to these residual plots. That is, the residuals seem to be spread uniformly above and below $e = 0$ for all values of X. Thus it may be concluded that the fitted model appears to be appropriate.

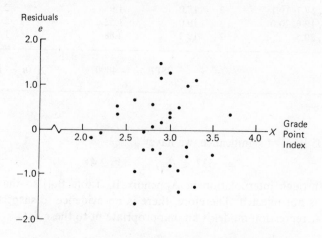

Figure 8.7 Plotting the residuals versus grade point index.

Nevertheless, it must be mentioned that had a pattern in the residual plots been observed—as often is the case—the fitted model would have been deemed inappropriate. As examples, note the scatter diagrams displayed in panels A and B of Figure 8.6. In both cases the *linear* relationship between Y and X was inadequate. However, such distinct patterns as shown in Figure 8.6 are sometimes hidden in the initial scatter diagrams. By plotting the residuals e against X, we "filter out" or remove the *linear* trend effect β_1 (which is positive in panel A and zero in panel B). It is therefore interesting to note that had we plotted the residuals e against the independent variable X for each of these cases, the two diagrams would be visually similar to panel B wherein the presence of a *second-degree polynomial* relationship (see Chapter 11) is clearly apparent.

8.11.2 Examining the Assumptions

Homoscedasticity.   The assumption of homoscedasticity (see Section 8.5.4) can also be evaluated from the plot of e with X. For the problem of interest to the Director, no apparent violation in this assumption is observed (see Figure 8.7). The residuals appear to fluctuate uniformly about zero for all values of X.

On the other hand, we can observe a *fanning effect* for the hypothetical plot of e with X depicted in Figure 8.8. The residuals seem to fan out as X increases—

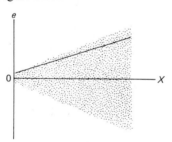

Figure 8.8   Violations in homoscedasticity.

thereby demonstrating a lack of homogeneity in the variances of the Y values at each level of X. Methods for treating this problem will be presented in Chapter 14.

Normality.   For the problem of interest to the Director, the normality assumption of Section 8.5.1 can now be evaluated from another viewpoint. The 30 residual values e_i may be tallied into a *frequency distribution* as shown in Table 8.7 with the results then plotted in a *histogram* as depicted in Figure 8.9 [see Chapter 3 of Berenson and Levine (1983)]. Since these data appear approximately bell-shaped, there would be no reason to suspect that the normality assumption has been violated.

**TABLE 8.7 Frequency Distribution
of 30 Residual Values**

Residual Values	No. of Observations
−2.0 but less than −1.2	3
−1.2 but less than −.4	8
−.4 but less than +.4	10
+.4 but less than +1.2	7
+1.2 but less than +2.0	2
Total	30

SOURCE: Table 8.1.

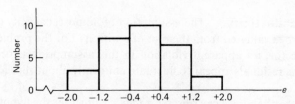

Figure 8.9   Histogram of 30 residual values.
Source: Table 8.7

Independence.   The independence assumption discussed in Section 8.5.3 can be evaluated by plotting the residuals in the order in which the observed data were obtained. Any resulting patterns which visually emerge through such a scatter plot would indicate a potential violation in the independence assumption.* Thus, for the problem of interest to the Director, the randomness and independence of our observed sequence of 30 observations can be evaluated from Figure 8.10.

A careful examination of Figure 8.10 reveals a possible pattern in these residual plots. In the first half of the sequence most of the residuals are positive, while in the second half the majority are negative. Thus the researcher might seriously be concerned that the independence assumption has been violated. Fortunately, however, this anomaly can be explained by noting from Table 8.1 that the first 15 individuals listed had majored in accountancy while the others had majored in marketing/management. Within these two groups the residual plot in Figure 8.10 does not indicate any pattern, and hence the assumption of independence no longer appears to be violated. Nevertheless, what we have

*For example, when data are collected over periods of time, there is often an *autocorrelation* effect among successive observations. That is, there exists a correlation between a particular observation and those values which precede and succeed it—thereby violating the assumption of independence. See Section 14.4 for a discussion of methods that may be used to remove such effects.

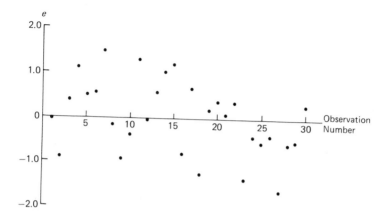

Figure 8.10 Studying the residuals in an observed sequence.

just observed is that the categorical variable "business major" may have an influence on starting salary. This interrelationship of business major and the grade point index on salary will be further investigated in Chapter 12.

8.12 ESTIMATING THE TRUE REGRESSION PARAMETERS

Now that a significant relationship between X and Y has been found and the appropriateness of the fitted simple linear regression model has been "verified," we may utilize the sample regression equation for prediction purposes. First we obtain estimates of its true parameters.

8.12.1 Estimating the True Slope β_1

A $100(1 - \alpha)\%$ confidence interval estimate of the true slope β_1 is given by

$$b_1 \pm t_{1-\frac{\alpha}{2}; n-2} S_{b_1} \tag{8.23}$$

For our problem, $b_1 = 3.92819$ and $S_{b_1} = .43045$. When using a 95% confidence coefficient, $t_{.975; 28} = 2.048$ and we have

$$3.92819 \pm 2.048(.43045)$$

$$3.92819 \pm .88156$$

$$+3.05 \leq \beta_1 \leq +4.81$$

With 95% confidence the Director of Career Development can interpret this to mean that for each additional achieved unit in the grade point index, the starting salaries offered to recent graduates majoring in business would increase by $3050 to $4810.

8.12.2 Estimating the True Intercept β_0

A $100(1 - \alpha)\%$ confidence interval estimate of the true intercept β_0 is given by

$$b_0 \pm t_{1-\frac{\alpha}{2}; n-2} S_{b_0} \tag{8.24}$$

where the standard error of the intercept is defined as

$$S_{b_0} = S_{Y|X} \sqrt{\frac{1}{n} + \frac{\bar{X}^2}{(n-1)S_X^2}} \tag{8.25}$$

where

$$(n-1)S_X^2 = \sum_{i=1}^{n} X_i^2 - \frac{\left(\sum_{i=1}^{n} X_i\right)^2}{n}$$

For our data, $b_0 = 6.41825$, $S_{Y|X} = .834675$, $\bar{X} = 2.90$, and $(n-1)S_X^2 = 3.76$. Thus, from (8.25),

$$S_{b_0} = .834675 \sqrt{\frac{1}{30} + \frac{(2.90)^2}{3.76}} = 1.25757$$

When using a 95% confidence coefficient, $t_{.975; 28} = 2.048$ and we have

$$6.41825 \pm 2.048(1.25757)$$

$$6.41825 \pm 2.57551$$

$$+3.84 \leq \beta_0 \leq +9.00$$

With 95% confidence we may conclude that the true intercept is between $3840 and $9000.

As in this example, confidence interval estimates of β_0 are often not directly interpretable in terms of practical use. Thus we are *not* concluding with 95% confidence that recent "graduates" who have achieved a grade point index of 0.0 could expect starting salaries in the range $3840–$9000. Such individuals would not have been graduated unless their index had been at least 2.0! Therefore, the purpose of obtaining such a confidence interval here is only to get an estimate of the magnitude of β_0—one of the two parameters in the simple linear regression model.

8.13 USING THE REGRESSION EQUATION FOR PREDICTION PURPOSES: ESTIMATING $\mu_{Y|X}$ AND $\hat{Y}_I$

The *primary purpose* of developing the regression equation is to be able to take advantage of the relationship between the predictor variable X and the response variable Y so that by using the former we can obtain a better estimate of mean response than if we had evaluated the sample Y values alone. However, as we have just observed in Section 8.12.2, to use the sample regression equation for prediction purposes, it is important that we consider only the *relevant* range of the predictor variable X. But this depends on whether the values of X are to be thought of as *fixed* or *random* (see Section 6.5).*

If the predictor (or independent) variable X is considered to be *random* (as in the case of a random sample taken in a survey), the relevant range encompasses all possible values from the smallest to the largest X used in developing the sample regression equation. On the other hand, if the predictor variable X is considered to be *fixed* or controlled at specific levels (as in the case of a designed experiment), the relevant range includes only those specifically selected values of X for prediction purposes. Any *interpolations* between such values of X for prediction purposes must be done with caution because the estimated responses would be valid only if the sample regression equation based on the controlled levels of X also holds for these intermediate values of X. In either situation, however, we should not *extrapolate* beyond the relevant range of X values. Hence, for the data in Table 8.1 the Director of Career Development can utilize the sample regression equation to predict expected starting salaries based on *any* grade point index from 2.1 through 3.7 [the predictor variable (GPI) is considered to be *random*]. Although in the population there may have been recent graduates whose grade point indexes either exceeded 3.7 or were at the minimum of 2.0 (for graduation purposes), such individuals were not observed in the particular sample listed in Table 8.1 These indexes then are inappropriate for the sample regression equation developed here. As discussed in Section 8.12.2, a prediction based on a grade point index of 0.0 is, of course, not only outside the relevant range but also an impossible level for a college graduate to obtain.

8.13.1 Estimating the True Mean Response $\mu_{Y|X}$ for a Given X

A $100(1 - \alpha)\%$ confidence interval estimate of the true *mean response* $\mu_{Y|X}$ at a particular value of X (say X_g) is given by

$$\hat{Y}_g \pm t_{1-\frac{\alpha}{2};n-2}\, S_{Y|X} \sqrt{\frac{1}{n} + \frac{(X_g - \bar{X})^2}{(n-1)S_X^2}} \tag{8.26}$$

where $\hat{Y}_g = b_0 + b_1 X_g$, the predicted value of Y when X is fixed at X_g. For example, suppose the Director wishes to estimate with 95% confidence the true mean starting salary for recent graduates who achieved a grade point index of 3.0. From (8.3) we compute

$$\hat{Y}_g = 6.42 + 3.93(3.0) = 18.21$$

The predicted value $\hat{Y}_g$, equal to 18.21 (thousands of dollars), is a point estimate of $\mu_{Y|X}$, the true mean starting salary for all recent graduates who have achieved

---

*In Section 8.5 we may recall that X was assumed to be fixed at various levels, and at each such level the random variable Y was assumed to be independent and normally distributed. Frequently, however, it is desirable to consider both Y and X as random variables. Under appropriate conditions, the alteration of this portion of the assumption does not affect our descriptive or inferential results [for further discussion, see Neter and Wasserman (1974)].

a 3.0 grade point index while majoring in business. To form the 95% confidence interval estimate, we have $S_{Y|X} = .835$, $n = 30$, $\bar{X} = 2.90$, $t_{.975;28} = 2.048$, and $(n - 1)S_X^2 = \sum_{i=1}^{n} X_i^2 - [(\sum_{i=1}^{n} X_i)^2/n] = 3.76$. Thus, from (8.26),

$$18.21 \pm 2.048(.835)\sqrt{\frac{1}{30} + \frac{(3.00 - 2.90)^2}{3.76}}$$

$$18.21 \pm .324$$

$$17.89 \leq \mu_{Y|X} \leq 18.53$$

Therefore, the Director estimates that for all recent graduates majoring in business who had achieved a 3.0 grade point index, the true mean starting salary is between \$17,890 and \$18,530. The Director has 95% confidence that this interval correctly estimates the true mean starting salary for such individuals.

     If the Director also wishes to obtain other 95% confidence interval estimates of the true mean starting salaries for various grade point indexes in the relevant range and plots those lower and upper limits for each of the obtained confidence intervals, the scatter diagram will demonstrate the presence of a *band* effect around the fitted regression line.

     As indicated in Figure 8.11, the width of the confidence interval varies at different levels of X. From (8.26) we note that for a given sample size n, desired amount of confidence $t_{1-\frac{\alpha}{2};n-2}$, sample mean $\bar{X}$, and total variation in the X values [that is, $(n - 1)S_X^2$], the closer the selected value of X is to $\bar{X}$, the narrower the confidence interval will be. Of course the *minimum width* occurs at values of X identical to the mean $\bar{X}$. On the other hand, for selected values of X which greatly differ from $\bar{X}$, we observe in Figure 8.11 that the interval estimate of the true mean response $\mu_{Y|X}$ becomes much wider. Therefore, we may conclude that since the interval estimate for $\mu_{Y|X}$ varies *hyperbolically* as a function of the *closeness* of the given X to $\bar{X}$, the much wider (and less useful) intervals obtained when making predictions with X values much different from $\bar{X}$ are the trade-off for predicting at such values of X.

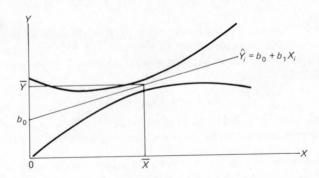

Figure 8.11   Interval estimates of $\mu_{Y|X}$ for different values of X.

8.13.2 Predicting the Individual Response $\hat{Y}_I$ for a Given X

By obtaining such interval estimates from (8.26), the Director of Career Development would be able to provide current seniors majoring in business with information pertaining to *average* starting salary based on achieved grade point index.

Much more important to an *individual* student, however, is information pertaining to the starting salary that he/she could expect based on an achieved grade point index. Thus, a $100(1 - \alpha)\%$ *prediction** interval of the *individual response* $\hat{Y}_I$ at a particular value of X, say X_g, is given by

$$\hat{Y}_g \pm t_{1-\frac{\alpha}{2};n-2}S_{Y|X}\sqrt{1 + \frac{1}{n} + \frac{(X_g - \bar{X})^2}{(n-1)S_X^2}} \tag{8.27}$$

where $\hat{Y}_g = b_0 + b_1 X_g$, the predicted value of Y when X is fixed at X_g. For example, suppose the Director wishes to inform a particular individual with an achieved 3.0 grade point index as to his/her expected starting salary. From (8.3), the point estimate is

$$\hat{Y}_g = 6.42 + 3.93(3.0) = 18.21$$

When using (8.27), the 95% prediction interval estimate would then be computed as

$$18.21 \pm 2.048(.835)\sqrt{1 + \frac{1}{30} + \frac{(3.00 - 2.90)^2}{3.76}}$$

$$18.21 \pm 1.74$$

$$16.47 \le \hat{Y}_I \le 19.95$$

Hence, with 95% confidence, the Director of Career Development would predict that for such an individual with a 3.0 grade point index the starting salary will be somewhere within the interval \$16,470 and \$19,950.

We should note that such a statement, of course, does not take into consideration any other variables that could perhaps also influence the prediction (see Chapter 10), nor does it make any adjustments for other effects such as inflation on starting salary.

Table 8.8 provides a summary of both 95% confidence interval estimates for the true mean starting salaries and 95% prediction interval estimates for individual starting salaries based on a selected set of grade point indexes.

Observing (8.26) and (8.27), we note that the only difference in these formulas is that a constant value (1) appears under the square root sign of the

---

*Although in *form* they appear to resemble one another, we distinguish between a prediction interval estimate and a confidence interval estimate. This latter uses the sampling distribution of the statistic to develop an estimate of the unknown parameter, whereas the former does not deal with estimating any parameter—merely predicting an *individual* observation drawn from some random variable.

TABLE 8.8   Summary Table of Confidence Interval Estimates for $\mu_{Y|X}$ and Prediction Interval Estimates for $\hat{Y}_I$ at Selected Values of X (95% Confidence Coefficient)

X	$\hat{Y}_i$ (in \$000)	Estimates for $\mu_{Y|X}$ (in \$000)	Estimates for $\hat{Y}_I$ (in \$000)
2.1	14.67	13.90–15.44	12.79–16.55
2.2	15.07	14.38–15.76	13.23–16.91
2.3	15.46	14.85–16.07	13.64–17.28
2.4	15.85	15.31–16.39	14.06–17.64
2.5	16.24	15.77–16.71	14.47–18.01
2.6	16.64	16.23–17.05	14.88–18.40
2.7	17.03	16.67–17.39	15.28–18.78
2.8	17.42	17.10–17.74	15.68–19.16
2.9	17.82	17.51–18.13	16.08–19.56
3.0	18.21	17.89–18.53	16.47–19.95
3.1	18.60	18.24–18.96	16.85–20.35
3.2	19.00	18.59–19.41	17.24–20.76
3.3	19.39	18.92–19.86	17.62–21.16
3.4	19.78	19.24–20.32	17.99–21.57
3.5	20.18	19.57–20.79	18.36–22.00
3.6	20.57	19.88–21.26	18.73–22.41
3.7	20.96	20.19–21.73	19.08–22.84

latter. Hence, it is obvious that for the same confidence coefficient the prediction interval must be wider than the corresponding confidence interval. Moreover, a band effect is also present in a set of prediction intervals. The prediction bands and confidence bands for our problem are displayed in Figure 8.12.

8.14 CORRELATION ANALYSIS

8.14.1 Introduction

To this point of the chapter we have extensively analyzed a simple linear relationship between X and Y using the least-squares method of regression. As such, we have formulated and developed appropriate prediction models. For completeness, however, we shall now examine this relationship from a slightly different perspective. That is, using methods of correlation, we shall focus our attention on measuring, evaluating, and understanding the strength of this association between X and Y.

The fundamental difference in the simple correlation and regression models lies in the manner in which the X and Y values are obtained. We may recall from Sections 8.5 and 8.13 that in regression analysis it is assumed that the Y values are random variables which are independent and normally distributed for fixed levels of X. However, in correlation analysis we shall assume that X

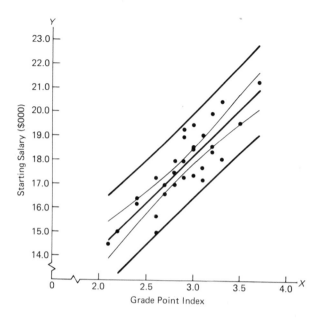

Figure 8.12 95% confidence interval estimates for $\mu_{X|Y}$ and 95% prediction interval estimates for $\hat{Y}_I$ at selected values of X. (Source: Table 8.8.)

and Y are each random variables that are *jointly normally distributed* and, in addition, that the obtained data consist of a random sample of n independent pairs of observations $(X_1, Y_1), (X_2, Y_2), \ldots, (X_n, Y_n)$ from an underlying *bivariate normal population* [see Draper and Smith (1981)].

8.14.2 Descriptive Measures

To measure the strength of the linear relationship between X and Y we return to the measures of variation considered in Section 8.8. From (8.11) we observed that SST (total variation) was partitioned into two additive components: SSR (variation explained by the relationship between X and Y) and SSE (unexplained variation).

The **coefficient of determination**, denoted by r^2, is now obtained from

$$r^2 = \frac{\text{SSR}}{\text{SST}} = \frac{\sum_{i=1}^{n}(\hat{Y}_i - \bar{Y})^2}{\sum_{i=1}^{n}(Y_i - \bar{Y})^2} = \frac{b_0 \sum_{i=1}^{n} Y_i + b_1 \sum_{i=1}^{n} X_i Y_i - \left[\left(\sum_{i=1}^{n} Y_i\right)^2 / n\right]}{\sum_{i=1}^{n} Y_i^2 - \left[\left(\sum_{i=1}^{n} Y_i\right)^2 / n\right]} \quad (8.28)$$

The coefficient of determination represents the portion of the total variation in Y that is "explained" by the variation in X. In regression terms, r^2 measures that portion of the total variability in Y that is "accounted for" by the fitted simple linear regression model.

Note that r^2 ranges from 0 to 1 inclusive. If a perfect linear relationship

233

between X and Y exists, all the variability in Y will be explained by the variation in X, and r^2 would be 1. At the other extreme, if no relationship exists, none of the variability in Y will be explained by the variation in X and r^2 would equal 0. Hence the **coefficient of correlation**, denoted by r, measures the strength of the linear relationship between X and Y. It is obtained as the square root of the coefficient of determination with the appropriate sign $(+ \text{ or } -)$ affixed, according to the sign of b_1. Thus

$$r = \pm\sqrt{r^2} \tag{8.29}$$

Note that r ranges from -1 (perfect negative correlation) to $+1$ (perfect positive correlation). Thus for any given sample size n, the closer r is to ±1, the stronger the linear relationship between X and Y. If $r = 0$, then the slope will equal zero, and X cannot be used to predict Y (see panel F of Figure 8.1 or panel B of Figure 8.6).

If the researcher is not interested in the regression equation, r can be computed directly from

$$r = \frac{\sum_{i=1}^{n} X_i Y_i - \left[\left(\sum_{i=1}^{n} X_i\right)\left(\sum_{i=1}^{n} Y_i\right)\big/ n\right]}{\sqrt{\sum_{i=1}^{n} X_i^2 - \left[\left(\sum_{i=1}^{n} X_i\right)^2\big/ n\right]}\sqrt{\sum_{i=1}^{n} Y_i^2 - \left[\left(\sum_{i=1}^{n} Y_i\right)^2\big/ n\right]}} \tag{8.30}$$

However, for the example of interest to the Director of Career Development we may use the results in Table 8.4 to compute

$$r^2 = \frac{\text{SSR}}{\text{SST}} = \frac{58.0199}{77.5270} = .7484$$

$$r = +\sqrt{.7484} = +.865$$

Thus, 74.8% of the variation in starting salaries is explained by the simple linear regression model based on the grade point index. On the other hand, 25.2% of the variability in starting salaries can be explained by other factors such as *chance* and/or other important predictor variables that had not been considered.

8.14.3 Inferential Methods

In Section 8.9 we tested for the existence of a significant linear relationship between X and Y by using either the ANOVA F test (8.14) or the t test (8.15). A third approach involves the sample correlation coefficient r.

To test the null hypothesis that there is no linear relationship

$$H_0: \quad \rho = 0 \quad \text{(the true correlation is zero)}$$

against the alternative

$$H_1: \quad \rho \neq 0 \quad \text{(the true correlation is not zero)}$$

we have

$$t = \frac{r - \rho}{S_r} \sim t_{n-2} \tag{8.31}$$

where the standard error of the correlation coefficient is given by

$$S_r = \sqrt{\frac{1 - r^2}{n - 2}} \qquad (8.32)$$

Using an α level of significance, H_0 may be rejected if $t \geq t_{1-\frac{\alpha}{2};n-2}$ or if $t \leq t_{\frac{\alpha}{2};n-2}$. For our example,

$$t = \frac{.865}{\sqrt{(1 - .7484)/28}} = +9.125$$

and since $t = +9.125 > t_{.975;28} = +2.048$, H_0 is rejected.

Except for rounding errors we should note that the two t tests (8.15) and (8.31) give identical results.

Now that a highly significant positive correlation between the grade point index and the starting salary has been found, the Director might be interested in obtaining a confidence interval estimate of the true correlation coefficient ρ. Analogous to our other interval estimates, we would expect a $100(1 - \alpha)\%$ confidence interval estimate of ρ to have the form $r \pm t_{1-\frac{\alpha}{2};n-2} S_r$. However, from the previous hypothesis test the Director believes that ρ significantly exceeds zero. Unfortunately, unless ρ equals zero, the sampling distribution of the sample correlation coefficient r is not normally distributed. To adjust for this, we utilize the Fisher Z transformation [Fisher (1959)]. The procedure for obtaining the confidence interval estimate of ρ takes the following six steps:

1. Compute r.
2. Transform r to Z_{F_r} using Table B.8 in Appendix B.
3. Instead of S_r, use $\hat{\sigma}_{Z_{F_r}}$ (the estimated standard error of the transformed correlation value), where

$$\hat{\sigma}_{Z_{F_r}} = \frac{1}{\sqrt{n - 3}} \qquad (8.33)$$

4. Instead of $t_{1-\frac{\alpha}{2};n-2}$, use $Z_{1-\frac{\alpha}{2}}$, the $100[1 - (\alpha/2)]$ percentile point of the standardized normal distribution (Appendix B, Table B.1).
5. Obtain the $100(1 - \alpha)\%$ confidence interval estimate of Z_{F_ρ} from

$$Z_{F_r} \pm Z_{1-\frac{\alpha}{2}}\hat{\sigma}_{Z_{F_r}} \qquad (8.34)$$

6. Using Table B.8 in Appendix B, reconvert the lower and upper limits of (8.34), the confidence interval estimate of Z_{F_ρ}, back to units of r and then obtain the $100(1 - \alpha)\%$ confidence interval estimate of ρ.

For the example of interest to the Director of Career Development, the 95% confidence interval estimate of the true correlation coefficient ρ is computed as follows:

1. Using (8.29), $r = +.865$.

2. Rounding r to two decimal places, we use Table B.8 in Appendix B to transform $r = +.87$ to $Z_{F_r} = +1.333$ as follows:

r	Z_{F_r}	r	Z_{F_r}	r	Z_{F_r}	r	Z_{F_r}	r	Z_{F_r}
.00	.000	.20	.203	.40	.424	.60	.693	.80	1.099
.	.	.	.	.	.	.	.	.	.
.	.	.	.	.	.	.	.	.	.
.	.	.	.	.	.	.	.	.	.
.07	.070	.27	.277	.47	.510	.67	.811	.87 $\rightarrow$	1.333
.	.	.	.	.	.	.	.	.	.
.	.	.	.	.	.	.	.	.	.
.	.	.	.	.	.	.	.	.	.
.19	.192	.39	.412	.59	.678	.79	1.071	.99	2.647

3. Using (8.33), $\hat{\sigma}_{Z_{F_r}} = 1/\sqrt{27} = .19245$.
4. For a 95% confidence coefficient, $Z_{.975} = \pm 1.96$ (see Appendix B, Table B.1).
5. From (8.34), the 95% confidence interval estimate of Z_{F_ρ} is

$$Z_{F_r} \pm Z_{.975}\hat{\sigma}_{Z_{F_r}}$$
$$+ 1.333 \pm 1.96(.19245)$$
$$+ 1.333 \pm .377$$
$$+ .956 \leq Z_{F_\rho} \leq +1.710$$

6. Using the closest Z_{F_r} values in Table B.8 of Appendix B for the lower and upper limits of the confidence interval, we reconvert back to the scale of r as follows:

r	Z_{F_r}	r	Z_{F_r}	r	Z_{F_r}	r	Z_{F_r}	r	Z_{F_r}
.00	.000	.20	.203	.40	.424	.60	.693	.80	1.099
.	.	.	.	.	.	.	.	.	.
.	.	.	.	.	.	.	.	.	.
.13	.131	.33	.343	.53	.590	.73	.929	.93	1.658
.14	.141	.34	.354	.54	.604	.74 $\leftarrow$	.950	.94 $\leftarrow$	1.738
.15	.151	.35	.365	.55	.618	.75	.973	.95	1.832
.	.	.	.	.	.	.	.	.	.
.	.	.	.	.	.	.	.	.	.
.19	.192	.39	.412	.59	.678	.79	1.071	.99	2.647

Thus, the 95% confidence interval estimate of the true correlation coefficient ρ is given by

$$+.74 \leq \rho \leq +.94$$

In the population there is evidence of a strong positive correlation between the grade point index and the starting salary.

8.15 COMPUTER PACKAGES AND SIMPLE LINEAR REGRESSION: USE OF SPSS, SAS, AND BMDP

8.15.1 Introduction

In this section we shall explain how to use either SPSS, SAS, or BMDP to analyze the data of interest to the Director of Career Development (see Table 8.1). The variables involved can be organized for data entry purposes with the identification number in columns 1 and 2, the grade point index in columns 4–6, the starting salary in columns 8–11, and the business major in column 13.

8.15.2 Using the SPSS Subprogram SCATTERGRAM for Simple Linear Regression

The SCATTERGRAM subprogram of SPSS can be used to obtain a scatter diagram for two variables and to perform a simple regression and correlation analysis. The basic setup for the SCATTERGRAM procedure statement is

1	16		
SCATTERGRAM	$\begin{Bmatrix} \text{dependent} \\ \text{variable } Y \end{Bmatrix}$	♭WITH♭	$\begin{Bmatrix} \text{independent} \\ \text{variable } X \end{Bmatrix}$ /
OPTIONS	6		
STATISTICS	ALL		

The SCATTERGRAM procedure statement identifies the dependent and independent variables to be analyzed. The use of ALL on the STATISTICS statement provides for the computation of r, r^2, the standard error of the estimate $(S_{Y|X})$, b_0, and b_1, while OPTIONS 6 allows for a two-tailed t test for the significance of the linear relationship.

If the Director of Career Development wanted to provide for separate regression analyses for various subgroups (such as different business majors—Table 8.1), the ∗SELECT IF control statement can be utilized:

1	16
∗SELECT♭IF	(expression)

This control statement permits *only* those observations that satisfy the ∗SELECT IF conditions to be processed. Moreover, this control statement will be applied only to the procedure that directly follows the ∗SELECT IF statement. Thus if we wish to analyze only accountancy majors, the following four statements would be needed:

```
1              16
*SELECTƀIF     (MAJORƀEQƀ1)
SCATTERGRAM    SALARYƀWITHƀGPI/
OPTIONS        6
STATISTICS     ALL
```

Figure 8.13 illustrates the complete SPSS program that was written for analyzing the starting salary data. Figure 8.14 represents annotated partial output.

8.15.3 Using the SAS PLOT and GLM Procedures for the Simple Regression Model

We may recall that in Section 3.6 we used PROC GLM for the ANOVA model. In this chapter we shall use the GLM procedure for simple linear regression and the PLOT procedure to obtain a scatter diagram of the two variables. The basic setup of these procedures is

PROCƀPLOT;

$$\text{PLOTƀ}\begin{Bmatrix}\text{dependent}\\ \text{variable } Y\end{Bmatrix} * \begin{Bmatrix}\text{independent}\\ \text{variable } X\end{Bmatrix};$$

PROCƀGLM;

$$\text{MODELƀ}\begin{Bmatrix}\text{dependent}\\ \text{variable } Y\end{Bmatrix} = \begin{Bmatrix}\text{independent}\\ \text{variable } X\end{Bmatrix}/\text{Pƀ}\begin{Bmatrix}\text{CLM}\\ \text{or}\\ \text{CLI}\end{Bmatrix};$$

Note that we have invoked options P and either CLM or CLI of PROC GLM (see Section 10.12 for additional options) for our regression analysis. Option P requests PROC GLM to print the observed, predicted, and residual values for each observation; option CLM allows PROC GLM to provide 95% confidence limits for the true mean response; while option CLI enables PROC GLM to give 95% prediction intervals.

If the Director of Career Development wanted to provide for separate regression analyses for various subgroups (such as different business majors—Table 8.1), we need to use

$$\text{DATAƀ}\begin{Bmatrix}\text{name of}\\ \text{new data set}\end{Bmatrix};$$

$$\text{SETƀ}\begin{Bmatrix}\text{name of}\\ \text{original data set}\end{Bmatrix};$$

IFƀ{expression};

The DATA statement provides a name for the *new* data set to be created, while the SET statement gives the name of the *original* data set to be accessed. The IF statement indicates the basis on which the new data set is to be formed.

```
{SYSTEM CARDS}
RUN NAME          REGRESSION OF STARTING SALARY
DATA LIST         FIXED(1)/1 ID 1-2, GPI 4-6,
                  SALARY 8-11, MAJOR 13
VAR LABELS        ID, STUDENT IDENTIFICATION NO/
                  GPI, GRADE POINT INDEX/
                  SALARY, STARTING SALARY IN $000/
                  MAJOR, UNDERGRADUATE MAJOR/
VALUE LABELS      MAJOR(1)ACCOUNTING(2)MARKETING/
READ INPUT DATA
01  2.7  17.0  1
 :    :    :    :
30  3.0  18.5  2
END INPUT DATA
SCATTERGRAM       SALARY WITH GPI/
OPTIONS           6
STATISTICS        ALL
*SELECT IF        (MAJOR EQ 1)
SCATTERGRAM       SALARY WITH GPI/
OPTIONS           6
STATISTICS        ALL
*SELECT IF        (MAJOR EQ 2)
SCATTERGRAM       SALARY WITH GPI/
OPTIONS           6
STATISTICS        ALL
FINISH
{SYSTEM CARDS}
```

Figure 8.13 SPSS program for starting salary example.

REGRESSION OF STARTING SALARY (ALL STUDENTS)

STATISTICS..

CORRELATION (R)-	0.86509	R SQUARED	–	0.74838	SIGNIFICANCE	–	0.00000
STD ERR OF EST - $S_{Y/x}$	0.83469	INTERCEPT (A) - b_0	6.41826	SLOPE (B) - b_1	3.92819
PLOTTED VALUES -	30	EXCLUDED VALUES-	0	MISSING VALUES -	0

STATISTICS.. (ACCOUNTING MAJORS)

CORRELATION (R)-	0.91650	R SQUARED	–	0.83997	SIGNIFICANCE	–	0.00000
STD ERR OF EST - $S_{Y/x}$	0.76248	INTERCEPT (A) - b_0	4.95719	SLOPE (B) - b_1	4.57746
PLOTTED VALUES -	15	EXCLUDED VALUES-	0	MISSING VALUES -	0

STATISTICS.. (MARKETING MAJORS)

CORRELATION (R)-	0.88036	R SQUARED	–	0.77504	SIGNIFICANCE	–	0.00001
STD ERR OF EST - $S_{Y/x}$	0.71918	INTERCEPT (A) - b_0	6.84790	SLOPE (B) - b_1	3.65576
PLOTTED VALUES -	15	EXCLUDED VALUES-	0	MISSING VALUES -	0

Figure 8.14 Partial SPSS output for starting salary data.

For example, if we wish to analyze only accountancy majors, the following five statements are needed:

> DATAbACCOUNT;
> SETbSALARYD;
> IFbMAJORbEQ 1;
> PROCbGLM;
> MODELbSALARY = GPI/PbCLM;

SAS accesses the *last* data set created when evaluating any PROC step. Thus if we wish to examine a data set that is *not* the last one created,

$$\text{DATA} = \begin{Bmatrix} \text{name of} \\ \text{data set} \end{Bmatrix}$$

must appear as part of the PROC step following the name of the procedure (i.e., PROC GLM DATA = {name of data set};).

Figure 8.15 illustrates the complete SAS program that has been written for

```
{SYSTEM CARDS}
DATA SALARYD;
    INPUT ID 1-2 GPI 4-6 SALARY 8-11 MAJOR 13;
LABEL ID=STUDENT IDENTIFICATION NO
        GPI=GRADE POINT INDEX
        SALARY=STARTING SALARY IN $000
        MAJOR=UNDERGRADUATE MAJOR;
CARDS;
01 2.7 17.0 1
   :     :     :
30 3.0 18.5 2
PROC FORMAT;
    VALUE MAJORF 1=ACCOUNTING
                 2=MARKETING;
PROC PLOT;
    PLOT SALARY*GPI;
PROC GLM;
    MODEL SALARY=GPI/P CLM;
DATA ACCOUNT;
    SET SALARYD;
    IF MAJOR EQ 1;
PROC GLM;
    MODEL SALARY=GPI/P CLM;
DATA MARKET;
    SET SALARYD;
    IF MAJOR EQ 2;
PROC GLM;
    MODEL SALARY=GPI/P CLM;
{SYSTEM CARDS}
```

Figure 8.15   SAS program for starting salary example.

analyzing the starting salary data. Figure 8.16 represents annotated partial output.

GENERAL LINEAR MODELS PROCEDURE

DEPENDENT VARIABLE SALARY   STARTING SALARY IN $000   (ALL STUDENTS)

SOURCE	DF	SUM OF SQUARES	MEAN SQUARE	F VALUE	PR > F	R-SQUARE
MODEL	1	58.01938830	58.01938830	83.28	0.0001	0.748377
ERROR	28	19.50761170	0.69670042		STD DEV	
CORRECTED TOTAL	29	77.52700000		$S_{y|x}$ 0.83468582		

SOURCE	DF	TYPE I SS	F VALUE	PR > F
GPI	1	58.01938830	83.28	0.0001

PARAMETER	ESTIMATE	T FOR H0 PARAMETER=0	PR > T	STD ERROR OF ESTIMATE
INTERCEPT	b_0 6.41824468	5.10	0.0001	S_{b_0} 1.25759081
GPI	b_1 3.92819149	9.13	0.0001	S_{b_1} 0.43045634

DEPENDENT VARIABLE SALARY   STARTING SALARY IN $000   (ACCOUNTING MAJORS)

SOURCE	DF	SUM OF SQUARES	MEAN SQUARE	F VALUE	PR > F	R-SQUARE
MODEL	1	39.67136150	39.67136150	68.24	0.0001	0.839973
ERROR	13	7.55797183	0.58138245		STD DEV	
CORRECTED TOTAL	14	47.22933333		$S_{y|x}$ 0.76248439		

SOURCE	DF	TYPE I SS	F VALUE	PR > F
GPI	1	39.67136150	68.24	0.0001

PARAMETER	ESTIMATE	T FOR H0 PARAMETER=0	PR > T	STD ERROR OF ESTIMATE
INTERCEPT	b_0 4.95718310	3.13	0.0079	S_{b_0} 1.58235080
GPI	b_1 4.57746479	8.26	0.0001	S_{b_1} 0.55413734

DEPENDENT VARIABLE SALARY   STARTING SALARY IN $000   (MARKETING MAJORS)

SOURCE	DF	SUM OF SQUARES	MEAN SQUARE	F VALUE	PR > F	R-SQUARE
MODEL	1	23.16539103	23.16539103	44.79	0.0001	0.775039
ERROR	13	6.72394231	0.51722633		STD DEV	
CORRECTED TOTAL	14	29.88933333		$S_{y|x}$ 0.71918449		

SOURCE	DF	TYPE I SS	F VALUE	PR > F
GPI	1	23.16539103	44.79	0.0001

PARAMETER	ESTIMATE	T FOR H0 PARAMETER=0	PR > T	STD ERROR OF ESTIMATE
INTERCEPT	b_0 6.84788462	4.20	0.00.⁻	S_{b_0} 1.63117492
GPI	b_1 3.65576923	6.69	0.0001	S_{b_1} 0.54625983

Figure 8.16   Partial SAS output for starting salary data.

8.15.4 Using BMDP Program 1R for Simple Linear Regression

The BMDP program 1R can be utilized for both regression and multiple regression analysis (see Section 10.12.3). The only paragraphs that refer specifically to BMDP program 1R are the /REGRESS and /PLOT paragraphs. The basic setup of these paragraphs is

$$\text{/REGRESS} \qquad \text{DEPENDENT\flat IS\flat} \begin{Bmatrix} \text{name of} \\ \text{dependent} \\ \text{variable} \end{Bmatrix}.$$

$$\text{INDEPENDENT\flat IS\flat} \begin{Bmatrix} \text{name of} \\ \text{independent variable(s)} \end{Bmatrix}.$$

$$\text{/PLOT} \qquad \text{VARIABLES} = \begin{Bmatrix} \text{name of} \\ \text{independent variable(s)} \end{Bmatrix}.$$

The DEPENDENT sentence of the /REGRESS paragraph names the dependent variable, while the INDEPENDENT sentence names the independent variable(s). Although several options are possible for the /PLOT paragraph, the VARIABLES = sentence provides plots of the observed and predicted values of Y (as well as the residuals) against each independent variable listed.

If the Director of Career Development wanted to provide for separate regression analyses for various subgroups (such as different business majors—Table 8.1), a GROUPING sentence must be presented as part of the /VARIABLE paragraph, and a /GROUP paragraph must be included.

Figure 8.17 illustrates the complete program using BMDP-1R that has been written for analyzing the starting salary data. Figure 8.18 represents annotated partial output.

```
{SYSTEM CARDS}
/PROBLEM    TITLE IS 'REGRESSION OF SALARY'.
/INPUT VARIABLES ARE 4.
      FORMAT IS '(F2.0,1X,F3.1,1X,F4.1,1X,F1.0)'.
/VARIABLE NAMES ARE ID, GPI, SALARY, MAJOR.
          LABEL IS ID.
          GROUPING IS MAJOR.
/GROUP    CODES(4) ARE 1,2.
          NAMES(4) ARE ACCOUNTING, MARKETING.
/REGRESS  DEPENDENT IS SALARY.
          INDEPENDENT IS GPI.
/END
01 2.7 17.0 1
 :    :    :    :
30 3.0 18.5 2
{SYSTEM CARDS}
```

Figure 8.17 Using BMDP program 1R for starting salary example.

8.16 SIMPLE LINEAR REGRESSION AND CORRELATION: A SUMMARY AND OVERVIEW

In this chapter we have examined the simple linear regression and correlation models both from a descriptive and inferential viewpoint. In Chapter 9 we shall review matrix algebra and then apply the matrix approach to the simple linear

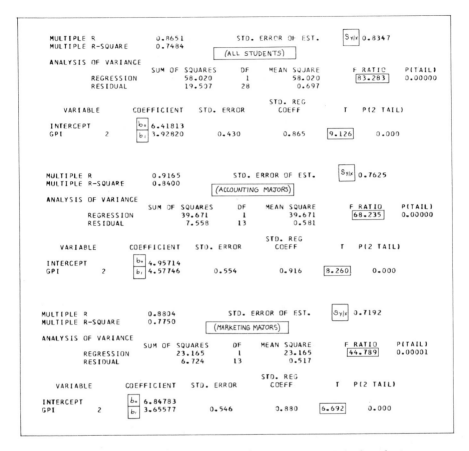

| MULTIPLE R | 0.8651 | STD. ERROR OF EST. | $S_{y\mid x}$ 0.8347 | | | |

MULTIPLE R 0.8651   STD. ERROR OF EST. $S_{y\mid x}$ 0.8347
MULTIPLE R-SQUARE 0.7484

(ALL STUDENTS)

ANALYSIS OF VARIANCE

	SUM OF SQUARES	DF	MEAN SQUARE	F RATIO	P(TAIL)
REGRESSION	58.020	1	58.020	83.283	0.00000
RESIDUAL	19.507	28	0.697		

VARIABLE		COEFFICIENT	STD. ERROR	STD. REG COEFF	T	P(2 TAIL)
INTERCEPT		b_0 6.41813				
GPI	2	b_1 3.92820	0.430	0.865	9.126	0.000

MULTIPLE R 0.9165   STD. ERROR OF EST. $S_{y\mid x}$ 0.7625
MULTIPLE R-SQUARE 0.8400

(ACCOUNTING MAJORS)

ANALYSIS OF VARIANCE

	SUM OF SQUARES	DF	MEAN SQUARE	F RATIO	P(TAIL)
REGRESSION	39.671	1	39.671	68.235	0.00000
RESIDUAL	7.558	13	0.581		

VARIABLE		COEFFICIENT	STD. ERROR	STD. REG COEFF	T	P(2 TAIL)
INTERCEPT		b_0 4.95714				
GPI	2	b_1 4.57746	0.554	0.916	8.260	0.000

MULTIPLE R 0.8804   STD. ERROR OF EST. $S_{y\mid x}$ 0.7192
MULTIPLE R-SQUARE 0.7750

(MARKETING MAJORS)

ANALYSIS OF VARIANCE

	SUM OF SQUARES	DF	MEAN SQUARE	F RATIO	P(TAIL)
REGRESSION	23.165	1	23.165	44.789	0.00001
RESIDUAL	6.724	13	0.517		

VARIABLE		COEFFICIENT	STD. ERROR	STD. REG COEFF	T	P(2 TAIL)
INTERCEPT		b_0 6.84783				
GPI	2	b_1 3.65577	0.546	0.880	6.692	0.000

Figure 8.18 Partial BMDP program 1R output for starting salary data.

regression problem of interest to the Director of Career Development. In Chapters 10–14 we extend our current discussions to include other inferential procedures and, even more importantly, to develop and interpret multiple regression models which include several predictor variables.

PROBLEMS

*8.1. The Director of Management Information Systems at a conglomerate must prepare his long-range forecasts for the company's 3-year budget. In particular, he must develop staffing ratios to predict the number of managers and project leaders based on the number of programmers. The results of a sample of the electronic data processing staffs of 10 companies within the industry are displayed in Table P8.1. *Hint:* First determine which is the independent variable and which is the dependent variable.

No. of Applications Programmers	No. of Managers and Project Leaders
15	6
7	2
20	10
12	4
16	7
20	8
10	4
9	6
18	7
15	9

(a) Use the least-squares method to find the regression coefficients b_0 and b_1. State the linear regression equation.

(b) Interpret the meaning of b_0 and b_1 in this problem.

(c) Compute the standard error of estimate.

(d) Compute the coefficient of determination, and interpret its meaning in this problem.

(e) Compute the coefficient of correlation.

(f) At the .05 level of significance, is there a linear relationship between the number of managers and the number of applications programmers?

(g) At the .05 level of significance, test for the appropriateness of the simple linear regression model.

(h) Set up a 95% confidence interval estimate of the true population slope.

(i) Set up a 95% confidence interval estimate of the true population intercept.

(j) Set up a 95% confidence interval estimate of the average number of managers at companies where there are 10 programmers.

(k) Set up a 95% prediction interval estimate of the number of managers for a particular company in which there are 10 programmers.

(l) Set up a 95% confidence interval estimate of the true population coefficient of correlation.

8.2. A random sample of 30 utility companies was selected in order to predict the net interest charges (in millions of dollars) from the net utility plant (the asset value of the physical plant in billions of dollars). The data are displayed in Table P8.2.

(a) Use the least-squares method to find the regression coefficients b_0 and b_1. State the linear regression equation.

(b) Interpret the meaning of b_0 and b_1 in this problem.

(c) Compute the standard error of estimate.

(d) Compute the coefficient of determination, and interpret its meaning in this problem.

(e) Compute the coefficient of correlation.

(f) At the .05 level of significance, is there a linear relationship between the net interest charges and the net utility plant?

(g) Evaluate the appropriateness of the simple linear regression model.

(h) Set up a 95% confidence interval estimate of the true population slope.

(i) Set up a 95% confidence interval estimate of the true population intercept.

(j) Set up a 95% confidence interval estimate of the mean net interest charges for all utilities with a net utility plant of 2.0 (billions of dollars).

(k) Set up a 95% prediction interval estimate of the net interest charge for a particular utility with a net utility plant of 2.0 (billions of dollars).

(l) Set up a 95% confidence interval estimate of the true population coefficient of correlation.

TABLE P8.2   Utility Study

Utility No.	Net Interest Charges (in millions of dollars)	Net Utility Plant (in billions of dollars)
1	21.793	.84925
2	49.151	1.38467
3	50.423	1.34101
4	49.859	1.56751
5	7.345	.25510
6	1.670	.04528
7	30.730	.92933
8	42.582	1.81214
9	11.087	.48214
10	47.111	1.31671
11	13.175	.43755
12	30.498	.90415
13	.741	.01749
14	109.720	4.07132
15	2.514	.00074
16	2.478	.04347
17	14.202	.38593
18	16.434	.63070
19	122.738	3.77211
20	78.032	2.42566
21	71.979	2.90627
22	26.095	.53251
23	71.855	2.96440
24	9.021	.04353
25	19.180	.73339
26	51.135	1.55539
27	2.500	.04690
28	166.755	4.63015
29	1.857	.03021
30	.173	.01208

SOURCE: *Statistics of Privately Owned Utilities in the United States*, Energy Information Administration, Washington, D.C., 1978.

8.3. The data in Table P8.3 represent the number of real estate brokers (in thousands) and the estimated number of households with $10,000 or higher after-tax income (as of February 1980) for selected Standard Metropolitan Areas (SMSAs) in

Texas. *Hint:* First determine which is the independent variable and which is the dependent variable.

TABLE P8.3 Real Estate Study

SMSA	Number of Households with $10,000+ (000)	Number of Brokers (000)
Abilene	30.9	.440
Amarillo	44.9	.681
Austin	113.5	2.638
Beaumont	96.5	.759
Brownsville	27.1	.408
Bryan/College Station	16.7	.246
Corpus Christi	64.6	1.014
El Paso	89.6	1.103
Galveston	50.3	.565
Killeen/Temple	30.5	.393
Lubbock	50.1	.700
McAllen	31.0	.417
Midland	20.8	.320
Odessa	29.3	.274
San Angelo	18.1	.298
San Antonio	223.5	2.876
Texarkana	15.9	.166
Waco	36.0	.407
Wichita Falls	32.6	.368

SOURCE: Texas Real Estate Research Center, Texas A&M University, College Station, Texas, 1980.

(a) Use the least-squares method to find the regression coefficients b_0 and b_1. State the linear regression equation.

(b) Interpret the meaning of b_0 and b_1 in this problem.

(c) Compute the standard error of estimate.

(d) Compute the coefficient of determination, and interpret its meaning in this problem.

(e) Compute the coefficient of correlation.

(f) At the .05 level of significance, is there a linear relationship between the number of brokers and the number of households with income above $10,000?

(g) Evaluate the appropriateness of the simple linear regression model.

(h) Set up a 95% confidence interval estimate of the true population slope.

(i) Set up a 95% confidence interval estimate of the true population intercept.

(j) Set up a 95% confidence interval estimate of the average number of brokers for all SMSAs which have 60,000 households with an after-tax income above $10,000.

(k) Set up a 95% prediction interval for a particular SMSA which has 60,000 households with an after-tax income above $10,000.

(l) Set up a 95% confidence interval estimate of the true population coefficient of correlation.

8.4. Referring to Table 2.1, for the 31 *technical* employees, we wish to predict the weekly salary based on the months of employment.

  (a) Use the least-squares method to find the regression coefficients b_0 and b_1. State the linear regression equation.

  (b) Interpret the meaning of b_0 and b_1 in this problem.

  (c) Compute the standard error of estimate.

  (d) Compute the coefficient of determination, and interpret its meaning in this problem.

  (e) Compute the coefficient of correlation.

  (f) At the .05 level of significance, is there a linear relationship between salary and length of employment?

  (g) Evaluate the appropriateness of the simple linear regression model.

  (h) Set up a 95% confidence interval estimate of the true population slope.

  (i) Set up a 95% confidence interval estimate of the true population intercept.

  (j) Set up a 95% confidence interval estimate of the average weekly salary for all technical employees with 5 years (60 months) of experience.

  (k) Set up a 95% prediction interval estimate of weekly salary for a particular technical employee with 5 years (60 months) of experience.

  (l) Set up a 95% confidence interval estimate of the true population coefficient of correlation.

8.5. Referring to Problem 3.5, an industrial psychologist wished to study the effects of alcoholic consumption on typing ability. The data are displayed in Table P3.5.

  (a) Treating this problem from a regression viewpoint, use the least-squares method to find the regression coefficients b_0 and b_1. State the linear regression equation.

  (b) Interpret the meaning of b_0 and b_1 in this problem.

  (c) Compute the standard error of estimate.

  (d) Compute the coefficient of determination, and interpret its meaning in this problem.

  (e) Compute the coefficient of correlation.

  (f) At the .05 level of significance, is there a linear relationship between the number of typing errors and alcoholic consumption?

  (g) At the .05 level of significance, test for the appropriateness of the simple linear regression model.

  (h) Set up a 95% confidence interval estimate of the true population slope.

  (i) Set up a 95% confidence interval estimate of the true population intercept.

  (j) Set up a 95% confidence interval estimate of the average number of typing errors made by all secretaries consuming 2 ounces of alcohol.

  (k) Set up a 95% prediction interval estimate of the number of typing errors made by a particular secretary who consumed 2 ounces of alcohol.

  (l) Set up a 95% confidence interval estimate of the true population coefficient of correlation.

8.6. Referring to Problem 3.8, a behavioral researcher wished to examine the effects of noise on driving performance. The data are displayed in Table P3.8.

  (a) Treating this problem from a regression viewpoint, use the least-squares method to find the regression coefficients b_0 and b_1. State the linear regression equation.

(b) Interpret the meaning of b_0 and b_1 in this problem.

(c) Compute the standard error of estimate.

(d) Compute the coefficient of determination, and interpret its meaning in this problem.

(e) Compute the coefficient of correlation.

(f) At the .05 level of significance, is there a linear relationship between driving performance and noise level?

(g) At the .05 level of significance, test for the appropriateness of the simple linear regression model.

(h) Set up a 95% confidence interval estimate of the true population slope.

(i) Set up a 95% confidence interval estimate of the true population intercept.

(j) Set up a 95% confidence interval estimate of average driving performance for all subjects exposed to a constant noise level of 100 decibels.

(k) Set up a 95% prediction interval estimate of the driving performance for a particular subject exposed to a constant noise level of 100 decibels.

(l) Set up a 95% confidence interval estimate of the true population coefficient of correlation.

8.7. Referring to Problem 3.9, a pharmaceutical statistician wished to investigate the effects of drug dosage levels on diastolic blood pressure. The data are displayed in Table P3.9.

(a) Treating this problem from a regression viewpoint, use the least-squares method to find the regression coefficients b_0 and b_1. State the linear regression equation.

(b) Interpret the meaning of b_0 and b_1 in this problem.

(c) Compute the standard error of estimate.

(d) Compute the coefficient of determination, and interpret its meaning in this problem.

(e) Compute the coefficient of correlation.

(f) At the .05 level of significance, is there a linear relationship between the diastolic blood pressure and drug dosage level?

(g) At the .05 level of significance, test for the appropriateness of the simple linear regression model.

(h) Set up a 95% confidence interval estimate of the true population slope.

(i) Set up a 95% confidence interval estimate of the true population intercept.

(j) Set up a 95% confidence interval estimate of the average diastolic blood pressure for all such rats under a dosage of 10 milligrams.

(k) Set up a 95% prediction interval estimate of the diastolic blood pressure for a particular rat under a dosage of 10 milligrams.

(l) Set up a 95% confidence interval estimate of the true population coefficient of correlation.

8.8. Using the real estate data base (see Appendix A), select a random sample of 30 houses. Develop a simple linear regression model to predict the selling price based on the lot size of a house. ∎

8.9. Using the real estate data base (see Appendix A), select a random sample of 30 houses. Develop a simple linear regression model to predict the assessed value based on the heating area of the dwelling. ∎

8.10. Using the real estate data base (see Appendix A), select a random sample of 30 houses. Develop a simple linear regression model to predict the selling price based on the assessed value.   ∎

8.11. Reviewing concepts:

   (a) Describe the simple linear regression model.

   (b) Discuss the assumptions of the simple linear regression model.

   (c) What is meant by measuring the aptness of a simple linear regression model?

   (d) Describe the partitioning of total variation.

9

A MATRIX APPROACH
TO SIMPLE LINEAR REGRESSION

9.1 INTRODUCTION

Researchers have found that a matrix algebra approach to regression and other multivariable methods is a useful means of compactly presenting lengthy and sometimes complex formulas. Hence the statistical literature (i.e., journals and advanced texts) often utilize matrix algebra terminology for pedagogical simplicity. As we shall see in Sections 9.8, 9.9, and 10.11, a matrix algebra formulation enables the researcher to more easily understand the interrelationship between simple and multiple regression analysis. In addition, a matrix approach also permits a better understanding of the interrelationship between regression analysis and the experimental design methods discussed in Chapters 3, 5, 6, and 7.

The following six sections are intended as a review of the fundamentals of matrix algebra. However, only those aspects of matrix algebra necessary for both simple and multiple regression analysis are presented here.* As such, depending on the background of the reader, these sections may be skimmed, used as a reference, or omitted. A matrix approach to regression analysis is presented in Section 9.8.

9.2 DESCRIBING A MATRIX

A matrix is a set of elements (i.e., numbers, variables, operators, etc.) arranged in a rectangular pattern into r rows and c columns. As is demonstrated a matrix is denoted by a boldface capital letter but possesses elements denoted by lower-

*Readers interested in a more complete discussion of the subject of matrix algebra should refer to such texts as Hadley (1961) or Noble and Daniel (1977).

case letters enclosed in brackets:

$$\underset{r \times c}{\mathbf{A}} = \begin{bmatrix} a_{11} & a_{12} & \cdots & a_{1j} & \cdots & a_{1c} \\ a_{21} & a_{22} & \cdots & a_{2j} & \cdots & a_{2c} \\ \cdot & \cdot & & \cdot & & \cdot \\ \cdot & \cdot & & \cdot & & \cdot \\ a_{i1} & a_{i2} & \cdots & a_{ij} & \cdots & a_{ic} \\ \cdot & \cdot & & \cdot & & \cdot \\ \cdot & \cdot & & \cdot & & \cdot \\ \cdot & \cdot & & \cdot & & \cdot \\ a_{r1} & a_{r2} & \cdots & a_{rj} & \cdots & a_{rc} \end{bmatrix}$$

The dimension of the matrix (denoted as "r by c") is determined by the number of rows and columns. The number of rows is always stated first. Thus element a_{ij} represents the observation in the ith row and the jth column (where $i = 1, 2, \ldots, r$ and $j = 1, 2, \ldots, c$) of this matrix.* If $r = c$, such a matrix having the same number of rows and columns is called a *square* matrix. Moreover, a square matrix possessing n rows and n columns is often referred to as an nth-order matrix.

9.3 MATRIX OPERATIONS

9.3.1 Addition

Matrix *addition* can be accomplished only if the matrices to be added have the same dimensions for rows and columns. Thus

$$\underset{r \times c}{\mathbf{C}} = \underset{r \times c}{\mathbf{A}} + \underset{r \times c}{\mathbf{B}} \tag{9.1}$$

where $c_{ij} = a_{ij} + b_{ij}$. As an example, suppose

$$\underset{2 \times 3}{\mathbf{A}} = \begin{bmatrix} 1 & 4 & 6 \\ 2 & 3 & 1 \end{bmatrix} \quad \text{and} \quad \underset{2 \times 3}{\mathbf{B}} = \begin{bmatrix} 0 & 1 & 10 \\ 5 & 2 & 6 \end{bmatrix}$$

Thus

$$\underset{2 \times 3}{\mathbf{C}} = \begin{bmatrix} c_{11} = a_{11} + b_{11} & c_{12} = a_{12} + b_{12} & c_{13} = a_{13} + b_{13} \\ c_{21} = a_{21} + b_{21} & c_{22} = a_{22} + b_{22} & c_{23} = a_{23} + b_{23} \end{bmatrix} = \begin{bmatrix} 1 & 5 & 16 \\ 7 & 5 & 7 \end{bmatrix}$$

9.3.2 Multiplication by a Scalar

The matrix $\mathbf{H}_{r \times c} = \lambda \mathbf{A}_{r \times c}$ is formed as the product of a constant λ with a matrix $\mathbf{A}_{r \times c}$ such that $h_{ij} = \lambda a_{ij}$ for all i and j. In matrix algebra terminology λ is called a *scalar*. Suppose, for example, $\lambda = 4$. Then for the particular matrix $\mathbf{A}_{2 \times 3}$ given in Section 9.3.1,

*Sometimes it is more convenient to merely describe the matrix by the boldface capital letter and not indicate its dimensionality. Thus $\mathbf{A}_{r \times c} = \mathbf{A}$.

$$4\mathbf{A}_{2\times 3} = \begin{bmatrix} 4 & 16 & 24 \\ 8 & 12 & 4 \end{bmatrix}$$

Moreover, if $\lambda = -1$, then for $\mathbf{B}_{2\times 3}$ given in Section 9.3.1,

$$(-1)\mathbf{B}_{2\times 3} = \begin{bmatrix} 0 & -1 & -10 \\ -5 & -2 & -6 \end{bmatrix}$$

9.3.3 "Subtraction"

Provided that the matrices involved have the same dimensions for rows and columns, matrix "subtraction" can be achieved through multiplication by a scalar as follows:

$$\mathbf{K}_{r\times c} = \mathbf{A}_{r\times c} + (-1)\mathbf{B}_{r\times c} = \mathbf{A}_{r\times c} - \mathbf{B}_{r\times c} \tag{9.2}$$

Using the two previously defined matrices $\mathbf{A}$ and $\mathbf{B}$, we have

$$\mathbf{K}_{2\times 3} = \begin{bmatrix} k_{11} = a_{11} + (-1)b_{11} & k_{12} = a_{12} + (-1)b_{12} & k_{13} = a_{13} + (-1)b_{13} \\ k_{21} = a_{21} + (-1)b_{21} & k_{22} = a_{22} + (-1)b_{22} & k_{23} = a_{23} + (-1)b_{23} \end{bmatrix}$$

$$= \begin{bmatrix} 1 & 3 & -4 \\ -3 & 1 & -5 \end{bmatrix}$$

9.3.4 Multiplication

Two matrices are conformable to multiplication only if the number of columns in the first matrix equals the number of rows in the second matrix. Thus, if $\mathbf{E}_{r\times n}$ is a matrix possessing r rows and n columns while $\mathbf{F}_{n\times c}$ is a matrix possessing n rows and c columns, the product $\mathbf{E}_{r\times n} \cdot \mathbf{F}_{n\times c} = \mathbf{G}_{r\times c}$ can be obtained. (Moreover, if $r = c$, then $\mathbf{F} \cdot \mathbf{E}$ can also be obtained, but, in general, $\mathbf{E} \cdot \mathbf{F} \neq \mathbf{F} \cdot \mathbf{E}$.) The element in the ith row and jth column of the product matrix $\mathbf{G}_{r\times c}$ would be computed from

$$g_{ij} = \sum_{k=1}^{n} e_{ik}f_{kj} = e_{i1}f_{1j} + e_{i2}f_{2j} + \cdots + e_{in}f_{nj} \tag{9.3}$$

where

$$\begin{bmatrix} g_{11} & g_{12} & \cdots & g_{1j} & \cdots & g_{1c} \\ g_{21} & g_{22} & \cdots & g_{2j} & \cdots & g_{2c} \\ \cdot & \cdot & & \cdot & & \cdot \\ \cdot & \cdot & & \cdot & & \cdot \\ \cdot & \cdot & & \cdot & & \cdot \\ g_{i1} & g_{i2} & \cdots & \boxed{g_{ij}} & \cdots & g_{ic} \\ \cdot & \cdot & & \cdot & & \cdot \\ \cdot & \cdot & & \cdot & & \cdot \\ \cdot & \cdot & & \cdot & & \cdot \\ g_{r1} & g_{r2} & \cdots & g_{rj} & \cdots & g_{rc} \end{bmatrix}$$

$$
= \begin{bmatrix} e_{11} & e_{12} & \cdots & e_{1n} \\ e_{21} & e_{22} & \cdots & e_{2n} \\ \cdot & \cdot & & \cdot \\ \cdot & \cdot & & \cdot \\ \cdot & \cdot & & \cdot \\ \boxed{e_{i1}} & e_{i2} & \cdots & e_{in} \\ \cdot & \cdot & & \cdot \\ \cdot & \cdot & & \cdot \\ \cdot & \cdot & & \cdot \\ e_{r1} & e_{r2} & \cdots & e_{rn} \end{bmatrix} \begin{bmatrix} f_{11} & f_{12} & \cdots & \boxed{f_{1j}} & \cdots & f_{1c} \\ f_{21} & f_{22} & \cdots & \boxed{f_{2j}} & \cdots & f_{2c} \\ \cdot & \cdot & & \cdot & & \cdot \\ \cdot & \cdot & & \cdot & & \cdot \\ \cdot & \cdot & & \cdot & & \cdot \\ f_{n1} & f_{n2} & \cdots & \boxed{f_{nj}} & \cdots & f_{nc} \end{bmatrix}
$$

That is, $\underset{r \times c}{\mathbf{G}} = \underset{r \times n}{\mathbf{E}} \cdot \underset{n \times c}{\mathbf{F}}$

In the preceding, $\mathbf{E}$ is called the *premultiplier* and $\mathbf{F}$ is called the *postmultiplier*. For example, suppose

$$
\underset{3 \times 2}{\mathbf{E}} = \begin{bmatrix} 3 & 0 \\ 5 & 4 \\ 1 & 1 \end{bmatrix} \quad \text{and} \quad \underset{2 \times 4}{\mathbf{F}} = \begin{bmatrix} 1 & 4 & 6 & -2 \\ 5 & 0 & 2 & 7 \end{bmatrix}
$$

Then

$$
\underset{3 \times 4}{\mathbf{G}} = \underset{3 \times 2}{\mathbf{E}} \underset{2 \times 4}{\mathbf{F}} = \begin{bmatrix} g_{11} = e_{11}f_{11} + e_{12}f_{21} & \cdots & g_{14} = e_{11}f_{14} + e_{12}f_{24} \\ g_{21} = e_{21}f_{11} + e_{22}f_{21} & \cdots & g_{24} = e_{21}f_{14} + e_{22}f_{24} \\ g_{31} = e_{31}f_{11} + e_{32}f_{21} & \cdots & g_{34} = e_{31}f_{14} + e_{32}f_{24} \end{bmatrix}
$$

$$
= \begin{bmatrix} g_{11} = 3(1) + 0(5) & \cdots & g_{14} = 3(-2) + 0(7) \\ g_{21} = 5(1) + 4(5) & \cdots & g_{24} = 5(-2) + 4(7) \\ g_{31} = 1(1) + 1(5) & \cdots & g_{34} = 1(-2) + 1(7) \end{bmatrix}
$$

$$
= \begin{bmatrix} 3 & 12 & 18 & -6 \\ 25 & 20 & 38 & 18 \\ 6 & 4 & 8 & 5 \end{bmatrix}
$$

9.4 SOME USEFUL MATRICES

The matrix $\mathbf{A}'_{c \times r}$ is the *transpose* of the matrix $\mathbf{A}_{r \times c}$ if the c elements in each of the various rows of the former identically correspond to the c elements in each of the various columns of the latter (that is, if $a_{ij} = a'_{ji}$ for all i and j). Thus, if

$$
\underset{2 \times 3}{\mathbf{A}} = \begin{bmatrix} a_{11} & a_{12} & a_{13} \\ a_{21} & a_{22} & a_{23} \end{bmatrix}
$$

then

$$
\underset{3 \times 2}{\mathbf{A}'} = \begin{bmatrix} a'_{11} = a_{11} & a'_{12} = a_{21} \\ a'_{21} = a_{12} & a'_{22} = a_{22} \\ a'_{31} = a_{13} & a'_{32} = a_{23} \end{bmatrix}
$$

so that for the particular 2×3 matrix described earlier,

$$\mathbf{A}_{2 \times 3} = \begin{bmatrix} 1 & 4 & 6 \\ 2 & 3 & 1 \end{bmatrix} \quad \text{and} \quad \mathbf{A}'_{3 \times 2} = \begin{bmatrix} 1 & 2 \\ 4 & 3 \\ 6 & 1 \end{bmatrix}.$$

The matrix

$$\mathbf{V}_{r \times 1} = \begin{bmatrix} V_{11} \\ V_{21} \\ \cdot \\ \cdot \\ \cdot \\ V_{r1} \end{bmatrix}$$

contains only one column and is called a *column vector*. Its transpose

$$\mathbf{V}'_{1 \times r} = [V'_{11} \quad V'_{12} \quad \cdots \quad V'_{1r}]$$

contains only one row and is called a *row vector*. The product of the column vector by the row vector yields an $r \times r$ (square) matrix, while the product of the row vector by the column vector yields a 1×1 (square) matrix containing but one element—a numerical result which we have previously referred to as a *scalar*. If, for example,

$$\mathbf{V}_{4 \times 1} = \begin{bmatrix} 2 \\ 7 \\ 1 \\ 3 \end{bmatrix}$$

then

$$\mathbf{S}_{4 \times 4} = \mathbf{V}_{4 \times 1} \cdot \mathbf{V}'_{1 \times 4} = \begin{bmatrix} 2 \\ 7 \\ 1 \\ 3 \end{bmatrix} [2 \quad 7 \quad 1 \quad 3] = \begin{bmatrix} 4 & 14 & 2 & 6 \\ 14 & 49 & 7 & 21 \\ 2 & 7 & 1 & 3 \\ 6 & 21 & 3 & 9 \end{bmatrix}$$

while

$$\mathbf{S}_{1 \times 1} = \mathbf{V}'_{1 \times 4} \cdot \mathbf{V}_{4 \times 1} = [2 \quad 7 \quad 1 \quad 3] \begin{bmatrix} 2 \\ 7 \\ 1 \\ 3 \end{bmatrix} = 2(2) + 7(7) + 1(1) + 3(3) = 63$$

Note that the square matrix $\mathbf{S}_{4 \times 4}$ is *symmetric*. A symmetric matrix can be defined as one which is equal to its transpose. Thus for all elements $s_{ij} = s'_{ij} = s_{ji}$. A symmetric matrix must be a square matrix.

It should also be mentioned that for any matrix $\mathbf{A}_{r \times c}$ the products $\mathbf{A}_{r \times c} \cdot \mathbf{A}'_{c \times r}$ and $\mathbf{A}'_{c \times r} \cdot \mathbf{A}_{r \times c}$ can be obtained and will always be square and symmetric.

Two other square and symmetric matrices of interest are the *diagonal* matrix and the *identity* matrix.

The diagonal matrix **D** is one whose elements off the main diagonal are all equal to zero, while those along the main diagonal are not all zero:

$$\mathbf{D}_{n \times n} = \begin{bmatrix} d_{11} & 0 & \cdots & 0 \\ 0 & d_{22} & \cdots & 0 \\ \cdot & \cdot & \cdot & \cdot \\ \cdot & \cdot & \cdot & \cdot \\ \cdot & \cdot & \cdot & \cdot \\ 0 & 0 & \cdots & d_{nn} \end{bmatrix}$$

The identity matrix **I** is a diagonal matrix in which all the elements along the main diagonal are 1:

$$\mathbf{I}_{n \times n} = \begin{bmatrix} 1 & 0 & \cdots & 0 \\ 0 & 1 & \cdots & 0 \\ \cdot & \cdot & \cdot & \cdot \\ \cdot & \cdot & \cdot & \cdot \\ \cdot & \cdot & \cdot & \cdot \\ 0 & 0 & \cdots & 1 \end{bmatrix}$$

Hence, for **A**, an $n \times n$ square matrix,

$$\underset{n \times n}{\mathbf{A}} \cdot \underset{n \times n}{\mathbf{I}} = \underset{n \times n}{\mathbf{I}} \cdot \underset{n \times n}{\mathbf{A}} = \underset{n \times n}{\mathbf{A}} \qquad (9.4)$$

More generally, for an $r \times c$ matrix **B**,

$$\underset{r \times c}{\mathbf{B}} \cdot \underset{c \times c}{\mathbf{I}} = \underset{r \times c}{\mathbf{B}} \qquad (9.5)$$

while

$$\underset{r \times r}{\mathbf{I}} \cdot \underset{r \times c}{\mathbf{B}} = \underset{r \times c}{\mathbf{B}} \qquad (9.6)$$

From this we see that the identity matrix is analogous to the number 1 in arithmetic operations. In multiplication, for example, $b \times 1 = b$ and $1 \times b = b$. Therefore, the identity matrix **I** can be included or omitted from a matrix expression as necessary.

9.5 DETERMINANTS

Associated with every square matrix is a numerical value called its *determinant*. The determinant of matrix **A** is denoted by $|\mathbf{A}|$. If **A** is a 1×1 matrix, the determinant is the element a_{11} itself. If **A** is a 2×2 matrix, the determinant is obtained from

$$|\mathbf{A}| = \left| \begin{bmatrix} a_{11} & a_{12} \\ a_{21} & a_{22} \end{bmatrix} \right| = a_{11}a_{22} - a_{12}a_{21} \qquad (9.7)$$

Suppose, for example,

$$\mathbf{A}_{2 \times 2} = \begin{bmatrix} 3 & 4 \\ 1 & 7 \end{bmatrix}$$

Then
$$|A| = 3(7) - 4(1) = 17$$
If A is a 3×3 matrix, the determinant can be computed from

$$|A| = \left| \begin{bmatrix} a_{11} & a_{12} & a_{13} \\ a_{21} & a_{22} & a_{23} \\ a_{31} & a_{32} & a_{33} \end{bmatrix} \right|$$
$$= a_{11}(a_{22}a_{33} - a_{23}a_{32}) - a_{12}(a_{21}a_{33} - a_{23}a_{31}) + a_{13}(a_{21}a_{32} - a_{22}a_{31})$$

$$(9.8)$$

Suppose, for example,

$$\underset{3 \times 3}{A} = \begin{bmatrix} 3 & 4 & 2 \\ 1 & 7 & 1 \\ -9 & 0 & 6 \end{bmatrix}$$

Then

$$|A| = 3(42 - 0) - 4[6 - (-9)] + 2[0 - (-63)] = 192$$

Indeed the computation of the determinant becomes more and more complex as the dimensions of the matrix increase. In these circumstances the computation of the determinant may be accomplished through *expansion of cofactors* [see, for example, Hadley (1961) or Noble and Daniel (1977)]. Nevertheless, regardless of its dimensions, the determinant of a diagonal matrix is merely the product of the elements along the main diagonal.

9.6 THE INVERSE OF A MATRIX

Only a square matrix whose determinant is not zero has an inverse. Such a matrix is called *nonsingular*. If any row (or column) of a square matrix is some multiple or *linear combination* of any other row(s) [or column(s)], the matrix will be *singular*, i.e., have a determinant of zero and have no inverse.

Since matrix inversion takes the place of "matrix division," we draw the following analogy to arithmetic: Given any real number a (not equal to zero), there exists another real number, the reciprocal $a^{-1} = 1/a$, such that

$$a \cdot a^{-1} = a^{-1} \cdot a = 1$$

Similarly, with square matrices A and B, if

$$A \cdot B = B \cdot A = I$$

then B must be the *inverse* of A. The inverse of A is denoted by A^{-1}. Thus

$$A \cdot A^{-1} = I \quad \text{and} \quad A^{-1} \cdot A = I \tag{9.9}$$

Therefore, to determine whether one matrix is the inverse of another, multiply the two matrices and observe whether or not the resulting product is equal to the identity matrix.

9.6.1 Obtaining the Inverse

The inverse of a 1×1 matrix $\mathbf{A}$ is merely the reciprocal of the element a_{11}. That is,

$$\underset{1 \times 1}{\mathbf{A}^{-1}} = \frac{1}{a_{11}} \qquad (\text{provided } a_{11} \neq 0) \tag{9.10}$$

The inverse of a 2×2 matrix $\mathbf{A}$ is obtained from

$$\underset{2 \times 2}{\mathbf{A}^{-1}} = \frac{\begin{bmatrix} a_{22} & -a_{12} \\ -a_{21} & a_{11} \end{bmatrix}}{|\mathbf{A}|} \tag{9.11}$$

As an example, if

$$\underset{2 \times 2}{\mathbf{A}} = \begin{bmatrix} 3 & 4 \\ 1 & 7 \end{bmatrix}$$

then

$$\underset{2 \times 2}{\mathbf{A}^{-1}} = \frac{\begin{bmatrix} 7 & -4 \\ -1 & 3 \end{bmatrix}}{17} = \begin{bmatrix} \frac{7}{17} & -\frac{4}{17} \\ -\frac{1}{17} & \frac{3}{17} \end{bmatrix}$$

The inverse of a 3×3 matrix $\mathbf{A}$ is computed from

$$\underset{3 \times 3}{\mathbf{A}^{-1}} = \frac{\begin{bmatrix} a_{22}a_{33} - a_{32}a_{23} & a_{32}a_{13} - a_{12}a_{33} & a_{12}a_{23} - a_{22}a_{13} \\ a_{31}a_{23} - a_{21}a_{33} & a_{11}a_{33} - a_{31}a_{13} & a_{21}a_{13} - a_{11}a_{23} \\ a_{21}a_{32} - a_{31}a_{22} & a_{31}a_{12} - a_{11}a_{32} & a_{11}a_{22} - a_{21}a_{12} \end{bmatrix}}{|\mathbf{A}|} \tag{9.12}$$

For example, as in Section 9.5, if

$$\underset{3 \times 3}{\mathbf{A}} = \begin{bmatrix} 3 & 4 & 2 \\ 1 & 7 & 1 \\ -9 & 0 & 6 \end{bmatrix}$$

then

$$\underset{3 \times 3}{\mathbf{A}^{-1}} = \frac{\begin{bmatrix} 42 - 0 & 0 - 24 & 4 - 14 \\ -9 - 6 & 18 - (-18) & 2 - 3 \\ 0 - (-63) & -36 - 0 & 21 - 4 \end{bmatrix}}{192} = \begin{bmatrix} \frac{42}{192} & -\frac{24}{192} & -\frac{10}{192} \\ -\frac{15}{192} & \frac{36}{192} & -\frac{1}{192} \\ \frac{63}{192} & -\frac{36}{192} & \frac{17}{192} \end{bmatrix}$$

For a higher-order square matrix the inverse may be computed through *expansion of cofactors* and by obtaining the corresponding *adjoint matrix* [see, for example, Hadley (1961)] or by employing an appropriate computer package [see SAS (1979)]. Nevertheless, regardless of its dimensions, the inverse of a diagonal matrix is easily obtained by replacing the elements along the main diagonal by their respective reciprocals.

9.6.2 Use of Matrix Inversion

A matrix inverse is especially important in solving simultaneous equations. For example, suppose we have the following two equations with two unknowns:

$$14v_{11} + 6v_{21} = 90$$
$$4v_{11} + 2v_{21} = 20$$

To solve for v_{11} and v_{21}, we can write this in matrix form:

$$\begin{bmatrix} 14 & 6 \\ 4 & 2 \end{bmatrix} \begin{bmatrix} v_{11} \\ v_{21} \end{bmatrix} = \begin{bmatrix} 90 \\ 20 \end{bmatrix}$$

$$\underset{2 \times 2}{\mathbf{A}} \cdot \underset{2 \times 1}{\mathbf{V}} = \underset{2 \times 1}{\mathbf{C}}$$

Now to solve for the elements of the column vector V, we have

$$\mathbf{A} \cdot \mathbf{V} = \mathbf{C}$$
$$\mathbf{A}^{-1} \cdot \mathbf{A} \cdot \mathbf{V} = \mathbf{A}^{-1} \cdot \mathbf{C}$$
$$\mathbf{I} \cdot \mathbf{V} = \mathbf{A}^{-1} \cdot \mathbf{C}$$
$$\mathbf{V} = \mathbf{A}^{-1} \cdot \mathbf{C}$$

Hence we need the inverse of the matrix $\mathbf{A}$. Since

$$\mathbf{A}^{-1} = \frac{\begin{bmatrix} 2 & -6 \\ -4 & 14 \end{bmatrix}}{4}$$

then

$$\mathbf{V} = \frac{\begin{bmatrix} 2 & -6 \\ -4 & 14 \end{bmatrix}}{4} \begin{bmatrix} 90 \\ 20 \end{bmatrix} = \begin{bmatrix} 15 \\ -20 \end{bmatrix} = \begin{bmatrix} v_{11} \\ v_{21} \end{bmatrix}$$

so that $v_{11} = 15$ and $v_{21} = -20$.

This matrix method for solving simultaneous equations is called *Cramer's rule*.

9.7 MATRIX PROPERTIES

Four second-order (i.e., 2×2) square matrices are presented:

$$\mathbf{A} = \begin{bmatrix} 8 & 1 \\ 3 & 5 \end{bmatrix} \quad \mathbf{B} = \begin{bmatrix} 2 & 7 \\ 1 & 10 \end{bmatrix} \quad \mathbf{C} = \begin{bmatrix} 5 & 4 \\ 0 & 3 \end{bmatrix} \quad \mathbf{I} = \begin{bmatrix} 1 & 0 \\ 0 & 1 \end{bmatrix}$$

Using these matrices as appropriate, the reader may now demonstrate the following properties of matrices which will be useful in many situations:

1. $(\mathbf{A} \cdot \mathbf{B})\mathbf{C} = \mathbf{A}(\mathbf{B} \cdot \mathbf{C}) = \mathbf{A} \cdot \mathbf{B} \cdot \mathbf{C}$ (9.13)
2. $\mathbf{A}(\mathbf{B} + \mathbf{C}) = \mathbf{A} \cdot \mathbf{B} + \mathbf{A} \cdot \mathbf{C}$ (9.14)
3. $\mathbf{A} \cdot \mathbf{B} \neq \mathbf{B} \cdot \mathbf{A}$ (9.15)

4. $\mathbf{I}^2 = \mathbf{I} \cdot \mathbf{I} = \mathbf{I}$ (9.16)

5. $\mathbf{C} \cdot \mathbf{C}^2 = \mathbf{C}^2 \cdot \mathbf{C}$ (9.17)

6. $\mathbf{A} \cdot \mathbf{I} = \mathbf{I} \cdot \mathbf{A}$ (9.4)

7. $\mathbf{B}^2 \cdot \mathbf{I} = \mathbf{I} \cdot \mathbf{B}^2$ (9.4)

8. $(\mathbf{A}')' = \mathbf{A}$ (9.18)

9. $(\mathbf{A} \cdot \mathbf{B})' = \mathbf{B}' \cdot \mathbf{A}'$ (9.19)

10. $|\mathbf{A} \cdot \mathbf{B}| = |\mathbf{A}| \cdot |\mathbf{B}|$ (9.20)

11. $\mathbf{A} \cdot \mathbf{A}^{-1} = \mathbf{I}$ (9.9)

12. $\mathbf{B}^{-1} \cdot \mathbf{B} = \mathbf{I}$ (9.9)

13. $(\mathbf{A} \cdot \mathbf{B})^{-1} = \mathbf{B}^{-1} \cdot \mathbf{A}^{-1}$ (9.21)

14. $(\mathbf{C}^{-1})^{-1} = \mathbf{C}$ (9.22)

15. $(\mathbf{C}')^{-1} = (\mathbf{C}^{-1})'$ (9.23)

The reader is also invited to demonstrate that the preceding properties hold in general for any 2×2 (or higher-order) nonsingular matrices.

9.8 USING MATRIX ALGEBRA IN REGRESSION

9.8.1 Developing the Regression Equation

Now that we have reviewed the fundamentals of matrix algebra, we are ready to reformulate the principles and methods of simple linear regression (Chapter 8) using a matrix approach. To fix ideas, we may express the simple linear regression model (8.2) in matrix form as follows*:

$$\mathbf{Y} = \mathbf{X}\boldsymbol{\beta} + \boldsymbol{\epsilon} \qquad (9.24)$$

where $\mathbf{Y} = N \times 1$ column vector of observations Y_i $(i = 1, 2, \ldots, N)$
$\mathbf{X} = N \times 2$ matrix of independent variables X_i
$\boldsymbol{\beta} = 2 \times 1$ column vector of parameters (β_0 and β_1) to be estimated
$\boldsymbol{\epsilon} = N \times 1$ column vector of residual terms ϵ_i

Drawing a sample of n observations from this population and using the principle of *least squares*, we wish to obtain $\hat{\boldsymbol{\beta}}$, a 2×1 column vector of sample regression coefficients (b_0 and b_1), so that the resulting sample regression equation provides the *best linear fit* to the observed data. To accomplish this, the *normal equations* (8.4) can be written as

$$\mathbf{X}'\mathbf{X}\hat{\boldsymbol{\beta}} = \mathbf{X}'\mathbf{Y} \qquad (9.25)$$

where, from the sample,

---

*Note that for the population the matrix $\mathbf{X}$ has dimensions $N \times 2$ so that it is conformable to multiplication with the 2×1 column vector $\boldsymbol{\beta}$. The matrix $\mathbf{X}$ consists of a column of 1's and a column of X_i values. For a sample the matrix $\mathbf{X}$ is similarly defined—except, of course, that its dimensions are $n \times 2$.

$\mathbf{Y} = n \times 1$ column vector of observations Y_i $(i = 1, 2, \ldots, n)$
$\mathbf{X} = n \times 2$ matrix of independent variables X_i
$\mathbf{X'} =$ transpose of $\mathbf{X}$
$\hat{\boldsymbol{\beta}} = 2 \times 1$ column vector of sample regression coefficients

Note that the product matrix $\mathbf{X'X}$ is square and symmetric:

$$\mathbf{X'X} = \begin{bmatrix} 1 & 1 & \cdots & 1 \\ X_1 & X_2 & \cdots & X_n \end{bmatrix} \begin{bmatrix} 1 & X_1 \\ 1 & X_2 \\ \cdot & \cdot \\ \cdot & \cdot \\ \cdot & \cdot \\ 1 & X_n \end{bmatrix} = \begin{bmatrix} n & \sum_{i=1}^{n} X_i \\ \sum_{i=1}^{n} X_i & \sum_{i=1}^{n} X_i^2 \end{bmatrix}$$

Also,

$$\mathbf{X'Y} = \begin{bmatrix} 1 & 1 & \cdots & 1 \\ X_1 & X_2 & \cdots & X_n \end{bmatrix} \begin{bmatrix} Y_1 \\ Y_2 \\ \cdot \\ \cdot \\ \cdot \\ Y_n \end{bmatrix} = \begin{bmatrix} \sum_{i=1}^{n} Y_i \\ \sum_{i=1}^{n} X_i Y_i \end{bmatrix}$$

For the problem of interest to the Director of Career Development (Chapter 8), the normal equations (9.25) can now be expressed as

$$\begin{bmatrix} 30 & 87.00 \\ 87.00 & 256.06 \end{bmatrix} \begin{bmatrix} b_0 \\ b_1 \end{bmatrix} = \begin{bmatrix} 534.30 \\ 1564.24 \end{bmatrix}$$

To solve for the elements of $\hat{\boldsymbol{\beta}}$, we first take the inverse of $\mathbf{X'X}$. Then, premultiplying both sides of (9.25) by this inverse, we obtain

$$(\mathbf{X'X})^{-1}\mathbf{X'X}\hat{\boldsymbol{\beta}} = (\mathbf{X'X})^{-1}\mathbf{X'Y}$$
$$\hat{\boldsymbol{\beta}} = (\mathbf{X'X})^{-1}\mathbf{X'Y} \tag{9.26}$$

By using (9.11), this inverse can be expressed as

$$(\mathbf{X'X})^{-1} = \left\{ \frac{1}{n\sum_{i=1}^{n} X_i^2 - \left(\sum_{i=1}^{n} X_i\right)^2} \right\} \begin{bmatrix} \sum_{i=1}^{n} X_i^2 & -\sum_{i=1}^{n} X_i \\ -\sum_{i=1}^{n} X_i & n \end{bmatrix} \tag{9.27}$$

where the scalar term is $|\mathbf{X'X}|$, the determinant of the product matrix. For our example,

$$(\mathbf{X'X})^{-1} = \left\{ \frac{1}{112.80} \right\} \begin{bmatrix} 256.06 & -87.00 \\ -87.00 & 30 \end{bmatrix}$$

In all operations involving $(\mathbf{X'X})^{-1}$, the multiplication by the scalar term would be performed last in order to reduce *rounding* errors. By using (9.26), the regression coefficients are computed:

$$\begin{bmatrix} b_0 \\ b_1 \end{bmatrix} = \left\{\frac{1}{112.80}\right\} \begin{bmatrix} 256.06 & -87.00 \\ -87.00 & 30 \end{bmatrix} \begin{bmatrix} 534.30 \\ 1564.24 \end{bmatrix} = \begin{bmatrix} 6.41825 \\ 3.92819 \end{bmatrix}$$

The sample regression equation (8.3) can now be stated as

$$\hat{\mathbf{Y}} = \mathbf{X}\hat{\boldsymbol{\beta}} \tag{9.28}$$

where $\hat{\mathbf{Y}}$ is an $n \times 1$ column vector of *fitted* values $\hat{Y}_i$. That is,

$$\begin{bmatrix} \hat{Y}_1 \\ \hat{Y}_2 \\ \cdot \\ \cdot \\ \cdot \\ \hat{Y}_n \end{bmatrix} = \begin{bmatrix} 1 & X_1 \\ 1 & X_2 \\ \cdot & \cdot \\ \cdot & \cdot \\ \cdot & \cdot \\ 1 & X_n \end{bmatrix} \begin{bmatrix} b_0 \\ b_1 \end{bmatrix} = \begin{bmatrix} b_0 + b_1 X_1 \\ b_0 + b_1 X_2 \\ \cdot \\ \cdot \\ \cdot \\ b_0 + b_1 X_n \end{bmatrix}$$

and for our example

$$\begin{bmatrix} \hat{Y}_1 \\ \hat{Y}_2 \\ \cdot \\ \cdot \\ \cdot \\ \hat{Y}_n \end{bmatrix} = \begin{bmatrix} 1 & 2.7 \\ 1 & 3.1 \\ \cdot & \cdot \\ \cdot & \cdot \\ \cdot & \cdot \\ 1 & 3.0 \end{bmatrix} \begin{bmatrix} 6.41825 \\ 3.92819 \end{bmatrix} = \begin{bmatrix} 17.03 \\ 18.60 \\ \cdot \\ \cdot \\ \cdot \\ 18.21 \end{bmatrix}$$

9.8.2 Preparation for Inference: Developing the ANOVA Table

Based on the principle of least squares, (9.28) is that equation for which the summation of squared differences between the observed and fitted Y values in the sample provides a *minimum total*. This minimum total, which represents the *unexplained variation* SSE in the data, can be computed from the identity in (8.8); in matrix form this is

$$\boldsymbol{\epsilon}'\boldsymbol{\epsilon} = (\mathbf{Y} - \hat{\mathbf{Y}})'(\mathbf{Y} - \hat{\mathbf{Y}}) = \mathbf{Y}'\mathbf{Y} - \hat{\boldsymbol{\beta}}'\mathbf{X}'\mathbf{Y} \tag{9.29}$$

where $\boldsymbol{\epsilon} = \mathbf{Y} - \hat{\mathbf{Y}}$ is an $n \times 1$ column vector of residuals e_i and $\boldsymbol{\epsilon}'$ is the transpose of $\boldsymbol{\epsilon}$. Thus

$$\boldsymbol{\epsilon} = \begin{bmatrix} e_1 \\ e_2 \\ \cdot \\ \cdot \\ \cdot \\ e_n \end{bmatrix} = \begin{bmatrix} Y_1 \\ Y_2 \\ \cdot \\ \cdot \\ \cdot \\ Y_n \end{bmatrix} + (-1) \begin{bmatrix} \hat{Y}_1 \\ \hat{Y}_2 \\ \cdot \\ \cdot \\ \cdot \\ \hat{Y}_n \end{bmatrix}$$

so that for our example

$$\boldsymbol{\epsilon} = \begin{bmatrix} 17.0 \\ 17.7 \\ \cdot \\ \cdot \\ \cdot \\ 18.5 \end{bmatrix} - \begin{bmatrix} 17.03 \\ 18.60 \\ \cdot \\ \cdot \\ \cdot \\ 18.21 \end{bmatrix} = \begin{bmatrix} -.03 \\ -.90 \\ \cdot \\ \cdot \\ \cdot \\ +.29 \end{bmatrix}$$

and therefore,

$$\boldsymbol{\epsilon}'\boldsymbol{\epsilon} = [e_1 \quad e_2 \quad \cdots \quad e_n] \begin{bmatrix} e_1 \\ e_2 \\ \cdot \\ \cdot \\ \cdot \\ e_n \end{bmatrix} = [-.03 \quad -.90 \quad \cdots \quad +.29] \begin{bmatrix} -.03 \\ -.90 \\ \cdot \\ \cdot \\ \cdot \\ +.29 \end{bmatrix} = 19.5071$$

By using (9.29), this unexplained variation SSE can also be obtained from

$$\text{SSE} = \mathbf{Y}'\mathbf{Y} - \hat{\boldsymbol{\beta}}'\mathbf{X}'\mathbf{Y} \tag{9.30}$$

$$= [Y_1 \quad Y_2 \quad \cdots \quad Y_n] \begin{bmatrix} Y_1 \\ Y_2 \\ \cdot \\ \cdot \\ \cdot \\ Y_n \end{bmatrix} - [b_0 \quad b_1] \begin{bmatrix} \sum_{i=1}^{n} Y_i \\ \sum_{i=1}^{n} X_i Y_i \end{bmatrix}$$

so that,

$$\text{SSE} = [17.0 \quad 17.7 \quad \cdots \quad 18.5] \begin{bmatrix} 17.0 \\ 17.7 \\ \cdot \\ \cdot \\ \cdot \\ 18.5 \end{bmatrix} - [6.41825 \quad 3.92819] \begin{bmatrix} 534.30 \\ 1564.24 \end{bmatrix}$$

$$= 19.5071$$

Note that $\mathbf{Y}'\mathbf{Y} = \sum_{i=1}^{n} Y_i^2$, the summation of squared Y values. Moreover, from (8.12) the *explained variation* SSR can be expressed as

$$\text{SSR} = \hat{\boldsymbol{\beta}}'\mathbf{X}'\mathbf{Y} - \frac{\left(\sum_{i=1}^{n} Y_i\right)^2}{n} \tag{9.31}$$

so that

$$\text{SSR} = [6.41825 \quad 3.92819] \begin{bmatrix} 534.30 \\ 1564.24 \end{bmatrix} - \frac{(534.30)^2}{30} = 58.0199$$

Furthermore, from (8.10) the *total variation* SST may be stated as

$$\text{SST} = \mathbf{Y}'\mathbf{Y} - \frac{\left(\sum_{i=1}^{n} Y_i\right)^2}{n} \tag{9.32}$$

and therefore,

$$\text{SST} = [17.0 \quad 17.7 \quad \cdots \quad 18.5] \begin{bmatrix} 17.0 \\ 17.7 \\ \cdot \\ \cdot \\ \cdot \\ 18.5 \end{bmatrix} - \frac{(534.30)^2}{30} = 77.5270$$

The complete ANOVA table (Table 9.1) can now be displayed using matrix notation and the appropriate tests of hypotheses made.

TABLE 9.1 ANOVA Table Using Matrix Notation

Source of Variation	Degrees of Freedom	Sum of Squares	Mean Square
Due to Regression	1	$SSR = \hat{\boldsymbol{\beta}}'X'Y - \dfrac{\left(\sum\limits_{i=1}^{n} Y_i\right)^2}{n}$	$MSR = \dfrac{SSR}{1}$
Error	$n - 2$	$SSE = Y'Y - \hat{\boldsymbol{\beta}}'X'Y$	$MSE = \dfrac{SSE}{n - 2}$
Lack of Fit	$n - 2 - \sum\limits_{j=1}^{l} n_j + l$	$SSLF = SSE - SSPE$	$MSLF = \dfrac{SSLF}{n - 2 - \sum\limits_{j=1}^{l} n_j + l}$
Pure Error	$\sum\limits_{j=1}^{l} n_j - l$	$SSPE$	$MSPE = \dfrac{SSPE}{\sum\limits_{j=1}^{l} n_j - l}$
Total	$n - 1$	$SST = Y'Y - \dfrac{\left(\sum\limits_{i=1}^{n} Y_i\right)^2}{n}$	

9.8.3 The Variance-Covariance Matrix S_b of the Regression Coefficients

To continue our discussion of inferential methods in simple linear regression using matrix algebra, it becomes necessary to develop S_b, *the variance-covariance matrix of the vector of sample regression coefficients.*

S_b is a symmetric matrix. The elements along the main diagonal represent the variances of the sample regression coefficients (i.e., the squares of the standard errors S_{b_0} and S_{b_1}), while the elements off the main diagonal ($S_{b_0 b_1}$) represent the covariances between b_0 and b_1. The variance-covariance matrix of the vector $\hat{\beta}$ is obtained from

$$S_b = (X'X)^{-1}\{MSE\} \tag{9.33}$$

where the scalar MSE, the mean square error term (i.e., the variance $S_{Y|X}^2$ around the sample regression line), is defined in the ANOVA table (Table 9.1) as

$$MSE = S_{Y|X}^2 = \frac{\{Y'Y - \hat{\beta}'X'Y\}}{(n-2)} \tag{9.34}$$

Thus

$$S_b = \begin{bmatrix} \dfrac{MSE\sum\limits_{i=1}^{n} X_i^2}{n\sum\limits_{i=1}^{n} X_i^2 - \left(\sum\limits_{i=1}^{n} X_i\right)^2} & \dfrac{-MSE\sum\limits_{i=1}^{n} X_i}{n\sum\limits_{i=1}^{n} X_i^2 - \left(\sum\limits_{i=1}^{n} X_i\right)^2} \\[4mm] \dfrac{-MSE\sum\limits_{i=1}^{n} X_i}{n\sum\limits_{i=1}^{n} X_i^2 - \left(\sum\limits_{i=1}^{n} X_i\right)^2} & \dfrac{MSE(n)}{n\sum\limits_{i=1}^{n} X_i^2 - \left(\sum\limits_{i=1}^{n} X_i\right)^2} \end{bmatrix} \tag{9.35}$$

For our example MSE = .69668 so that

$$S_b = \begin{bmatrix} \dfrac{.69668(256.06)}{112.80} & \dfrac{-.69668(87.00)}{112.80} \\[4mm] \dfrac{-.69668(87.00)}{112.80} & \dfrac{.69668(30)}{112.80} \end{bmatrix}$$

$$= \begin{bmatrix} 1.581488 & -.537333 \\ -.537333 & .185287 \end{bmatrix}$$

9.8.4 Estimating the True Intercept β_0 and True Slope β_1

To obtain *separate* $100(1 - \alpha)\%$ confidence interval estimates for β_0 and β_1, we merely extract the appropriate elements from the $\hat{\beta}$ vector and its corresponding variance-covariance matrix S_b. Thus, as in (8.24) and (8.23), we have

$$b_0 \pm t_{1-\frac{\alpha}{2};\, n-2}\sqrt{S_{b_0}^2} \tag{9.36}$$

and

$$b_1 \pm t_{1-\frac{\alpha}{2};\, n-2}\sqrt{S_{b_1}^2} \tag{9.37}$$

and the estimates

$$+3.84 \leq \beta_0 \leq +9.00$$

and

$$+3.05 \leq \beta_1 \leq +4.81$$

are obtained as in Chapter 8.

9.8.5 Estimating the True Mean Response $\mu_{Y|X}$ for a Given X

To obtain a $100(1 - \alpha)\%$ confidence interval estimate for the *mean response* $\mu_{Y|X}$ at a given value of X (say X_g), we must develop the corresponding matrix algebra expression for (8.26). First, we define the column vector

$$\mathbf{X}_g = \begin{bmatrix} 1 \\ X_g \end{bmatrix}$$

with transpose

$$\mathbf{X}_g' = [1 \quad X_g]$$

The point estimate $\hat{Y}_g$ is then found from

$$\hat{Y}_g = \mathbf{X}_g'\hat{\boldsymbol{\beta}} = [1 \quad X_g]\begin{bmatrix} b_0 \\ b_1 \end{bmatrix}$$

In matrix terminology the confidence interval statement (8.26) can now be written as

$$\mathbf{X}_g'\hat{\boldsymbol{\beta}} \pm \{t_{1-\frac{\alpha}{2};\, n-2}\}\{(\mathbf{X}_g'(\mathbf{X}'\mathbf{X})^{-1}\mathbf{X}_g)\{MSE\}\}^{1/2} \qquad (9.38)$$

As in Chapter 8, if $X_g = 3.0$, then

$$\hat{Y}_g = [1 \quad 3.0]\begin{bmatrix} 6.41825 \\ 3.92819 \end{bmatrix} = 18.21$$

Using (9.38), we obtain

$$18.21 \pm \{2.048\}\left\{[1 \quad 3.0]\left\{\frac{1}{112.80}\right\}\begin{bmatrix} 256.06 & -87.00 \\ -87.00 & 30 \end{bmatrix}\begin{bmatrix} 1 \\ 3.0 \end{bmatrix}\{.69668\}\right\}^{1/2}$$

$$18.21 \pm .32$$

so that, as before,

$$17.89 \leq \mu_{Y|X} \leq 18.53$$

9.8.6 Predicting the Individual Response $\hat{Y}_I$ for a Given X

Similarly, to obtain a $100(1 - \alpha)\%$ prediction interval for an *individual response* $\hat{Y}_I$ at a given value of X (say X_g), we may re-express (8.27) in matrix form as follows:

$$\mathbf{X}_g'\hat{\boldsymbol{\beta}} \pm \{t_{1-\frac{\alpha}{2};\, n-2}\}\{[1 + (\mathbf{X}_g'(\mathbf{X}'\mathbf{X})^{-1}\mathbf{X}_g)]\{MSE\}\}^{1/2} \qquad (9.39)$$

Once again, if $X_g = 3.0$, the prediction interval becomes

$$16.47 \leq \hat{Y}_I \leq 19.95$$

9.9 A MATRIX APPROACH TO REGRESSION: AN OVERVIEW

Now that the fundamental aspects of matrix algebra necessary for both simple and multiple regression have been reviewed, the reader is better able to appreciate its value as a method for compactly presenting lengthy and sometimes complex formulas in as simple a manner as possible. Regardless of whether we are dealing with a simple linear regression model having but one predictor variable or a complex multiple regression model with numerous predictor variables, the matrix algebra terminology for

1. Describing the regression model (9.24)
2. Solving the normal equations [(9.25) and (9.26)]
3. Expressing the sample regression equation (9.28)
4. Developing the sources of variation [(9.30), (9.31), and (9.32)]
5. Stating the variance-covariance matrix (9.34)
6. Making inferential statements [(9.38) and (9.39)]

is the same. Therefore, once we have mastered such terminology, we shall not only have a better understanding of the relationship between simple and multiple regression, but, even more importantly, we shall be able to concentrate on the ideas and concepts underlying the more complex multiple regression models without becoming bogged down with the corresponding complexity in notation and formulas.

PROBLEMS

9.1. Do Problem 8.1 using a matrix algebra approach.

9.2. Do Problem 8.2 using a matrix algebra approach.

9.3. Do Problem 8.3 using a matrix algebra approach.

9.4. Do Problem 8.4 using a matrix algebra approach.

9.5. Do Problem 8.5 using a matrix algebra approach.

9.6. Do Problem 8.6 using a matrix algebra approach.

9.7. Do Problem 8.7 using a matrix algebra approach.

10

MULTIPLE REGRESSION: AN INTRODUCTION AND BASIC CONCEPTS

10.1 INTRODUCTION

In our discussion of regression and correlation in Chapters 8 and 9, we have focused upon a linear relationship between a single quantitative independent (explanatory) variable X and a quantitative dependent (response) variable Y. In this chapter we begin to extend this discussion in order to explore the basic principles of multiple regression. In multiple regression at least two independent variables are used to predict the value of a dependent variable.

To develop our ideas, let us suppose that the Marketing Department for a large manufacturer of electronic games would like to measure the effectiveness of different types of advertising media in the promotion of its products. Specifically, two types of media are to be considered: radio and television advertising and newspaper advertising (including the cost of discount coupons). A sample of 22 cities with approximately equal populations is selected for study during a test period of 1 month. Each city is to be allocated a specific expenditure level for both radio and television advertising as well as newspaper advertising. The sales (in millions of dollars) for electronic games during the test month are recorded in Table 10.1 along with the levels of media expenditure.

With two independent variables in the multiple regression problem, a scatter diagram of the points can be plotted on a three-dimensional graph as shown in Figure 10.1.

For a particular investigation, when there are several independent variables present, the simple linear regression model of Chapter 8 can be extended if we assume a linear relationship between each independent variable and the depen-

TABLE 10.1 Monthly Sales, Radio and Television Advertising Expenditures, and Newspaper Advertising Expenditures in a Sample of 22 Cities

City	Y Sales (\$ Millions)	X_1 Radio and TV Advertising (\$000)	X_2 Newspaper Advertising (\$000)
1	9.73	0	20
2	11.19	0	20
3	8.75	5	5
4	6.25	5	5
5	9.10	10	10
6	9.71	10	10
7	9.31	15	15
8	11.77	15	15
9	8.82	20	5
10	9.82	20	5
11	16.28	25	25
12	15.77	25	25
13	10.44	30	0
14	9.14	30	0
15	13.29	35	5
16	13.30	35	5
17	14.05	40	10
18	14.36	40	10
19	15.21	45	15
20	17.41	45	15
21	18.66	50	20
22	17.17	50	20

dent variable. Thus, for our example, the multiple linear regression model is expressed as

$$Y_i = \beta_0 + \beta_1 X_{1_i} + \beta_2 X_{2_i} + \epsilon_i \qquad (10.1)$$

where $\beta_0 = Y$ intercept

$\beta_1 =$ slope of Y with variable X_1 holding X_2 constant

$\beta_2 =$ slope of Y with variable X_2 holding X_1 constant

$\epsilon_i =$ random error in Y for observation i

This multiple linear regression model can be compared to the simple linear regression model (8.2) expressed as

$$Y_i = \beta_0 + \beta_1 X_i + \epsilon_i \qquad (8.2)$$

In the latter we should note that the slope β_1 represents the unit change in Y per unit change in X and does not take into account any other variables besides the single independent variable that is included in the model. Thus β_1 may be considered as a *gross regression coefficient*. On the other hand, in the multiple linear regression model (10.1), the slope β_1 represents the unit change

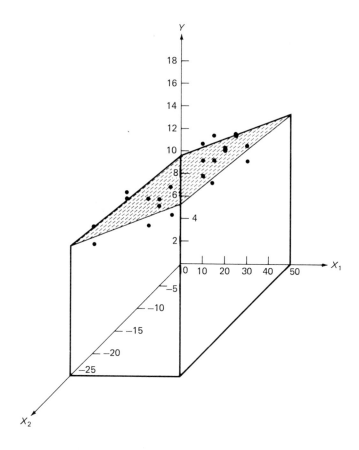

Figure 10.1 Scatter diagram for the advertising media problem.

in Y per unit change in X_1, taking into account the effect of X_2, and is referred to as either a *net* or *partial regression coefficient*.

As in the case of simple linear regression, when sample data are analyzed, the sample regression coefficients (b_0, b_1, and b_2) are used as estimates of the true parameters (β_0, β_1, and β_2). Thus the sample regression equation for the multiple linear regression model with two independent variables would be

$$\hat{Y}_i = b_0 + b_1 X_{1_i} + b_2 X_{2_i} \qquad (10.2)$$

10.2 FINDING THE REGRESSION COEFFICIENTS

If the least-squares method (see Section 8.3) is utilized to compute the regression coefficients (b_0, b_1, and b_2), the following three normal equations would be developed:

$$\text{I.} \quad \sum_{i=1}^{n} Y_i = nb_0 + b_1 \sum_{i=1}^{n} X_{1_i} + b_2 \sum_{i=1}^{n} X_{2_i}$$

$$\text{II.} \quad \sum_{i=1}^{n} X_{1_i} Y_i = b_0 \sum_{i=1}^{n} X_{1_i} + b_1 \sum_{i=1}^{n} X_{1_i}^2 + b_2 \sum_{i=1}^{n} X_{1_i} X_{2_i} \qquad (10.3)$$

$$\text{III.} \quad \sum_{i=1}^{n} X_{2_i} Y_i = b_0 \sum_{i=1}^{n} X_{2_i} + b_1 \sum_{i=1}^{n} X_{1_i} X_{2_i} + b_2 \sum_{i=1}^{n} X_{2_i}^2$$

By using an appropriate computer package, the regression coefficients are easily obtained from this set of simultaneous equations. Thus Figure 10.2 presents the partial output from the SAS GLM and CORR procedures for the data in Table 10.1.

From this computer output we observe that the regression coefficients are

$$b_0 = 5.2573824 \qquad b_1 = +.1621127 \qquad b_2 = +.2488677$$

so that the multiple regression equation can be expressed as

$$\hat{Y}_i = 5.2573824 + .1621127 X_{1_i} + .2488677 X_{2_i}$$

where $\hat{Y}_i$ = predicted sales (in millions of dollars) of electronic games per month for city i

X_{1_i} = radio and television advertising expenditures (in thousands of dollars) for city i

X_{2_i} = newspaper advertising expenditures (in thousands of dollars) for city i

The interpretation of the regression coefficients is analogous to that of the simple linear regression model. The Y intercept b_0, computed as 5.2574, represents the predicted sales ($5,257,400) for a city in which there has been no radio and television advertising and no newspaper advertising. The slope of radio and television advertising with sales (b_1, computed as .1621) can be interpreted to mean that for a city with a given amount of newspaper advertising, sales will increase by .1621 millions of dollars ($162,100) for each increase of one thousand dollars in radio and television advertising expenditures. Furthermore, the slope of newspaper advertising with sales (b_2, computed as .2489) can be interpreted to mean that for a city with a given amount of radio and television advertising, sales will increase by .2489 millions of dollars ($248,900) for each increase of one thousand dollars in newspaper advertising.

10.3 TESTING FOR THE SIGNIFICANCE OF THE RELATIONSHIP EXPRESSED BY THE FITTED MODEL

Once a regression model has been fitted to a set of data, we can determine whether there is a significant relationship between the dependent variable Y and the set of independent variables. Since there are two independent variables in the media expenditures problem, the null and alternative hypotheses can be set up as follows:

SALES AND MEDIA EXPENDITURES

GENERAL LINEAR MODELS PROCEDURE

DEPENDENT VARIABLE   SALES   SALES IN DOLLARS 000

SOURCE	DF	SUM OF SQUARES	MEAN SQUARE	F VALUE	PR > F	R-SQUARE	C.V.
MODEL	2	232.65758686	116.32879343	121.97	0.0001	0.927739	7.9715
ERROR	19	18.12167223	0.95377222			STD DEV	SALES MEAN
CORRECTED TOTAL	21	250.77925909				0.97661263	12.25136364

$S_{y|x_{12}} = 0.97661263$

SOURCE	DF	TYPE I SS	F VALUE	PR > T		DF	TYPE IV SS	F VALUE	PR > F
RADTVDOL	1	156.89965500	164.50	0.0001		1	144.06108962	151.04	0.0001
NEWSDOL	1	75.75793186	79.43	0.0001		1	75.75793186	79.43	0.0001

PARAMETER		ESTIMATE	T FOR H0 PARAMETER=0	PR > T	STD ERROR OF ESTIMATE	
INTERCEPT	b_0	5.25738235	10.55	0.0001	S_{b_0}	0.49843716
RADTVDOL	b_1	0.16211270	12.29	0.0001	S_{b_1}	0.01319064
NEWSDOL	b_2	0.24896771	8.91	0.0001	S_{b_2}	0.02792395

CORRELATION COEFFICIENTS / PROB   R UNDER H0 RHO=0 / N = 22

	CITY	SALES	RADTVDOL	NEWSDOL
CITY NUMBER OF CITY	1.00000 0.0000	0.79091 0.0001	0.99689 0.0001	0.05756 0.7992
SALES SALES IN DOLLARS 000	0.79091 0.0001	1.00000 0.0000	0.79098 0.0001	0.59438 0.0035
RADTVDOL RADIO AND TV EXPENDITURE DOL 000	0.99689 0.0001	0.79098 0.0001	1.00000 0.0000	0.05774 0.7986
NEWSDOL NEWSPAPER EXPENDITURES DOLLARS 000	0.05756 0.7992	0.59438 0.0035	0.05774 0.7986	1.00000 0.0000

Figure 10.2 Partial SAS output for multiple regression model.

H_0: $\beta_1 = \beta_2 = 0$ (there is no linear relationship between the dependent variable and the independent variables)

H_1: $\beta_1 \neq \beta_2 \neq 0$ (at least one regression coefficient is not equal to zero)

This null hypothesis is tested by subdividing the total variation in the Y values (SST) into two components, variation due to *regression* (SSR) and variation due to *error* (SSE). Table 10.2 provides the ANOVA table for a multiple regression model containing p independent variables. The appropriate F test is given by*

$$F = \frac{\text{MSR}}{\text{MSE}} \tag{10.4}$$

where

$$F \sim F_{p,\,n-p-1}$$

and, using an α level of significance, the null hypothesis may be rejected if $F \geq F_{1-\alpha;\,p,\,n-p-1}$ (see Appendix B, Table B.4).

For the media expenditures problem, the ANOVA table is provided as part of the GLM output of SAS (see Figure 10.2). Thus we have

$$\text{MSR} = 116.32879343 \qquad \text{and} \qquad \text{MSE} = .95377222$$

so that from (10.4)

$$F = \frac{116.32879343}{.95377222} = 121.97$$

Using an .05 level of significance, we note that $F = 121.97 > F_{.95;\,2,\,19} = 3.52$, so we can reject H_0 and conclude that *at least* one of the independent variables (radio and television and/or newspaper advertising) is related to sales of electronic games.

The next step is to measure the contribution of each variable to the regression model to determine whether the multiple regression model or merely a simple regression model is the better choice. Once this is accomplished, the selected model can be tested for its aptness. Based on this, the researcher finally settles on the particular fitted model to be used for prediction purposes and/or other inferential endeavors. These procedures will be discussed in Sections 10.5–10.12. Prior to this, however, it is useful to focus on measures of association.

10.4 MEASURING ASSOCIATION IN THE MULTIPLE REGRESSION MODEL

As in Table 10.2, once the total variation is partitioned into its two components, the coefficient of multiple determination may be computed. The *coefficient of multiple determination* ($r^2_{Y.12\cdots p}$) represents the proportion of the variation in Y

---

*Note that in (10.4) the error variance (MSE) is merely the square of the standard error of estimate ($S_{Y|X_{12}}$) described in Section 8.6.

TABLE 10.2 Analysis of Variance Table for Testing the Significance of a Set of Regression Coefficients in Multiple Linear Regression Containing $p = 2$ Independent Variables

Source	Degrees of Freedom	Sum of Squares	Mean Square	F
Regression	p	$\text{SSR} = b_0 \sum_{i=1}^{n} Y_i + b_1 \sum_{i=1}^{n} X_{1,i} Y_i + b_2 \sum_{i=1}^{n} X_{2,i} Y_i - \dfrac{\left(\sum_{i=1}^{n} Y_i \right)^2}{n}$	$\text{MSR} = \dfrac{\text{SSR}}{p}$	$F = \dfrac{\text{MSR}}{\text{MSE}}$
Error	$n - p - 1$	$\text{SSE} = \sum_{i=1}^{n} Y_i^2 - b_0 \sum_{i=1}^{n} Y_i - b_1 \sum_{i=1}^{n} X_{1,i} Y_i - b_2 \sum_{i=1}^{n} X_{2,i} Y_i$	$\text{MSE} = \dfrac{\text{SSE}}{n - p - 1}$	
Total	$n - 1$	$\text{SST} = \sum_{i=1}^{n} Y_i^2 - \dfrac{\left(\sum_{i=1}^{n} Y_i \right)^2}{n}$		

that is explained by the set of (p) independent variables selected. That is,

$$r_{Y \cdot 12 \cdots p}^2 = \frac{\text{SSR}}{\text{SST}} \tag{10.5}$$

In our example containing two independent variables, the coefficient of multiple determination ($r_{Y \cdot 12}^2$) is given by

$$r_{Y \cdot 12}^2 = \frac{\text{SSR}}{\text{SST}} = \frac{b_0 \sum_{i=1}^{n} Y_i + b_1 \sum_{i=1}^{n} X_{1i} Y_i + b_2 \sum_{i=1}^{n} X_{2i} Y_i - \left[\left(\sum_{i=1}^{n} Y_i\right)^2 \Big/ n\right]}{\sum_{i=1}^{n} Y_i^2 - \left[\left(\sum_{i=1}^{n} Y_i\right)^2 \Big/ n\right]} \tag{10.6}$$

Thus, as displayed in the SAS output (Figure 10.2),

$$r_{Y \cdot 12}^2 = \frac{\text{SSR}}{\text{SST}} = \frac{232.6575869}{250.7792591} = .9277$$

This coefficient of multiple determination, computed as .9277, can be interpreted to mean that 92.77% of the variation in monthly sales can be explained by the variation in both radio and television advertising as well as newspaper advertising expenditures.

To further study the relationship among the variables, it is often important to examine the correlation between each pair of variables included in the model. Referring to Figure 10.2, we can observe that a correlation matrix has been obtained (by using the CORR procedure). This matrix, displayed in Table 10.3,

TABLE 10.3   Correlation Matrix for the Advertising Media Problem

	Y	X_1	X_2
Y (Sales)	$r_{YY} = 1.0$	$r_{Y1} = .79098$	$r_{Y2} = .59438$
X_1 (Radio and Television Advertising)	$r_{Y1} = .79098$	$r_{11} = 1.0$	$r_{12} = .05774$
X_2 (Newspaper Advertising)	$r_{Y2} = .59438$	$r_{12} = .05774$	$r_{22} = 1.0$

indicates the coefficient of correlation between each pair of variables. From Table 10.3, we observe that the correlation between sales and radio and television advertising expenditures is $+.79098$, indicating a strong positive association. We may also observe that the correlation between sales and newspaper advertising expenditures is $+.59438$, indicating a moderate positive correlation. Furthermore, we also note that there is only a small correlation ($+.05774$) between the two independent variables, radio and television advertising and newspaper advertising. Finally, we may note that the correlation coefficients along the main diagonal of the matrix (r_{YY}, r_{11}, r_{22}) are each 1.0 since there will always be perfect correlation between a variable and itself.

10.5 EVALUATING THE CONTRIBUTION OF EACH INDEPENDENT VARIABLE TO A MULTIPLE REGRESSION MODEL

In developing a multiple regression model, the objective is to include only those independent variables that are useful in predicting the value of a dependent variable. If an independent variable is not helpful in making this prediction, then it could be deleted from the multiple regression model, and a simpler model with fewer independent variables could be utilized in its place.

10.5.1 Determining the Contribution of an Independent Variable by Comparing Different Regression Models

Although several approaches in choosing a model will be considered (see Chapter 13), one method for determining the contribution of an independent variable is called the *partial F test criterion*. This involves determining the contribution to the regression sum of squares (SSR) made by each independent variable after all other independent variables have been included in a model. The new independent variable would be included only if it significantly improved the model. To apply this partial F criterion in our advertising media problem containing two independent variables, we need to evaluate the contribution of radio and television advertising (X_1) once newspaper advertising (X_2) has been included in a simple model, and, conversely, we also must evaluate the contribution of newspaper advertising (X_2) once radio and television advertising (X_1) has been included in a simple model.

This contribution can be determined in two ways, depending on the information provided by the particular computer package utilized. In this section we shall evaluate the contribution of each independent variable by determining the regression sum of squares for a model that includes all independent variables except the one of interest (say variable k)—that is, SSR (*slopes of all variables except k*). Thus, in general, to determine the contribution of variable k given that all other variables are already included, we would have

$$\text{SSR}(b_k | \textit{slopes of all variables except k})$$
$$= \text{SSR}(\textit{slopes of all variables including k})$$
$$- \text{SSR}(\textit{slopes of all variables except k}) \qquad (10.7)$$

If, as in the advertising media problem, there are two independent variables, the contribution of each can be determined from (10.8) and (10.9).

Contribution of Variable X_1 Given X_2 Has Been Included:
$$\text{SSR}(b_1 | b_2) = \text{SSR}(b_1 \text{ and } b_2) - \text{SSR}(b_2) \qquad (10.8)$$

Contribution of Variable X_2 Given X_1 Has Been Included:

$$\text{SSR}(b_2 \mid b_1) = \text{SSR}(b_1 \text{ and } b_2) - \text{SSR}(b_1) \qquad (10.9)$$

The term $\text{SSR}(b_2)$ represents the regression sum of squares for a model that includes all independent variables except X_1 (which in this case consists only of X_2). Similarly, the term $\text{SSR}(b_1)$ represents the regression sum of squares for a model that includes all independent variables except X_2 (which in this case consists only of X_1). Computer output obtained from the SAS GLM procedure for these two models is presented in Figures 10.3 and 10.4.

DEPENDENT VARIABLE SALES		SALES IN DOLLARS 000				
SOURCE	DF	SUM OF SQUARES	MEAN SQUARE	F VALUE	PR > F	R-SQUARE
MODEL	1	156.89965500	156.89965500	33.43	0.0001	0.625648
ERROR	20	93.87960409	4.69398020		STD DEV	
CORRECTED TOTAL	21	250.77925909		$S_{Y\mid X}$	2.16655953	
SOURCE	DF	TYPE I SS	F VALUE	PR > F	DF	TYPE IV SS	F VALUE
RADTVDOL	1	156.89965500	33.43	0.0001	1	156.89965500	33.43

Figure 10.3 Partial SAS output for radio and television expenditure model.

DEPENDENT VARIABLE SALES		SALES IN DOLLARS 000				
SOURCE	DF	SUM OF SQUARES	MEAN SQUARE	F VALUE	PR > F	R-SQUARE
MODEL	1	88.59649724	88.59649724	10.93	0.0035	0.353285
ERROR	20	162.18276185	8.10913809		STD DEV	
CORRECTED TOTAL	21	250.77925909		$S_{Y\mid X}$	2.84765484	
SOURCE	DF	TYPE I SS	F VALUE	PR > F	DF	TYPE IV SS	F VALUE
NEWSDOL	1	88.59649724	10.93	0.0035	1	88.59649724	10.93

Figure 10.4 Partial SAS output for newspaper expenditure model.

We can observe from Figure 10.4 that

$$\text{SSR}(b_2) = 88.59649724$$

and therefore from (10.8) we have

$$\text{SSR}(b_1 \mid b_2) = \text{SSR}(b_1 \text{ and } b_2) - \text{SSR}(b_2)$$
$$= 232.65758686 - 88.59649724$$
$$= 144.06108962$$

We should note that this value is also provided as the Type IV SS obtained from the SAS GLM procedure for the regression model with two independent variables (Figure 10.2).

Now that $\text{SSR}(b_1 \mid b_2)$ has been computed, we can subdivide the regression sum of squares into its two component parts as shown in Table 10.4.

To test for the contribution of X_1 to the model, the null and alternative hypotheses would be

TABLE 10.4 **Analysis of Variance Table for Determining the Contribution of Variable** X_1

Source	Degrees of Freedom	Sum of Squares	Mean Square	F
Regression	2	232.65758686	116.32879343	
b_2	1	88.59649724	88.59649724	
$b_1 \mid b_2$	1	144.06108962	144.06108962	151.04
Error	19	18.12167223	.95377222	
Total	21	250.77925909		

H_0: variable X_1 (radio and television advertising) does not significantly improve the simple model once variable X_2 (newspaper advertising) has been included

H_1: variable X_1 significantly improves the model already containing X_2

The partial F test criterion is expressed by

$$F = \frac{\text{SSR}(b_k \mid slopes\ of\ all\ variables\ except\ k)}{\text{MSE}} \qquad (10.10)$$

where

$$F \sim F_{1,\,n-p-1}$$

and, by using an α level of significance, the null hypothesis may be rejected if $F \geq F_{1-\alpha;\ 1,n-p-1}$. Thus, from Table 10.4 we have

$$F = \frac{144.06108962}{.95377222} = 151.04$$

By using a level of significance of .05, since $F = 151.04 > F_{.95;\ 1,19} = 4.38$, our decision is to reject H_0 and conclude that the addition of variable X_1 (radio and television advertising) significantly improves a regression model that already contains variable X_2 (newspaper advertising).

To evaluate the contribution of variable X_2 (newspaper advertising) to a simple model in which variable X_1 has already been included, we need to compute (10.9). From Figure 10.3 we determine that

$$\text{SSR}(b_1) = 156.899655$$

Therefore, by using (10.9),

$$\text{SSR}(b_2 \mid b_1) = \text{SSR}(b_1\ and\ b_2) - \text{SSR}(b_1)$$
$$= 232.65758686 - 156.899655$$
$$= 75.75793186$$

Now that $\text{SSR}(b_2 \mid b_1)$ has been computed, we can subdivide the regression sum of squares into its two component parts as shown in Table 10.5.

TABLE 10.5 Analysis of Variance Table for Determining the Contribution of Variable X_2

Source	Degrees of Freedom	Sum of Squares	Mean Square	F
Regression	2	232.65758686	116.32879343	
b_1	1	156.899655	156.899655	
$b_2 \mid b_1$	1	75.75793186	75.75793186	79.43
Error	19	18.12167223	.95377222	
Total	21	250.77925909		

The null and alternative hypotheses to test for the contribution of X_2 to the model would be

H_0: variable X_2 (newspaper advertising) does not significantly improve the simple model once variable X_1 (radio and television advertising) has been included

H_1: variable X_2 significantly improves the model already containing variable X_1

Thus, from (10.10) and Table 10.5, we have

$$F = \frac{75.75793186}{.95377222} = 79.43$$

and again, if a level of significance of .05 is selected, then

$$F = 79.43 > F_{.95; \, 1, 19} = 4.38$$

so that our decision is to reject H_0 and conclude that the addition of variable X_2 (newspaper advertising) significantly improves a regression model that already contains variable X_1 (radio and television advertising).

By testing for the contribution of each independent variable—after the other had been included in the model—we have determined that each of the variables contributed significantly toward improving the overall model. Therefore, based on the partial F test criterion, the *multiple* regression model should include radio and television advertising (X_1) and newspaper advertising (X_2) expenditures in predicting the monthly sales of electronic games.

10.5.2 Determining the Contribution of an Independent Variable Based on the Standard Error of Its Regression Coefficient

A second approach to evaluating the contribution made by an independent variable is based on the standard error of its regression coefficient. Thus the contribution of a particular variable (say X_k) can be determined in the following manner:

$$\text{SSR}(b_k \,|\, slopes\ of\ all\ variables\ except\ k) = \frac{b_k^2 \text{MSE}}{S_{b_k}^2} \qquad (10.11)$$

For our problem the standard errors of the regression coefficients for each independent variable (S_{b_1} and S_{b_2}) are available as part of the SAS output (Figure 10.2). Thus, for example, if we wanted to determine the contribution of variable X_2 after X_1 has been included, from (10.11) we have

$$\text{SSR}(b_2 \,|\, b_1) = \frac{b_2^2 \text{MSE}}{S_{b_2}^2}$$

$$= \frac{(.24886771)^2(.95377222)}{(.02792395)^2} = 75.75793186$$

As indicated in Table 10.5, this result is the same as that obtained using (10.9).

10.5.3 Determining the Contribution of an Independent Variable Based on the t Test for the Slope

A third approach to evaluating the contribution made by an independent variable involves the test of hypothesis for the slope. Recall from Section 1.6.4 the identity

$$(t_{1-\frac{\alpha}{2};\,\nu})^2 = F_{1-\alpha;\,1,\nu} \qquad (1.20)$$

so that whenever F has but one degree of freedom in the numerator, there is an equivalence between it and the appropriate t (see Section 8.9.2).

To test a hypothesis regarding the contribution of a specific regression coefficient, say β_k, (8.15) is generalized for multiple regression as follows:

$$t = \frac{b_k}{S_{b_k}} \qquad (10.12)$$

where $t \sim t_{n-p-1}$ and where p is the number of independent variables in the regression equation, and S_{b_k} is the standard error of the regression coefficient b_k. Using an α level of significance, the decision rule is to reject the null hypothesis if $t > t_{1-\frac{\alpha}{2};\,n-p-1}$ or if $t \leq t_{\frac{\alpha}{2};\,n-p-1}$.

Since the formulas for the standard errors of the regression coefficients are unwieldy and are expressed better by using a matrix approach (see Section 10.11), it is fortunate that the results are almost always provided as part of the output from a computer package (see Figure 10.2).

Thus if we wish to determine whether variable X_2 (newspaper advertising) has a significant effect on the monthly sales of electronic games, taking into account X_1 (radio and television advertising), the null and alternative hypotheses would be

H_0: $\beta_2 = 0$ (X_2 does not significantly contribute to a model once X_1 has been included)

H_1: $\beta_2 \neq 0$ [X_2 significantly improves a (simple) model already containing X_1]

From (10.12) we have

$$t = \frac{b_2}{S_{b_2}}$$

and from the data of this problem (see Figure 10.2),

$$b_2 = .24886771$$

$$S_{b_2} = .02792395$$

so that

$$t = \frac{.24886771}{.02792395} = + 8.912$$

If an .05 level of significance is selected, the decision rule then is to reject H_0 since

$$t = 8.912 > t_{.975;\ 19} = +2.093$$

Hence, we may again conclude that variable X_2 (newspaper advertising) significantly contributes to a model already containing radio and television advertising (X_1).

10.6 COEFFICIENTS OF PARTIAL DETERMINATION AND CORRELATION

In Section 10.4 we described the coefficient of multiple determination ($r^2_{Y.12}$) as a measure of the proportion of the variation in Y that was explained by variation in the two independent variables. Now that we have evaluated the contribution of each independent variable to a multiple regression model, we can also examine the coefficients of partial determination ($r^2_{Y1.2}$ and $r^2_{Y2.1}$) and partial correlation ($r_{Y1.2}$ and $r_{Y2.1}$). The *coefficient of partial determination* measures the proportion of the variation in the dependent variable that is explained by each independent variable while controlling for, or holding constant, the other independent variable(s). Thus for a multiple regression model with two independent variables we have

$$r^2_{Y1.2} = \frac{\text{SSR}(b_1 \mid b_2)}{\text{SST} - \text{SSR}(b_1 \text{ and } b_2) + \text{SSR}(b_1 \mid b_2)} \qquad (10.13)$$

and also

$$r^2_{Y2.1} = \frac{\text{SSR}(b_2 \mid b_1)}{\text{SST} - \text{SSR}(b_1 \text{ and } b_2) + \text{SSR}(b_2 \mid b_1)} \qquad (10.14)$$

where $\text{SSR}(b_1 \mid b_2) = $ sum of squares of the contribution of variable X_1 to the regression model given that variable X_2 has already been included

$$SST = \text{total sum of squares for } Y$$
$$SSR(b_1 \text{ and } b_2) = \text{regression sum of squares when variables } X_1 \text{ and } X_2 \text{ are both included in the multiple regression model}$$
$$SSR(b_2 | b_1) = \text{sum of squares of the contribution of variable } X_2 \text{ to the regression model given that variable } X_1 \text{ has already been included}$$

In the general multiple regression model containing p independent variables, we have

$$r^2_{Yk \cdot (\text{all variables except } k)} = \frac{SSR(b_k | \text{slopes of all variables except } k)}{SST - SSR(\text{slopes of all variables including } k) + SSR(b_k | \text{slopes of all variables except } k)} \quad (10.15)$$

For our advertising media problem we can compute

$$r^2_{Y1 \cdot 2} = \frac{144.06108962}{250.77925909 - 232.65758686 + 144.06108962}$$
$$= .8883$$

and

$$r^2_{Y2 \cdot 1} = \frac{75.75793186}{250.77925909 - 232.65758686 + 75.75793186}$$
$$= .8070$$

The coefficient of partial determination $r^2_{Y1 \cdot 2}$ can be interpreted to mean that for a fixed or constant amount of newspaper advertising, 88.83% of the variation in monthly sales of electronic games can be explained by the variation in radio and television advertising expenditures from city to city. Moreover, the coefficient of partial determination $r^2_{Y2 \cdot 1}$ can be interpreted to mean that for a given or constant amount of radio and television advertising, 80.70% of the variation in the monthly sales of electronic games can be explained by variation in newspaper advertising expenditures from city to city.

The *coefficient of partial correlation* measures the strength of the association between Y and an independent variable while controlling for, or holding constant, the other independent variable(s). It is often utilized as an aid in selecting the variables to be included in a regression model. For our data, then, we have

$$r_{Y1 \cdot 2} = \sqrt{r^2_{Y1 \cdot 2}} = \sqrt{.8883} = .9425$$

and

$$r_{Y2 \cdot 1} = \sqrt{r^2_{Y2 \cdot 1}} = \sqrt{.8070} = .8983$$

We should, of course, realize that each of the coefficients of partial correlation is significant since in the previous section we had determined that each variable made a significant contribution to a model already including the other.

10.7 TESTING FOR THE APPROPRIATENESS OF THE MULTIPLE REGRESSION MODEL

Now that we have determined the contribution of each independent variable, we need to evaluate the aptness of the particular multiple regression model that has been fit to the data.

In our discussion of the simple linear regression model, we observed in Section 8.10 that if there existed at least two independent measures Y for at least one level of X, the error sum of squares (SSE) could be subdivided into its two components, *pure error* (SSPE) and *lack of fit* (SSLF). The pure error component measures the inherent variability of Y, while the lack of fit component represents all other departures from the predicted values.

For the multiple regression model, the possibility of performing a test of lack of fit becomes less likely as more independent variables and/or additional levels of the independent variables are introduced. To test for lack of fit, at least two independent measurements of Y would be needed for at least one specific combination of levels of the independent variables. Since this might require an excessively large sample size and since the levels of the independent variables often cannot be experimentally controlled, it is frequently not possible to test for lack of fit in the multiple regression model. In such cases, residual analysis procedures (to be discussed in Section 10.8) are particularly useful. However, in our advertising media problem, we have obtained two measurements for each combination of X_1 (radio and television advertising) and X_2 (newspaper advertising), and hence we shall be able to test for lack of fit by extending the procedures of Section 8.10. In the simple linear regression model, the pure error sum of squares (SSPE) was computed as

$$\text{SSPE} = \sum_{j=1}^{l} \sum_{k=1}^{n_j} (Y_{jk} - \bar{Y}_j)^2 \tag{8.18}$$

where $\bar{Y}_j$ is the mean of Y for level j of variable X.

Similarly, for the multiple regression model with two independent variables we have

$$\text{SSPE} = \sum_{j=1}^{l} \sum_{k=1}^{n_j} (Y_{jk} - \bar{Y}_j)^2 \tag{10.16}$$

where $\bar{Y}_j$ is the mean of Y for the jth combination of levels of X_1 and X_2. From (8.17) and (8.20) we have

$$\text{SSE} = \text{SSPE} + \text{SSLF} \tag{10.17}$$

and

$$\text{SSLF} = \text{SSE} - \text{SSPE} \tag{10.18}$$

For our advertising media problem, we obtained two observations for each level of radio and television and newspaper advertising. Hence, the computations of the pure error sum of squares (SSPE) are summarized in Table 10.6. Since

TABLE 10.6 Computation of the Pure Error Sum of Squares for the Advertising Media Problem

X	Y_{jk}	$\bar{Y}_j$	n_j	$Y_{jk} - \bar{Y}_j$	$(Y_{jk} - \bar{Y}_j)^2$
$X_1 = 0$	9.73			$-.730$	.532900
$X_2 = 20$	11.19	10.460	2	$+.730$	.532900
$X_1 = 5$	8.75			$+1.250$	1.562500
$X_2 = 5$	6.25	7.500	2	-1.250	1.562500
$X_1 = 10$	9.10			$-.305$	.093025
$X_2 = 10$	9.71	9.405	2	$+.305$	.093025
$X_1 = 15$	9.31			-1.230	1.512900
$X_2 = 15$	11.77	10.540	2	$+1.230$	1.512900
$X_1 = 20$	8.82			$-.500$	.250000
$X_2 = 5$	9.82	9.320	2	$+.500$	.250000
$X_1 = 25$	16.28			$+.255$	.065025
$X_2 = 25$	15.77	16.025	2	$-.255$	.065025
$X_1 = 30$	10.44			$+.650$	.422500
$X_2 = 0$	9.14	9.790	2	$-.650$	.422500
$X_1 = 35$	13.29			$-.005$	.000025
$X_2 = 5$	13.30	13.295	2	$+.005$	.000025
$X_1 = 40$	14.05			$-.155$	.024025
$X_2 = 10$	14.36	14.205	2	$+.155$	.024025
$X_1 = 45$	15.21			-1.100	1.210000
$X_2 = 15$	17.41	16.310	2	$+1.100$	1.210000
$X_1 = 50$	18.66			$+.745$	.555025
$X_2 = 20$	17.17	17.915	2	$-.745$	.555025
				SSPE =	12.455850

$$SSPE = 12.455850$$

from (10.18) we have

$$SSLF = 18.121672 - 12.455850$$

$$= 5.665822$$

The pure error component has $\sum_{j=1}^{l} (n_j - 1)$ degrees of freedom

where n_j = number of observations Y for combination j of variables X_1 and X_2
l = number of different combinations of levels of variables X_1 and X_2 with repeat measurements.

Thus for our example there are 11 degrees of freedom for the pure error component since there are two observations at each of the 11 different combinations of levels of variables X_1 and X_2. Therefore, since

$$\text{MSPE} = \frac{\text{SSPE}}{\sum_{j=1}^{l}(n_j - 1)} \tag{10.19}$$

we have

$$\text{MSPE} = \frac{12.455850}{11}$$

$$= 1.13235$$

The variance due to lack of fit can be computed from

$$\text{MSLF} = \frac{\text{SSLF}}{(n - p - 1) - \sum_{j=1}^{l}(n_j - 1)} \tag{10.20}$$

The degrees of freedom for the lack of fit mean square represents the difference in the degrees of freedom of the error and pure error components. Thus for our data we have

$$\text{MSLF} = \frac{5.665822}{19 - 11} = \frac{5.665822}{8}$$

$$= .708228$$

The null and alternative hypotheses for testing for lack of fit would be

H_0:   there is no lack of fit in the regression model

H_1:   there is lack of fit in the regression model

The test statistic for these hypotheses is

$$F = \frac{\text{MSLF}}{\text{MSPE}} \tag{10.21}$$

where

$$F \sim F_{n-p-1-\sum_{j=1}^{l}(n_j-1),\,\sum_{j=1}^{l}(n_j-1)}$$

and using a level of significance of α, H_0 may be rejected if

$$F \geq F_{1-\alpha;\,n-p-1-\sum_{j=1}^{l}(n_j-1),\,\sum_{j=1}^{l}(n_j-1)}$$

Thus for the advertising media problem we have

$$F = \frac{.708228}{1.13235} = .625$$

If a level of significance of .05 is selected, $F = .625 < F_{.95;\,8,11} = 2.95$, and so we do not reject H_0. We therefore conclude that there is no evidence of any lack of fit in the multiple regression model.

10.8 RESIDUAL ANALYSIS IN MULTIPLE REGRESSION MODELS

In Section 8.11 we also utilized residual analysis to evaluate whether the simple linear regression model was appropriate for the set of data being analyzed. Since tests of lack of fit are often not possible, residual analysis is a particularly useful tool for evaluating the fit of a multiple regression model. When examining a multiple linear regression model with two independent variables, the following residual plots are of particular interest:

1. Residuals $(Y_i - \hat{Y}_i)$ versus predicted values $\hat{Y}_i$
2. Residuals $(Y_i - \hat{Y}_i)$ versus X_1
3. Residuals $(Y_i - \hat{Y}_i)$ versus X_2
4. Residuals $(Y_i - \hat{Y}_i)$ versus X_1 for different levels of X_2

The first residual plot examines the pattern of residuals for the predicted values of Y. If the residuals appear to vary for different levels of the predicted Y value, it provides evidence of a possible curvilinear effect in at least one independent variable and/or the need to transform the dependent variable (see Section 11.6). The second and third residual plots involve the independent variables. Patterns in the plot of the residuals versus an independent variable may indicate the existence of a curvilinear effect in the relationship of Y and the independent variable or indicate the need for a possible transformation of the independent variable. The last plot is utilized to investigate the possible interaction of the two independent variables. Separate residual plots can be obtained for variable X_1 at low values of X_2 and high values of X_2. If the pattern of residuals were different between the two plots, it would lead to the possible inclusion of an interaction term (such as $X_1 X_2$) in the regression model (see Section 11.6).

The residual plots discussed here can be obtained as part of the output of the SPSS NEW REGRESSION subprogram, SAS GLM procedure, or BMDP program 1R (see Section 10.12). Figure 10.5 consists of the residual output obtained by using the appropriate options of the SAS GLM procedure.

We can observe from Figure 10.5 that there appears to be very little or no pattern in the relationship between the residuals and either the predicted value $\hat{Y}_i$, the value of X_1 (radio and television advertising), and the value of X_2 (newspaper advertising). Thus one may conclude that the multiple linear regression model is an appropriate model for the advertising media problem.

10.9 INFERENCES CONCERNING THE POPULATION REGRESSION COEFFICIENTS

Now that we have developed our multiple regression model, we are interested in estimating its true population parameters. In multiple regression, a confidence interval estimate of the true regression coefficient β_k can be obtained from

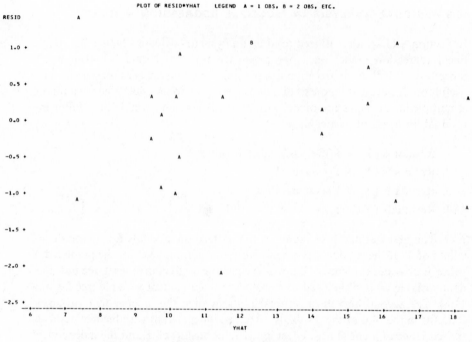

Figure 10.5 Residual plots for advertising media problem.

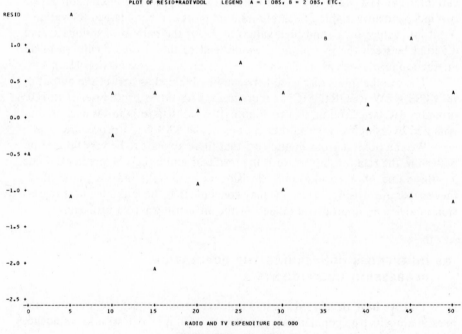

Figure 10.5 (Continued)

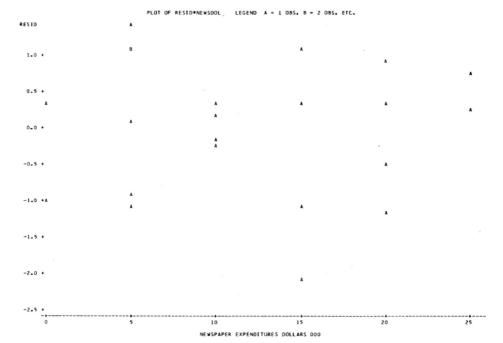

Figure 10.5 (Continued)

$$b_k \pm t_{1-\frac{\alpha}{2}; n-p-1} S_{b_k} \qquad (10.22)$$

For example, if we wish to obtain a 95% confidence interval estimate of the true slope β_1 (that is, the effect of radio and television advertising on monthly sales of electronic games, holding constant the effect of newspaper advertising X_2), we would have, from (10.22) and Figure 10.2,

$$b_1 \pm t_{.975; 19} S_{b_1}$$

so that

$$.1621127 \pm 2.093(.0131908)$$
$$.1621127 \pm .0276083$$

and

$$.134504 \le \beta_1 \le .189720$$

Thus, taking into account the effect of newspaper advertising, we estimate that the true effect of radio and television advertising is to increase monthly sales by between \$134,504 and \$189,720 for each increase of one thousand dollars in expenditures. Furthermore, we have 95% confidence that this interval correctly estimates the true relationship between these variables. Of course, from a hypothesis testing perspective, since this confidence interval did not include zero, the regression coefficient β_1 would be considered to have a significant effect on the overall model.

10.10 INFERENCE ABOUT THE MEAN RESPONSE $\mu_{Y|X_{12\cdots p}}$ AND INDIVIDUAL RESPONSE $\hat{Y}_I$

Now that the multiple regression model has been developed, it can be used for prediction purposes. For example, suppose that we wanted to predict the monthly sales for a city in which radio and television advertising expenditures are $20,000 and newspaper advertising expenditures are also $20,000. By using our multiple regression equation,

$$\hat{Y}_i = 5.2574 + .1621X_{1_i} + .2489X_{2_i}$$

with $X_{1_i} = 20$ and $X_{2_i} = 20$, we have

$$\hat{Y}_i = 5.2574 + .1621(20) + .2489(20)$$

and thus

$$\hat{Y}_i = 13.4774$$

Therefore, we would predict that monthly sales in that city would be approximately 13.4774 millions of dollars ($13,477,400).

Since this is merely the point estimate, it becomes important to develop a confidence interval estimate of the true average predicted response $\mu_{Y|X_{12}}$ as well as a prediction interval for an individual response $\hat{Y}_I$. However, as previously mentioned in our discussion of the standard error of the regression coefficients, the formulas used are rather complex and are usually expressed in matrix notation (see Sections 10.11.5 and 10.11.6). Nevertheless, it should be noted that one of the options of the GLM procedure of SAS provides these confidence and prediction interval estimates for each observation in the sample (see Figure 10.6). On the other hand, if either confidence or prediction interval estimates are desired for a combination of levels of X_1 and X_2 that are not

OBSERVATION	OBSERVED VALUE	PREDICTED VALUE	RESIDUAL	LOWER 95% CL FOR MEAN	UPPER 95% CL FOR MEAN
1	9.73000000	10.23473650	-0.50473650	9.26878078	11.20069221
2	11.19000000	10.23473650	0.95526350	9.26878078	11.20069221
3	8.75000000	7.31228438	1.43771562	6.51970191	8.10486685
4	6.25000000	7.31228438	-1.06228438	6.51970191	8.10486685
5	9.10000000	9.36718641	-0.26718641	8.76086593	9.97350690
6	9.71000000	9.36718641	0.34281359	8.76086593	9.97350690
7	9.31000000	11.42208844	-2.11208844	10.86833052	11.97584637
8	11.77000000	11.42208844	0.34791156	10.86833052	11.97584637
9	8.82000000	9.74397487	-0.92397487	9.14279774	10.34515199
10	9.82000000	9.74397487	0.07602513	9.14279774	10.34515199
11	16.28000000	15.53189250	0.74810750	14.64676443	16.41702057
12	15.77000000	15.53189250	0.23810750	14.64676443	16.41702057
13	10.44000000	10.12076332	0.31923668	9.28585731	10.95566933
14	9.14000000	10.12076332	-0.98076332	9.28585731	10.95566933
15	13.29000000	12.17566535	1.11433465	11.51412563	12.83720507
16	13.30000000	12.17566535	1.12433465	11.51412563	12.83720507
17	14.05000000	14.23056738	-0.18056738	13.61592333	14.84521143
18	14.36000000	14.23056738	0.12943262	13.61592333	14.84521143
19	15.21000000	16.28546941	-1.07546941	15.56607691	17.00486191
20	17.41000000	16.28546941	1.12453059	15.56607691	17.00486191
21	18.66000000	18.34037144	0.31962856	17.41470975	19.26603313
22	17.17000000	18.34037144	-1.17037144	17.41470975	19.26603313

Figure 10.6 Confidence intervals obtained from SAS PROC GLM for the media expenditures model.

present in the observed data, then the matrix approach discussed in Sections 10.11.5 and 10.11.6 can be utilized.

10.11 A MATRIX APPROACH TO MULTIPLE REGRESSION

In Chapter 9, we developed the matrix algebra approach to the formulation of the linear regression model. Since the matrix approach does not distinguish between models with one or more independent variables, we can apply the matrix terminology to the multiple regression model. In fact, the matrix approach is particularly appropriate for the multiple regression model because it offers a concise representation of a set of procedures that otherwise could be expressed only with complex formulas.

10.11.1 Developing the Multiple Regression Equation

To fix ideas, let us generalize the multiple linear regression model (10.1) to one containing p predictor variables, that is,

$$Y_i = \beta_0 + \beta_1 X_{1_i} + \beta_2 X_{2_i} + \cdots + \beta_p X_{p_i} + \epsilon_i \qquad (10.23)$$

where $\beta_0 = Y$ intercept

$\beta_1 = $ slope of Y with X_1, holding constant $X_2, X_3, \ldots, X_p$
$\beta_2 = $ slope of Y with X_2, holding constant $X_1, X_3, \ldots, X_p$
$\beta_p = $ slope of Y with X_p, holding constant $X_1, X_2, \ldots, X_{p-1}$
$\epsilon_i = $ random error in Y for observation i

Expressing (10.23) in matrix form, we have

$$\mathbf{Y} = \mathbf{X}\boldsymbol{\beta} + \boldsymbol{\epsilon} \qquad (10.24)$$

where $\mathbf{Y} = N \times 1$ column vector of observations Y_i $(i = 1, 2, \ldots, N)$
$\mathbf{X} = N \times (p + 1)$ matrix of independent variables
$\boldsymbol{\beta} = (p + 1) \times 1$ column vector of parameters $(\beta_0, \beta_1, \beta_2, \ldots, \beta_p)$
$\boldsymbol{\epsilon} = N \times 1$ column vector of residual terms ϵ_i

Drawing a sample of n observations from this population and using the principle of *least squares*, we wish to obtain $\hat{\boldsymbol{\beta}}$, a $(p + 1) \times 1$ column vector of sample regression coefficients (that is, $b_0, b_1, b_2, \ldots,$ and b_p), so that the resulting sample regression equation provides the *best linear fit* to the observed data. To accomplish this, the *normal equations* can be written as

$$\mathbf{X}'\mathbf{X}\hat{\boldsymbol{\beta}} = \mathbf{X}'\mathbf{Y} \qquad (10.25)$$

where, from the sample,

$\mathbf{Y} = n \times 1$ column vector of observations Y_i $(i = 1, 2, \ldots, n)$
$\mathbf{X} = n \times (p + 1)$ matrix of independent variables
$\mathbf{X}' = $ transpose of $\mathbf{X}$
$\hat{\boldsymbol{\beta}} = (p + 1) \times 1$ column vector of sample regression coefficients

As in (9.27), solving for the elements of $\hat{\boldsymbol{\beta}}$, we obtain

$$\hat{\boldsymbol{\beta}} = (\mathbf{X}'\mathbf{X})^{-1}\mathbf{X}'\mathbf{Y} \tag{10.26}$$

In general, for a sample containing information on p predictor variables, the $\mathbf{X}$ matrix takes the form

$$\underset{n \times (p+1)}{\mathbf{X}} = \begin{bmatrix} 1 & X_{1_1} & X_{2_1} & \cdots & X_{p_1} \\ 1 & X_{1_2} & X_{2_2} & \cdots & X_{p_2} \\ \cdot & \cdot & \cdot & & \cdot \\ \cdot & \cdot & \cdot & \cdot & \cdot \\ \cdot & \cdot & \cdot & & \cdot \\ 1 & X_{1_n} & X_{2_n} & \cdots & X_{p_n} \end{bmatrix}$$

More specifically, for our advertising media problem involving but two independent variables, this $\mathbf{X}$ matrix is of dimension $n \times 3$. Thus we may obtain the $\mathbf{X}'\mathbf{X}$ product matrix as

$$\mathbf{X}'\mathbf{X} = \begin{bmatrix} n & \sum_{i=1}^{n} X_{1_i} & \sum_{i=1}^{n} X_{2_i} \\ \sum_{i=1}^{n} X_{1_i} & \sum_{i=1}^{n} X_{1_i}^2 & \sum_{i=1}^{n} X_{1_i}X_{2_i} \\ \sum_{i=1}^{n} X_{2_i} & \sum_{i=1}^{n} X_{1_i}X_{2_i} & \sum_{i=1}^{n} X_{2_i}^2 \end{bmatrix}$$

and also

$$\mathbf{X}'\mathbf{Y} = \begin{bmatrix} \sum_{i=1}^{n} Y_i \\ \sum_{i=1}^{n} X_{1_i}Y_i \\ \sum_{i=1}^{n} X_{2_i}Y_i \end{bmatrix}$$

From Table 10.1 the following summary computations are obtained:

$$\sum_{i=1}^{n} X_{1_i} = 550 \qquad \sum_{i=1}^{n} X_{1_i}^2 = 19{,}250 \qquad \sum_{i=1}^{n} X_{1_i}Y_i = 7667.20$$

$$\sum_{i=1}^{n} X_{2_i} = 260 \qquad \sum_{i=1}^{n} X_{2_i}^2 = 4300 \qquad \sum_{i=1}^{n} X_{2_i}Y_i = 3515.10$$

$$\sum_{i=1}^{n} Y_i = 269.53 \qquad \sum_{i=1}^{n} Y_i^2 = 3552.8893 \qquad \sum_{i=1}^{n} X_{1_i}X_{2_i} = 6650$$

$$n = 22$$

Thus

$$\mathbf{X}'\mathbf{X} = \begin{bmatrix} 22 & 550 & 260 \\ 550 & 19{,}250 & 6650 \\ 260 & 6650 & 4300 \end{bmatrix}$$

and, from Section 9.6, its inverse is

$$(\mathbf{X'X})^{-1} = \begin{bmatrix} .26048106 & -.00429715 & -.00910442 \\ -.00429715 & .00018243 & -.00002230 \\ -.00910442 & -.00002230 & .00081754 \end{bmatrix}$$

By using (10.26), the regression coefficients are computed:

$$\hat{\boldsymbol{\beta}} = \begin{bmatrix} b_0 \\ b_1 \\ b_2 \end{bmatrix} = \begin{bmatrix} .26048106 & -.00429715 & -.00910442 \\ -.00429715 & .00018243 & -.00002230 \\ -.00910442 & -.00002230 & .00081754 \end{bmatrix} \begin{bmatrix} 269.53 \\ 7667.20 \\ 3515.10 \end{bmatrix}$$

$$= \begin{bmatrix} 5.2573824 \\ .1621127 \\ .2488677 \end{bmatrix}$$

and the sample regression equation (10.2) can be stated:

$$\hat{Y}_i = 5.2573824 + .1621127X_{1_i} + .2488677X_{2_i}$$

Fortunately the researcher does not have to compute the regression coefficients manually. They are simply interpreted from the output of a computer package. In fact, in the case of the SAS package, the use of the I option of the GLM procedure (see Section 10.12.2) provides for the computation and printout of the matrix $(\mathbf{X'X})^{-1}$.

10.11.2 Preparation for Inference: Developing the ANOVA Table

In Section 10.3 the existence of a relationship between the dependent variable and the set of p independent variables was investigated through the F test in the ANOVA table (see Table 10.2). From Section 9.8.2, the corresponding ANOVA table using matrix notation is presented as Table 10.7. Using matrix notation, we may write

$$\text{SSR} = \hat{\boldsymbol{\beta}}'\mathbf{X'Y} - \frac{\left(\sum_{i=1}^{n} Y_i \right)^2}{n} \tag{10.27}$$

$$\text{SSE} = \mathbf{Y'Y} - \hat{\boldsymbol{\beta}}'\mathbf{X'Y} \tag{10.28}$$

and

$$\text{SST} = \mathbf{Y'Y} - \frac{\left(\sum_{i=1}^{n} Y_i \right)^2}{n} \tag{10.29}$$

For our advertising media problem we compute

$$\text{SSR} = [5.2573824 \quad .1621127 \quad .2488677] \begin{bmatrix} 269.53 \\ 7667.20 \\ 3515.10 \end{bmatrix} - \frac{(269.53)^2}{22} = 232.6575869$$

TABLE 10.7 ANOVA Table Using Matrix Notation

Source of Variation	Degrees of Freedom	Sum of Squares	Mean Square	F
Due to Regression	p	$\text{SSR} = \hat{\boldsymbol{\beta}}'\mathbf{X}'\mathbf{Y} - \dfrac{\left(\sum\limits_{i=1}^{n} Y_i\right)^2}{n}$	$\text{MSR} = \dfrac{\text{SSR}}{p}$	$F = \dfrac{\text{MSR}}{\text{MSE}}$
Error	$n - p - 1$	$\text{SSE} = \mathbf{Y}'\mathbf{Y} - \hat{\boldsymbol{\beta}}'\mathbf{X}'\mathbf{Y}$	$\text{MSE} = \dfrac{\text{SSE}}{n - p - 1}$	
Total	$n - 1$	$\text{SST} = \mathbf{Y}'\mathbf{Y} - \dfrac{\left(\sum\limits_{i=1}^{n} Y_i\right)^2}{n}$	—	

$$\text{SSE} = [9.73 \quad 11.19 \quad \cdots \quad 17.17] \begin{bmatrix} 9.73 \\ 11.19 \\ \cdot \\ \cdot \\ \cdot \\ 17.17 \end{bmatrix}$$

$$- [5.2573824 \quad .1621127 \quad .2488677] \begin{bmatrix} 269.53 \\ 7667.20 \\ 3515.10 \end{bmatrix} = 18.1216722$$

and

$$\text{SST} = [9.73 \quad 11.19 \quad \cdots \quad 17.17] \begin{bmatrix} 9.73 \\ 11.19 \\ \cdot \\ \cdot \\ \cdot \\ 17.17 \end{bmatrix} - \frac{(269.53)^2}{22} = 250.7792591$$

From the ANOVA table the mean square terms are obtained, and then F is computed using (10.4) with the same results as in Section 10.3.

10.11.3 The Variance-Covariance Matrix S_b

As described in Section 9.8.3, S_b, the variance-covariance matrix of the vector of sample regression coefficients, is useful for purposes of inference.

S_b is a symmetric matrix. The elements along the main diagonal represent the variances of the sample regression coefficients (i.e., the squares of the standard errors $S_{b_0}, S_{b_1}, S_{b_2}, \ldots, S_{b_p}$), while the elements off the main diagonal represent the covariances between pairs of variables.

The variance-covariance matrix of the vector $\hat{\beta}$ is obtained from

$$S_b = (X'X)^{-1}\text{MSE} \qquad (10.30)$$

where the scalar MSE, the mean square error term (i.e., the variance $S^2_{Y|X_{12}\cdots p}$ around the sample regression surface), is defined in Table 10.7 as

$$\text{MSE} = S^2_{Y|X_{12}\cdots p} = \frac{Y'Y - \hat{\beta}'X'Y}{n - p - 1} \qquad (10.31)$$

Thus for our advertising media problem we have

$$\text{MSE} = .95377222$$

and

$$S_b = \begin{bmatrix} .26048106 & -.00429715 & -.00910442 \\ -.00429715 & .00018243 & -.00002230 \\ -.00910442 & -.00002230 & .00081754 \end{bmatrix} (.95377222)$$

$$= \begin{bmatrix} .248439599 & -.0040985023 & -.0086835429 \\ -.0040985023 & .0001739967 & -.0000212691 \\ -.0086835429 & -.0000212691 & .0007797469 \end{bmatrix}$$

so that from the main diagonal

$$S_{b_0}^2 = .248439599$$
$$S_{b_1}^2 = .0001739967$$
$$S_{b_2}^2 = .0007797469$$

Once again, these values could also be obtained from the GLM procedure of the SAS package (see Section 10.12.2).

10.11.4 Estimating the True Regression Parameters

To obtain *separate* $100(1 - \alpha)\%$ confidence interval estimates for the $p + 1$ regression parameters, we merely extract the appropriate elements from the $\hat{\boldsymbol{\beta}}$ vector and its corresponding variance-covariance matrix S_b. Thus, as in (10.22), we have

$$b_k \pm t_{1-\frac{\alpha}{2};\, n-p-1} S_{b_k} \tag{10.22}$$

and for the advertising media example the following 95% confidence interval estimates are obtained:

$$+4.21415 \leq \beta_0 \leq +6.30061 \quad \text{(in millions of dollars)}$$
$$+.13450 \leq \beta_1 \leq +.18972 \quad \text{(in millions of dollars)}$$
$$+.19042 \leq \beta_2 \leq +.30731 \quad \text{(in millions of dollars)}$$

10.11.5 Estimating the True Mean Response $\mu_{Y|X_{12...p}}$

In Section 10.10 we noted that the computations involved in obtaining a confidence interval estimate of an average predicted value (i.e., mean response) at a given combination of $X_1, X_2, \ldots, X_p$ levels were laborious. Moreover, the CLM option of the GLM procedure of the SAS package (see Section 10.12.2) will provide these estimates only for the observed combinations of the $X_1, X_2, \ldots, X_p$ levels (i.e., the observations in the sample). Thus for a given combination of $X_1, X_2, \ldots, X_p$ levels in the respective relevant ranges, the following $100(1 - \alpha)\%$ confidence interval statement for $\mu_{Y|X_{12...p}}$ can be expressed in matrix terms as

$$\mathbf{X}_g'\hat{\boldsymbol{\beta}} \pm t_{1-\frac{\alpha}{2};\, n-p-1}\{[\mathbf{X}_g'(\mathbf{X}'\mathbf{X})^{-1}\mathbf{X}_g]\text{MSE}\}^{1/2} \tag{10.32}$$

where the column vector $\mathbf{X}_g$ is defined as

$$\mathbf{X}_g = \begin{bmatrix} 1 \\ X_{1_g} \\ X_{2_g} \\ \cdot \\ \cdot \\ \cdot \\ X_{p_g} \end{bmatrix}$$

with the transpose

$$\mathbf{X}'_g = [1 \quad X_{1_g} \quad X_{2_g} \quad \cdots \quad X_{p_g}]$$

and whose elements represent the *given* combination of $X_1, X_2, \ldots, X_p$ levels necessary for the desired prediction. Moreover,

$$\mathbf{X}'_g \hat{\boldsymbol{\beta}} = [1 \quad X_{1_g} \quad X_{2_g} \quad \cdots \quad X_{p_g}] \begin{bmatrix} b_0 \\ b_1 \\ b_2 \\ \cdot \\ \cdot \\ \cdot \\ b_p \end{bmatrix}$$

is simply the point estimate $\hat{Y}_g$, the predicted value of Y for the given combination of $X_1, X_2, \ldots, X_p$ levels.

For example, in the advertising media problem we might be interested in obtaining a 95% confidence interval estimate of the average monthly sales for all cities in which radio and television advertising is \$20,000 ($X_1 = 20$) and newspaper advertising is \$20,000 ($X_2 = 20$). Since this combination of X_1 and X_2 did not appear among the 22 sample observations, this confidence interval would not have been estimated as part of the output of the GLM procedure of the SAS package.

For these data, $X_{1_g} = 20$ and $X_{2_g} = 20$ so that

$$\hat{Y}_g = \mathbf{X}'_g \hat{\boldsymbol{\beta}} = [1 \quad 20 \quad 20] \begin{bmatrix} 5.2574 \\ .1621 \\ .2489 \end{bmatrix} = 13.4774 \qquad \text{(in millions of dollars)}$$

This, of course, is the same result obtained in Section 10.10. However, by using (10.32), the following 95% confidence interval estimate of the mean sales is obtained:

$$13.4774 \pm 2.093 \left\{ [1 \quad 20 \quad 20] \begin{bmatrix} .26048106 & -.00429715 & -.00910442 \\ -.00429715 & .00018243 & -.00002230 \\ -.00910442 & -.00002230 & .00081754 \end{bmatrix} \right.$$

$$\left. \times \begin{bmatrix} 1 \\ 20 \\ 20 \end{bmatrix} (.95377222) \right\}^{1/2}$$

$$13.4774 \pm 2.093(.31881)$$

$$13.4774 \pm .6673$$

so that

$$12.8101 \leq \mu_{Y|X_{12}} \leq 14.1447 \qquad \text{(in millions of dollars)}$$

or

$$\$12,810,100 \leq \mu_{Y|X_{12}} \leq \$14,144,700$$

10.11.6 Predicting the Individual Response $\hat{Y}_I$

In Section 10.10 we also noted that the computations involved in obtaining a prediction interval estimate for an individual response at a given combination of $X_1, X_2, \ldots, X_p$ levels were laborious. Furthermore, the CLI option of the GLM procedure of the SAS package (see Section 10.12.2) will only provide these estimates for the observed sample values of $X_1, X_2, \ldots, X_p$. Thus for a given (desired) combination of $X_1, X_2, \ldots, X_p$ levels in the respective relevant ranges, the $100(1 - \alpha)\%$ prediction interval statement for $\hat{Y}_I$ can be written as follows using matrix notation:

$$\mathbf{X}_g'\hat{\boldsymbol{\beta}} \pm t_{1-\frac{\alpha}{2}; n-p-1}(\{1 + [\mathbf{X}_g'(\mathbf{X}'\mathbf{X})^{-1}\mathbf{X}_g]\}MSE)^{1/2} \tag{10.33}$$

We note that the only difference between (10.32) and (10.33) is the scalar 1 added to the value of $[\mathbf{X}_g'(\mathbf{X}'\mathbf{X})^{-1}\mathbf{X}_g]$ in (10.33). Therefore, by using (10.33) in the advertising media example, if a 95% prediction interval estimate is desired for monthly sales for a particular city in which \$20,000 is spent on radio and television advertising ($X_1 = 20$) and \$20,000 is spent on newspaper advertising ($X_2 = 20$), we have

$$13.4774 \pm 2.093\left(\left\{1 + [1\ 20\ 20]\begin{bmatrix} .26048106 & -.00429715 & -.00910442 \\ -.00429715 & .00018243 & -.00002230 \\ -.00910442 & -.00002230 & .00081754 \end{bmatrix} \times \begin{bmatrix} 1 \\ 20 \\ 20 \end{bmatrix}\right\}(.95377222)\right)^{1/2}$$

$13.4774 \pm 2.093(1.02733)$

13.4774 ± 2.1502

so that

$$11.3272 \le \hat{Y}_I \le 15.6276 \qquad \text{(in millions of dollars)}$$

or

$$\$11,327,200 \le \hat{Y}_I \le \$15,627,600$$

10.11.7 A Matrix Approach: An Overview

If we compare the various matrix expressions in Section 10.11 to those given in Section 9.8, we shall observe that they are the *same*—regardless of the number of independent variables in the model. Hence we can appreciate the use of matrix notation as a way of compactly presenting lengthy and sometimes complex formulas in as simple a manner as possible. Nevertheless, once we understand the fundamental concepts and know how to use the various inferential procedures, we would greatly facilitate our research efforts if we utilize a computer package. This will be the subject of Section 10.12.

10.12 COMPUTER PACKAGES AND MULTIPLE REGRESSION: USE OF SPSS, SAS, AND BMDP

In this section we shall explain how to use either SPSS, SAS, or BMDP to analyze the multiple regression data of the advertising media problem (see Table 10.1). The four variables involved can be organized for data entry purposes with the city number in columns 1 and 2, monthly sales in columns 4–8, radio and television advertising expenditures in columns 10 and 11, and newspaper advertising expenditures in columns 13 and 14.

10.12.1 Using the SPSS Subprogram NEW REGRESSION for the Multiple Regression Model

The NEW REGRESSION subprogram of SPSS [see Hull and Nie (1981)] can be used to perform a regression analysis. The basic setup of the NEW REGRESSION procedure statements is:

```
1                    16
NEWbREGRESSIONbDESCRIPTIVES/
                    VARIABLES = {list of all variables}/
                    STATISTICS = DEFAULTS, HISTORY/
                    DEPENDENT = {name of dependent variable}/
                    STEPWISE/
                    RESIDUALS/
                    SCATTERPLOT = (var₁, var₂), . . ./
```

The DESCRIPTIVES statement provides the mean, standard deviation and correlations for all specified variables. The VARIABLES = statement provides a list of all the variables to be included in the model. The STATISTICS = statement provides various statistics for the regression equation. The DEFAULTS command specifies the multiple correlation coefficients, the analysis of variance table, and the regression coefficients and statistics for those variables that are included in the model as well as those that have not yet been entered into the model. The HISTORY command provides a history of the step by step entry of variables into the model. The DEPENDENT = statement names the dependent variable while the STEPWISE statement indicates that the variables are being entered into the model one at a time (see Chapter 13). The RESIDUALS statement provides a histogram and normal probability plot for the standardized residuals as well as a list of the ten largest residuals (i.e., possible outliers). The SCATTERPLOT = statement provides for residual plots. The standardized residuals are given the name *ZRESID and the predicted Y values ($\hat{Y}_i$) are given the name *PRED. Thus for the advertising media problem the SCATTERPLOT statement would be

```
16
SCATTERPLOT = (*ZRESID, SALES), (*ZRESID, RADTVDOL),
(*ZRESID, NEWSDOL)/
```

The SCATTERPLOT statement must follow the last regression method statement (i.e., the STEPWISE statement). Figure 10.7 illustrates the complete SPSS program that was written for the advertising media problem. Figure 10.8 represents annotated partial output of the SPSS subprogram NEW REGRESSION for the advertising media problem.

```
{SYSTEM CARDS}
RUN NAME              SALES AND MEDIA EXPENDITURES
DATA LIST             FIXED(1)/1 CITY 1-2, SALES 4-8,
                      RADTVDOL 10-11, NEWSDOL 13-14,
VAR LABELS            CITY, NUMBER OF CITY/
                      SALES, SALES IN $MILLIONS/
                      RADTVDOL, RADIO AND TV $000/
                      NEWSDOL, NEWSPAPER $000/
READ INPUT DATA
01  09.73  00  20
02  11.19  00  20
:     :     :   :
22  17.17  50  20
END INPUT DATA
NEW REGRESSION        DESCRIPTIVES/VARIABLES=SALES
                      TO NEWSDOL/
                      STATISTICS=DEFAULTS, HISTORY/
                      DEPENDENT=SALES/STEPWISE/
                      RESIDUALS/
                      SCATTERPLOT=((*ZRESID,SALES)),
                      (*ZRESID,RADTVDOL)),
                      (*ZRESID,NEWSDOL)/
```

Figure 10.7 SPSS program for media expenditures example.

10.12.2 Using the SAS GLM and CORR Procedures for the Multiple Regression Model

We may recall that in Chapter 8 we used the SAS GLM procedure to perform a simple linear regression analysis and the PLOT procedure to obtain a two-dimensional plot or scatter diagram of the dependent and independent variables.

The setup for the GLM and PLOT procedures were

PROCƀGLM;

$$\text{MODEL}ƀ \left\{ \begin{matrix} \text{dependent} \\ \text{variable} \end{matrix} \right\} = \left\{ \begin{matrix} \text{list of} \\ \text{independent} \\ \text{variables} \end{matrix} \right\} / \{\text{options}\};$$

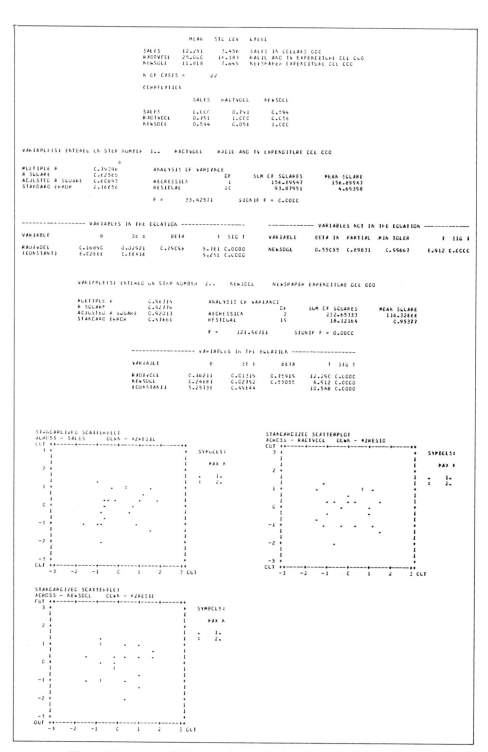

Figure 10.8 Partial SPSS output for media expenditures problem.

PROC♭PLOT;

$$\text{PLOT♭} \begin{Bmatrix} \text{dependent} \\ \text{variable} \end{Bmatrix} * \begin{Bmatrix} \text{independent} \\ \text{variable} \end{Bmatrix};$$

The options involved in these procedures will be expanded upon to enable us to obtain residuals and other output for the multiple regression model. Although many options are available for use with the MODEL statement [see SAS (1979)], we shall focus upon the options entitled P, I, CLM, and CLI. The P option provides the observed, predicted, and residual values for each observation and also computes the Durbin-Watson statistic (see Section 14.4). The I (or INVERSE) option provides the inverse of the $\mathbf{X'X}$ matrix (see Section 10.11.1), which is particularly useful when employing the matrix approach to multiple regression. The CLM option provides confidence limits for each average response $\mu_{Y|X_{12\ldots P,}}$ while the CLI option provides for prediction intervals for $\hat{Y}_I$ (see Sections 10.10, 10.11.5, and 10.11.6).

In addition to the options available for the MODEL statement, we can also obtain residual plots as part of the GLM procedure. This can be achieved by using an OUTPUT statement directly following the MODEL statement of GLM. The form of the OUTPUT statement is

$$\text{OUTPUT♭OUT} = \begin{Bmatrix} \text{name of} \\ \text{new data set} \end{Bmatrix} \text{♭PREDICTED}$$

$$= \begin{Bmatrix} \text{name of} \\ \text{predicted } Y \text{ values} \end{Bmatrix} \text{♭RESIDUAL} = \begin{Bmatrix} \text{name of} \\ \text{residuals} \end{Bmatrix};$$

Figure 10.9 illustrates the SAS program that has been written to analyze the advertising media problem (see Figures 10.2–10.6). We can observe that in the OUTPUT statement the new data set has been assigned the name NEW, the predicted Y values are called YHAT, and the residuals are called RESID. This statement is immediately followed by a PLOT statement of the form

$$\text{PROC♭PLOT♭DATA} = \begin{Bmatrix} \text{name of} \\ \text{new data set} \end{Bmatrix};$$

$$\text{PLOT♭} \begin{Bmatrix} \text{name of} \\ \text{residual} \end{Bmatrix} * \{\text{variable}\};$$

In the PLOT statement for this procedure, the residuals are plotted against the predicted Y values and the independent variables.

The CORR procedure

PROC♭CORR;
VAR♭{list of variables};

provides for a correlation matrix of all variables of interest. In Figure 10.9 the statements

PROC♭CORR;
VAR♭SALES♭RADTVDOL♭NEWSDOL;

produce the matrix displayed in Table 10.3.

```
{SYSTEM CARDS}
DATA SALES;
    INPUT CITY 1-2 SALES 4-8 RADTVDOL 10-11
    NEWSDOL 13-14;
LABEL CITY=NUMBER OF CITY
      SALES=SALES IN $MILLIONS
      RADTVDOL=RADIO AND TV $000
      NEWSDOL=NEWSPAPER $000;
CARDS;
01 09.73 00 20
  :     :  :  :
22 17.17 50 20
PROC PRINT;
TITLE SALES AND MEDIA EXPENDITURES;
PROC CORR;
    VAR SALES RADTVDOL NEWSDOL;
PROC GLM;
    MODEL SALES=RADTVDOL NEWSDOL/P I CLM;
    OUTPUT OUT=NEW PREDICTED=YHAT RESIDUAL=RESID;
PROC PLOT DATA=NEW;
    PLOT RESID*YHAT RESID*RADTVDOL RESID*NEWSDOL;
PROC GLM;
    MODEL SALES=RADTVDOL;
PROC GLM;
    MODEL SALES=NEWSDOL;
{SYSTEM CARDS}
```

Figure 10.9 SAS program for media expenditures example.

10.12.3 Using the BMDP Program 1R for the Multiple Regression Model

The BMDP program 1R can be used for both simple regression analysis and multiple regression analysis. The only paragraphs of BMDP (see Section 2.7) that refer specifically to P-1R are the /REGRESS, /PRINT, and /PLOT paragraphs. The form of the /REGRESS paragraph is

$$/REGRESS\flat DEPENDENT\flat IS\flat \left\{ \begin{array}{l} \text{name of} \\ \text{dependent variable} \end{array} \right\} .$$

$$INDEPENDENT\flat ARE\flat \left\{ \begin{array}{l} \text{names of} \\ \text{independent variables} \end{array} \right\} .$$

The DEPENDENT sentence of the /REGRESS paragraph names the dependent variable, while the INDEPENDENT sentence names the set of independent variables. Although various options are available from P-1R [see Dixon et al. (1981)] for the /PRINT and /PLOT paragraphs, we shall focus on the /PRINT options DATA, CORR, and COV and the /PLOT options RESID and VARIABLES.

For the /PRINT paragraph the DATA option provides the observed, predicted, and residual values along with the values of the independent variables. The CORR option provides the correlation matrix, while the COV option provides the variance-covariance matrix (see Section 10.11.3) for the variables.

The /PLOT paragraph provides various residual and other plots of the dependent variable. The RESID option plots residuals and squared residuals separately against the predicted values of Y. The "VARIABLES = option" names the independent variables that are to be plotted against the observed, predicted, and residual values.

Figure 10.10 illustrates the program (using BMDP-1R) that has been written to analyze the advertising media problem, while Figure 10.11 represents annotated partial output from the BMDP program 1R analysis of these data.

```
{SYSTEM CARDS}
/PROBLEM  TITLE IS 'SALES AND MEDIA EXPENDITURES'.
/INPUT  VARIABLES ARE 4.
     FORMAT IS '(F2.0,1X,F5.0,1X,F2.0,1X,F2.0)'.
     CASES ARE 22.
/VARIABLE  NAMES ARE CITY,SALES,RADTVDOL,NEWSDOL.
           LABEL IS CITY.
/REGRESS   DEPENDENT IS SALES.
           INDEPENDENT ARE RADTVDOL,NEWSDOL.
/PRINT     DATA.
           CORR.
           COV.
/PLOT      RESID.
           VARIABLE ARE RADTVDOL,NEWSDOL.
/END
01  09.73  00  20
  :      :      :      :
22  17.17  50  20
{SYSTEM CARDS}
```

Figure 10.10   Using BMDP program 1R for media expenditures example.

PROBLEMS

*10.1. A statistician for a large American automobile manufacturer would like to develop a statistical model for predicting delivery time (the number of days between the ordering and actual delivery) of custom-ordered new automobiles. Two independent variables—number of options ordered (X_1) and hundreds of shipping miles (X_2)—are to be considered in predicting the delivery time. By assuming each independent variable is linearly related to the delivery time, a random sample of 16 custom-ordered cars is selected, and the computer output obtained from SAS is provided in Table P10.1.

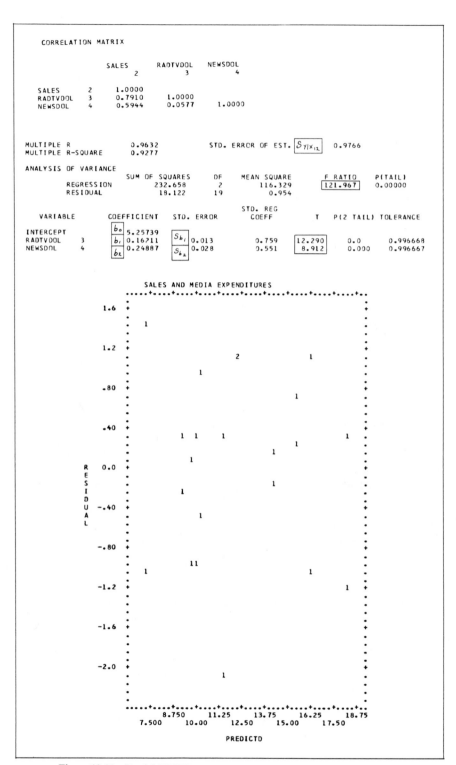

Figure 10.11 Partial BMDP—IR output for media expenditures data.

TABLE P10.1 Statistical Analysis System General Linear Models Procedure

DEPENDENT VARIABLE: TIME (IN DAYS)

SOURCE	DF	SS	MS	F
MODEL	2	2985.287	1492.643	270.584
ERROR	13	71.713	5.516	
CORRECTED TOTAL	15	3057.000	STD DEV =	2.349

SOURCE	DF	TYPE I SS	F	DF	TYPE IV SS	F
OPTIONS	1	2927.230	530.68	1	2808.669	509.19
MILES	1	58.057	10.53	1	58.057	10.53

PARAMETER	ESTIMATE	T FOR $H0$ PARAMETER $= 0$	STD ERROR OF ESTIMATE (S_{b_i})
INTERCEPT	16.196	7.53	2.151
OPTIONS	2.038	22.64	.090
MILES	.563	3.25	.173

(a) State the multiple regression equation. Interpret the meaning of the slopes in this problem.

(b) If a car was ordered with 10 options and had to be shipped 800 miles, what would you predict the delivery time to be?

(c) Determine whether there is a significant relationship between the delivery time and the two independent variables at the .05 level of significance.

(d) Interpret the meaning of the coefficient of multiple determination $r^2_{Y.12}$ in this problem.

(e) At the .05 level of significance, determine whether each independent variable makes a contribution to the regression model. Based on these results, indicate the regression model that should be utilized in this problem.

(f) Set up a 95% confidence interval estimate of the true population slope between the delivery time and the number of options.

(g) Compute the coefficients of partial determination $r^2_{Y1.2}$ and $r^2_{Y2.1}$ and interpret their meaning in this problem.

10.2. Referring to Problem 8.3, if we wished to consider the average sales price per home as well as the number of higher-income households in predicting the number of brokers (in thousands) use an available computer package (such as SPSS, SAS, or BMDP) and a matrix algebra approach as appropriate to perform a multiple linear regression analysis. The average sales price (as of February 1980) for each of the selected SMSAs are displayed in Table P10.2.

(a) State the multiple regression equation. Interpret the meaning of the slopes in this problem.

(b) Determine whether there is a significant relationship between the number of brokers and the two independent variables at the .05 level of significance.

(c) Interpret the meaning of the coefficient of multiple determination $r^2_{Y.12}$ in this problem.

TABLE P10.2 Average Sales Price for Selected SMSAs

SMSA	Average Sales Price ($000)
Abilene	39.110
Amarillo	39.853
Austin	65.002
Beaumont	45.768
Brownsville	46.510
Bryan/College Station	47.938
Corpus Christi	49.891
El Paso	49.579
Galveston	51.922
Killeen/Temple	47.632
Lubbock	47.242
McAllen	48.646
Midland	57.806
Odessa	44.174
San Angelo	33.294
San Antonio	48.427
Texarkana	35.882
Waco	40.089
Wichita Falls	33.501

(d) At the .05 level of significance, determine whether each independent variable makes a contribution to the regression model. Based on these results, indicate the regression model that should be utilized in this problem.

(e) Evaluate the appropriateness of the developed model.

(f) Set up a 95% confidence interval estimate of the true population slope between the number of brokers and the average sales price.

(g) Compute the coefficients of partial determination $r^2_{Y1.2}$ and $r^2_{Y2.1}$, and interpret their meaning in this problem.

(h) Predict the average number of brokers for an SMSA which has 60,000 households with above $10,000 after-tax income and an average sales price per home of $52,000.

(i) Set up a 95% confidence interval estimate of the mean number of brokers for all SMSA having 60,000 households with above $10,000 after-tax income and an average sales price per home of $52,000.

(j) Set up a 95% prediction interval estimate of the expected number of brokers for a particular SMSA having 60,000 households with above $10,000 after-tax income and an average sales price per home of $52,000.

10.3. The Fortune 500 consists of the 500 largest industrial corporations ranked by sales. We would like to develop a model to predict the net income of a corporation based on sales and assets. A random sample of 33 corporations was selected from the Fortune 500 for the year 1979 with the results displayed in Table P10.3. Use an available computer package (such as SPSS, SAS, or BMDP) and a matrix algebra approach as appropriate to perform a multiple linear regression analysis.

(a) State the multiple regression equation. Interpret the meaning of the slopes in this problem.

TABLE P10.3 Predicting Net Income Based on Sales and Assets

Corporation	Net Income ($ Millions)	Sales ($ Millions)	Assets ($ Millions)
Mobil	2,007	44,721	27,506
DuPont	939	12,572	8,940
Caterpillar Tractor	492	7,613	5,403
National Steel	126	4,234	3,160
Eaton	154	3,360	2,355
American Cyanamid	169	3,187	2,827
Boise Cascade	175	2,917	2,309
Burroughs	306	2,785	3,387
Central Soya	34	2,448	558
Interco	93	1,851	1,003
National Distillers	136	1,773	1,665
Emhart	54	1,573	965
U.S. Gypsum	124	1,525	1,048
Staley Mfg.	24	1,435	642
Timken	102	1,282	943
Smith Kline	234	1,351	1,196
Pitney-Bowes	64	1,025	902
Baker Int.	99	1,169	1,123
Witco Chemical	43	967	566
Champion Spark Plug	57	807	611
Joy Mfg.	50	782	676
AM Int.	12	754	544
Hoover	39	754	491
Outboard Marine	17	741	560
Morton-Norwich	46	732	587
Sybron	22	723	513
Bemis	27	648	364
Con Agra	21	645	250
Dan River	22	579	351
Avery Int.	21	562	348
United Refining	17	498	175
Consolidated Papers	55	445	342
Dorsey	10	421	203

SOURCE: "The Fortune Directory of the 500 Largest Industrial Corporations," *Fortune* (May 5, 1980), pp. 276–299. © Time Inc. All rights reserved.

(b) Determine whether there is a significant relationship between net income and the two independent variables at the .05 level of significance.

(c) Interpret the meaning of the coefficient of multiple determination $r^2_{Y.12}$ in this problem.

(d) At the .05 level of significance, determine whether each independent variable makes a contribution to the regression model. Based on these results, indicate the regression model that should be utilized in this problem.

(e) Evaluate the appropriateness of the developed model.

(f) Set up a 95% confidence interval estimate of the true population slope between net income and sales.

(g) Compute the coefficients of partial determination $r^2_{Y1.2}$ and $r^2_{Y2.1}$, and interpret their meaning in this problem.

(h) Predict the net income (millions of dollars) for a corporation with sales of $1 billion and assets of $1 billion.

(i) Set up a 95% confidence interval estimate of the mean net income (millions of dollars) for all corporations with sales of $1 billion and assets of $1 billion.

(j) Set up a 95% prediction interval estimate of the expected net income (millions of dollars) for a particular corporation with sales of $1 billion and assets of $1 billion.

10.4. Referring to Table 2.1 and Problem 8.4, if we wished to consider the age of employees as well as their length of employment in predicting salary, use an available computer package (such as SPSS, SAS, or BMDP) and a matrix algebra approach as appropriate to perform a multiple linear regression analysis.

(a) State the multiple regression equation. Interpret the meaning of the slopes in this problem.

(b) Determine whether there is a significant relationship between the weekly salary and the two independent variables at the .05 level of significance.

(c) Interpret the meaning of the coefficient of multiple determination $r^2_{Y.12}$ in this problem.

(d) At the .05 level of significance, determine whether each independent variable makes a contribution to the regression model.

(e) Compute the coefficients of partial determination $r^2_{Y1.2}$ and $r^2_{Y2.1}$, and interpret their meaning in this problem.

(f) Set up a 95% confidence interval estimate of the true population slope between the weekly salary and the length of employment. Compare the result to that of problem 8.4(h). Discuss.

(g) Evaluate the appropriateness of the developed model. Suggest a possible course of action in light of your findings in problem 8.4.

10.5. Using the real estate data base (see Appendix A), select a random sample of 30 houses. Develop a multiple linear regression model to predict the assessed value based on the age and heating area of the house. ■

10.6. Using the real estate data base (see Appendix A), select a random sample of 30 houses. Develop a multiple linear regression model to predict the selling price based on the assessed value and time period. ■

10.7. Using the real estate data base (see Appendix A), select a random sample of 30 houses. Develop a multiple regression model to predict the asking price based on the age and assessed value. ■

10.8. Reviewing concepts:

(a) Describe the multiple regression model.

(b) Describe the process by which we may evaluate the contribution of a variable to the multiple regression model.

(c) What is meant by the coefficients of partial determination and correlation?

(d) Discuss why residual analysis is often more appropriate for evaluating the aptness of a multiple regression model than tests of lack of fit.

(e) In what ways are a matrix approach to regression and correlation analysis useful?

11

POLYNOMIAL
AND OTHER
REGRESSION MODELS

11.1 DEVELOPING THE POLYNOMIAL REGRESSION MODEL

In our discussions of simple regression (Chapter 8) and multiple regression (Chapter 10), we have focused upon linear relationships between variables. However, as mentioned in those chapters, not all relationships are linear in nature. One of the more common nonlinear relationships between two variables is a curvilinear polynomial wherein Y increases (or decreases) at a *changing* rate for various values of X.

The second-degree polynomial relationship between X and Y can be expressed by the following model:

$$Y_i = \beta_0 + \beta_1 X_{1_i} + \beta_2 X_{1_i}^2 + \epsilon_i \tag{11.1}$$

where $\beta_0 = Y$ intercept

$\beta_1 =$ linear effect on Y

$\beta_2 =$ curvilinear effect on Y

$\epsilon_i =$ random error in Y_i for observation i

Note that (11.1) is similar to the multiple regression model with two independent variables [see (10.1)] except that the second "independent" variable in this instance is merely the square of the first independent variable.

As in the case of multiple linear regression, when sample data are analyzed the sample regression coefficients (b_0, b_1, and b_2) are used as estimates of the true parameters. Thus the regression equation for the second-degree polynomial model having one independent variable (X_1) and a dependent variable (Y) is

$$\hat{Y}_i = b_0 + b_1 X_{1_i} + b_2 X_{1_i}^2 \tag{11.2}$$

11.2 FINDING THE REGRESSION COEFFICIENTS AND PREDICTING Y

To illustrate the second-degree curvilinear model, we shall examine the following: An auditor for a county government would like to develop a model to predict the county taxes based on the age of single family homes. A random sample of 19 single-family homes has been selected with the results presented in Table 11.1.

TABLE 11.1 Taxes and Age of a Random Sample of 19 Homes

County Taxes ($)	Age (Years)
925	1
870	2
809	4
720	4
694	5
630	8
626	10
562	10
546	12
523	15
480	20
486	22
462	25
441	25
426	30
368	35
350	40
348	50
322	50

To investigate the type of relationship that exists between county taxes and the age of houses, a scatter diagram can be plotted as in Figure 11.1.

An examination of Figure 11.1 seems to indicate not only a negative relationship but also a *leveling off* in the decrease of county taxes for an increasing age of houses. Therefore, it appears that a curvilinear polynomial model may be appropriate for predicting county taxes based on the age of the houses.

If the least-squares method is utilized to compute the sample regression coefficients (b_0, b_1, and b_2), we will have the following three normal equations:

$$\text{I} \quad \sum_{i=1}^{n} Y_i = nb_0 + b_1 \sum_{i=1}^{n} X_{1_i} + b_2 \sum_{i=1}^{n} X_{1_i}^2$$

$$\text{II} \quad \sum_{i=1}^{n} X_{1_i} Y_i = b_0 \sum_{i=1}^{n} X_{1_i} + b_1 \sum_{i=1}^{n} X_{1_i}^2 + b_2 \sum_{i=1}^{n} X_{1_i}^3 \qquad (11.3)$$

$$\text{III} \quad \sum_{i=1}^{n} X_{1_i}^2 Y_i = b_0 \sum_{i=1}^{n} X_{1_i}^2 + b_1 \sum_{i=1}^{n} X_{1_i}^3 + b_2 \sum_{i=1}^{n} X_{1_i}^4$$

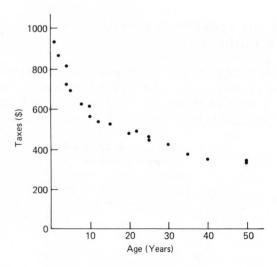

Figure 11.1 Scatter diagram of age of houses X_1 and county taxes Y.

Figure 11.2 represents a partial output of the SAS GLM procedure that has been utilized to solve these equations for the data of Table 11.1. The computed values of the regression coefficients are

$$b_0 = 857.5884 \qquad b_1 = -24.7221 \qquad b_2 = +.2935$$

Therefore, by using (11.2), the sample curvilinear regression equation can be expressed as

$$\hat{Y}_i = 857.5884 - 24.7221 X_{1_i} + .2935 X_{1_i}^2$$

where $\hat{Y}_i$ is the predicted county taxes for house i and X_{1_i} is the age of house i (in years).

From this equation, the Y intercept ($b_0 = 857.5884$) can be interpreted to mean that for a brand-new house ($X_1 = 0$) there would be an expected county tax assessment of \$857.59. However, to interpret the meaning of b_1 and b_2, it would be useful to predict county taxes for different-aged houses. For example, for $X_{1_i} = 10$ (years) we have

$$\hat{Y}_i = 857.5884 - 24.7221(10) + .2935[(10)^2] = \$639.72$$

For $X_{1_i} = 30$ (years) we have

$$\hat{Y}_i = 857.5884 - 24.7221(30) + .2935[(30)^2] = \$380.08$$

For $X_{1_i} = 50$ (years) we have

$$\hat{Y}_i = 857.5884 - 24.7221(50) + .2935[(50)^2] = \$355.23$$

Thus we observe that although the assessed county taxes decrease with aging houses, the rate of this decrease levels off for older houses.

DEPENDENT VARIABLE TAXES   ANNUAL CONTY TAXES

SOURCE	DF	SUM OF SQUARES	MEAN SQUARE	F VALUE	PR > F	R-SQUARE	C.V.
MODEL	2	544220.94753411	272110.47376705	116.12	0.0001	0.935548	8.6867
ERROR	16	37492.73667642	2343.29604228			STD DEV	TAXES MEAN
CORRECTED TOTAL	18	581713.68421053				48.40760315	557.26315789

$S_{y|x}$ = 48.40760315

SOURCE	DF	TYPE I SS	F VALUE	PR > F		DF	TYPE IV SS	F VALUE	PR > F
AGE	1	467247.68910221	199.40	0.0001		1	208041.52370436	88.78	0.0001
AGESQ	1	76973.45843190	32.85	0.0001		1	76973.45843190	32.85	0.0001

PARAMETER		ESTIMATE	T FOR H0 PARAMETER=0	PR > T	STD ERROR OF ESTIMATE	
INTERCEPT	b_0	857.58837433	34.05	0.0001	25.18944815	S_{b_0}
AGE	b_1	-24.72228658	-9.42	0.0001	2.62375748	S_{b_1}
AGESQ	b_2	0.29353893	5.73	0.0001	0.05121637	S_{b_2}

Figure 11.2 Partial SAS output for county taxes data.

11.3 INFERENCES FOR THE POLYNOMIAL REGRESSION MODEL

Now that we have fit the curvilinear model to the data, we can determine whether there is a significant relationship between county taxes Y and the age of houses X_1. Similar to multiple regression (see Section 10.3), the null and alternative hypotheses can be set up as follows:

H_0: $\beta_1 = \beta_2 = 0$ (there is no relationship between X_1 and Y)

H_1: $\beta_1 \neq \beta_2 \neq 0$ (at least one regression coefficient is not equal to zero)

As indicated in the analysis of variance table for the curvilinear polynomial model (Table 11.2), this null hypothesis can be tested by subdividing the total variation (SST) into its two components: regression variation (SSR) and error variation (SSE).

For the county taxes data, the ANOVA table is provided as part of the GLM output of SAS (see Figure 11.2). From this we have

$$\text{MSR} = 272,110.47376705 \quad \text{and} \quad \text{MSE} = 2343.29604228$$

Thus

$$F = \frac{\text{MSR}}{\text{MSE}} = \frac{272,110.47376705}{2343.29604228} = 116.12$$

If a level of significance of .05 is chosen, since $F = 116.12 > F_{.95;\,2,16} = 3.63$, we can reject H_0 and conclude that there is a significant relationship between county taxes and the age of houses.

We may recall from Section 10.4 that once the ANOVA table for the multiple regression model was developed, we could compute the coefficient of multiple determination ($r_{Y\cdot 12}^2$) given by

$$r_{Y\cdot 12}^2 = \frac{\text{SSR}}{\text{SST}} \tag{11.4}$$

From Figure 11.1 we have

$$r_{Y\cdot 12}^2 = \frac{544,220.94753411}{581,713.68421053} = .9356$$

Thus we can state that 93.56% of the variation in county taxes can be explained by the curvilinear relationship between county taxes and the age of houses.

11.4 COMPARING THE CURVILINEAR POLYNOMIAL MODEL TO THE LINEAR MODEL

In using a regression model to examine a relationship between two variables, we wish to choose the simplest as well as the most accurate model. Therefore, it becomes important to examine whether there is a significant difference between the simple curvilinear model

TABLE 11.2 ANOVA Table for Testing the Significance of a Simple Curvilinear Relationship

Source	Degrees of Freedom	Sum of Squares	Mean Square	F
Regression	2	$\text{SSR} = b_0 \sum\limits_{i=1}^{n} Y_i + b_1 \sum\limits_{i=1}^{n} X_{1_i} Y_i + b_2 \sum\limits_{i=1}^{n} X_{1_i}^2 Y_i - \dfrac{\left(\sum\limits_{i=1}^{n} Y_i\right)^2}{n}$	$\text{MSR} = \dfrac{\text{SSR}}{2}$	$F = \dfrac{\text{MSR}}{\text{MSE}}$
Error	$n - 3$	$\text{SSE} = \sum\limits_{i=1}^{n} Y_i^2 - b_0 \sum\limits_{i=1}^{n} Y_i - b_1 \sum\limits_{i=1}^{n} X_{1_i} Y_i - b_2 \sum\limits_{i=1}^{n} X_{1_i}^2 Y_i$	$\text{MSE} = \dfrac{\text{SSE}}{n - 3}$	
Total	$n - 1$	$\text{SST} = \sum\limits_{i=1}^{n} Y_i^2 - \dfrac{\left(\sum\limits_{i=1}^{n} Y_i\right)^2}{n}$		

$$Y_i = \beta_0 + \beta_1 X_{1_i} + \beta_2 X_{1_i}^2 + \epsilon_i \tag{11.1}$$

and the simple linear model

$$Y_i = \beta_0 + \beta_1 X_{1_i} + \epsilon_i \tag{8.2}$$

When we evaluated the multiple linear regression model in Section 10.5, we determined the contribution of each independent variable using (10.9) or (10.11). Similarly, the contribution of $X_1^2 \mid X_1$ can be computed using either

$$SSR(b_2 \mid b_1) = SSR(b_1 \text{ and } b_2) - SSR(b_1) \tag{11.5}$$

or

$$SSR(b_2 \mid b_1) = \frac{b_2^2 MSE}{S_{b_2}^2} \tag{11.6}$$

From Figure 11.2 we note that the *Type I SS* indicates the contribution of the variable age, $SS(b_1)$, and the age squared, $SS(b_2 \mid b_1)$. Thus

$$SS(b_1) = 467,247.48910221 \quad \text{and} \quad SS(b_2 \mid b_1) = 76,973.45843190$$

The null and alternative hypotheses to test for the contribution of the curvilinear effect to the regression model are

H_0: $\beta_2 = 0$ (the curvilinear model does not represent an improvement over the linear model)

H_1: $\beta_2 \neq 0$ (the curvilinear model is a better fit than the linear model)

From Figure 11.2

$$F = \frac{MSR(b_2 \mid b_1)}{MSE} = \frac{76,973.45843190}{2343.29604228} = 32.85$$

By using an .05 level of significance, since $F = 32.85 > F_{.95; 1, 16} = 4.49$, our decision would be to reject H_0 and conclude that the curvilinear polynomial model is significantly better than the linear model in representing the relationship between county taxes and the age of houses.

Now that we have determined that $X_1^2 \mid X_1$ is significant, we should also determine whether $X_1 \mid X_1^2$ is significant, that is, whether the contribution of the linear effect (b_1) is significant given that the curvilinear effect (b_2) has already been included. This $SSR(b_1 \mid b_2)$ can be computed using either

$$SSR(b_1 \mid b_2) = SSR(b_1 \text{ and } b_2) - SSR(b_2) \tag{11.7}$$

or

$$SSR(b_1 \mid b_2) = \frac{b_1^2 MSE}{S_{b_1}^2} \tag{11.8}$$

From Figure 11.2, we note that the *Type IV SS* indicates the contribution of the variable age given the age squared, $SS(b_1 \mid b_2)$, as well as the contribution of the variable age squared given the age, $SS(b_2 \mid b_1)$.

The null and alternative hypotheses to test for the contribution of the linear effect to the regression model are

H_0: $\beta_1 = 0$ (including the linear effect does not improve the curvilinear effect model)

H_1: $\beta_1 \neq 0$ (including the linear effect does improve the curvilinear effect model)

From Figure 11.2

$$F = \frac{MSR(b_1 | b_2)}{MSE} = \frac{208,041.52370436}{2343.29604228} = 88.78$$

By using an .05 level of significance, since $F = 88.78 > F_{.95; 1,16} = 4.49$, our decision would be to reject H_0 and conclude that the polynomial model ($\hat{Y}_i = b_0 + b_1 X_{1_i} + b_2 X_{1_i}^2$) is a significantly better fit than one which includes only the curvilinear effect ($\hat{Y}_i = b_0 + b_2 X_{1_i}^2$).

11.5 TESTING FOR THE APPROPRIATENESS OF THE SIMPLE CURVILINEAR REGRESSION MODEL

We may recall from Sections 8.10 and 10.7 that once a regression model was fitted to the data, if repeated measurements were available for at least one level of X, a test for *lack of fit* could be performed to evaluate the appropriateness of the model. Since in the county taxes model two measurements on Y have been observed at each of four levels of X_1 (the age of houses), the aptness of our selected curvilinear model can be evaluated using the test for lack of fit.

From (8.18) and (8.20) we have

$$SSPE = \sum_{j=1}^{l} \sum_{k=1}^{n_j} (Y_{jk} - \bar{Y}_j)^2 \tag{8.18}$$

and

$$SSLF = SSE - SSPE \tag{8.20}$$

Table 11.3 presents the computation of SSPE for the county taxes example. Since SSPE = 6567.00 and, from Figure 11.2, SSE = 37,492.736676,

TABLE 11.3   Obtaining Pure Error Estimates from Repeated Observations on the Age of Houses

Y_{jk}	X_j	$\bar{Y}_j$	$Y_{jk} - \bar{Y}_j$	$(Y_{jk} - \bar{Y}_j)^2$
809	4	764.5	+44.5	1980.25
720	4		−44.5	1980.25
626	10	594.0	+32.0	1024.00
562	10		−32.0	1024.00
462	25	451.5	+10.5	110.25
441	25		−10.5	110.25
348	50	335.0	+13.0	169.00
322	50		−13.0	169.00
				SSPE = 6567.00

$$\text{SSLF} = 37{,}492.736676 - 6567.00$$
$$= 30{,}925.736676$$

Since there are two measurements of Y_i at each of four levels of X_1, the degrees of freedom for *pure error* are $\sum_{j=1}^{l}(n_j - 1) = 4$; by subtraction, the degrees of freedom for *lack of fit* are $(n - p - 1) - \sum_{j=1}^{l}(n_j - 1) = 12$. By using (8.21),

$$\text{MSLF} = \frac{\text{SSLF}}{(n - p - 1) - \sum_{j=1}^{l}(n_j - 1)} = \frac{30{,}925.736676}{12} = 2577.144723$$

and from (8.19)

$$\text{MSPE} = \frac{\text{SSPE}}{\sum_{j=1}^{l}(n_j - 1)} = \frac{6567.00}{4} = 1641.75$$

To test the hypothesis

H_0:   there is no lack of fit in the selected curvilinear model

against the alternative

H_1:   the selected curvilinear model is not the appropriate model

we have (8.22)

$$F = \frac{\text{MSLF}}{\text{MSPE}} = \frac{2577.144723}{1641.75} = 1.57$$

Using an .05 level of significance, since $F = 1.57 < F_{.95; \, 12,4} = 5.91$, our decision is not to reject H_0, and thus we may conclude that there is no evidence of a lack of fit in the selected curvilinear model.

11.6 OTHER TYPES OF REGRESSION MODELS

In our discussion of multiple regression models we have thus far examined the linear model,

$$Y_i = \beta_0 + \beta_1 X_{1_i} + \beta_2 X_{2_i} + \epsilon_i \qquad (10.1)$$

and the curvilinear polynomial model,

$$Y_i = \beta_0 + \beta_1 X_{1_i} + \beta_2 X_{1_i}^2 + \epsilon_i \qquad (11.1)$$

11.6.1 Interaction Terms in Regression Models

We should be aware, however, that interaction terms involving the product of independent variables sometimes also contribute to the multiple regression model (see Section 12.5). When two such independent variables are involved, this model would be

$$Y_i = \beta_0 + \beta_1 X_{1_i} + \beta_2 X_{2_i} + \beta_3 X_{1_i} X_{2_i} + \epsilon_i \qquad (11.9)$$

For example, we may recall that in Chapter 10 we utilized a multiple linear model to predict sales based on radio and television advertising (X_1) and newspaper advertising (X_2). It is theoretically possible that there could exist an interaction effect of these two types of advertising media that would greatly affect sales when various combinations of X_1 and X_2 were applied. If this were the case, a regression model that includes the interaction term might provide a better fit than the model selected in Chapter 10.

11.6.2 Using Transformations in Regression Models

We may recall that in Section 3.4 we discussed *data transformations* as a vehicle for overcoming the effects of the violation of assumptions in statistical inference. Clearly, these transformations can be generalized to the regression model as well. Depending on the situation, it may be appropriate to transform the values of the independent variables, the dependent variable, or both. For example, if a reciprocal transformation were applied to the values of each of two independent variables, the multiple regression model would be

$$Y_i = \beta_0 + \beta_1\left(\frac{1}{X_{1_i}}\right) + \beta_2\left(\frac{1}{X_{2_i}}\right) + \epsilon_i \tag{11.10}$$

However, if a logarithmic transformation had been applied, the model would be

$$Y_i = \beta_0 + \beta_1 \ln X_{1_i} + \beta_2 \ln X_{2_i} + \epsilon_i \tag{11.11}$$

Furthermore, if a square root transformation had been applied, the model would be

$$Y_i = \beta_0 + \beta_1\sqrt{X_{1_i}} + \beta_2\sqrt{X_{2_i}} + \epsilon_i \tag{11.12}$$

Interestingly, in some situations we may note that transformations can be applied to change what appear to be nonlinear models into linear models. For example, the multiplicative model

$$Y_i = \beta_0 X_{1_i}^{\beta_1} X_{2_i}^{\beta_2}\epsilon_i \tag{11.13}$$

can be transformed (by taking natural logarithms of both the dependent and independent variables) to

$$\ln Y_i = \ln \beta_0 + \beta_1 \ln X_{1_i} + \beta_2 \ln X_{2_i} + \ln \epsilon_i \tag{11.14}$$

Hence (11.14) is *linear in the logarithms*. Similarly, the exponential model

$$Y_i = e^{\beta_0 + \beta_1 X_{1_i} + \beta_2 X_{2_i}}\epsilon_i \tag{11.15}$$

can also be transformed to one of linear form (by taking natural logarithms of both sides of the equation) as follows

$$\ln Y_i = \beta_0 + \beta_1 X_{1_i} + \beta_2 X_{2_i} + \ln \epsilon_i \tag{11.16}$$

From the preceding discussion, two important questions arise:

1. Will a transformation of the variables be useful for a particular set of data?
2. If so, which transformation should be applied?

Choosing the appropriate transformation: Looking at the bulge. By using the transformed variables, we are often able to obtain a simpler regression model than had we maintained the original variables. That is, by re-expressing the original X or Y values (or both), we may simplify the relationship between the variables to one which is linear in the transformation. The purpose of such re-expressions then is to facilitate data interpretation—and the simpler the model, the more readily this is achieved.

However, the choice of the "best" transformation is not always an easy one to make. Therefore, to aid in the selection of appropriate transformations in order to "straighten out" or *linearize* the relationship in a set of data, Mosteller and Tukey (1977) suggested that we examine the X,Y scatter plot and move in the direction of the *bulge*. Four possible patterns are exhibited in Figure 11.3.

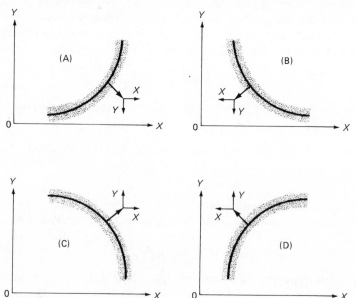

Figure 11.3 Patterns in curvilinear relationships indicating the direction of (power) re-expressions for X and Y.

In panel A the bulge in the scatter plot points *southeast*—suggesting either higher powers of X or lower powers of Y (or both) to linearize the relationship—while in panel B the bulge points *southwest*—indicating either lower powers of X or lower powers of Y (or both) to straighten out the fit. In a similar manner, we observe from panel C that a bulge in the *northeast* direction suggests possibly higher powers for both X and/or Y, while in panel D a *northwest* bulge indicates lower powers of X and/or higher powers of Y.

When dealing with simple (bivariate) regression problems, whether we choose to re-express either X or Y or both is often dependent on the nature of the data. Occasionally, a particular transformation for one of the variables will be suggested by the manner in which the data are measured. At other times, researchers may choose to re-express the variable whose values have displayed greater amounts of scatter or variability in the observed data. Moreover, some researchers prefer to re-express only Y because they consider X as the explanatory or predictor variable used to describe Y.

On the other hand, when dealing with multiple regression problems, there is far less flexibility in the choice. Since a separate scatter plot is needed to study the individual relationships of Y against each predictor variable X, it is simplest to search only for the appropriate transformations of the various X which would linearize the individual fits.

Choosing the appropriate transformation: Examining the ladder of powers. As to the particular powers to select, Tukey (1977) introduces a *ladder of powers for transformations*, as displayed in Table 11.4.

TABLE 11.4   Ladder of Powers for Transformations
$$\begin{cases} X \longrightarrow X^P \\ Y \longrightarrow Y^P \end{cases}$$

Power P	Re-expressions X	Y	Name	Comments[a, b]
p	X^p	Y^p	pth power	p is a positive number
·	·	·	·	·
·	·	·	·	·
·	·	·	·	·
3	X^3	Y^3	Cube	Infrequently used power
2	X^2	Y^2	Square	Highest commonly used power
1	X^1	Y^1	Original data	No re-expression at all
$\frac{1}{2}$	$\sqrt{X}$	$\sqrt{Y}$	Square root	Commonly used power
"0"	$\ln X$ or $\log_{10}(X)$	$\ln Y$ or $\log_{10}(Y)$	Logarithm	Commonly used power
$-\frac{1}{2}$	$-1/\sqrt{X}$	$-1/\sqrt{Y}$	Reciprocal root	Commonly used power
-1	$-1/X$	$-1/Y$	Reciprocal	Lowest commonly used power
-2	$-1/X^2$	$-1/Y^2$	Reciprocal square	Infrequently used power
·	·	·	·	·
·	·	·	·	·
$-p$	$-1/X^p$	$-1/Y^p$	Reciprocal pth power	p is a positive number

[a]Tukey (1977) suggests using logarithms to the base 10. However, regardless of whether the common or natural logarithm is used, this type of transformation holds the place of the *zero power* in the ladder of powers.

[b]The negative sign for all reciprocal-type transformations preserves the order in the original data.

For example, if in a multiple regression study the scatter plot of Y against the first independent variable X_1 appeared as in panel A of Figure 11.3, we would want to express X_1 with a higher power to straighten out the relationship. The best starting point is likely to be as close as possible to but higher than $P = 1$, the *base power* (which, of course, is the original data). Therefore, we would fit Y against X_1^2. Hence, based on the direction of the bulge in the scatter plot, an "educated guess" is obtained regarding the direction on the ladder of powers (higher or lower) the transformed variable should take.

Evaluating the results of the transformation.   At this point two alternative procedures may be employed to evaluate the results of the transformation. The first is a systematic approach described by Tukey (1977). Basically it is an iterative "paper and pencil" procedure wherein easily obtained summary measures are compared after the re-expression to determine whether a linearization has been achieved or whether further re-expression is necessary. A more detailed discussion of this *exploratory data analysis* technique is outside the scope of this text. An excellent presentation, however, is given in Velleman and Hoaglin (1981).

The second procedure outlined here is more classical in nature. A least-squares linear regression analysis is obtained for Y and the transformed $X_1^* = X_1^2$. By using the methods of Sections 8.9 and 8.10, respectively, both the model (i.e., slope) and the aptness of the linear fit may be tested for significance. Based on the results, a determination may be made as to whether a linearization has been achieved or whether further refinement through re-expression is necessary.

Regardless of which procedure is used, in multiple regression studies the process would be repeated for each independent variable until we are satisfied that we have attained a linearization in the fit between Y and each X (see Problem 11.6). In summary, searching for appropriate data transformations may substantially enhance a regression analysis.

11.7 POLYNOMIAL AND OTHER REGRESSION MODELS AND COMPUTER PACKAGES: USE OF SPSS, SAS, AND BMDP

The regression programs of SPSS and SAS which have been utilized for multiple regression can be modified for use with polynomial and other regression models. However, the BMDP package has a separate program (BMDP program 5R) for polynomial regression models.

11.7.1 Using SPSS for Polynomial and Other Regression Models

The SPSS subprogram NEW REGRESSION can be utilized for polynomial regression by defining the second independent variable as the square of the

first independent variable. This is accomplished through the use of the COM-
PUTE statement:

1   **16**

COMPUTE   {variable = expression}

Referring to the county taxes problem of Section 11.2, we may define
AGE^2 by using

1   **16**

COMPUTE   AGESQ = AGE ∗ AGE

and then performing a (multiple) regression analysis to predict TAXES from
AGE and AGESQ.

The COMPUTE statement can also be utilized to transform the dependent
variable and/or any of the independent variables in developing a regression
model. For example, if we wish to transform the exponential model (11.15) to
linear form (11.16), we would require the following COMPUTE statement

1   **16**

COMPUTE   YTRANS = LN(Y)

as part of the set of SPSS control statements.

11.7.2 Using SAS for Polynomial and Other Regression Models

The SAS GLM procedure can be utilized for polynomial regression by
defining the second independent variable as the square of the first independent
variable. This may be accomplished by defining

{new variable} = {expression};

Referring to the county taxes problem of Section 11.2, we may define the
new variable AGE^2 by using

AGESQ = AGE ∗ AGE;

and then perform a (multiple) regression analysis to predict TAXES from AGE
and AGESQ.

If we wish to transform the dependent variable and/or any of the independent variables in developing a regression model, the new variable can be
expressed as a mathematical function of the original variable. Thus, for example,
if we wish to transform the exponential model (11.15) to linear form (11.16), we
would require that the statement

YTRANS = LOG(Y);

be included as part of the SAS data step.

11.7.3 Using BMDP for Polynomial and Other Regression Models

The P-5R program of BMDP provides for a polynomial regression analysis
of the form

$$Y_i = \beta_0 + \beta_1 X_{1_i} + \beta_2 X_{1_i}^2 + \cdots + \beta_p X_{1_i}^p + \epsilon_i \qquad (11.17)$$

where p is the degree of the polynomial.

The only paragraph of BMDP unique to program P-5R is the/REGRESS paragraph, which takes the form

/REGRESSβDEPENDENTβISβ{name of dependent variable}.
INDEPENDENTβISβ{name of independent variable}.
DEGREE = {integer}.

The DEGREE sentence specifies the degree of the polynomial relationship. Thus a curvilinear model of the form $Y_i = \beta_0 + \beta_1 X_{1_i} + \beta_2 X_{1_i}^2 + \epsilon_i$ is considered a parabola or second-degree polynomial. Since the X^2 term is the highest-order term in the model for the county taxes problem of Section 11.2, we would have

/REGRESSβDEPENDENTβISβTAXES.
INDEPENDENTβISβAGE.
DEGREE = 2.

and a regression analysis to predict TAXES from AGE and AGESQ would be performed.

If we desire to transform the dependent and/or the independent variables in a regression analysis, we need to use the /TRANSFORM paragraph in conjunction with BMDP—IR. The format for this paragraph is

/VARIABLES   ADD = {number of transformed variables}.

/TRANSFORM $\begin{Bmatrix} \text{new} \\ \text{variable} \end{Bmatrix}$ = {expression}.

For example, if we wish to use the reciprocal transformation (11.10) to transform X_1 and X_2, we would have the following:

/VARIABLES   ADD = 2.
/TRANSFORM   X1TRANS = 1/X1.
                 X2TRANS = 1/X2.

PROBLEMS

*11.1. The marketing department of a large supermarket chain would like to develop a regression model to predict sales of disposable razors at various stores in the chain. A random sample of 15 stores with equivalent store traffic and product placement (i.e., at the checkout counter) was selected. Five stores were randomly assigned to each of three different price levels (79 cents, 99 cents, and $1.19) for a package of three razors. The results (number of packages sold in a test period of 1 week) are presented in Table P11.1. Assuming a second-degree polynomial model between price and sales, use an available computer package (such as SPSS, SAS, or BMDP) to perform a regression analysis.

(a) State the regression equation.

(b) Explain the meaning of the regression coefficients.

TABLE P11.1 Sales of Disposable Razors in 15 Stores

Price (Cents)	Number of Packages Sold
79	142
79	151
79	163
79	168
79	176
99	91
99	100
99	107
99	115
99	126
119	77
119	86
119	95
119	100
119	106

(c) Predict the expected sales if the price is set at $.99 per package.

(d) Determine whether there is a significant second-degree polynomial relationship between price and sales ($\alpha = .05$).

(e) Compute the coefficient of multiple determination, and interpret its meaning.

(f) Determine the contribution of each effect to the regression model at the .05 level of significance. Based on these results, indicate the regression model that should be used in this problem.

(g) Evaluate the aptness of the second-degree polynomial relationship for these data.

11.2. Referring to Problem 11.1, use a square root transformation on the total number of packages sold. Fit a *linear* relationship between the price and the square root of the number of packages sold. For this model, $\sqrt{\hat{Y}_i} = b_0 + b_1 X_{1_i}$:

(a) Determine the regression equation.

(b) Interpret the meaning of the slope.

(c) Predict the expected number of packages sold if the price is set at $.99 per package.

(d) Determine whether there is a significant relationship between price and the square root of sales. Use $\alpha = .05$.

(e) Compute the coefficient of determination, and interpret its meaning in this problem.

(f) Evaluate the aptness of this model.

(g) Compare the results of this model to the second-degree polynomial fit in Problem 11.1. Which model do you think is more appropriate? Why?

11.3. For the data of Problem 8.4, fit a second-degree polynomial model to predict the salary of *technical* workers based on length of employment. Use a computer package (such as SPSS, SAS, or BMDP) to perform a regression analysis.

(a) State the regression equation.

(b) Explain the meaning of the regression coefficients.

(c) Predict the expected weekly salary of a technical worker who has been employed for 5 years (i.e., 60 months).

(d) Determine whether there is a significant second-degree polynomial relationship between the length of employment and salary ($\alpha = .05$).

(e) Compute the coefficient of multiple determination, and interpret its meaning.

(f) Determine the contribution of each effect to the regression model ($\alpha = .05$). Based on these results, indicate the regression model that should be used in this problem.

(g) Evaluate the aptness of the second-degree polynomial relationship for these data.

11.4. Referring to Problem 8.4, using the natural logarithm of salary (instead of salary), fit a *linear* relationship between the length of employment and the natural logarithm of salary. For this model, $\ln(\hat{Y}_i) = b_0 + b_1 X_i$:

(a) Determine the regression equation.

(b) Interpret the meaning of the slope.

(c) Predict the expected weekly salary of a technical worker who has been employed for 5 years (i.e., 60 months).

(d) Determine whether there is a significant relationship between the length of employment and the natural logarithm of salary at the .05 level of significance.

(e) Compute the coefficient of determination, and interpret its meaning in this problem.

(f) Evaluate the aptness of this model.

(g) Compare the results of this model to the simple linear model of Problem 8.4 and the second-degree polynomial model of Problem 11.3. Which model do you think is most appropriate? Why?

11.5. Referring to the data of Table 11.1, the number of rooms for each house was also available. The results are the following:

House	1	2	3	4	5	6	7	8	9	10	11	12	13	14	15	16	17	18	19
Number of Rooms	9	10	10	7	9	8	7	6	9	8	7	8	8	6	9	6	7	8	7

Fit a regression model of the form

$$\text{Taxes} = b_0 + b_1(\text{Age}) + b_2(\text{Age})^2 + b_3(\text{Rooms})$$

using a computer package (such as SPSS, SAS, or BMDP).

(a) State the regression equation.

(b) Explain the meaning of the regression coefficients.

(c) Using this model, predict the county taxes for a house that is 10 years old and has 8 rooms.

(d) Compute the coefficient of multiple determination, and interpret its meaning.

(e) Determine the contribution of each regression coefficient to the regression model ($\alpha = .05$). Based on these results, indicate the regression model that should be used in this problem. (In particular, should the curvilinear effect of age be included once the number of rooms is included?)

11.6. Referring to Problem 11.5, reformulate the regression model by studying the separate scatter plots of Y against age and Y against rooms, and by making the appropriate data transformation(s). Analyze completely. Compare your results to those obtained in Problem 11.5(e).

11.7. Using the real estate data base (see Appendix A), select a random sample of 30 houses. Develop a second-degree polynomial regression model to predict the heating area based on the lot size of a house. Is this curvilinear model a better fit than the linear model? ∎

11.8. Using the real estate data base (see Appendix A), select a random sample of 30 houses. Develop a second-degree polynomial regression model to predict the selling price based on the asking price. Is this curvilinear model a better fit than the linear model? ∎

11.9. Using the real estate data base (see Appendix A), select a random sample of 30 houses. Develop a second-degree polynomial regression model to predict the assessed value based on the age. Is this curvilinear model a better fit than the linear model? ∎

11.10. Reviewing concepts:
   (a) What is meant by polynomial regression?
   (b) Describe the differences and similarities among simple linear regression, polynomial regression, and multiple regression.
   (c) How may data transformations be useful in regression analysis?

12

THE INTERRELATIONSHIP
BETWEEN REGRESSION
AND EXPERIMENTAL DESIGN:
A COMPARISON OF FITTED MODELS

12.1 INTRODUCTION

In Sections 7.2 and 7.8 we stated that there are occasions when it is useful to examine certain types of designed experiments from a regression viewpoint (see also Problems 8.5, 8.6, and 8.7). In this chapter we shall expand on the interrelationships between regression and experimental design and focus on the following two problems:

1. How can we compare two separately fitted simple linear regression models in order to evaluate their similarities? That is, how can we determine the plausibility that both samples came from the same population or identical populations? To answer this, we introduce tests for the equality of the model parameters. As examples, in Section 12.3 a test for *parallelism* will enable us to conclude whether or not both populations may have the same slope—regardless of the intercept—while in Section 12.4 a test for the *equality of intercepts* will enable us to determine whether the populations may have the same origin (at $X = 0$)—regardless of slope. Moreover, by developing a multiple regression model containing *dummy* variables (Section 12.5), we shall then be able to test for *coincidence* among the two population models.

The application of dummy variables is most important for three reasons: First, it permits a test for coincidence based on the *full regression model*; second, it provides us with a fresh approach to comparing the two fitted models through the highly useful ANACOVA (i.e., **analysis of covariance**) procedures described in Section 12.7; third, it allows us to consider qualitative variables in a regression model.

2. How can we develop a regression approach to the solution of unbalanced factorial experiments when an approximate solution based on the method of unweighted means (Section 7.6.1) is inappropriate? This important interrelationship between regression analysis and the design of experiments will be the subject of Sections 12.9 and 12.10.

12.2 A COMPARISON OF TWO FITTED SIMPLE LINEAR REGRESSION MODELS

To compare two fitted simple linear regression models, we must consider the five possibilities pertaining to the true population models as illustrated in Figure 12.1. Some form of equality between the two population models (I and II) is depicted

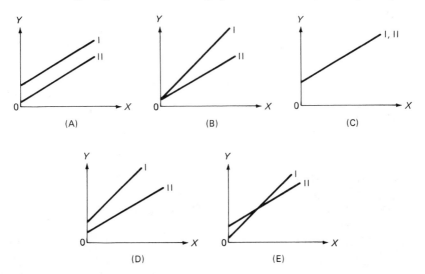

Figure 12.1   Comparison of simple linear regression models.

in panels A, B, and C. In panel A the models have equal slopes but different intercepts, while in panel B the models have equal intercepts but different slopes. Moreover, in panel C the two models are identical (i.e., *coincident*). On the other hand, panels D and E demonstrate situations in which the models possess both unequal intercepts and unequal slopes. In panel D the two models do not intersect within the relevant range of X so that $\mu_{Y_1|X} > \mu_{Y_{II}|X}$ for all X. In panel E, however, the two models do intersect; $\mu_{Y_1|X} > \mu_{Y_{II}|X}$ at high levels of X, but $\mu_{Y_1|X} < \mu_{Y_{II}|X}$ at low levels of X.

In the following three sections we shall describe methods for testing the equality of the models' parameters based on estimates obtained from two sample regression equations. In making these comparisons, it must be assumed that for each fitted regression model the population slope is not zero (i.e., there is a linear relationship between X and Y) and that the simple linear fit is adequate. Such assumptions have been tested as indicated in Table 12.1 for the example which follows.

TABLE 12.1   Summary of Simple Linear Regression Analyses Performed for Two Groups

Accountancy		Marketing/Management	
$\sum_{i=1}^{n_a} X_{a_i} = 42.5$	$\bar{X}_a = 2.833$	$\sum_{i=1}^{n_m} X_{m_i} = 44.5$	$\bar{X}_m = 2.967$
$\sum_{i=1}^{n_a} Y_{a_i} = 268.9$	$\bar{Y}_a = 17.927$	$\sum_{i=1}^{n_m} Y_{m_i} = 265.4$	$\bar{Y}_m = 17.693$
$\sum_{i=1}^{n_a} X_{a_i} Y_{a_i} = 770.55$	$\sum_{i=1}^{n_a} X_{a_i}^2 = 122.31$	$\sum_{i=1}^{n_m} X_{m_i} Y_{mi} = 793.69$	$\sum_{i=1}^{n_m} X_{m_i}^2 = 133.75$
$n_a = 15$	$\sum_{i=1}^{n_a} Y_{a_i}^2 = 4867.71$	$n_m = 15$	$\sum_{i=1}^{n_m} Y_{m_i}^2 = 4725.70$
$\hat{Y}_{a_i} = 4.957183 + 4.577465 X_{a_i}$		$\hat{Y}_{m_i} = 6.847885 + 3.655769 X_{m_i}$	

Accountancy	Marketing/Management
$S_{X_a}^2 = .135238$	$S_{X_m}^2 = .123810$
$S_{Y_a}^2 = 3.373524$	$S_{Y_m}^2 = 2.134952$
$S_{Y_a \mid X_a}^2 = .581382$	$S_{Y_m \mid X_m}^2 = .517226$
$\text{SSR}_a = 39.67136$	$\text{SSR}_m = 23.16539$
$\text{SSE}_a = 7.55797$	$\text{SSE}_m = 6.72394$
$\text{SST}_a = 47.22933$	$\text{SST}_m = 29.88933$
$r_a^2 = .83997; r_a = +.917$	$r_m^2 = .77504; r_m = +.880$
$F_a = \dfrac{\text{MSR}_a}{\text{MSE}_a} = 68.24; F_{.95; \, 1, \, 3} = 4.67$	$F_m = \dfrac{\text{MSR}_m}{\text{MSE}_m} = 44.79; F_{.95; \, 1, \, 3} = 4.67$
(slope β_{1_a} is significant)	(slope β_{1_m} is significant)
$\text{SSPE}_a = 2.41000; \text{d.f.}_{\text{PE}_a} = 4$	$\text{SSPE}_m = 3.62000; \text{d.f.}_{\text{PE}_m} = 4$
$\text{SSLF}_a = 5.14797; \text{d.f.}_{\text{LF}_a} = 9$	$\text{SSLF}_m = 3.10394; \text{d.f.}_{\text{LF}_m} = 9$
$F_a = \dfrac{\text{MSLF}_a}{\text{MSPE}_a} = .949; F_{.95; \, 9, \, 4} = 6.00$	$F_m = \dfrac{\text{MSLF}_m}{\text{MSPE}_m} = .381; F_{.95; \, 9, \, 4} = 6.00$
(simple linear model is adequate)	(simple linear model is adequate)

Source: Table 8.1.

To motivate the discussion for the next three sections, we recall that in Chapters 8 and 9 the concepts and methods of simple linear regression were highlighted through an examination of the relationship between the starting salary and the achieved grade point index for a sample of 30 recent college graduates who had received their baccalaureate in business administration. Continuing with this problem, we again note from Table 8.1 that the first 15 individuals listed had majored in accountancy, while the other 15 had majored in marketing/management.

Suppose that in an effort to provide guidance to the current senior class the Director of Career Development had fitted *two separate* linear regression models—one for accountancy majors and the other for marketing/management majors. By using the procedures of Chapter 8, the simple linear regression analyses performed separately for the two groups are summarized in Table 12.1. These results were part of the output obtained using SAS PROC GLM (see Figure 8.16). The scatter diagram illustrating the developed sample regression equations is presented in Figure 12.2.

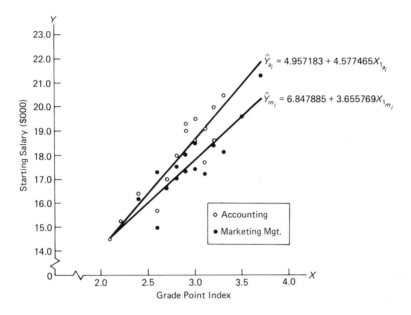

$$\hat{Y}_{a_j} = 4.957183 + 4.577465X_{1_{a_j}}$$

$$\hat{Y}_{m_j} = 6.847885 + 3.655769X_{1_{m_j}}$$

Figure 12.2   Comparison of regression equations. (Source: Tables 8.1 and 12.1).

12.3 TESTING FOR PARALLELISM

To test for the equality of the two population slopes (i.e., the *parallelism* of the slopes β_{1_a} and β_{1_m}) without regard to the level of the intercepts requires the pooling of the two $S^2_{Y|X}$ values. However, such pooling is appropriate only if the assumption of homoscedasticity in the response variable Y holds over the two models. Recall from Figure 8.10 that the residual plot for the 15 accountancy majors indicates a uniformity of scatter over increasing values of X—as does the plot for the 15 marketing/management majors. Thus we have no reason to suspect that the homoscedasticity assumption is violated. More formally, however, we may test

$$H_0: \quad \sigma^2_{Y_a} = \sigma^2_{Y_m}$$

against

$$H_1: \quad \sigma^2_{Y_a} \neq \sigma^2_{Y_m}$$

using the F test defined under H_0 by

$$F = \frac{S^2_{Y_a|X_a}}{S^2_{Y_m|X_m}} \sim F_{n_a-2,\,n_m-2} \tag{12.1}$$

and we may reject H_0 if, at the α level of significance, $F \geq F_{1-\frac{\alpha}{2}; n_a-2, n_m-2}$ or $F \leq F_{\frac{\alpha}{2}; n_a-2, n_m-2}$. From Table 12.1 we have

$$S^2_{Y_a|X_a} = .581382 \qquad S^2_{Y_m|X_m} = .517226$$

so that from (12.1)

$$F = \frac{S^2_{Y_a|X_a}}{S^2_{Y_m|X_m}} = 1.12$$

Using an .05 level of significance, since $F_{.025; 13,13} \cong .321 < F = 1.12 < F_{.975; 13,13} \cong 3.12$ (obtained through interpolation in Appendix B, Table B.4), the null hypothesis cannot be rejected. Again, there is no evidence that the assumption of homogeneity of variances has been violated.

The pooled estimate $S^2_{P|X}$ is given by

$$S^2_{P|X} = \frac{SSE_a + SSE_m}{n_a + n_m - 4} = \frac{(n_a - 2)S^2_{Y_a|X_a} + (n_m - 2)S^2_{Y_m|X_m}}{n_a + n_m - 4} \qquad (12.2)$$

Using the information summarized in Table 12.1, we compute

$$S^2_{P|X} = \frac{7.55797 + 6.72394}{15 + 15 - 4} = .549304$$

To study the possible parallelism between the two population slopes (without considering the levels of the intercepts), we may test

$$H_0: \quad \beta_{1_a} = \beta_{1_m} \quad \text{or} \quad \beta_{1_a} - \beta_{1_m} = 0$$

against

$$H_1: \quad \beta_{1_a} \neq \beta_{1_m} \quad \text{or} \quad \beta_{1_a} - \beta_{1_m} \neq 0$$

by the following t test using the sample slopes from the two independently developed regression equations:

$$t = \frac{(b_{1_a} - b_{1_m}) - (\beta_{1_a} - \beta_{1_m})}{S_{b_{1_a}-b_{1_m}}} \sim t_{n_a+n_m-4} \qquad (12.3)$$

where $S_{b_{1_a}-b_{1_m}}$, the standard error of the difference in the two slopes, is given by

$$S_{b_{1_a}-b_{1_m}} = \sqrt{S^2_{P|X}\left[\frac{1}{(n_a - 1)S^2_{X_a}} + \frac{1}{(n_m - 1)S^2_{X_m}}\right]} \qquad (12.4)$$

Using an α level of significance, the null hypothesis is rejected if $t \geq t_{1-\frac{\alpha}{2}; n_a+n_m-4}$ or if $t \leq t_{\frac{\alpha}{2}; n_a+n_m-4}$. From Table 12.1 we have

$$b_{1_a} = 4.577465 \qquad b_{1_m} = 3.655769$$

$$(n_a - 1)S^2_{X_a} = \sum_{i=1}^{n_a} X^2_{a_i} - \frac{\left(\sum_{i=1}^{n_a} X_{a_i}\right)^2}{n_a} = 1.89333$$

and

$$(n_m - 1)S^2_{X_m} = \sum_{i=1}^{n_m} X^2_{m_i} - \frac{\left(\sum_{i=1}^{n_m} X_{m_i}\right)^2}{n_m} = 1.73333$$

so that using (12.4), we obtain

$$S_{b_{1_a} - b_{1_m}} = \sqrt{.549304 \left(\frac{1}{1.89333} + \frac{1}{1.73333} \right)} = .779123$$

Thus from (12.3) the t statistic is now computed:

$$t = \frac{4.577465 - 3.655769}{.779123} = 1.183$$

By choosing an .05 level of significance, since $t = 1.183 < t_{.975; \, 26} = 2.056$, the null hypothesis cannot be rejected.

Since there is no evidence to suspect that the two population slopes are different, we may pool the two separate sample estimates to obtain a combined estimate of the common slope as follows:

$$\hat{\beta}_1 = \frac{(n_a - 1)S_{X_a}^2 b_{1_a} + (n_m - 1)S_{X_m}^2 b_{1_m}}{(n_a - 1)S_{X_a}^2 + (n_m - 1)S_{X_m}^2} \tag{12.5}$$

so that

$$\hat{\beta}_1 = \frac{1.89333(4.577465) + 1.73333(3.655769)}{1.89333 + 1.73333} = 4.136949$$

12.4 TESTING FOR THE EQUALITY OF INTERCEPTS

To examine the possible equality of the two population intercepts (without regard to the slopes), we may test

$$H_0: \quad \beta_{0_a} = \beta_{0_m} \quad \text{or} \quad \beta_{0_a} - \beta_{0_m} = 0$$

against

$$H_1: \quad \beta_{0_a} \neq \beta_{0_m} \quad \text{or} \quad \beta_{0_a} - \beta_{0_m} \neq 0$$

by the following t test using the sample intercepts from the two independently developed regression equations:

$$t = \frac{(b_{0_a} - b_{0_m}) - (\beta_{0_a} - \beta_{0_m})}{S_{b_{0_a} - b_{0_m}}} \sim t_{n_a + n_m - 4} \tag{12.6}$$

where $S_{b_{0_a} - b_{0_m}}$, the standard error of the difference in the two intercepts, is defined as

$$S_{b_{0_a} - b_{0_m}} = \sqrt{S_{P|X}^2 \left\{ \frac{1}{n_a} + \frac{1}{n_m} + \frac{\bar{X}_a^2}{(n_a - 1)S_{X_a}^2} + \frac{\bar{X}_m^2}{(n_m - 1)S_{X_m}^2} \right\}} \tag{12.7}$$

By using an α level of significance, the null hypothesis is rejected if $t \geq t_{1 - \frac{\alpha}{2}; \, n_a + n_m - 4}$ or if $t \leq t_{\frac{\alpha}{2}; \, n_a + n_m - 4}$.

Note that as in (12.3) this test for the equality of intercepts also requires the pooling of the two $S_{Y|X}^2$ values. Again, as was the case in our example, such pooling is appropriate only if the homoscedasticity assumption holds over each of the models. Thus from (12.2) we calculated $S_{P|X}^2 = .549304$, and from Table 12.1 we have

$$b_{0_a} = 4.957183 \qquad b_{0_m} = 6.847885 \qquad \bar{X}_a = 2.833 \qquad \bar{X}_m = 2.967$$

$$(n_a - 1)S_{X_a}^2 = \sum_{i=1}^{n_a} X_{a_i}^2 - \frac{\left(\sum_{i=1}^{n_a} X_{a_i}\right)^2}{n_a} = 1.89333$$

and

$$(n_m - 1)S_{X_m}^2 = \sum_{i=1}^{n_m} X_{m_i}^2 - \frac{\left(\sum_{i=1}^{n_m} X_{m_i}\right)^2}{n_m} = 1.73333$$

so that using (12.7), we obtain

$$S_{b_{0_a} - b_{0_m}} = \sqrt{.549304 \left[\frac{1}{15} + \frac{1}{15} + \frac{(2.833)^2}{1.89333} + \frac{(2.967)^2}{1.73333} \right]} = 2.278473$$

Thus from (12.6)

$$t = \frac{4.957183 - 6.847885}{2.278473} = -.830$$

By choosing an .05 level of significance, since $t = -.830 > t_{.025;\, 26} = -2.056$, the null hypothesis is not rejected.

Since there is no evidence that the two population intercepts differ, we may pool the two separate sample estimates to obtain a combined estimate of the common intercept as follows:

$$\hat{\beta}_0 = \frac{n_a b_{0_a} + n_m b_{0_m}}{n_a + n_m} \qquad\qquad (12.8)$$

and hence

$$\hat{\beta}_0 = \frac{15(4.957183) + 15(6.847885)}{15 + 15} = 5.902534$$

12.5 USING DUMMY VARIABLES IN REGRESSION

12.5.1 Problems with Separate Tests of Regression Parameters

Had either of the separate tests for parallelism (Section 12.3) or for equality of intercepts (Section 12.4) led to a rejection of the null hypothesis, we could have concluded that the population models were different. On the other hand, the reverse is not true; although neither test caused a rejection of the null hypothesis, we thus far are unable to conclude that the two sample regression equations are coming from the same or identical populations. There is a twofold problem here:

1. The test for equality of slopes (12.3) was performed regardless of whether there were differences in the intercepts, and, likewise, the test for equality of intercepts (12.6) was conducted irrespective of whether there were differences in the slopes. A researcher may legitimately argue that when testing for the

coincidence of the two fitted regression models the respective regression coefficients on which a model depends should not be looked at as separate from each other.

2. When deciding on the coincidence of the two fitted regression models, the fact that the two tests were conducted separately affects the α level of significance. As discussed in Section 4.2, a researcher would argue that when studying the coincidence of two fitted regression models an *experimentwise* error rate would be more appropriate.

To resolve these dilemmas and obtain a more adequate test for the equality of the fitted regression equations, we must develop a single multiple regression model which includes as a variable (and thus takes into account) the classification factor (i.e., "business major") previously used to separate the different groups.

12.5.2 Using Dummy Variables

Up to this point in our discussion of regression models we have utilized only those variables which were measured on some quantitative scale. There are frequent occasions, however, where variables are needed to identify a category as opposed to representing a measurement from some physical, biological, or social process. For example, in the problem of interest to the Director, the grade point indexes and salary data were obtained from two groups of individuals —those who had majored in accountancy and those who had majored in marketing/management. If, instead, we were to develop a multiple regression model to predict the starting salary based on both the grade point index and the business major, we could not establish a natural scale to differentiate the two categories of business major. We must, however, assign some *values* to the categories of this qualitative variable so that we can address the possibility that the different majors will have separate effects upon the response variable (salary).

The use of *dummy variables* is the vehicle which permits the researcher to handle such situations. If a given qualitative predictor variable to be included in a multiple regression model has c categories, then $c - 1$ dummy variables will be needed to uniquely represent all these categories. For our example, the two types of business majors are designated as "accountancy" and "marketing/management." Let D_1 be a dichotomous dummy variable defined by

$$D_1 = \begin{cases} 0 & \text{if the individual majored in accountancy} \\ 1 & \text{if the individual majored in marketing/management} \end{cases}$$

Entering the dummy variable D_1 into a regression model will require a parameter β_2 (say) to be estimated in conjunction with the other parameters, β_0 and β_1. Moreover, in stating the full regression model, an *interaction* parameter β_3 is needed to express the potential relationship between the two explanatory variables X_1 and D_1.

To fix ideas, consider the sample of $n = 30$ recent business school graduates. As displayed in Table 8.1, there are three *measurements* recorded for each of the individuals: grade point index (X_1), business major (D_1), and salary (Y). The complete (i.e., full) multiple regression model to predict starting salary is given by

$$Y_i = \beta_0 + \beta_1 X_{1_i} + \beta_2 D_{1_i} + \beta_3 X_{1_i} D_{1_i} + \epsilon_i \qquad (12.9)$$

where Y_i = value of Y at observation i

β_0 = Y intercept

β_1 = slope of Y with variable X_1, holding variable D_1 constant

β_2 = incremental effect of variable D_1 on Y, holding variable X_1 constant

β_3 = incremental interaction effect of variables X_1 and D_1 on Y

ϵ_i = random error at observation i

The corresponding fitted sample regression equation is

$$\hat{Y}_i = b_0 + b_1 X_{1_i} + b_2 D_{1_i} + b_3 X_{1_i} D_{1_i} \qquad (12.10)$$

where $\hat{Y}_i$ is the predicted value of Y at observation i and b_0, b_1, b_2, and b_3 are estimates of the parameters β_0, β_1, β_2, and β_3.

12.5.3 Testing for Coincidence

In testing for coincidence the hypotheses are

H_0: $\beta_2 = \beta_3 = 0$

H_1: at least one of the two parameters is not zero

Note that if the null hypothesis is not rejected, the full-fitted model (12.10) reduces to the simple linear regression equation

$$\hat{Y}_{a_i} = b_{0_a} + b_{1_a} X_{1_{a_i}} \qquad (12.11)$$

fitted for accountancy majors in Table 12.1. This is so because if β_3 were zero, there would be no interaction effect between the grade point index and the business major so that the slopes would be parallel, while if β_2 were zero, there would be no incremental effect on the salary for either category of business major so that the intercepts would be equal. Hence the full-fitted multiple regression model would reduce accordingly. Of course, if coincidence were observed, then the business major would not be a useful predictor of the salary; the Director would then obtain the "best" simple linear model by developing the sample regression equation for all 30 individuals—regardless of major—as was achieved in Chapter 8.

Figure 12.3 presents a partial output for the full model (12.10) using the SAS GLM procedure. From Figure 12.3 the full model (12.10) fitted to the sample data is

$$\hat{Y}_i = 4.957183 + 4.577465 X_{1_i} + 1.890702 D_{1_i} - .921696 X_{1_i} D_{1_i}$$

Note that for any accountancy major, D_1 is zero, and the model is re-expressed as

FULL MODEL (12.10)

DEPENDENT VARIABLE: SALARY   STARTING SALARY IN $000

SOURCE	DF	SUM OF SQUARES	MEAN SQUARE	F VALUE	PR > F	R-SQUARE	C.V.
MODEL	3	63.24508586	21.08169529	38.38	0.0001	0.815781	4.1614
ERROR	26	14.28191414	0.54930439			STD DEV	SALARY MEAN
CORRECTED TOTAL	29	77.52700000				0.74115072	17.81000000

SOURCE	DF	TYPE I SS	F VALUE	PR > F
GPI	1	58.01938830	105.62	0.0001
MAJORD	1	4.45696281	8.11	0.0085
GPI*MAJORD	1	0.76873476	1.40	0.2475

SOURCE	DF	TYPE IV SS	F VALUE	PR > F
GPI	1	39.67136150	72.22	0.0001
MAJORD	1	0.37824380	0.69	0.4142
GPI*MAJORD	1	0.76873476	1.40	0.2475

PARAMETER	ESTIMATE	T FOR H0: PARAMETER=0	PR > |T|	STD ERROR OF ESTIMATE
INTERCEPT	4.95718310	3.22	0.0034	1.53807796
GPI	4.57746479	8.50	0.0001	0.53863305
MAJORD	1.89070152	0.83	0.4142	2.27847149
GPI*MAJORD	-0.92169556	-1.18	0.2475	0.77912254

REDUCED MODEL (12.16)

DEPENDENT VARIABLE: SALARY   STARTING SALARY IN $000

SOURCE	DF	SUM OF SQUARES	MEAN SQUARE	F VALUE	PR > F	R-SQUARE	C.V.
MODEL	2	62.47635110	31.23817555	56.04	0.0001	0.805866	4.1921
ERROR	27	15.05064890	0.55743144			STD DEV	SALARY MEAN
CORRECTED TOTAL	29	77.52700000				0.74661331	17.81000000

SOURCE	DF	TYPE I SS	F VALUE	PR > F
GPI	1	58.01938830	104.08	0.0001
MAJORD	1	4.45696281	8.00	0.0087

SOURCE	DF	TYPE IV SS	F VALUE	PR > F
GPI	1	62.06801777	111.35	0.0001
MAJORD	1	4.45696281	8.00	0.0087

PARAMETER	ESTIMATE	T FOR H0: PARAMETER=0	PR > |T|	STD ERROR OF ESTIMATE
INTERCEPT	6.20531250	5.50	0.0001	1.12741289
GPI	4.13694853	10.55	0.0001	0.39205041
MAJORD	-0.78492647	-2.83	0.0087	0.27759088

Figure 12.3 Partial output for the full model (12.10) and reduced model (12.16) using PROC GLM of SAS.

$$\hat{Y}_i = 4.957183 + 4.577465 X_{1_i}$$

as shown in Table 12.1. Similarly, for any marketing/management major, D_1 equals 1, and the model becomes

$$\hat{Y}_i = 6.847885 + 3.655769 X_{1_i}$$

as presented in Table 12.1.

More formally, when $D_{1_i} = 0$, then (12.10) reduces to

$$\hat{Y}_i = b_0 + b_1 X_{1_i} \tag{12.12}$$

where $b_0 = b_{0_a}$ and $b_1 = b_{1_a}$ as in Table 12.1; also, when $D_{1_i} = 1$, then (12.10) reduces to

$$\hat{Y}_i = (b_0 + b_2) + (b_1 + b_3)X_{1_i} \tag{12.13}$$

where $b_0 + b_2 = b_{0_m}$ and $b_1 + b_3 = b_{1_m}$ as in Table 12.1. Thus the full multiple regression equation containing the dummy variable D_1 combines the two separate regression equations in such a manner that differences in slopes (as estimated by the effect of b_3) and/or differences in intercepts (as estimated by the effect of b_2) are *simultaneously* taken into account.

Using the full model (12.10), we may now test for coincidence in the two separate regression lines. The test simultaneously considers the equality of the corresponding regression coefficients comprising the models. Again, the hypotheses are

$$H_0 : \quad \beta_2 = \beta_3 = 0$$

$$H_1 : \quad \text{at least one of the two parameters is not zero}$$

The test statistic is given by

$$F = \frac{[\text{SSR}(b_1, b_2, b_3) - \text{SSR}(b_1)]/(p-1)}{\text{SSE}(b_1, b_2, b_3)/(n-p-1)} \sim F_{p-1, n-p-1} \tag{12.14}$$

where $p = 3$ is the *total* number of predictor variables (including dummy variables), and the null hypothesis of coincidence is rejected if, at the α level of significance, $F \geq F_{1-\alpha; \, p-1, n-p-1}$.

For our data, the regression sums of squares [$\text{SSR}(b_1)$] using the single predictor X_1 is obtained from the *Type I SS* in Figure 12.3 (it was previously displayed in Figure 8.16). Of course, the $\text{SSR}(b_1, b_2, b_3)$ and $\text{SSE}(b_1, b_2, b_3)$ are also obtained from Figure 12.3. Using (12.14), we compute

$$F = \frac{(63.245086 - 58.019388)/2}{14.281914/26} = \frac{2.612849}{.549304} = 4.76$$

By using an .05 level of significance, since $F = 4.76 > F_{.95; \, 2, 26} = 3.37$, we may reject the null hypothesis and conclude the two fitted models are not coincident. Hence the (dummy) variable pertaining to the business major is a significant contributor to a model predicting the starting salary.

Although this result differs from that obtained by the separate tests for equality of slopes (12.3) and equality of intercepts (12.6), we again emphasize that the conclusions drawn from this *simultaneous* test (12.14) are more appropriate.

12.5.4 Testing for Parallelism Using Dummy Variables

The question that still needs to be answered is: How does the variable business major relate to starting salary? In other words, is there essentially an equivalent incremental difference in the starting salaries of individuals who had majored in accountancy versus those who had majored in marketing/management—*regardless* of the grade point index? If this is so, we can then interpret this to mean that there is no significant interaction parameter β_3 (and the full model could be reduced to one having separate regression equations but common slopes). If this is not so, we may conclude that there is an interaction effect of the major and the grade point index on the salary. We would then conclude that the slopes were not parallel and that salary differences between individuals who had majored in accountancy and those who had majored in marketing/management would change with changing levels of the grade point index.

The test procedure.   To test for parallelism using the complete multiple regression model (12.9), we have

$$H_0: \quad \beta_3 = 0$$

and

$$H_1: \quad \beta_3 \neq 0$$

and the test statistic F simply measures the contribution of the *product variable* $X_1 D_1$ into a model already containing their separate effects. Thus we have

$$F = \frac{[\text{SSR}(b_1,b_2,b_3) - \text{SSR}(b_1,b_2)]/1}{\text{SSE}(b_1,b_2,b_3)/(n - p - 1)} \sim F_{1,\,n-p-1} \qquad (12.15)$$

and the null hypothesis of parallelism is rejected if, at the α level of significance, $F \geq F_{1-\alpha;\,1,\,n-p-1}$.

For our data all the terms in (12.15) can be found in the SAS output (Figure 12.3). From the *Type I* SS we obtain $\text{SSR}(b_1,b_2)$ as the sum of the two *separate* regression contributions, $\text{SSR}(b_1)$ and $\text{SSR}(b_2|b_1)$. Using (12.15), we compute

$$F = \frac{(63.245086 - 62.476351)/1}{14.281914/26} = \frac{.768735}{.549304} = 1.40$$

By using an .05 level of significance, since $F = 1.40 < F_{.95;\,1,\,26} = 4.23$, the null hypothesis is not rejected, and we may conclude that there is no interaction effect present. Hence the fitted model (12.10) reduces to

$$\hat{Y}_i = b'_0 + b'_1 X_{1_i} + b'_2 D_{1_i} \qquad (12.16)$$

where for accountancy majors we have

$$\hat{Y}_i = b'_0 + b'_1 X_{1_i} \qquad (12.17)$$

while for marketing/management majors we have

$$\hat{Y}_i = (b_0' + b_2') + b_1'X_{1_i} \tag{12.18}$$

so that the two separately fitted regression equations have the common slope b_1' but differing intercepts.

Note that the estimates (b_0', b_1', b_2') obtained by least-squares methods for this (reduced) fitted regression model (12.16) are not the same as those generated in (12.10) for the full model.

From Figure 12.3 the reduced model (12.16) for these data is

$$\hat{Y}_i = 6.205313 + 4.136949X_{1_i} - .784926D_{1_i}$$

so that for accounting majors

$$\hat{Y}_i = 6.205313 + 4.136949\,X_{1_i}$$

while for marketing/management majors

$$\hat{Y}_i = 5.420387 + 4.136949X_{1_i}$$

Thus, with common slope, the coefficient $b_2' = -.784926$ measures the reduction in the starting salary for individuals who had majored in marketing/management as compared to those who had majored in accountancy.

Some interesting relationships. From the preceding, several interesting relationships are worth mentioning. First, it is important to note that the common slope b_1' developed for the multiple regression equation (12.16) is *identical* to that which we obtained $(\hat{\beta}_1)$ when we pooled the separate slopes using (12.5). Moreover, once the common slope is estimated, the intercept b_0' for accountancy majors would have been obtained in the usual manner with ordinary least-squares methods by using (8.6). Thus

$$b_0' = \frac{\sum\limits_{i=1}^{n_a} Y_{a_i}}{n_a} - b_1'\frac{\sum\limits_{i=1}^{n_a} X_{a_i}}{n_a} \tag{12.19}$$

which for accountancy majors (Table 12.1) is

$$b_0' = \frac{268.9}{15} - 4.136949\left(\frac{42.5}{15}\right) = 6.205313$$

while for marketing/management majors the intercept $b_0' + b_2'$ could similarly have been computed,

$$b_0' + b_2' = \frac{\sum\limits_{i=1}^{n_m} Y_{m_i}}{n_m} - b_1'\frac{\sum\limits_{i=1}^{n_m} X_{m_i}}{n_m} \tag{12.20}$$

so that (see Table 12.1)

$$b_0' + b_2' = \frac{265.4}{15} - 4.136949\left(\frac{44.5}{15}\right) = 5.420387$$

Finally, it is interesting to observe from the identity between $F_{1-\alpha;\,1,\nu}$ and

$(t_{1-\frac{\alpha}{2};\,\nu})^2$ [see (1.20)] that the partial F test for parallelism (i.e., $H_0 : \beta_3 = 0$) gives, except for rounding errors, the same result as the (square of the) t test for parallelism. Using (12.15), we computed $F = 1.40$, and the critical value was $F_{.95;\,1.26} = 4.23$. From (12.3) we note that $t^2 = (1.183)^2 = 1.40$, and the squared critical value is $(t_{.975;\,26})^2 = (2.056)^2 \cong 4.23$. Thus, unlike the F test for coincidence (12.14), we may again describe this F test for parallelism (12.15) as a *separate* test.

The reduced multiple regression model (12.16) with common slope b'_1 provides the basis for yet another approach to comparing two (or more) regression equations. The important and useful *analysis of covariance* (ANACOVA) *procedures* will be described in Section 12.7. Prior to this, however, it is important to demonstrate to the researcher (1) how to interpret multiple regression models containing more than one qualitative variable, particularly where the number of categories for a specific variable is not necessarily dichotomous (Section 12.5.5), and (2) how to use computer packages for multiple regression models containing dummy variables (Section 12.6).

12.5.5 Interpreting the Parameter Estimates for Dummy Variables

As an example, suppose a qualitative predictor variable has *three* possible categories of response (d_1, d_2, d_3). The *two* required dummy variables D_1 and D_2 may be defined by

$$(D_1, D_2) = \begin{cases} (1,0) & \text{for observations from } d_1 \\ (0,1) & \text{for observations from } d_2 \\ (0,0) & \text{for observations from } d_3 \end{cases}$$

Whatever model has been decided upon for the remaining variables, it will be necessary to include at least the additional term $\beta_j D_1 + \beta_k D_2$. More complicated models would, of course, necessitate the inclusion of some interaction parameters between the dummy variables and the other explanatory variables.

To illustrate some of these ideas, we shall consider part of an analysis presented by Chatterjee and Price (1977). In their discussion the issue was the identification and quantification of those factors that determine salary differentials for computer professionals in a large corporation. The variables included in the study were

1. Experience (X_1)—measured in years
2. Education—coded using dummy variables:

$$(D_1, D_2) = \begin{cases} (1,0) & \text{for high school graduate} \\ (0,1) & \text{for bachelor's degree} \\ (0,0) & \text{for advanced degree} \end{cases}$$

3. Management responsibility—coded using a dummy variable:

$$D_3 = \begin{cases} 1 & \text{if individual has management responsibility} \\ 0 & \text{if not} \end{cases}$$

4. Salary (Y)—measured in dollars per annum

The reduced model considered by the authors is

$$Y_i = \beta_0 + \beta_1 X_{1_i} + \beta_2 D_{1_i} + \beta_3 D_{2_i} + \beta_4 D_{3_i} + \epsilon_i \qquad (12.21)$$

so that the corresponding fitted model becomes

$$\hat{Y}_i = b_0 + b_1 X_{1_i} + b_2 D_{1_i} + b_3 D_{2_i} + b_4 D_{3_i} \qquad (12.22)$$

The three dummy variables help to determine the base salary level as a function of education and management status after adjusting for years of experience.

Using a sample of 46 individuals, the model given in (12.21) was estimated. The resulting estimated parameters, standard errors, and t values are displayed in Table 12.2 along with $r^2_{Y.1234}$ and $S^2_{Y|X1234}$.

TABLE 12.2   Summary of Results for Fitted Model

Variable Name	Parameter Estimate	Standard Error	t
Experience (X_1)	546.16	30.52	17.90
High School Graduate (D_1)	−2,996.00	411.73	−7.28
Bachelor's Degree (D_2)	147.98	387.64	0.38
Management Responsibility (D_3)	6,883.50	313.94	21.93
(Constant)	11,032.00	383.20	28.79

$$r^2_{Y.1234} = .9568 \qquad S^2_{Y|X1234} = 1027.39$$

SOURCE: S. Chatterjee and B. Price, *Regression Analysis by Example*, Wiley, New York, 1977.

As usual, we should exercise caution when interpreting results without first looking at pertinent residual plots and/or testing for the adequacy of the fit. However, we assume that this has been done and that the researcher is satisfied with the secondary analysis. By using Table 12.2, the following conclusions can be made while taking into account or adjusting for the effects of the other predictor variables:

1. Each additional year of job experience is estimated to be worth an annual salary increment of $546.16.

2. The average incremental value in annual salary associated with a management position is $6883.50.

3. b_2 measures the salary differential for the high school graduate relative to the advanced degree recipient, while b_3 measures the salary differential for

the bachelor's degree recipient relative to the advanced degree recipient. The difference $b_3 - b_2$ measures the salary differential for the bachelor's degree recipient relative to the high school graduate. Hence, an advanced degree is worth \$2996.00 more than a high school diploma, a bachelor's degree is worth \$147.98 more than an advanced degree (but this is not statistically significant since $t = .38$), and a bachelor's degree is worth \$3143.98 more than a high school diploma.

12.6 COMPUTER PACKAGES AND DUMMY VARIABLES: USE OF SPSS, SAS, AND BMDP

In Section 12.5 we developed the idea of utilizing qualitative (i.e., dummy) variables as explanatory variables in regression analysis. In this section, we shall describe how dummy variables can be created using either SPSS, SAS, or BMDP.

12.6.1 Forming Dummy Variables Using SPSS

We may recall from Section 11.7.1 that the COMPUTE statement of SPSS was used to transform variables. To create dummy variables in SPSS, the COMPUTE and RECODE statements can be utilized. The format is

1	16
COMPUTE	{variable} = {arithmetic expression}
RECODE	(variable name or list)
	(value list$_1$ = new value$_1$)
	(value list$_2$ = new value$_2$) etc.

If we refer to the problem of interest to the Director of Career Development (see Sections 8.4 and 12.5.2), we may note that accountancy has been coded 1 and marketing/management has been coded 2. This may be converted into a 0-1 dummy variable as follows:

1	16
COMPUTE	MAJORD = MAJOR
RECODE	MAJORD(1 = 0) (2 = 1)
VARϸLABELS	MAJORD, MAJORϸRECODED/
VALUEϸLABELS	MAJORD(0) ACCOUNTANCY
	(1) MARKETINGϸMANAGEMENT/

Refer next to the example concerning the salaries of computer professionals (Section 12.5.5). If the variable education (named EDUC) is coded as

(1) HS GRAD   (2) BACHELOR   (3) ADVANCED

we need to create two dummy variables with HS GRAD coded as (1,0), BACHELOR coded as (0,1), and ADVANCED coded as (0,0). The necessary control statements would be

```
1                16
COMPUTE          EDUC1 = EDUC
COMPUTE          EDUC2 = EDUC
RECODE           EDUC1 (1 = 1) (2,3 = 0)
RECODE           EDUC2 (1,3 = 0) (2 = 1)
VALUEƀLABELS     EDUC1 (0) NONƀHSƀGRAD (1) HSƀGRAD/
                 EDUC2 (0) NONƀBACHELOR (1) BACHELOR/
```

12.6.2 Forming Dummy Variables Using SAS

We may recall from Section 11.7.2 that the "variable = expression" statement of SAS was used to transform variables. To create dummy variables in SAS, the {variable =} and the IF THEN statements can be provided. The format of these control statements is

$$\{variable\} = \{expression\};$$
$$IFƀ\{expression\}ƀTHENƀ\{statement\};$$

If we refer to the problem of interest to the Director of Career Development (see Sections 8.4 and 12.5.2), we may note that accountancy has been coded 1 and marketing/management has been coded 2. This may be recoded into a 0-1 dummy variable as follows:

```
MAJORD = 0;        ← needed to initialize MAJORD
IFƀMAJORƀEQƀ1ƀTHENƀMAJORD = 0;
IFƀMAJORƀEQƀ2ƀTHENƀMAJORD = 1;
```

Refer next to the example concerning the salaries of computer professionals (Section 12.5.5). If the variable education (named EDUC) is coded as

(1) HS GRAD   (2) BACHELOR   (3) ADVANCED

we need to create two dummy variables with HS GRAD coded as (1,0), BACHELOR coded as (0,1), and ADVANCED coded as (0,0). The necessary SAS control statements would be

```
EDUC1 = 0;
EDUC2 = 0;
IF ƀEDUCƀEQƀ1ƀTHENƀEDUC1 = 1;
IF ƀEDUCƀNEƀ1THENƀEDUC1 = 0;
IF ƀEDUCƀEQƀ2THENƀEDUC2 = 1;
IF ƀEDUCƀNEƀ2ƀTHENƀEDUC2 = 0;
```

12.6.3 Forming Dummy Variables Using BMDP

We may recall from Section 11.7.3 that the /TRANSFORM paragraph of BMDP was used to transform variables. To create dummy variables in BMDP, a *logical* transformation can be utilized with the following format:

$$/\text{TRANSFORM}\flat \begin{Bmatrix} \text{new} \\ \text{variable} \end{Bmatrix} = \begin{Bmatrix} \text{old} \\ \text{variable} \end{Bmatrix} \flat \begin{pmatrix} \text{EQ} \\ \text{NE} \\ \text{LT} \\ \text{LE} \\ \text{GT} \\ \text{GE} \\ \text{OR} \\ \text{IF} \end{pmatrix} \flat \begin{Bmatrix} \text{value} \\ \text{or} \\ \text{variable} \end{Bmatrix}.$$

The result of a logical operation in BMDP is equal to 1 if the expression is true and equal to 0 if the expression is false. The logical operations equal (EQ), not equal (NE), less than (LT), less than or equal (LE), greater than (GT), greater than or equal (GE), OR, and IF may be utilized.

If we refer to the problem of interest to the Director of Career Development (see Sections 8.4 and 12.5.5), we may note that accountancy has been coded 1 and marketing/management has been coded 2. This may be recoded into a 0-1 dummy variable as follows:

/VARIABLESþADD = 1.
/TRANSFORMþMAJORD = MAJORþEQþ2.

Once again we note that an ADD sentence must be provided as part of the /VARIABLE paragraph to indicate the number of variables added by transformation.

Refer next to the example concerning the salaries of computer professionals (Section 12.5.5). If the variable education (named EDUC) is coded as

(1) HS GRAD   (2) BACHELOR   (3) ADVANCED

we need to create two dummy variables with HS GRAD coded as (1,0), BACHELOR coded as (0,1), and ADVANCED coded as (0,0). The necessary BMDP paragraphs would be

/VARIABLESþADD = 2.
/TRANSFORMþEDUC1 = EDUCþEQþ1.
        EDUC2 = EDUCþEQþ2.

12.7 THE ANALYSIS OF COVARIANCE (ANACOVA)

12.7.1 Introduction

One approach to describing the integration of regression with methods of experimental design (Chapters 3–7) is through a study of the **analysis of covariance** (ANACOVA). The ANACOVA procedures are invaluable to the researcher when planning experiments because they attempt to reduce experimental error and thereby provide for a more powerful analysis of treatment effects. Such procedures then are particularly useful for completely randomized experimental design models (Chapter 3) wherein other methods for reducing experimental error such as *blocking* (Chapter 5) cannot be applied.

12.7.2 ANACOVA Versus ANOVA

To develop ideas, we return to the problem of interest to the Director of Career Development. From Table 8.1, neglecting the grade point index, we extract the starting salaries of the 15 recent graduates who had majored in accountancy as well as the 15 who had majored in marketing/management. Treating these data as the outcomes of a completely randomized experiment having but two levels of the (qualitative) classification factor "business major," we may use the ANOVA procedure of Chapter 3 to determine whether or not there are significant differences in the starting salaries among the two groups. From the information summarized in Table 12.1, we first test the assumption of homoscedasticity. Thus

$$H_0: \quad \sigma^2_{Y_a} = \sigma^2_{Y_m}$$
$$H_1: \quad \sigma^2_{Y_a} \neq \sigma^2_{Y_m}$$

and using Table 1.3 (formula G), we have*

$$F = \frac{S^2_{Y_a}}{S^2_{Y_m}} = \frac{3.37352}{2.13495} = 1.58$$

If an .05 level of significance is chosen, $F = 1.58$ falls between $F_{.025; 14, 14} \cong .334$ and $F_{.975; 14, 14} \cong 2.99$; hence there is no reason to suspect a violation in this assumption. Continuing with the analysis using the summary information obtained from Table 12.1, we note that the total sample size $(n..)$ is 30, the grand total $(Y..)$ is 534.3, and the sum of the squares $(\sum_{i=1}^{n_a} Y^2_{a_i} + \sum_{i=1}^{n_m} Y^2_{m_i})$ is 9593.41.

To test the hypotheses

$$H_0: \quad \mu_{Y_a} = \mu_{Y_m}$$
$$H_1: \quad \mu_{Y_a} \neq \mu_{Y_m}$$

we compute F as in (3.1). Table 12.3 presents the ANOVA table for these data. Using an .05 level of significance, since $F = .148 < F_{.95; 1, 28} = 4.20$, we could not reject the null hypothesis. Hence we would conclude that there is no evidence of any real differences in the starting salaries among the two groups of majors.

On the other hand, however, if we utilize the known sample information pertaining to the grade point index in our efforts to compare the starting salaries among the two groups, we shall observe that the analysis and resulting conclusions will change drastically. Thus ANACOVA permits the researcher to employ a *covariate* (a quantitative independent variable such as the grade point index) along with the qualitative groupings (i.e., the business major) in order to study

---

*Note that this test of homoscedasticity uses the two sample variances $S^2_{Y_a}$ and $S^2_{Y_m}$, whereas the test in Section 12.3 employed the two sample variances about the respective regression lines $S^2_{Y_a|X_a}$ and $S^2_{Y_m|X_m}$.

the responses (i.e., the starting salary). The covariate is also known as a *concomitant* variable. If the chosen concomitant variable is related to the response (i.e., dependent) variable, ANACOVA methods will be successful in reducing the experimental error in the data and thereby yield a more powerful analysis. If, of course, the concomitant variable does not relate significantly to the response variable, nothing is really gained by ANACOVA methods that could not have been obtained more simply by ANOVA procedures. Thus the selection of the covariate is most important. In a designed experiment the concomitant variable should, if possible, be studied prior to conducting the experiment. If such measurements are obtained during the experiment or are obtained as part of a survey, these should not in any way be affected by the imposed treatment conditions.* For the problem of interest to the Director, the choice of the grade

TABLE 12.3   ANOVA of Salary Data

Source of Variation	Degrees of Freedom	Sum of Squares	Mean Square	F
Among Groups (Majors)	1	SSα = .408333	MSα = .408333	$F = \dfrac{\text{MS}\alpha}{\text{MSW}} = .148$
Within Groups	28	SSW = 77.118667	MSW = 2.754238	
Total	29	SST = 77.527000		

point index as the concomitant variable seems to meet these criteria. The grade point index is a good predictor of starting salary (see Chapter 8). Moreover, the grade point index is not strongly related to the treatments (i.e., groupings of business major)—that is, from Table 12.1 there do not appear to be any real differences in the distribution of the grade point indexes achieved by individuals who had majored in accountancy versus those who had majored in marketing/management. Hence we should expect ANACOVA methods to be useful here.

12.7.3 The Analysis of Covariance Model

From a regression viewpoint, the ANACOVA model (12.16) was already developed in Section 12.5.4. The ANACOVA procedures are based on the assumption that the separate regression equations evolving from the full-fitted model have equal slopes. Thus for our data, the full-fitted model (12.10) contains

*If the selected covariate is highly related to the treatment, ANACOVA procedures will confound much of the effect which the treatments had on the response variable (see the problem of multicollinearity in Section 14.5)—yielding perplexing results to an unsophisticated researcher. For example, suppose the accountancy majors had all achieved high grade point indexes, while the marketing/management majors had all achieved low ones. If such a relationship between the two independent variables (the grade point index and major) were to exist, ANACOVA procedures would hide actual salary differences when controlling for the grade point index.

estimates b_0, b_1, b_2, and b_3 of the respective true parameters β_0, β_1, β_2, and β_3. Testing for parallelism (H_0: $\beta_3 = 0$), we compute the partial F statistic (12.15) to evaluate the contribution of the product variable $X_1 D_1$ (i.e., the interaction) to a model already containing X_1 and D_1. If the null hypothesis is rejected, we conclude that the separate regression models are different, and the thrust of our data analysis and interpretations involves predictions for each (separate) group. No ANACOVA would be performed. On the other hand, if the null hypothesis is not rejected, we can conclude that, in the populations from which the (fitted) simple regression equations were developed, there is no evidence of any differences in the slopes. Therefore, by removing the β_3 term from (12.9), we can then fit the reduced model

$$Y_i = \beta_0 + \beta_1 X_{1_i} + \beta_2 D_{1_i} + \epsilon_i \tag{12.23}$$

having no interaction term present. Such a model is known as the ANACOVA model. From (12.23) the fitted model becomes

$$Y_i = b'_0 + b'_1 X_{1_i} + b'_2 D_{1_i} \tag{12.16}$$

where, as previously mentioned, the resulting estimates b'_0, b'_1, and b'_2 differ numerically from those obtained using the full model.

Recall from Figure 12.3 that the ANACOVA model for these data is

$$\hat{Y}_i = 6.205313 + 4.136949 X_{1_i} - .784926 D_{1_i}$$

From this the two separate regression equations are

$$\begin{aligned} \hat{Y}_i &= b'_0 + b'_1 X_{1_i} \\ &= 6.205313 + 4.136949 X_{1_i} \end{aligned} \tag{12.17}$$

for accountancy majors and

$$\begin{aligned} \hat{Y}_i &= (b'_0 + b'_2) + b'_1 X_{1_i} \\ &= (6.205313 - .784926) + 4.136949 X_{1_i} \\ &= 5.420387 + 4.136949 X_{1_i} \end{aligned} \tag{12.18}$$

for marketing/management majors.

Since the two models have common slope (b'_1), we can interpret the dummy variable regression coefficient b'_2 to represent the incremental difference in starting salaries between marketing/management majors and accountancy· majors—having *adjusted* for the grade point index. That is, for those individuals who had achieved the *same* grade point index, starting salaries were approximately \$785 lower for marketing/management majors. This is illustrated in Figure 12.4.

To determine whether such differences in the adjusted effects are significant, we must compute the adjusted mean salary for accountancy majors $\bar{Y}_{a(\text{ADJ})}$ and for marketing/management majors $\bar{Y}_{m(\text{ADJ})}$. These adjusted means are simply the predicted values of Y when evaluating the separate regression equations using $\bar{X}$ (the combined mean of the covariate), given by

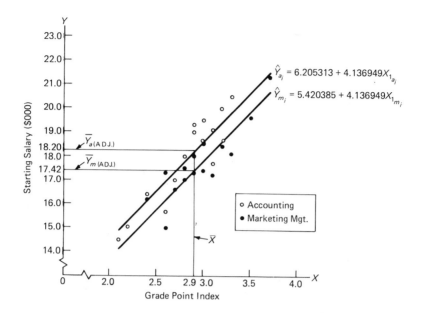

Figure 12.4 Using ANACOVA procedures.

$$\bar{X} = \frac{n_a \bar{X}_a + n_m \bar{X}_m}{n_a + n_m} \tag{12.24}$$

Thus $\bar{X}$ represents the mean grade point index in the pooled data—without regard to the major. By using Table 12.1, for our data

$$\bar{X} = \frac{15(2.833333) + 15(2.966667)}{15 + 15} = 2.90$$

Hence the adjusted salary means are as follows:

1. *Accountancy*:

$$\bar{Y}_{a(\text{ADJ})} = b_0' + b_1' \bar{X}$$
$$= 6.205313 + 4.136949(2.90) = 18.20247$$

   (in thousands of dollars)

2. *Marketing/Management*:

$$\bar{Y}_{m(\text{ADJ})} = (b_0' + b_2') + b_1' \bar{X}$$
$$= 5.420387 + 4.136949(2.90) = 17.41754$$

   (in thousands of dollars)

Note that the "unadjusted" sample means of starting salaries displayed in Table 12.1 are

1. *Accountancy*:

$$\bar{Y}_a = 17.92667 \qquad \text{(in thousands of dollars)}$$

2. *Marketing/Management*:

$$\bar{Y}_m = 17.69333 \qquad \text{(in thousands of dollars)}$$

Thus without regard to a covariate such as the grade point index, the (unadjusted) mean difference in starting salaries between accountancy majors and marketing/management majors was approximately \$233; such an observed difference was deemed insignificant when performing the ANOVA F test (see Table 12.3). Now, however, by adjusting for the grade point index, the mean difference in starting salaries is approximately \$785. Hence, by taking into account differences in the covariate (the grade point index), the salary gap between accountancy and marketing/management majors has widened.

12.7.4 The Test Procedure

To determine whether these adjusted means are significantly different, we test the hypotheses

$$H_0: \quad \beta_2 = 0$$

$$H_1: \quad \beta_2 \neq 0$$

by use of a partial F test to measure the contribution of β_2 to a simple linear model already containing β_1. If the null hypothesis is not rejected, then the variable "business major" is not a significant contributor to the prediction of salary once a model containing the covariate (the grade point index) has already been fitted. In such a case we would merely use the simple linear relationship between the salary and grade point index taken over all individuals, regardless of major, to complete our analysis (see Chapter 8). On the other hand, if the null hypothesis is rejected, we may conclude that there is a significant difference in the mean starting salaries (adjusted for the grade point index) between the two business major groups.

The partial F test is given by

$$F = \frac{[\text{SSR}(b_1, b_2) - \text{SSR}(b_1)]/(p - 1)}{\text{SSE}(b_1, b_2)/(n - p - 1)} \sim F_{p-1, n-p-1} \qquad (12.25)$$

where p is the total number of predictor variables in the ANACOVA model and the null hypothesis may be rejected at an α level of significance if $F \geq F_{1-\alpha; \, p-1, n-p-1}$.

Using the computer output (for the reduced model) in Figure 12.3, we have

$$F = \frac{(62.476351 - 58.019388)/1}{15.050649/27} = \frac{4.456963}{.557431} \cong 8.00$$

Using an .05 level of significance, since $F = 8.00 > F_{.95;\,1,27} = 4.21$, we can reject the null hypothesis and conclude that the adjusted mean salaries are significantly different for the two majors.

We should now appreciate the value of the ANACOVA procedure in enhancing an analysis. The reduction in the experimental error through the introduction of the concomitant variable grade point index has led us to uncover significant differences in the adjusted mean salaries—differences which heretofore were not indicated by the less sophisticated ANOVA methods.

12.7.5 Extending ANACOVA Procedures in the Completely Randomized Experiment

In general, a completely randomized experiment will contain c (qualitative) levels or categories of the classification factor. If, at each level, the researcher is able to obtain a set of measurements for the covariate X and the response variable Y, a more general ANACOVA model can be written as

$$Y_i = \beta_0 + \beta_1 X_{1_i} + \beta_2 D_{1_i} + \beta_3 D_{2_i} + \beta_4 D_{3_i} + \cdots + \beta_c D_{c-1_i} + \epsilon_i \qquad (12.26)$$

Such a model contains $c + 1$ regression parameters to be estimated and $c - 1$ dummy variables. The fitted ANACOVA model becomes

$$\hat{Y}_i = b_0 + b_1 X_{1_i} + b_2 D_{1_i} + b_3 D_{2_i} + b_4 D_{3_i} + \cdots + b_c D_{c-1_i} \qquad (12.27)$$

Note that no interaction terms are permitted, and thus it is thereby assumed that the c separately evolving regression equations all have equal slopes.*

Defining the $c - 1$ dummy variables $(D_1, D_2, D_3, \ldots, D_{c-1})$, we obtain

$$(D_1, D_2, D_3, \ldots, D_{c-1}) = \begin{cases} (1, 0, 0, \ldots, 0) & \text{for observations from category 1} \\ (0, 1, 0, \ldots, 0) & \text{for observations from category 2} \\ (0, 0, 1, \ldots, 0) & \text{for observations from category 3} \\ \quad \vdots & \\ (0, 0, 0, \ldots, 0) & \text{for observations from category } c \end{cases}$$

Thus, the c separate regression equations are

*The parallelism assumption could be checked by fitting the full model containing the interaction terms for the product variables $X_1 D_1$, $X_1 D_2$, $X_1 D_3, \ldots, X_1 D_{c-1}$ and then testing that the corresponding regression coefficients are all zero.

$$\text{Category 1:} \quad \hat{Y}_i = (b_0 + b_2) + b_1 X_{1_i} \qquad (12.28a)$$

$$\text{Category 2:} \quad \hat{Y}_i = (b_0 + b_3) + b_1 X_{1_i} \qquad (12.28b)$$

$$\text{Category 3:} \quad \hat{Y}_i = (b_0 + b_4) + b_1 X_{1_i} \qquad (12.28c)$$

$$\vdots \qquad \qquad \vdots$$

$$\text{Category } c - 1: \quad \hat{Y}_i = (b_0 + b_c) + b_1 X_{1_i} \qquad (12.28d)$$

$$\text{Category } c: \quad \hat{Y}_i = b_0 + b_1 X_{1_i} \qquad (12.28e)$$

From these equations we note that b_1 is the *common* estimate of β_1. To proceed with the ANACOVA, the c adjusted means ($\bar{Y}_{1(\text{ADJ})}, \bar{Y}_{2(\text{ADJ})}, \bar{Y}_{3(\text{ADJ})}, \ldots, \bar{Y}_{c(\text{ADJ})}$) could be obtained by evaluating the separate regression equations (12.28a–e) using the pooled covariate mean $\bar{X}$, where

$$\bar{X} = \frac{n_1 \bar{X}_1 + n_2 \bar{X}_2 + n_3 \bar{X}_3 + \cdots + n_c \bar{X}_c}{n_1 + n_2 + n_3 + \cdots + n_c} \qquad (12.29)$$

To determine whether there are significant differences among these c adjusted means, we test

$$H_0: \quad \beta_2 = \beta_3 = \beta_4 = \cdots = \beta_c = 0$$

against

$$H_1: \quad \text{not all } \beta_j \text{ are zero} \qquad (j = 2, 3, 4, \ldots, c)$$

using the F test computed from

$$F = \frac{[\text{SSR}(b_1, b_2, b_3, \ldots, b_c) - \text{SSR}(b_1)]/(c - 1)}{\text{SSE}(b_1, b_2, b_3, \ldots, b_c)/(n - c - 1)} \sim F_{c-1, n-c-1} \qquad (12.30)$$

where the *numerator* degrees of freedom $(c - 1)$ are obtained as the difference in the degrees of freedom for the reduced model $(n - 2)$ and the ANACOVA model $[n - (c + 1)]$.

All the necessary information to perform the test can be obtained from computer output such as displayed in Figure 12.3. Using an α level of significance, if $F \geq F_{1-\alpha; \, c-1, n-c-1}$, we may reject the null hypothesis and conclude that the adjusted means are not all the same. A posteriori multiple comparison procedures (Chapter 4) such as the Scheffé S method can then be used to determine which of the adjusted means are significantly different from the others [see Neter and Wasserman (1974)].

12.8 COMPUTER PACKAGES AND THE ANALYSIS
OF COVARIANCE: USE OF SPSS, SAS, AND BMDP

In Sections 3.6, 5.6, and 6.9 we discussed the use of SPSS, SAS, or BMDP in studying experimental designs involving either the one-factor or two-factor analysis of variance model. In this section we shall describe how to utilize these packages to perform an analysis of covariance on the problem of interest to the Director of Career Development (see Sections 8.4 and 12.2).

12.8.1 Using the SPSS Subprogram ANOVA for the Analysis of Covariance

The ANOVA subprogram (see Sections 5.6.2 and 6.9.2) can perform the analysis of covariance by using the following format:

1   16

ANOVA $\begin{Bmatrix} \text{dependent or} \\ \text{response variable} \end{Bmatrix}$ ƀBYƀ $\begin{Bmatrix} \text{independent or} \\ \text{qualitative variable(s)} \end{Bmatrix}$

(min. value, max. value)ƀWITHƀ $\begin{Bmatrix} \text{name of} \\ \text{covariate} \end{Bmatrix}$ /

For the starting salary data, the following statements can be included as part of the program displayed as Figure 8.13:

1   16

ANOVA   SALARYƀBYƀMAJOR(1,2)

WITHƀGPI/

Figure 12.5 represents annotated partial output of the SPSS ANOVA subprogram for the starting salary data.

```
✧ ✧ ✧ ✧ ✧ ✧ ✧ ✧ A N A L Y S I S   O F   V A R I A N C E ✧ ✧ ✧ ✧ ✧ ✧ ✧ ✧ ✧
              SALARY      STARTING SALARY IN $000
          BY MAJOR    UNDERGRADUATE MAJOR
          WITH GPI    GRADE POINT INDEX
✧ ✧ ✧ ✧ ✧ ✧ ✧ ✧ ✧ ✧ ✧ ✧ ✧ ✧ ✧ ✧ ✧ ✧ ✧ ✧ ✧ ✧ ✧ ✧ ✧ ✧ ✧ ✧ ✧ ✧ ✧ ✧ ✧
```

SOURCE OF VARIATION	SUM OF SQUARES	DF	MEAN SQUARE	F	SIGNIF OF F
COVARIATES	58.019	1	58.019	1C4.C83	0.0D0
GPI	58.019	1	58.019	1G4.C83	0.C00
MAIN EFFECTS	4.457	1	4.457	7.995	0.0D9
MAJOR	4.457	1	4.457	[7.995]	0.009
EXPLAINED	62.476	2	31.238	56.C39	0.000
RESIDUAL	15.051	27	0.557		
TOTAL	77.527	29	2.673		

Figure 12.5 Partial SPSS output for ANACOVA analysis.

12.8.2 Using SAS PROC GLM for the Analysis of Covariance

The GLM procedure (see Sections 3.6.3 and 6.9.3) can perform the analysis of covariance by using the following format:

PROCþGLM;

$$\text{CLASSESþ} \begin{cases} \text{name of} \\ \text{independent or} \\ \text{qualitative variable(s)} \end{cases};$$

$$\text{MODELþ} \begin{cases} \text{dependent or} \\ \text{response variable} \end{cases} = \{\text{covariate}\} \text{þ} \begin{cases} \text{list of} \\ \text{qualitative} \\ \text{variable(s)} \end{cases};$$

For the starting salary data, the following statements can be included as part of the program depicted as Figure 8.15:

        PROCþGLM;
        CLASSESþMAJOR;
        MODELþSALARY = GPIþMAJOR;

Figure 12.6 represents annotated partial output of SAS PROC GLM for the starting salary data.

12.8.3 Using BMDP Program 1V for the Analysis of Covariance

The BMDP program 1V (see Section 3.6.4) can perform an analysis of covariance by including the following paragraphs:

$$\text{/VARIABLE}\quad\text{GROUPING} = \begin{cases} \text{name of} \\ \text{qualitative variable(s)} \end{cases}.$$

$$\text{/DESIGN}\quad\text{DEPENDENT} = \begin{cases} \text{name of dependent} \\ \text{or response variable} \end{cases}.$$

$$\text{INDEPENDENT} = \begin{cases} \text{name of} \\ \text{covariate} \end{cases}.$$

For the starting salary data, the following statements can be substituted for the /REGRESS paragraph (and P-1V used instead of P-1R) depicted in Figure 8.17:

      /VARIABLE   GROUPING = MAJOR.
      /DESIGN   DEPENDENT = SALARY.
                      INDEPENDENT = GPI.

Figure 12.7 represents annotated partial output from BMDP program 1V for the starting salary data. Contained therein is the test for the equality of adjusted means (12.25) and the test for parallelism (12.15).

12.9 A REGRESSION APPROACH TO THE NONORTHOGONAL TWO-FACTOR EXPERIMENT

12.9.1 Introduction

In Chapter 7 we described the nonorthogonal two-factor experimental design as one in which the cells contained an unequal number of observations. In such experiments the nonorthogonal structure of the design format prohibits

Figure 12.6 Partial SAS output for ANACOVA analysis.

```
ANALYSIS OF VARIANCE

SOURCE OF VARIANCE              D.F.     SUM OF SQ.      MEAN SQ.        F-VALUE

EQUALITY OF ADJ. CELL MEANS      1         4.4568         4.4568          7.9952
ZERO SLOPE                       1        62.0678        62.0678        111.3465
    ERROR                       27        15.0506         0.5574

EQUALITY OF SLOPES               1         0.7687         0.7687          1.3994
    ERROR                       26        14.2819         0.5493
```

Figure 12.7 Partial BMDP program 1V output for ANACOVA analysis.

the decomposition of the total sums of squares (SST) in the ANOVA table into a series of components which measure the separate effects. Although the *method of unweighted means* (see Section 7.6.1) was shown to give excellent approximate solutions provided that certain conditions were met, it was, however, noted that there would be many occasions when the method would yield unsatisfactory results and should be avoided. For example, if cell samples differed markedly or if empty cells existed in the $r \times c$ table, one of the many (more complex but flexible) least-squares regression methods would be better suited. In fact, Winer (1971) had argued that a least-squares approach is always more appropriate if, as in survey research, the data are classified within their natural strata. In addition, for more complex nonorthogonal experiments concerning three or more classification factors a least-squares approach is most appropriate.

In this section, we now provide yet another link in the relationship between experimental design and regression analysis.*

12.9.2 Development

The general layout for data obtained in a two-factor experiment with unequal numbers of observations in the cells was presented in Table 7.1. The corresponding summary of totals and means in the cells, rows, and columns was displayed in Table 7.2.

When the cell frequencies are both unequal and disproportionate, the usual ANOVA procedures cannot be used because there is a correlation among the classification factors. It then becomes difficult to determine the magnitude of the separate effects that each factor has on the response variable. In such situations, however, least-squares methods can be used to estimate the independent effect of each factor while controlling for possible relationships with other factors.

Nevertheless, in dealing with nonorthogonal experiments, the researcher is faced with a dilemma. It is extremely important to note that the various least-

---

*The least-squares methods could not have been discussed sooner because they are dependent on multiple regression analysis and the use of qualitative variables.

squares methods that have been devised yield different solutions which, of course, may result in different conclusions.* Thus, which of the many least-squares regression approaches available should be selected? The answer is not easy. Unfortunately, although much has been written on this subject, to date there is no consensus among statisticians as to which of the least-squares approaches is "best." In fact, it has been argued [see Herr and Gaebelein (1978)] that there is no optimal method and that the choice depends on the focus of the particular problem of interest.

The initial impetus into the selection problem was sparked by the work of Overall and Spiegel (1969). The authors described and compared three least-squares approaches to the nonorthogonal design problem:

Method 1: a *generalized linear model analysis* [see Scheffé (1959) or Graybill (1961)]

Method 2: an *experimental design method* [see Rao (1973), Snedecor and Cochran (1980), or Winer (1971)]

Method 3: a *hierarchical model analysis* [see Cohen (1968)]

After comparisons of applications in which each method would be appropriate, the authors summarize by suggesting that Method 2 should probably be used whenever the original problem had been formulated as a factorial experiment, which, if not for the orthogonality problem, could have been analyzed by the methods of Chapter 6. Moreover, they suggest that Method 1 is most appropriate if the researcher is developing a model whose goal is to explain the underlying structure in the data. Furthermore, they suggest Method 3 if a logical a priori ordering exists among hypotheses to be tested. On the whole, however, the authors seemed to favor Method 2 because of its likeness to experimental design situations.

The Overall and Spiegel (1969) article stimulated much interest in the nonorthogonality problem. For example, Carlson and Timm (1974) disagreed with the previous suggestions and strongly advocated the use of Method 1 for most research situations. They concluded that Method 2 should be used only if the researcher could assume no interaction between the classification factors.† On the other hand, Appelbaum and Cramer (1974) strongly disagreed with all the approaches. Instead, these writers advocate the use of a series of *model comparisons*—a step-by-step exploratory evaluation process which guides the researcher to an appropriate solution [see also Cramer and Appelbaum (1980)].

In this text we take the position that this *model comparison* procedure is one which provides the most insight to the researcher in analyzing the data.

---

*The least-squares methods yield identical results for (balanced) orthogonal factorial designs—experiments wherein the cell sample sizes are equal (see Problems 12.7 and 12.8).

†Perhaps these findings, among others, led Overall et al. (1975) to advocate Method 1 over Method 2.

Unfortunately, however, unless the researcher is quite familiar with the capabilities of available computer packages, it is a difficult procedure to employ. Two recent articles have addressed the problem of nonorthogonal factorial experiments and the use of computer packages [see Herr and Gaebelein (1978) and Hosking and Hamer (1979)]. A summary of their findings are presented in Table 12.4. Since the computer packages differ so markedly in their capabilities, we

TABLE 12.4   Using Computer Packages in Nonorthogonal
Factorial Experiments[a]

	Computer Package		
Least-Squares Approaches	SAS	SPSS	BMDP
Method 1	Yes	Yes	Yes
Method 2	Yes	Yes	No
Method 3	Yes	Yes[b]	No
Model comparisons	Yes	No	No

[a]Capabilities as of January 1982.
[b]Can be used only for two-factor experiments.

shall limit our discussions to the use of Method 1—the only least-squares regression approach available in SAS, SPSS, and BMDP—as a means of analyzing the nonorthogonal factorial design problem. However, the researcher should realize that each of these methods would lead to tests of different sets of hypotheses with corresponding differences in the obtained results. The researcher is cautioned that such differences would not be readily apparent from the summary tables displayed in the computer output. It is, therefore, essential today, as it was more than a decade ago [Overall and Spiegel (1969)], that in this era of computer packages the researcher (1) be aware of the differences in the methods, (2) know what method has been used, and (3) define the method sufficiently so that a reader would know how to interpret the results.

12.9.3 A Generalized Least-Squares Analysis: Method 1

The model for the nonorthogonal two-factor fixed-effects ANOVA was presented in (7.2). Use of the *generalized least-squares analysis* results in the ANOVA table shown as Table 12.5 in lieu of Table 7.9 for the method of unweighted means. From Table 12.5 we observe that each effect, whether it be a main effect or interaction, is adjusted for relationships to all other effects in the model. Hence $SS\alpha$ measures the contribution of factor α to a model already including factor β and $\alpha\beta$ interaction; $SS\beta$ measures the contribution of factor β to a model already containing factor α and $\alpha\beta$ interaction; and $SS\alpha\beta$ measures the contribution of interaction to a model already involving the main effects α and β.

TABLE 12.5 ANOVA Table for a Generalized Least-Squares Analysis (Method 1)

Source of Variation	Degrees of Freedom	Sum of Squares[a]	Mean Square
α	$r - 1$	$SS\alpha = SSR(\hat{\alpha}_i, \hat{\beta}_j, \widehat{\alpha\beta}_{ij}) - SSR(\hat{\beta}_j, \widehat{\alpha\beta}_{ij})$	$MS\alpha = \dfrac{SS\alpha}{r - 1}$
β	$c - 1$	$SS\beta = SSR(\hat{\alpha}_i, \hat{\beta}_j, \widehat{\alpha\beta}_{ij}) - SSR(\hat{\alpha}_i, \widehat{\alpha\beta}_{ij})$	$MS\beta = \dfrac{SS\beta}{c - 1}$
$\alpha\beta$	$(r - 1)(c - 1)$	$SS\alpha\beta = SSR(\hat{\alpha}_i, \hat{\beta}_j, \widehat{\alpha\beta}_{ij}) - SSR(\hat{\alpha}_i, \hat{\beta}_j)$	$MS\alpha\beta = \dfrac{SS\alpha\beta}{(r - 1)(c - 1)}$
Error	$n.. - rc$	$SSE = SST - SSR(\hat{\alpha}_i, \hat{\beta}_j, \widehat{\alpha\beta}_{ij})$	$MSE = \dfrac{SSE}{n.. - rc}$
Total	$n.. - 1$	SST	

[a] $\hat{\alpha}_i$, $\hat{\beta}_j$, $\widehat{\alpha\beta}_{ij}$ are the estimated effects from the model in (7.2).

12.9.4 Application

To compare the results using Method 1 with those obtained using the method of unweighted means, we return to the example of Section 7.6.2. Forty-two upperclassmen majoring in business were classified according to sex and intention to attend graduate school, and their grade point indexes were recorded in Table 7.10.

Figure 12.8 represents the partial output from BMDP program 2V.

```
              ANALYSIS OF INTENTIONS

ANALYSIS OF VARIANCE FOR    1-ST
DEPENDENT VARIABLE - GPI

       SOURCE           SUM OF      DEGREES OF    MEAN          F          TAIL
                        SQUARES     FREEDOM       SQUARE                   PROBABILITY

       MEAN             335.91726      1        335.91726    2242.70       0.0000
       SEX      (𝒶)       0.08672      1          0.08672       0.58       0.4517
       INTENT   (𝐵)       1.88083      2          0.94041       6.28       0.0046
       S I    (𝒶 × 𝐵)     0.03557      2          0.01779       0.12       0.8884
       ERROR              5.39216     36          0.14978
```

Figure 12.8 Partial output from BMDP program 2V.

The appropriate hypotheses tests for a fixed-effects model were presented in Table 7.3. The following hypotheses are again tested:

$$H_{0\alpha}: \quad \mu_{1.} = \mu_{2.}. \quad \text{(or no ``sex'' effect)}$$
$$H_{1\alpha}: \quad \mu_{1.} \neq \mu_{2.}.$$
$$H_{0\mathcal{B}}: \quad \mu_{.1} = \mu_{.2} = \mu_{.3} \quad \text{(or no ``intent'' effect)}$$
$$H_{1\mathcal{B}}: \quad \text{not all } \mu_{.j} \text{ are equal} \quad (j = 1, 2, 3)$$

and

$$H_{0\alpha\mathcal{B}}: \quad \mu_{ij} \equiv \text{for all } rc \text{ cells} \quad \text{(or no ``interaction'' effect)}$$
$$H_{1\alpha\mathcal{B}}: \quad \text{not all } \mu_{ij} \text{ are equal} \quad (i = 1, 2 \text{ and } j = 1, 2, 3)$$

At the .05 level of significance, from Figure 12.8 we observe that there is no evidence of an interaction effect, nor is there any reason to suspect the existence of an effect due to sex. However, since $F_{\mathcal{B}} = \text{MS}\mathcal{B}/\text{MSE} = 6.28 > F_{.95; 2, 36} = 3.26$, we may reject $H_{0\mathcal{B}}$ and conclude that there are significant differences in the grade point index based on graduate school intent (controlling for sex and interaction).

These findings are very similar to those displayed in Table 7.13 based on the method of unweighted means.

12.10 COMPUTER PACKAGES AND THE GENERALIZED LEAST-SQUARES ANALYSIS FOR UNBALANCED FACTORIAL DESIGNS: USE OF SPSS, SAS, AND BMDP

In our analysis of nonorthogonal factorial experiments in Section 7.7, we described how to use SPSS, SAS, or BMDP in order to obtain results based on the method of unweighted means. In this section we shall explain how to use these computer packages in order to obtain results based on a generalized least-squares approach.

12.10.1 Organizing the Data

Using as our example the study of the effects of sex and graduate school intention on the grade point index, the raw data are organized as in Table 7.14. For data entry purposes, the identification number was placed in columns 1 and 2, the GPI in columns 4–7, SEX in column 9, and INTENT in column 11.

12.10.2 Using SPSS for the Generalized Least-Squares Analysis

Figure 7.1 illustrates the program that has been written using the ANOVA subprogram of SPSS to perform a two-way ANOVA on the raw data to obtain the MSE term and the cell means. For a generalized least-squares analysis (Method 1), however, we merely replace the STATISTICS 3 statement by an OPTIONS 9 statement. Thus, to run a generalized least-squares analysis, the procedure statement would then appear as

```
1              16
ANOVA          GPIƀBYƀSEX(1,2)ƀINTENT(1,3)/
OPTIONS        9
```

Of course the RUN NAME statement would also be replaced by one with a more appropriate title such as

```
1              16
RUNƀNAME       GENERALƀLEASTƀSQUARESƀANALYSISƀOFƀGPI
```

Annotated partial output obtained from this SPSS program is displayed in Figure 12.9.

```
* * * * * * * * * * * * A N A L Y S I S   O F   V A R I A N C E * * * * * * * * * * *
            GPI       GRADE POINT INDEX OF STUDENT
         BY SEX       SEX OF STUDENT
            INTENT    GRAD SCHOOL INTENTIONS
* * * * * * * * * * * * * * * * * * * * * * * * * * * * * * * * * * * * * * * * * * * *
```

SOURCE OF VARIATION	SUM OF SQUARES	DF	MEAN SQUARE	F	SIGNIF OF F
MAIN EFFECTS	1.904	3	0.635	4.238	0.012
SEX (𝒜)	0.087	1	0.087	0.579	0.452
INTENT (ℬ)	1.881	2	0.940	6.278	0.005
2-WAY INTERACTIONS	0.036	2	0.018	0.119	0.888
SEX INTENT (𝒜 × ℬ)	0.036	2	0.018	0.119	0.888
EXPLAINED	1.932	5	0.386	2.580	0.043
RESIDUAL	5.392	36	0.150		
TOTAL	7.324	41	0.179		

Figure 12.9 Partial SPSS output for graduate school example.

12.10.3 Using SAS for the Generalized Least-Squares Analysis

Figure 7.3 illustrates the program that has been written using PROC GLM of SAS to perform a two-way ANOVA on the raw data to obtain the MSE term and the cell means. For a generalized least-squares analysis (Method 1), however, we simply replace all of the statements which follow the last *raw data* entry by

PROCþGLM;
CLASSESþSEXþINTENT;
MODELþGPI = SEXþINTENTþSEX * INTENT/SS4;

Moreover, the "DATA UNWEIGHT;" statement at the beginning of the program would also be replaced by one with a more appropriate name such as

DATAþMETHOD1;

We note that in the MODEL statement the two classification factors are listed along with the interaction term SEX * INTENT and followed immediately by /SS4;. This SS4 option yields the generalized least-squares analysis. Annotated partial output obtained from this program is displayed in Figure 12.10.

GENERAL LINEAR MODELS PROCEDURE

DEPENDENT VARIABLE GPI   GRADE POINT INDEX

SOURCE	DF	SUM OF SQUARES	MEAN SQUARE	F VALUE	PR F
MODEL	5	1.93188492	0.38637698	2.58	0.0429
ERROR	36	5.39216270	0.14978230		STD DEV
CORRECTED TOTAL	41	7.32404762			0.38701718

SOURCE	DF	TYPE IV SS	F VALUE	PR > F
SEX (α)	1	0.08672031	0.58	0.4517
INTENT (β)	2	1.88082934	6.28	0.0046
SEX*INTENT (α × β)	2	0.03557319	0.12	0.8884

Figure 12.10   SAS output for graduate school example.

12.10.4 Using BMDP for the Generalized Least-Squares Analysis

Figure 7.5 illustrates the program that has been written using BMDP program 2V to perform a two-way ANOVA on the raw data in order to extract the MSE term and the cell means for an *unweighted means* analysis (Section 7.6.1). More generally, however, we note that this program yields the desired generalized least-squares analysis. Of course, the /PROBLEMþTITLE paragraph should be replaced by one more appropriate such as

/PROBLEMþTITLEþISþ'AþGENERALIZEDþLEASTþSQUARESþANALYSIS'.

Annotated partial output obtained for this program was displayed in Figure 12.8.

PROBLEMS

12.1. For the data of Table 2.1, develop a regression model to predict the weekly salary based on the length of employment, age, sex, and job classification.

(a) State the regression equation.

(b) Interpret the meaning of the regression coefficients.

(c) Compute the coefficient of multiple determination, and interpret its meaning in this problem.

(d) Predict the expected weekly salary for a female technical worker who has been employed for 5 years and is 45 years old.

(e) Determine the contribution of each independent variable after all other independent variables have been included in the model ($\alpha = .05$). What model would you choose based on this approach?

(f) Compute the coefficients of partial determination, and interpret their meaning in this problem.

(g) Determine the adequacy of the fit for the selected model.

*12.2. A local real estate association in a metropolitan area would like to study the relationship between the size of a single-family house (as measured by the number of rooms) and the selling price of the house. Separate analyses are to be carried out for two different neighborhoods within the locality. A random sample of 10 recently sold houses in each neighborhood revealed the information displayed in Table P12.2.

TABLE P12.2 Real Estate Association Study

Neighborhood I		Neighborhood II	
Selling Price ($000)	Number of Rooms	Selling Price ($000)	Number of Rooms
89.6	7	88.5	6
87.4	8	161.3	13
120.3	9	117.4	10
126.5	12	126.2	10
78.2	6	122.4	9
117.8	9	103.7	8
104.1	10	109.6	8
93.2	8	123.6	9
107.8	9	140.7	11
105.3	8	128.3	9

(a) Assuming a linear relationship between the number of rooms and the selling price, fit a simple linear regression model for each neighborhood.

(b) Test for the significance of each fitted model ($\alpha = .05$).

(c) Test for the adequacy of each fitted model ($\alpha = .05$).

(d) Test for the parallelism of the two regression lines ($\alpha = .05$).

(e) Test for the equality of the intercepts for the two regression lines ($\alpha = .05$).

(f) What can be said about coincidence?

12.3. For the *entire* set of data displayed in Table P12.2 (see Problem 12.2), develop the full multiple regression model to predict the selling price based on the number of rooms and neighborhood (i.e., treat the neighborhood as a dummy variable).

(a) State the full regression equation.

(b) Interpret the meaning of the regression coefficients.

(c) Compute the coefficient of multiple determination, and interpret its meaning in this problem.

(d) Determine the contribution of each independent variable after the other independent variable has been included in the model ($\alpha = .05$). What model would you choose based on this approach?

(e) Test for coincidence between the two fitted regression equations (of Problem 12.2).

(f) Compute the coefficients of partial determination, and interpret their meaning.

(g) Compare the results obtained here to those of Problem 12.2.

*12.4. For the *entire* set of data displayed in Table P12.2 (see Problem 12.2),

(a) State the ANACOVA model.

(b) Determine whether there is a difference in the selling price of the houses in the two neighborhoods after adjusting for the number of rooms (use $\alpha = .05$).

(c) Compare the results in part (b) to those obtained using ANOVA methods. Discuss.

12.5. A professor wishes to compare three different teaching methods. She divides 30 students into three randomly assigned groups of 10. Before the start of the experiment each student is given an IQ test. At the end of the course, all students take the same examination. The results are summarized in Table P12.5. At the .05 level of significance, is there a difference among methods after adjusting for the IQ score?

TABLE P12.5 Study of Teaching Methods

Method I		Method II		Method III	
IQ Score	Exam	IQ Score	Exam	IQ Score	Exam
94	14	80	38	92	55
96	19	84	34	96	53
98	17	90	43	99	55
100	38	97	43	101	52
102	40	97	61	102	35
105	26	112	63	104	46
109	41	115	93	107	57
110	28	118	74	110	55
111	36	120	76	111	42
130	66	120	79	118	81

12.6. The developer of a combined diet and exercise program would like to study the effects of the program on men and women. A random sample of 12 men and 12 women was selected. Their weight was recorded both prior to the program's inception and 6 months thereafter. The results are displayed in Table P12.6.

TABLE P12.6   Study of Diet and Exercise Program

Men		Women	
Before	After	Before	After
250	213	163	149
196	179	184	155
174	162	152	139
193	183	137	126
164	150	128	120
227	192	154	142
237	216	132	120
208	189	118	111
173	162	174	146
185	170	193	158
244	225	127	119
275	232	160	141

(a) Perform separate regression analyses for the men and women to predict their weight "after the program" based on their weight "prior to the program."

(b) Test for the parallelism of the slopes ($\alpha = .05$).

(c) Based on your results in part (b), is it appropriate to develop a single multiple regression equation to predict the weight 6 months after the program based on the initial weight and sex? Discuss completely, and carry out all relevant analyses.

12.7. Perform a generalized least-squares analysis on the programming hours data in Table P6.4 of Problem 6.4. Compare your results to those obtained previously in parts (a), (b), and (c) of Problem 6.4.

12.8. Perform a generalized least-squares analysis on the restaurant ratings data in Table P6.5 of Problem 6.5. Compare your results to those obtained previously in parts (a), (b), and (c) of Problem 6.5.

12.9. Perform a generalized least-squares analysis on the earnings-per-share data in Table P7.1 of Problem 7.1. Compare your results to those obtained previously.

12.10. Perform a generalized least-squares analysis on the diastolic blood pressure data in Table P7.2 of Problem 7.2. Compare your results to those obtained previously.

12.11. Perform a generalized least-squares analysis on the television viewing data in Table P7.3 of Problem 7.3. Compare your results to those obtained previously.

*12.12. Perform a generalized least-squares analysis on the postsurgical hospitalization data in Table P7.5 of Problem 7.5. Compare your results to those obtained previously.

12.13. Perform a generalized least-squares analysis on the price-to-earnings ratio data in Table P7.6 of Problem 7.6. Compare your results to those obtained previously.

12.14. Referring to the data of Problem 10.6, use the qualitative variables "type of garage" and "presence of a fireplace" along with "age" and "size of the heating area" to predict "assessed value."   ∎

12.15. Using the real estate data base (see Appendix A), select a random sample of 30 houses. At the .05 level of significance, is there a difference in the selling price for houses with various garage types after adjusting for lot size?   ∎

12.16. Use the generalized least-squares procedure to analyze the data of Problem 7.8. Compare the results obtained to those of the unweighted means procedure.   ∎

12.17. Referring to Problem 10.7, use the qualitative variables of "the presence of a fireplace" and "the rating category of built-in features" along with "the assessed value" and "time period" to predict "the selling price."   ∎

12.18. Using the real estate data base (see Appendix A), select a random sample of 30 houses. Is there a difference ($\alpha = .01$) in the assessed value based on the presence or absence of a fireplace after adjusting for heating area?   ∎

12.19. Use the generalized least-squares procedure to analyze the data of Problem 7.9. Compare the results obtained to those based on the method of unweighted means.   ∎

12.20. Reviewing concepts:
  (a) When comparing two fitted simple linear regression models, what is the difference between parallelism and coincidence?
  (b) What are dummy variables and how are they used in regression analysis?
  (c) What are the major purposes of ANACOVA? How do they differ from those of ANOVA?
  (d) Describe a regression approach to the unbalanced two-factor experiment.

13

SELECTION
OF AN APPROPRIATE
REGRESSION MODEL

13.1 INTRODUCTION

In Chapters 8–12 we have considered both simple linear and curvilinear regression models having but one independent variable and then extended the developed ideas to the multiple regression model having two independent variables. We learned to test for the significance of the model as well as evaluate its aptness (1) through tests for *lack of fit*, (2) through study of residuals, and (3) through examination of the assumptions. These *exploratory* endeavors often led to such adjustments in the initially developed model as (1) data transformations, (2) alterations of functional relationships, (3) the inclusion of new predictor variables, and/or (4) the removal of ineffective independent variables. Nevertheless, in all instances heretofore, our objective was to seek and develop the "best"-fitting model for the data at hand—not only for purposes of prediction but also so that we could better explain the underlying structure inherent in the model.

For pedagogical reasons, however, it was necessary to describe models that contained no more than two predictor variables. In practice, this is indeed not the typical case. In developing multiple regression models, researchers usually have access to information regarding many (say p) independent variables. The problem then becomes one of formulating an appropriate model using these p predictor variables (or some subset thereof). In some instances, the researcher may be able to screen out variables that are either not fundamental to the problem or whose information is essentially duplicated by other independent variables. Nevertheless, the criterion of model building in regression is *parsimony*, that is, we would like to develop that multiple regression model which includes the *fewest* number of (independent) variables that permits an adequate interpretation of the responses. Multiple regression models that have fewer independent variables are inherently easier to analyze and interpret. Indeed, more complex regression models (with many independent variables) not only become difficult to analyze, but at the same time they are more likely to contain intercorrelated independent variables whose presence interferes with their interpretation [see Chapter 14 and also Belsley et al. (1980)].

Thus in this chapter we shall examine several selection procedures for finding the "best" regression model. Although some quantitative evaluation of competing models can be performed, there may not exist a *uniquely* best model but merely several equally appropriate models. The variable selection process thus requires a certain amount of judgment that makes the development of an appropriate model an "art as well as a science."

13.2 A SAMPLE DATA SET

To examine the different types of model selection procedures, we shall focus upon data collected from a sample of 30 large utility companies. The variables studied are

Y = net income before extraordinary items (millions of dollars)

X_1 = number of customers

X_2 = ratio of electric sales revenue to kilowatt-hour sales (price per 100 kilowatt-hours)

X_3 = net energy generated and received (billions of kilowatt-hours)

X_4 = total transmission system structure miles

X_5 = non-revenue-producing environmental protection expenses (thousands of dollars)

The data are summarized in Table 13.1. The dual objective of the multiple regression analysis is to develop a parsimonious model that can be used to predict net income for utility companies as well as to permit an explanation of the underlying structure inherent in the data.

13.3 EVALUATING ALL POSSIBLE REGRESSIONS

13.3.1 Introduction

In searching for the best subset out of the p independent variables to include in the model, the first preliminary step should involve obtaining a matrix of the correlation coefficients among the $p + 1$ variables in the study. Table 13.2 presents the correlation matrix (for the utility company data) obtained by using the SAS CORR procedure (see Sections 8.15 and 13.8).

TABLE 13.1 Electric Utility Company Data

Name of Utility	Y	X_1	X_2	X_3	X_4	X_5
Tucson G&E	46.964	164,625	3.81691	5.9863	1352	0
San Diego G&E	66.807	700,713	5.06130	10.3270	1096	806
Conn. L&P	61.475	574,438	4.00983	11.4788	838	1628
Florida Power	81.491	1,967,364	4.02469	18.1331	3312	16,594
Savannah E&P	11.399	85,195	4.15411	2.3526	76	0
Maui Elec.	2.223	25,608	6.80663	.4169	0	0
Idaho Elec.	33.498	11,102	1.54959	13.2484	0	1123
Boston Edison	49.432	590,852	4.86249	12.5886	296	20,697
Mass. Elec.	22.917	738,233	4.82930	11.4801	144	13
New Eng. Power	47.614	106	2.98368	17.4670	1911	2526
Ed. Slt. Elec.	.687	15,200	2.28075	.4990	268	0
Detroit Edison	146.870	1,714,902	4.08278	39.9302	4167	31,847
Cliffs Elec.	3.260	1	3.28203	2.2452	0	0
Missouri Utilities	2.002	55,988	3.79227	1.0445	581	0
Nevada Power	16.688	155,516	2.98573	5.1978	626	826
Alt. City Elec.	30.064	343,358	4.74226	5.7560	1236	852
Pbl. Service E&G	228.786	1,663,762	5.32623	31.6289	852	3815
Lilco	141.993	883,373	5.54267	14.5098	886	0
Carolina P&L	142.743	702,052	3.20089	29.7017	4793	1338
Ohio Elec.	28.778	1	2.39202	14.6370	0	0
Penna. P&L	149.036	955,589	3.58145	23.4297	945	2904
UGI Corp	14.153	53,527	4.83739	.6697	98	657
SW Elec. Power	38.083	34,112	3.79297	12.1197	2983	895
Utah P&L	71.111	397,116	2.85036	15.7864	6566	7606
Grn. Mt. Power	3.589	57,867	3.29273	1.3502	274	1
Virginia E&P	203.864	1,247,125	3.79522	39.9683	4627	0
Wheeling Elec.	.818	39,531	2.59182	1.9929	189	0
Public Service of Indiana	87.708	516,030	3.06638	18.6244	5713	4128
Iowa P&L	28.607	221,590	3.46828	4.7615	1495	1403
Wisconsin River Power	.073	3	.83751	.2433	0	0

TABLE 13.2 Correlation Matrix for the Utility Example

	Y	X_1	X_2	X_3	X_4	X_5
Y	1.0	.80336	.25656	.90410	.51539	.30609
X_1		1.0	.35680	.77031	.42722	.60078
X_2			1.0	.08937	−.08217	.12942
X_3				1.0	.63075	.48405
X_4					1.0	.36180
X_5						1.0

A cursory examination of Table 13.2 reveals that income (Y) appears to be highly correlated with the number of customers (X_1) and the net energy generated (X_3), and moderately correlated with the other three independent variables $(X_2, X_4, \text{and } X_5)$. In addition, there appears to exist some moderate intercorrelations among some of the independent variables.*

In selecting the best regression model, the first approach that we shall consider is to evaluate *all possible linear regression models*. This approach may be particularly appropriate when there are a relatively small (i.e., less than 10) number of independent variables present since both the number of possible models and the corresponding computer time required to develop the models increase rapidly as additional independent variables are considered.

In the examination of all possible linear regression models, we shall focus upon two selection procedures: the $R_{p^*}^2$ criterion and Mallows C_{p^*}. Table 13.3 presents the $R_{p^*}^2$ and C_{p^*} values obtained from all possible regressions by using the SAS RSQUARE procedure (see Section 13.8).

13.3.2 The $R_{p^*}^2$ Criterion

The $R_{p^*}^2$ criterion involves the comparison of the coefficients of determination for all possible multiple regression models. We can define $R_{p^*}^2$ as

$$R_{p^*}^2 = \frac{\text{SSR}_{p^*}}{\text{SST}} = 1 - \frac{\text{SSE}_{p^*}}{\text{SST}} \tag{13.1}$$

where p = number of independent variables included in the particular regression model

p^* = number of parameters included in the regression model in which there are p independent variables

SSR_{p^*} = regression sum of squares for a model that includes p^* parameters

SST = total sum of squares

SSE_{p^*} = error sum of squares for a model that includes p^* parameters

However, we must realize that since SST will be constant for all regression models, $R_{p^*}^2$ *cannot* decrease as additional independent variables are introduced into a regression model. Thus the highest $R_{p^*}^2$ value will occur when all possible independent variables are included in the model. With this in mind, our goal in utilizing $R_{p^*}^2$ is to compare alternative models so that we may determine when the introduction of additional independent variables does not produce a commensurate increase in $R_{p^*}^2$.

For the utility company example, from Table 13.3 and Figure 13.1 we can observe that there is great disparity among the $R_{p^*}^2$ values for the various regression models. In Figure 13.1 the maximum $R_{p^*}^2$ values for each p^* (where $p^* = 2$, 3, 4, 5, 6) have been connected with dashed lines. For the two-parameter model (i.e., an intercept parameter β_0 and one independent variable) the best fit clearly occurs when X_3 is included ($R_2^2 = .8174$). For the three-parameter model (i.e.,

---

*When independent variables are intercorrelated, the problem of multicollinearity may exist. See Section 14.5 for further discussion.

TABLE 13.3 $R_{p^*}^2$ and C_{p^*} Values for All Possible Regression Models for the Utility Example

Independent Variables Included in Regression Model	Number of Parameters, p^*	$R_{p^*}^2$	C_{p^*}
X_1	2	.64539425	69.31
X_2	2	.06582088	225.09
X_3	2	.81739669†	23.08
X_4	2	.26562484	171.39
X_5	2	.09369209	217.60
X_1, X_2	3	.64643109	71.03
X_1, X_3	3	.84551531	17.52
X_1, X_4	3	.68165717	61.57
X_1, X_5	3	.69417235	58.20
X_2, X_3	3	.84853649†	16.71
X_2, X_4	3	.35557749	149.21
X_2, X_5	3	.14155798	206.74
X_3, X_4	3	.82239767	23.74
X_3, X_5	3	.83999266	19.01
X_4, X_5	3	.28209024	168.96
X_1, X_2, X_3	4	.85821334	16.11
X_1, X_2, X_4	4	.68212114	63.44
X_1, X_2, X_5	4	.69750104	59.31
X_1, X_3, X_4	4	.84813691	18.82
X_1, X_3, X_5	4	.90298226†	4.08
X_1, X_4, X_5	4	.74477274	46.60
X_2, X_3, X_4	4	.85011489	18.29
X_2, X_3, X_5	4	.87695906	11.07
X_2, X_4, X_5	4	.36167688	149.57
X_3, X_4, X_5	4	.84340506	20.09
X_1, X_2, X_3, X_4	5	.85947932	17.77
X_1, X_2, X_3, X_5	5	.91068091†	4.01
X_1, X_2, X_4, X_5	5	.74477740	48.60
X_1, X_3, X_4, X_5	5	.90325678	6.00
X_2, X_3, X_4, X_5	5	.87746405	12.94
X_1, X_2, X_3, X_4, X_5	6	.91070942†	6.00

†Model with highest R^2 for a given number of parameters p^*.

an intercept parameter β_0 and two independent variables) several combinations of variables (X_2, X_3; X_1, X_3; X_3, X_5) appear to yield similar R_3^2 values. Once four parameters are considered, the X_1, X_3, X_5 model has the highest R_4^2 value (.903) followed by the X_2, X_3, X_5 model ($R_4^2 = .877$). For the five-parameter model, the X_1, X_2, X_3, X_5 model has the highest R_5^2 value (.9107) followed by the X_1, X_3, X_4, X_5 model (.9033). Finally, with all six parameters considered, $R_6^2 = .9107$. From Figure 13.1 we can observe that $R_{p^*}^2$ steadily rises until four parameters are included in the model and then does not exhibit any real further

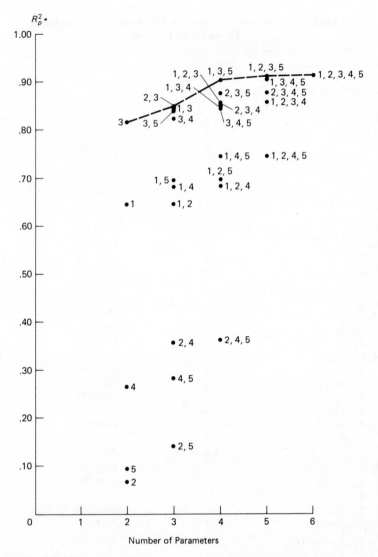

Figure 13.1 Plot of R_p^2. versus p^* for all regression models for the utility data. (Source: Table 13.3.)

increase as additional parameters are added. Note that the best four-parameter model (X_1, X_3, X_5) has $R_4^2 = .903$, while the best five-parameter model (X_1, X_2, X_3, X_5) has $R_5^2 = .9107$. Therefore, in the interest of simplicity, the researcher would decide to select that regression model which contains only X_1, X_3, and X_5 since inclusion of X_2 leads to such a negligible increase in R_p^2.

13.3.3 The C_{p*} Criterion

In contrast to the R_{p*}^2 criterion which approaches 1.000 as additional independent variables are included in the model, the C_{p*} criterion developed by Mallows (1964, 1966, 1973) measures the *total squared error* (TSE) of a regression model with p^* parameters. The total squared error consists of a *bias* component and a *random error* component. The bias component represents the difference in the predicted Y values obtained from the fitted regression model and the "true" regression model. The random error component represents the variability around the fitted line of regression. The C_{p*} statistic is defined as

$$C_{p*} = \frac{SSE_{p*}}{MSE_T} - (n - 2p^*) \tag{13.2}$$

where $p^* =$ number of parameters included in a particular model with p independent variables

$SSE_{p*} =$ error sum of squares for a regression model with p^* parameters

$T =$ total number of parameters to be considered for inclusion in the regression model

$MSE_T =$ mean square error (variance) of a regression model containing all T parameters

$n =$ sample size

C_{p*} can also be expressed in terms of R_{p*}^2

$$C_{p*} = \frac{(1 - R_{p*}^2)(n - T)}{1 - R_T^2} - (n - 2p^*) \tag{13.3}$$

where R_T^2 is the coefficient of multiple determination for a model containing all T parameters.

For example, using (13.3), if we wanted to compute C_{p*} for the X_1, X_3, X_5 model, we would have

$$p^* = 4 \qquad T = 6 \qquad n = 30$$

and from Table 13.3†

$$R_{p*}^2 = .90298226 \qquad R_T^2 = .91070942$$

Thus

$$C_4 = \frac{(1 - .90298226)(30 - 6)}{1 - .91070942} - [30 - 2(4)]$$

$$= 26.08 - 22$$

$$= 4.08$$

When a regression equation with p independent variables does not contain any bias component, the average value of C_{p*} is p^*, the number of parameters.

---

†It is particularly important to keep a large number of decimal places for R_{p*}^2 in order to prevent rounding errors in the computation of C_{p*}.

Thus the goal of the researcher is to identify those regression models for which C_{p^*} is close to p^*.

An extremely useful aid involves plotting C_{p^*} versus p^* for all possible regression models. For the utility example, such a plot is obtained (Figure 13.2) using the data of Table 13.3.

Since the average value of C_{p^*} for regression models with no bias is p^*,

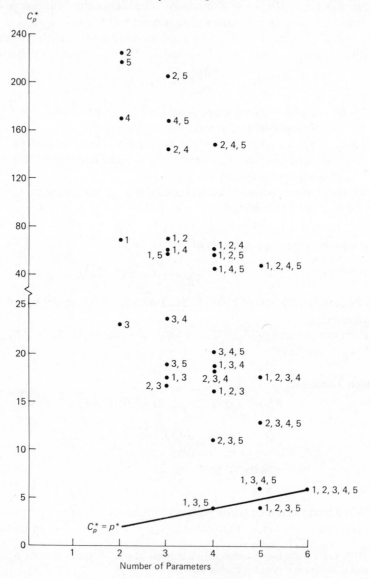

Figure 13.2 Plot of C_{p^*} versus p^* for all regression models for the utility data. (Source: Table 13.3.)

those models with substantial bias will plot far above the line of $C_{p^*} = p^*$, while those models with little or no bias will plot close to the $C_{p^*} = p^*$ line. Because C_{p^*} is a measure of *total squared error*, the researcher should attempt to find the set of independent variables that leads to the smallest C_{p^*} value while at the same time minimizing the bias component [see Daniel and Wood (1980) for a more detailed discussion]. If we refer back to Table 13.3 and Figure 13.2, we can observe that only four of the possible models have C_{p^*} values that are close to p^*: the X_1, X_3, X_5 model ($C_4 = 4.08$); the X_1, X_2, X_3, X_5 model ($C_5 = 4.01$); the X_1, X_3, X_4, X_5 model ($C_5 = 6.00$); and, of course, the complete X_1, X_2, X_3, X_4, X_5 model where $C_6 = 6.00$. Since C_{p^*} is lowest for the X_1, X_3, X_5 model ($C_4 = 4.08$) and the X_1, X_2, X_3, X_5 model ($C_5 = 4.01$), it would appear that the researcher should select one of these two models as containing the most appropriate subset of independent variables. Perhaps we should select the X_1, X_2, X_3, X_5 model ($C_5 = 4.01$) since its C_{p^*} value is slightly lower (4.01 as compared to 4.08) and it appears to contain less bias (in fact, its C_{p^*} value is even less than p^*, the average total squared error when no bias exists).

13.4 STEPWISE REGRESSION

In Section 13.3 we discussed two selection criteria, $R_{p^*}^2$ and C_{p^*}, that can be utilized when evaluating all possible regression models. Examining all possible regressions, however, becomes a laborious and expensive task once a large number of independent variables are available for consideration. Thus other *search* procedures have been developed in order to find the "best" set of independent variables without examining all possible regressions. Once a best model has been found, a thorough residual analysis should be performed to evaluate the aptness of the model.

Perhaps the most widely used search procedure is *stepwise regression*. We may recall from Section 10.5 that the contribution of each independent variable could be examined after the other independent variables were already included in the regression model. Using the partial F test criterion, if the variable made a significant contribution to the fit of the model, it would be included in the multiple regression equation; otherwise it would not. Stepwise regression extends this partial F test criterion to a model with p independent variables. In stepwise regression, variables are either added to or deleted from the regression model at each *step* of the model-building process. The steps involved can be summarized as follows:

1. The "stepwise" process begins by including in the model the independent variable having the highest simple correlation with the dependent variable Y. If the F statistic (MSR/MSE) for this model equals or exceeds the critical value of F needed to enter* the model (that is, $F \geq F_E = F_{1-\alpha;\, 1,\nu}$), the variable

---

*We should realize that since $t_\nu^2 = F_{1,\nu}$, we could alternatively express the partial F values as t values (see Section 10.5).

would be included. If F is less than F_E, the process terminates with no independent variables included in the model.

2. Once the first independent variable (say X_1, for example) is included in the model, the contribution of each of the remaining $p - 1$ independent variables is determined for the model that already includes variable X_1. A partial F test $[\text{MSR}(X_i \mid X_1)/\text{MSE}(X_i, X_1)]$ would be computed for each of these variables, given that variable X_1 was already included in the regression model. If the largest partial F value equals or exceeds the critical value of F needed to enter the model (F_E), then a second variable (say X_2, for example) would be included. On the other hand, if the partial F value was less than F_E, the process terminates, and only the first independent variable would have been included in the model.

3. A most important feature of this stepwise process is that an independent variable which has been entered into the model at an earlier stage may subsequently be removed once other independent variables have been evaluated. Presuming that the model contains two independent variables (such as X_1 and X_2), we may now determine whether any of the variables already included (for example, X_1) are no longer important, given that others have subsequently been added. If so, then they could be deleted from the model. Thus the partial F value for those variables already included in the model [in this case $\text{MSR}(X_1 \mid X_2)/\text{MSE}(X_1, X_2)$] could be computed. If the smallest of these partial F values is less than or equal to the critical value of F for removal from the model (that is, $F \leq F_R = F_{1-\alpha; 1-\nu'}$), then the particular variable which resulted in that partial F value could be removed from the regression model. However, if the smallest partial F value is greater than the critical F_R value, then the corresponding variable should not be eliminated from the model.

We may now observe this stepwise regression process by using the utility example. Figure 13.3 represents a partial output obtained from the SAS STEPWISE procedure for the utility company data.

The first variable entered into the model is X_3 (net energy generated), which provides a regression sum of squares (SSR) of 94,399.855 out of a total sum of squares (SST) of 115,488.424. Since the F value of 125.34 is clearly greater than the critical F value to enter the model ($F_E = F_{.95; 1,28} = 4.20$) at the .05 level of significance, X_3 is included in the regression model.

The next step involves the evaluation of the second variable to be included in the model. The variable making the largest contribution at this stage is X_2 (ratio of electric revenue to sales). From Figure 13.3, we can observe that SAS provides a *Type II Sum of Squares* for each independent variable included in the model. This represents the sum of squares contributed by a variable after the other independent variables are already included in the model. Thus the *Type II* SS for X_2 represents its contribution once variable X_3 has already been entered in the model. The regression sum of squares for the two independent variables model can be subdivided (as in Table 13.4) into contributions of variable X_3 as well as X_2 given X_3 (that is, $X_2 \mid X_3$). Since the F value for $X_2 \mid X_3$ is 5.55, which is greater than $F_E = F_{.95 \ 1,27} = 4.21$, X_2 is entered into the regres-

```
                        STEPWISE REGRESSION PROCEDURE FOR [EPENDENT VARIABLE INCOME

STEP 1   VARIABLE ENERGY ENTERED      R SQUARE = 0.81739745      C(P) =   23.08085431

                                   DF        SUM OF SQUARES        MEAN SQUARE          F       PROB>F

                   REGRESSION       1         94400.07299756      94400.07299756     125.34    0.0001
                   ERROR           28         21088.50997581        753.16107056
                   TOTAL           29        115488.58297337

                                B VALUE       STD ERROR          TYPE II SS          F       PROB>F

                   INTERCEPT   -2.61380465
                   ENERGY       5.00890196    0.44740445       94400.07299756      125.34    0.0001
-----------------------------------------------------------------------------------------------------
STEP 2   VARIABLE RVSALES ENTERED     R SQUARE = 0.84853754      C(P) =   16.71085930

                                   DF        SUM OF SQUARES        MEAN SQUARE          F       PROB>F

                   REGRESSION       2         97996.39840241      48998.19920120      75.63    0.0001
                   ERROR           27         17492.18457096        647.85868781
                   TOTAL           29        115488.58297337

                                B VALUE       STD ERROR          TYPE II SS          F       PROB>F

                   INTERCEPT  -35.4954703
                   RVSALES      9.10847279    3.86594987        3596.32540485        5.55    0.0260
                   ENERGY       4.92118086    0.41661769       90394.81517988      139.53    0.0001
-----------------------------------------------------------------------------------------------------
STEP 3   VARIABLE EPAEXP ENTERED      R SQUARE = 0.87696006      C(P) =   11.07130777

                                   DF        SUM OF SQUARES        MEAN SQUARE          F       PROB>F

                   REGRESSION       3        101278.87506306      33759.62502102      61.77    0.0001
                   ERROR           26         14209.70791030        546.52722732
                   TOTAL           29        115488.58297337

                                B VALUE       STD ERROR          TYPE II SS          F       PROB>F

                   INTERCEPT  -39.33835405
                   RVSALES      9.97295123    3.56824366        4259.23622973        7.81    0.0096
                   ENERGY       5.43207259    0.43575225       84930.49716463      155.40    0.0001
                   EPAEXP      -0.00169767    0.00069272        3282.47666066        6.01    0.0213
-----------------------------------------------------------------------------------------------------
STEP 4   VARIABLE CUST ENTERED        R SQUARE = 0.91068087      C(P) =    4.00765542

                                   DF        SUM OF SQUARES        MEAN SQUARE          F       PROB>F

                   REGRESSION       4        105173.24335835      26293.31083959      63.72    0.0001
                   ERROR           25         10315.33961502        412.61358460
                   TOTAL           29        115488.58297337

                                B VALUE       STD ERROR          TYPE II SS          F       PROB>F

                   INTERCEPT  -21.97834203
                   CUST         0.00003959    0.00001289        3894.36829528        9.44    0.0051
                   RVSALES      5.11009370    3.48110170         889.13524874        2.15    0.1546
                   ENERGY       4.22139572    0.54649004       24620.17629265       59.67    0.0001
                   EPAEXP      -0.00252887    0.00065991        6059.31911276       14.69    0.0008
-----------------------------------------------------------------------------------------------------
STEP 5   VARIABLE RVSALES REMOVED     R SQUARE = 0.90298197      C(P) =    4.07700587

                                   DF        SUM OF SQUARES        MEAN SQUARE          F       PROB>F

                   REGRESSION       3        104284.10810961      34761.36936987      80.66    0.0001
                   ERROR           26         11204.47486376        430.94134091
                   TOTAL           29        115488.58297337

                                B VALUE       STD ERROR          TYPE II SS          F       PROB>F

                   INTERCEPT   -3.54012493
                   CUST         0.00004819    0.00001173        7274.46927627       16.88    0.0004
                   ENERGY       3.97360187    0.53118468       24115.47240021       55.96    0.0001
                   EPAEXP      -0.00263167    0.00067060        6636.68925291       15.40    0.0006
-----------------------------------------------------------------------------------------------------
NO OTHER VARIABLES MET THE 0.0500 SIGNIFICANCE LEVEL FOR ENTRY INTO THE MODEL.
```

Figure 13.3 Stepwise regression using SAS for the utility data.

sion model. Now that X_2 has been included in the model, we can determine whether X_3 is still an important contributing variable or whether it may be eliminated from the model. From Figure 13.3, since the *Type II* SS for X_3 in the two independent variables model is 90,394.5915, the computed F value is 139.53. This exceeds the critical value, $F_R = F_{.95;\,1,\,27} = 4.21$. Therefore, since $X_3 \,|\, X_2$ makes a significant contribution to the model, X_3 cannot be removed.

TABLE 13.4   Contributions of Variables X_2 and X_3
to the Stepwise Regression Model

Source	Degrees of Freedom	Sum of Squares	Mean Square	F
Regression	2	97,996.1411	48,998.0705	75.63
$X_2 \mid X_3$	1	3,596.2856	3,596.2856	5.55
X_3	1	94,399.8555	94,399.8555	
Error	27	17,492.2825	647.8623	
Total	29	115,488.4236		

Now that a two independent variables model has been fitted, the stepwise procedure evaluates the remaining three variables to determine which, if any, should be next included in the model. From Figure 13.3 the third variable included is X_5 (non-revenue-producing EPA expenses). The contribution of $X_5 \mid X_2, X_3$ leads to an F value of 6.01, which is greater than $F_E = F_{.95; \, 1,26} = 4.23$ (see Table 13.5). Therefore X_5 makes a significant contribution and

TABLE 13.5   Contributions of Variables X_5, X_2, and X_3
to the Stepwise Regression Model

Source	Degrees of Freedom	Sum of Squares	Mean Square	F
Regression	3	101,278.6197	33,759.5399	61.77
$X_5 \mid X_2, X_3$	1	3,282.4786	3,282.4786	6.01
$X_2 \mid X_3$	1	3,596.2856	3,596.2856	
X_3	1	94,399.8555	94,399.8555	
Error	26	14,209.8039	546.5309	
Total	29	115,488.4236		

should be included in the model. Once X_5 has been entered into the regression model, we need to determine whether either X_2 or X_3 can be removed. From Figure 13.3, since the *Type II* SS for X_2 [i.e., SS($X_2 \mid X_3, X_5$)] is less than that for X_3 [i.e., SS($X_3 \mid X_2, X_5$)], the contribution of $X_2 \mid X_3, X_5$ can be tested to determine whether X_2 can be removed from the model. Since $F = 7.81 > F_R = F_{.95; \, 1,26} = 4.23$, variable X_2 still makes a significant contribution and cannot be deleted from the regression model.

Once the three-variable model (X_2, X_3, X_5) has been fitted, the stepwise procedure evaluates which, if any, of the two remaining independent variables (X_1 and X_4) should be included in the model. From Figure 13.3, the fourth variable included is X_1 (number of customers). The contribution of $X_1 | X_5$, X_2, X_3 leads to an F value of 9.44, which is greater than the $F_E = F_{.95; \, 1,25}$ = 4.24 (see Table 13.6). Therefore X_1 makes a significant contribution and

**TABLE 13.6   Contribution of Variables X_1, X_5, X_2, and X_3
to the Stepwise Regression**

Source	Degrees of Freedom	Sum of Squares	Mean Square	F
Regression	4	105,173.1025	26,293.2756	63.72
$X_1 \| X_5, X_2, X_3$	1	3,894.4828	3,894.4828	9.44
$X_5 \| X_2, X_3$	1	3,282.4786	3,282.4786	
$X_2 \| X_3$	1	3,596.2856	3,596.2856	
X_3	1	94,399.8555	94,399.8555	
Error	25	10,315.3211	412.6128	
Total	29	115,488.4236		

should be included. Once X_1 has been entered into the model, we then need to determine whether either X_2, X_3, or X_5 can be removed. From Figure 13.3, the Type II SS for X_2 (given the others) is less than that for either X_3 or X_5; hence variable X_2 becomes a possible candidate for removal. Testing the contribution of $X_2 | X_1, X_3, X_5$, we observe that $F = 2.15 < F_R = F_{.95; \, 1,25} = 4.24$. Thus variable X_2 no longer makes a significant contribution to the regression model and can be removed. The regression model now contains independent variables X_1, X_3, and X_5. From Figure 13.3, we can observe that the only remaining variable (X_4) does not make a significant contribution. Therefore, no other variables can be added to the model. The question then becomes one of determining whether or not others may yet be eliminated from the model. Since the Type II SS for X_5 (given the others) is smaller than that for either X_1 or X_3, the contribution of X_5 given X_1 and X_3 can now be tested to determine whether X_5 is to be removed from the model. Since $F = 15.40 > F_R = F_{.95; \, 1,26} = 4.23$, variable X_5 makes a significant contribution and cannot be eliminated. Thus, since no other variables are to be entered or removed, the stepwise process terminates with a final model that contains the variables X_1, X_3, and X_5.

13.5 FORWARD SELECTION

In contrast to the stepwise approach which allows independent variables to be entered and removed from the regression model at different steps in the process, the *forward selection* procedure does not permit the removal of a variable from the model once it has been entered. Although this simplifies the model selection process, it unfortunately leads to the inclusion of variables that do not make a significant contribution once other independent variables are entered in the regression model. If we apply the forward selection procedure to the utility example, from Figure 13.3 we can observe that, except for the final step, the forward selection procedure produces results identical to those obtained from stepwise regression. Since variable X_2 cannot be removed from the model as part of the forward selection procedure and since X_4 does not make a significant contribution given that X_1, X_2, X_3, and X_5 are included in the model, the forward selection procedure leads to a model that contains the variables X_1, X_2, X_3, and X_5.

13.6 BACKWARD ELIMINATION

In contrast to the stepwise regression and forward selection approaches that begin with the simple regression model and build upward, the *backward elimination* procedure starts with the complete regression model, one that includes all possible independent variables, and attempts to eliminate them from the model one at a time. Although the backward elimination approach evaluates fewer models than either the stepwise regression or forward selection approaches, the models examined are more complex and require more computer time.

Figure 13.4 illustrates a partial output (obtained from the SAS STEP-WISE procedure) for the utility example using the backward elimination approach. We observe that the first model fitted to the data contains all p independent variables—X_1, X_2, X_3, X_4, and X_5. Since the backward elimination approach seeks only to remove variables from the model, the variable with the smallest incremental contribution is tested at each step to determine whether it can be eliminated from the model. From Figure 13.4, since X_4 (transmission system structure miles) makes the smallest contribution (*Type II* SS), given all other variables are already included in the model, the F value associated with X_4 is compared to the critical F value for removing a variable (F_R). Since $F = .01 < F_R = F_{.95; 1,24} = 4.26$, X_4 is then eliminated from the model. The resulting model contains X_1, X_2, X_3, and X_5. Of these four remaining variables, we observe from Figure 13.4 that X_2 (ratio of electric revenue to sales) makes the smallest contribution to a model which includes the others. Since the F value associated with X_2 is less than the critical F needed to remove a variable from the model (that is, $F = 2.15 < F_R = F_{.95; 1,25} = 4.24$), X_2 can be eliminated. The resulting model contains only three independent variables—X_1, X_3, and X_5. The next step in the backward elimination process would again be to determine which of these remaining independent variables could possibly be removed from the model. Since X_5 (non-revenue-producing EPA expenses) makes the smallest contribution, its associated F value is compared to the critical F value (F_R) needed to remove a variable from the model. In this case, since $F = 15.40 > F_R = F_{.95; 1,26} = 4.23$, X_5 does contribute significantly and

BACKWARD ELIMINATION PROCEDURE FOR DEPENDENT VARIABLE INCOME

STEP 0   ALL VARIABLES ENTERED   R SQUARE = 0.91070935   C(P) = 6.00000000

	DF	SUM OF SQUARES	MEAN SQUARE	F	PROB>F
REGRESSION	5	105176.53265372	21035.30653074	48.96	0.0001
ERROR	24	10312.05031965	429.66876332		
TOTAL	29	115488.58297337			

	B VALUE	STD ERROR	TYPE II SS	F	PROB>F
INTERCEPT	-21.77601977				
CUST	0.00003948	0.00001321	3839.33816271	8.94	0.0064
RVSALES	5.06982718	3.58200503	860.73152568	2.00	0.1698
ENERGY	4.24817047	0.63611347	19163.21917831	44.60	0.0001
XSYS	-0.00023583	0.00269534	3.28929537	0.01	0.9310
EPAEXP	-0.00252101	0.00067938	5916.41362299	13.77	0.0011

STEP 1   VARIABLE XSYS REMOVED   R SQUARE = 0.91068687   C(P) = 4.00765542

	DF	SUM OF SQUARES	MEAN SQUARE	F	PROB>F
REGRESSION	4	105173.24335835	26293.31083959	63.72	0.0001
ERROR	25	10315.33961502	412.61358460		
TOTAL	29	115488.58297337			

	B VALUE	STD ERROR	TYPE II SS	F	PROB>F
INTERCEPT	-21.97834203				
CUST	0.00003959	0.00001289	3894.36829528	9.44	0.0051
RVSALES	5.11009370	3.48110170	889.13524874	2.15	0.1546
ENERGY	4.22139572	0.54649004	24620.17622655	59.67	0.0001
EPAEXP	-0.00252887	0.00065991	6059.31911276	14.69	0.0008

STEP 2   VARIABLE RVSALES REMOVED   R SQUARE = 0.90298197   C(P) = 4.37700587

	DF	SUM OF SQUARES	MEAN SQUARE	F	PROB>F
REGRESSION	3	104284.10809961	34761.36936987	80.66	0.0001
ERROR	26	11204.47486376	430.94134091		
TOTAL	29	115488.58297337			

	B VALUE	STD ERROR	TYPE II SS	F	PROB>F
INTERCEPT	-3.54012493				
CUST	0.00004819	0.00001173	7274.46927627	16.88	0.0004
ENERGY	3.97360187	0.53318468	24115.47240021	55.96	0.0001
EPAEXP	-0.00263167	0.00067060	6636.68925291	15.40	0.0006

ALL VARIABLES IN THE MODEL ARE SIGNIFICANT AT THE 0.0500 LEVEL.

Figure 13.4 Backward elimination using SAS for utility data.

cannot be removed. With this result, the backward elimination process is terminated, and the model containing variables X_1, X_3, and X_5 is selected.

13.7 SELECTION OF MODELS: AN OVERVIEW

In the chapter we have examined several approaches to selecting the set of independent variables to be included in a regression model. We have focused upon two types of procedures, those involving the evaluation of all possible regression models (R_p^2. and C_{p^*}) and those involving only a limited number of regression models (stepwise regression, forward selection, and backward elimination).

It is important to emphasize that these approaches do not necessarily result in the selection of the same set of variables for the "best" regression model. In particular, from our utility example we have observed that different regression models were selected using these various approaches. The forward selection and C_{p^*} procedures selected a model containing variables X_1, X_2, X_3, and X_5, while the $R_{p^*}^2$, stepwise regression, and backward elimination procedures developed a model containing variables X_1, X_3, and X_5.

Thus the researcher must be aware of the fact that there often is no "uniquely superior" or "best" regression model for a set of p independent variables. Nevertheless, several approaches (or combinations thereof) could be utilized in attempting to find a "best" model. For example, the stepwise regression approach could be used to determine the *number* of independent variables that should be included in the model. This could be followed by a C_{p^*} analysis of all possible regression models containing that number of independent variables. Finally, a residual analysis could be performed to study the aptness of the particular model fitted. In conclusion, although the various approaches provide guidelines for variable selection, the model ultimately developed would take into consideration such factors as the simplicity, interpretability, and the usefulness of the variables.

13.8 COMPUTER PACKAGES AND MODEL SELECTION: USE OF SPSS, SAS, AND BMDP

13.8.1 Introduction

In this section we shall explain how to use either SPSS, SAS, or BMDP in order to select the appropriate variables for a regression model. For the utility data (see Table 13.1) the variables involved can be organized for data entry purposes with the company code in columns 1 and 2, the net income (Y) in columns 4–10, the number of customers (X_1) in columns 11–17, the ratio of electric revenue to sales (X_2) in columns 18–24, the net energy generated and received (X_3) in columns 25–31, the total transmission system structure miles (X_4) in columns 32–35, and non-revenue-producing EPA expenses (X_5) in columns 36–40.

13.8.2 Using the SPSS Subprogram NEW REGRESSION
for Model Selection

In Section 10.12.1 we used the SPSS subprogram NEW REGRESSION to perform a multiple regression analysis. This NEW REGRESSION subprogram can also be used for model selection by providing a CRITERIA = statement. The setup of this statement is:

$$\text{CRITERIA} = \text{PIN}\binom{\alpha \text{ level}}{\text{to enter}} \text{POUT}\binom{\alpha \text{ level}}{\text{to remove}}/ \qquad \mathbf{16}$$

PIN represents the α level to enter a variable into a stepwise regression model. The default value is .05. POUT represents the α level to remove a variable from a stepwise regression model. The default value is .10. The STEPWISE statement indicates that stepwise regression with forward entry and backward elimination of variables is being utilized. If only forward inclusion is desired a FORWARD statement would be substituted for the STEPWISE statement. If only backward elimination is desired a BACKWARD statement would be substituted for the STEPWISE statement. Figure 13.5 represents the SPSS program that has been written to perform a stepwise regression analysis for the utility data while Figure 13.6 represents a partial output.

```
{SYSTEM CARDS}
RUN NAME            MODEL SELECTION FOR UTILITIES
DATA LIST           FIXED(1)// COCODE 1-2, INCOME
                    4-10, CUST 11-17, RVSALES 18-24,
                    ENERGY 25-31, XSYS 32-35, EPAEXP
                    36-40
VAR LABELS          COCODE, COMPANY CODE NUMBER/
                    INCOME, NET INCOME IN MILLIONS/
                    CUST, NUMBER OF CUSTOMERS/
                    RVSALES, RATIO REVENUE TO SALES/
                    ENERGY, NET ENERGY GENERATED/
                    XSYS, SYSTEM STRUCTURE MILES/
                    EPAEXP, NON REVENUE EPA EXP/
READ INPUT DATA
01    46.964   164625 3.81691   5.98631 352      0
:      :        :      :         :       :        :
30    .073    30.83751   .2483   0       0
END INPUT DATA
NEW REGRESSION      DESCRIPTIVES/VARIABLES=INCOME TO
                    EPAEXP/CRITERIA=PIN(.05)POUT(.05)/
                    STATISTICS=DEFAULTS,HISTORY/
                    DEPENDENT=INCOME/STEPWISE/RESIDUALS/
FINISH
{SYSTEM CARDS}
```

Figure 13.5 SPSS program for utility data.

```
DEPENDENT VARIABLE.. INCCME

BEGINNING BLOCK NUMBER  1.  METHOD: STEPWISE

VARIABLE(S) ENTERED CN STEP NUMBER  1..    ENERGY    NET ENERGY GENERATED

MULTIPLE R         C.9C410      ANALYSIS CF VARIANCE
R SQUARE           C.81740                       CF     SUM CF SCUARES      MEAN SCUARE
ADJUSTED R SQUARE  C.81688      REGRESSICN        1      94400.C6768       94400.C6768
STANDARD ERROR     27.44378     RESIDUAL         28      21088.51332         753.16115

                                F =     125.33847    SIGNIF F = C.COCO

------------- VARIABLES IN THE EQUATION -------------    ------------- VARIABLES NOT IN THE EQUATION -------------

VARIABLE          B        SE B     BETA      T  SIG T    VARIABLE   BETA IN  PARTIAL  MIN TCLER     T  SIG T

ENERGY        5.C0890    C.44740   0.9C410   11.155 C.0000   CUST     0.26296  0.39241   C.40663   2.217 0.C352
(CONSTANT)   -2.61381    7.42669            -0.352 C.7275    RVSALES  0.17717  0.41296   C.99201   2.356 0.026C
                                                            XSYS    -0.05113 -0.16549   C.6021S  -C.872 0.3905
                                                            EPAEXP  -0.17179 -C.35177   C.7657C  -1.953 0.C613

              * * * * * * * * * * * * * * * * * * * * * * * * *

VARIABLE(S) ENTERED CN STEP NUMBER  2..    RVSALES    RATIC OF REVENUE TO SALES

MULTIPLE R         0.92116      ANALYSIS CF VARIANCE
R SQUARE           C.84854                       CF     SUM CF SCUARES      MEAN SCUARE
ADJUSTED R SQUARE  C.83732      REGRESSICN        2      97996.39167       48998.19583
STANDARD ERROR     25.45307     RESIDUAL         27      17492.18533         647.85886

                                F =     75.63097    SIGNIF F = 0.00CO

------------- VARIABLES IN THE EQUATION -------------    ------------- VARIABLES NOT IN THE EQUATION -------------

VARIABLE          B        SE B     BETA      T  SIG T    VARIABLE   BETA IN  PARTIAL  MIN TCLER     T  SIG T

ENERGY        4.92118    C.41662   C.88627   11.812 C.0000   CUST     0.17307  0.2527E   0.32304   1.332 C.1544
RVSALES       9.10847    3.86555   0.17717    2.356 C.0260   XSYS    -0.05204 -C.10208   C.58208  -C.523 0.6052
(CONSTANT)  -35.45547   15.56333            -2.281 C.C307    EPAEXP  -0.19361 -0.43319   C.75822  -2.451 0.C213

              * * * * * * * * * * * * * * * * * * * * * * * * *

VARIABLE(S) ENTERED CN STEP NUMBER  3..    EPAEXP    NON REVENUE PRODUCING EPA EXPENSES

MULTIPLE R         C.93646      ANALYSIS CF VARIANCE
R SQUARE           C.87696                       CF     SUM CF SCUARES      MEAN SCUARE
ADJUSTED R SQUARE  C.86276      REGRESSICN        3     101278.86764       33759.62255
STANDARD ERROR     23.37793     RESIDUAL         26      14205.71335         546.52744

                                F =     61.77114    SIGNIF F = C.COCO

------------- VARIABLES IN THE EQUATION -------------    ------------- VARIABLES NOT IN THE EQUATION -------------

VARIABLE          B        SE B     BETA      T  SIG T    VARIABLE   BETA IN  PARTIAL  MIN TCLER     T  SIG T

ENERGY        5.43207    0.43575   C.98048   12.466 C.0000   CUST     0.35423  0.52351   C.26874   3.072 0.CC51
RVSALES       9.57295    3.56824   0.19359    2.755 C.CC96   XSYS    -0.02959 -C.06406  C.5176C  -C.321 0.7505
EPAEXP       -C.CC17C  C.65270-C3 -C.15361   -2.451 C.0213
(CONSTANT)  -35.33835   14.38C23             -2.736 C.C111

VARIABLE(S) ENTERED CN STEP NUMBER  4..    CUST    NUMBER OF CUSTOMERS

MULTIPLE R         C.95430      ANALYSIS CF VARIANCE
R SQUARE           C.91068                       CF     SUM CF SCUARES      MEAN SCUARE
ADJUSTED R SQUARE  C.89635      REGRESSICN        4     105173.24112       26293.31028
STANDARD ERROR     2C.31285     RESIDUAL         25      10315.33987         412.61359

                                F =     63.72381    SIGNIF F = 0.00CO

------------- VARIABLES IN THE EQUATION -------------    ------------- VARIABLES NOT IN THE EQUATION -------------

VARIABLE          B        SE B     BETA      T  SIG T    VARIABLE   BETA IN  PARTIAL  MIN TCLER     T  SIG T

ENERGY        4.2214C    0.54549   0.76156    7.725 C.C000   XSYS    -0.CC766 -0.01786   0.26635  -C.087 0.9310
RVSALES       5.11C05    3.4811C   C.C994C    1.468 C.1546
EPAEXP       -0.CC253  C.67060-C3 -C.28841   -3.832 C.CC08
CUST        0.395920-C4 0.12890-C4 0.35423    3.072 C.CC51
(CONSTANT)  -21.97833   13.71321             -1.6C3 C.1216

              * * * * * * * * * * * * * * * * * * * * * * * * *

VARIABLE(S) REMOVED ON STEP NUMBER  5..    RVSALES    RATIO OF REVENUE TO SALES

MULTIPLE R         C.95025      ANALYSIS CF VARIANCE
R SQUARE           C.9C298                       CF     SUM CF SCUARES      MEAN SCUARE
ADJUSTED R SQUARE  C.89179      REGRESSICN        3     104284.10707       34761.36502
STANDARD ERROR     2C.75513     RESIDUAL         26      11204.47393         430.94130

                                F =     80.66381    SIGNIF F = 0.0000
------------- VARIABLES IN THE EQUATION -------------    ------------- VARIABLES NOT IN THE EQUATION -------------

VARIABLE          B        SE B     BETA      T  SIG T    VARIABLE   BETA IN  PARTIAL  MIN TCLER     T  SIG T

ENERGY        3.9736C   C.531l8   C.71723     7.461 C.C000   RVSALES  0.05940  0.28170   C.26874   1.468 C.1546
EPAEXP      -0.CC263  C.67060-C3 -0.30C13    -3.524 C.CC06    XSYS    -0.C2173 -0.05318  C.29481  -C.266 0.7922
CUST        C.481540-C4 0.11730-C4 C.43119    4.109 C.0004
(CONSTANT)  -3.54C13    5.62419             -0.629 C.5345

FOR BLOCK NUMBER  1  PIN = C.050 LIMITS REACHED.
```

Figure 13.6 Partial SPSS output for stepwise regression analysis of utility data.

13.8.3 Using SAS PROC RSQUARE and PROC STEPWISE
for Model Selection

The RSQUARE and STEPWISE procedures are available to facilitate model selection using SAS. The RSQUARE procedure performs all possible regressions for a set of dependent and independent variables and computes R_p^2. for each of these models. The setup for this procedure step is

PROCⱀRSQUAREⱀCP;

$$\text{MODEL}ⱀ \begin{Bmatrix} \text{dependent} \\ \text{variable} \end{Bmatrix} = \begin{Bmatrix} \text{list of} \\ \text{independent variables} \end{Bmatrix} / \{\text{options}\};$$

The options available include

$$\text{START} = \text{value}$$
$$\text{STOP} = \text{value}$$

The START = option indicates the minimum number of independent variables to be evaluated, while the STOP = option indicates the maximum number of variables to be evaluated. The output of PROC RSQUARE for the utility data consists of R_p^2. and C_p. for each regression model (see Table 13.3).

In addition to using PROC RSQUARE for evaluating all possible regression models, we can use the STEPWISE procedure for a stepwise regression approach. The setup of the procedure step is

PROCⱀSTEPWISE;

$$\text{MODEL}ⱀ \begin{Bmatrix} \text{dependent} \\ \text{variable} \end{Bmatrix} = \begin{Bmatrix} \text{list of} \\ \text{independent variables} \end{Bmatrix} / \{\text{options}\};$$

Among the options available are

FORWARD (or *F*)
BACKWARD (or *B*)
STEPWISE
MAXR
SLE = .*value*
SLS = .*value*

The first four options listed refer to a particular model selection process FORWARD (or *F*) is used for forward selection (see Section 13.5), BACKWARD (or *B*) is for backward elimination (see Section 13.6), and STEPWISE is for stepwise selection (see Section 13.4). MAXR refers to the maximum R^2 improvement technique [see SAS (1979)] in which the selection of the "best" model is based on the maximum improvement in R^2 obtained by entering and switching variables in the regression model. SLE = .*value* represents the desired significance level for entering a variable into a model while using the forward selection or stepwise options. SLS = .*value* represents the desired significance level for removing a variable from the model while using the backward elimination or stepwise options.

```
{SYSTEM CARDS}
DATA UTILITY;
    INPUT COCODE 1-2 INCOME 4-10 CUST 11-17 RVSALES
    18-24 ENERGY 25-31 XSYS 32-35 EPAEXP 36-40;
LABEL COCODE=COMPANY CODE NUMBER
      INCOME=NET INCOME IN MILLIONS
      CUST=NUMBER OF CUSTOMERS
      RVSALES=RATIO OF REVENUE TO SALES
      ENERGY=NET ENERGY GENERATED
      XSYS=TOTAL TRANSMISSION SYSTEM STRUCTURE MLES
      EPAEXP=NON REVENUE PRODUCING EPA EXPENSES;
CARDS;
01   46.964  1646253.81691  5.98631352        0
  .                                   .    .
  .                                   .    .
30   .073         30.83751   .2433    0    0
PROC PRINT;
PROC CORR;
VAR INCOME CUST RVSALES ENERGY XSYS EPAEXP;
PROC RSQUARE CP;
MODEL INCOME=CUST RVSALES ENERGY XSYS EPAEXP;
PROC STEPWISE;
MODEL INCOME=CUST RVSALES ENERGY XSYS EPAEXP/
STEPWISE SLE=.05 SLS=.05;
PROC STEPWISE;
MODEL INCOME=CUST RVSALES ENERGY XSYS
EPAEXP/F SLE=.05;
PROC STEPWISE;
MODEL INCOME=CUST RVSALES ENERGY XSYS
EPAEXP/B SLS=.05;
{SYSTEM CARDS}
```

Figure 13.7 SAS program for utility data.

Figure 13.7 is the SAS program that has been written using model selection procedures for the utility data. Figure 13.3 (see Section 13.4) represents a partial output obtained from PROC RSQUARE and PROC STEPWISE for these data.

13.8.4 Using the BMDP Programs P-2R and P-9R for Model Selection

There are two BMDP programs that are available for application to model selection in multiple regression. The P-2R program can be used for stepwise regression, while the P-9R program is used for all possible regression models.

P-2R. The only paragraphs of BMDP (see Section 2.7) that refer specifically to P-2R are /REGRESS, /PRINT, and /PLOT. The setup for these paragraphs is similar to that used in P-1R (see Section 10.12.3):

$$/REGRESS\flat DEPENDENT\flat IS\flat \left\{\begin{array}{l}\text{name of}\\ \text{dependent variable}\end{array}\right\}.$$

$$INDEPENDENT\flat ARE\flat \left\{\begin{array}{l}\text{names of}\\ \text{independent variables}\end{array}\right\}.$$

384

$$\text{METHOD} = F.$$
$$\text{ENTER} = F_E, F_E.$$
$$\text{REMOVE} = F_R, F_R.$$
/PRINTÞCORR.
    FRATIO.
    PARTIAL.
    DATA.

As compared to P-1R, there are three new sentences within the /REGRESS paragraph. The METHOD sentence (in which we use F) is used to indicate how variables are to be entered or removed from the model. The ENTER and REMOVE sentences, respectively, indicate the F criteria for entering and removing variables from the model. The F_R value needs to be set lower than the F_E value to prevent variables from being entered and removed at alternate steps. If we wish to employ the forward selection procedure, we can assign very low values to the first entry for F_E and F_R. If we wish to use the backward elimination approach, we can assign very high values to the second pair set of entries.

In addition to the DATA and CORR sentences of the /PRINT paragraph (see Section 10.12.3), the FRATIO sentence provides a table of F to enter and remove values, while PARTIAL provides a table of partial correlations.

Figure 13.8 illustrates the program (using P-2R) that has been written to perform stepwise regression for the utility data, while Figure 13.9 represents partial output.

```
⌠SYSTEM CARDS⌡
/PROBLEM    TITLE IS 'STEPWISE REGRESSION FOR
            UTILITY DATA'.
/INPUT  VARIABLES ARE 7.
        FORMAT IS '(F2.0,1X,F7.3,F7.0,F7.5,F7.4,
        F4.0,F5.0)'.
        CASES ARE 30.
/VARIABLE  NAMES ARE COCODE,INCOME,CUST,RVSALES,
           ENERGY,XSYS,EPAEXP.
           LABEL IS COCODE.
/REGRESS   DEPENDENT IS INCOME.
        INDEPENDENT ARE CUST,RVSALES,ENERGY,XSYS,EPAEXP.
               METHOD=F.
               ENTER=4.25,4.25.
               REMOVE=4.2,4.2.
/PRINT  CORR.
        FRATIO.
        PARTIAL.
        DATA.
/END
01     46.964  164.6253.81691  5.98631352      0
30       .073       30.83751   .2433     0      0
⌠SYSTEM CARDS⌡
```

Figure 13.8 Using BMDP program 2R for utility data.

```
STEP NO.   1
VARIABLE ENTERED     5 ENERGY

MULTIPLE R                0.9041
MULTIPLE R-SQUARE         0.8174
ADJUSTED R-SQUARE         0.8109
STD. ERROR OF EST.       27.4437

ANALYSIS OF VARIANCE
                  SUM OF SQUARES     DF    MEAN SQUARE     F RATIO
    REGRESSION       94399.813        1      94399.81       125.34
    RESIDUAL         21088.398       28       753.1570

                VARIABLES IN EQUATION
                            STD. ERROR   STD REG                F TO
    VARIABLE    COEFFICIENT  OF COEFF     COEFF    TOLERANCE    REMOVE
(Y-INTERCEPT     -2.614 )
ENERGY    5        5.009      0.447       0.904    1.00000      125.34
```

```
                                      VARIABLES NOT IN EQUATION
                                        PARTIAL                 F TO
LEVEL.      VARIABLE                    CORR.     TOLERANCE     ENTER   LEVEL
  1   .   COCODE      1    0.08854    0.99217      0.21      0
      .   CUST        3    0.39241    0.40663      4.91      1
      .   RVSALES     4    0.41296    0.99201      5.55      1
      .   XSYS        6   -0.16549    0.60215      0.76      1
      .   EPAEXP      7   -0.35176    0.76570      3.81      1
```

```
STEP NO.   2
VARIABLE ENTERED     4 RVSALES

MULTIPLE R                0.9212
MULTIPLE R-SQUARE         0.8485
ADJUSTED R-SQUARE         0.8373
STD. ERROR OF EST.       25.4530

ANALYSIS OF VARIANCE
                  SUM OF SQUARES     DF    MEAN SQUARE     F RATIO
    REGRESSION       97996.125        2      48998.06        75.63
    RESIDUAL         17492.082       27       647.8547

                VARIABLES IN EQUATION
                            STD. ERROR   STD REG                F TO
    VARIABLE    COEFFICIENT  OF COEFF     COEFF    TOLERANCE    REMOVE
(Y-INTERCEPT    -35.495 )
RVSALES   4        9.108      3.866       0.177    0.99201        5.55
ENERGY    5        4.921      0.417       0.888    0.99201      139.53
```

```
                                      VARIABLES NOT IN EQUATION
                                        PARTIAL                 F TO
LEVEL.      VARIABLE                    CORR.     TOLERANCE     ENTER   LEVEL
  1   .   COCODE      1    0.28540    0.85613      2.31      0
      .   CUST        3    0.25276    0.32304      1.77      1
      .   XSYS        6   -0.10208    0.58280      0.27      1
      .   EPAEXP      7   -0.43318    0.75822      6.01      1
```

```
STEP NO.   3
VARIABLE ENTERED     7 EPAEXP

MULTIPLE R                0.9365
MULTIPLE R-SQUARE         0.8770
ADJUSTED R-SQUARE         0.8628
STD. ERROR OF EST.       23.3780
```

Figure 13.9 Partial BMDP program 2R output for utility data.

ANALYSIS OF VARIANCE

	SUM OF SQUARES	DF	MEAN SQUARE	F RATIO
REGRESSION	101278.44	3	33759.48	61.77
RESIDUAL	14209.777	26	546.5298	

VARIABLES IN EQUATION

VARIABLE		COEFFICIENT	STD. ERROR OF COEFF	STD REG COEFF	TOLERANCE	F TO REMOVE
(Y-INTERCEPT		-39.338)				
RVSALES	4	9.973	3.568	0.194	0.98232	7.81
ENERGY	5	5.432	0.436	0.980	0.76498	155.40
EPAEXP	7	-0.002	0.001	-0.194	0.75822	6.01

VARIABLES NOT IN EQUATION

LEVEL	VARIABLE		PARTIAL CORR.	TOLERANCE	F TO ENTER	LEVEL
1	COCODE	1	0.20769	0.80711	1.13	0
1	CUST	3	0.52351	0.26874	9.44	1
1	XSYS	6	-0.06406	0.57661	0.10	1

STEP NO. 4
VARIABLE ENTERED 3 CUST

MULTIPLE R 0.9543
MULTIPLE R-SQUARE 0.9107
ADJUSTED R-SQUARE 0.8964
STD. ERROR OF EST. 20.3130

ANALYSIS OF VARIANCE

	SUM OF SQUARES	DF	MEAN SQUARE	F RATIO
REGRESSION	105172.75	4	26293.19	63.72
RESIDUAL	10315.465	25	412.6184	

VARIABLES IN EQUATION

VARIABLE		COEFFICIENT	STD. ERROR OF COEFF	STD REG COEFF	TOLERANCE	F TO REMOVE
(Y-INTERCEPT		-21.978)				
CUST	3	0.000	0.000	0.354	0.26874	9.44
RVSALES	4	5.110	3.481	0.099	0.77922	2.15
ENERGY	5	4.221	0.546	0.762	0.36719	59.67
EPAEXP	7	-0.003	0.001	-0.288	0.63077	14.68

VARIABLES NOT IN EQUATION

LEVEL	VARIABLE		PARTIAL CORR.	TOLERANCE	F TO ENTER	LEVEL
1	COCODE	1	0.28573	0.80346	2.13	0
1	XSYS	6	-0.01785	0.57158	0.01	1

STEP NO. 5
VARIABLE REMOVED 4 RVSALES

MULTIPLE R 0.9503
MULTIPLE R-SQUARE 0.9030
ADJUSTED R-SQUARE 0.8918
STD. ERROR OF EST. 20.7592

ANALYSIS OF VARIANCE

	SUM OF SQUARES	DF	MEAN SQUARE	F RATIO
REGRESSION	104283.63	3	34761.21	80.66
RESIDUAL	11204.590	26	430.9456	

VARIABLES IN EQUATION

VARIABLE		COEFFICIENT	STD. ERROR OF COEFF	STD REG COEFF	TOLERANCE	F TO REMOVE
(Y-INTERCEPT		-3.540)				
CUST	3	0.000	0.000	0.431	0.33878	16.88
ENERGY	5	3.974	0.531	0.717	0.40592	55.96
EPAEXP	7	-0.003	0.001	-0.300	0.63795	15.40

VARIABLES NOT IN EQUATION

LEVEL	VARIABLE		PARTIAL CORR.	TOLERANCE	F TO ENTER	LEVEL
1	COCODE	1	0.17772	0.88182	0.82	0
1	RVSALES	4	0.28170	0.77922	2.15	1
1	XSYS	6	-0.05318	0.58117	0.07	1

* * * * * F-LEVELS(4.250, 4.200) OR TOLERANCE INSUFFICIENT FOR FURTHER STEPPING

Figure 13.9 (Continued)

P-9R. BMDP-9R provides a regression model for varying subsets of independent variables. The setup for the /REGRESS paragraph is

$$/\text{REGRESS}\flat\text{DEPENDENT}\flat\text{IS}\flat \begin{Bmatrix} \text{name of} \\ \text{dependent variable} \end{Bmatrix}.$$

$$\text{INDEPENDENT}\flat\text{IS}\flat \begin{Bmatrix} \text{names of} \\ \text{independent variables} \end{Bmatrix}.$$

$$\text{METHOD} = \begin{Bmatrix} \text{CP or} \\ \text{RSQ} \end{Bmatrix}.$$

$$\text{NUMBER} = \begin{Bmatrix} \text{value} \\ \text{up to 10} \end{Bmatrix}.$$

In the P-9R program, the METHOD sentence of the /REGRESS paragraph indicates the process to be used in selecting subsets of independent variables. CP refers to Mallows C_{p^*} (see Section 13.3), and RSQ refers to $R_{p^*}^2$. The NUMBER sentence indicates the number of models to be presented. For CP this number (which must be ≤ 10) represents the *total* number of models provided (for all p^* values), while for RSQ the number represents the number of models provided for *each* value of p^*.

As mentioned in Section 10.12.3, a /PLOT paragraph is also available. The /PLOT paragraph of P-9R differs from the corresponding paragraph in P-1R and P-2R in that the predicted values (named PREDICTD), the residuals

```
{SYSTEM CARDS}
/PROBLEM TITLE IS 'MODEL SELECTION FOR UTILITY DATA'.
/INPUT VARIABLES ARE 7.
   FORMAT IS '(F2.0,1X,F7.3,F7.0,F7.5,F7.4,F4.0,F5.0)'
   CASES ARE 30.
/VARIABLE NAMES ARE COCODE,INCOME,CUST,RVSALES,
          ENERGY,XSYS,EPAEXP.
          LABEL IS COCODE.
/REGRESS  DEPENDENT IS INCOME.
   INDEPENDENT ARE CUST,RVSALES,ENERGY,XSYS,EPAEXP.
          METHOD=CP.
          NUMBER=10.
/PRINT    RESI.
/PLOT YVAR ARE RESIDUAL,RESIDUAL,RESIDUAL,RESIDUAL,
          RESIDUAL,RESIDUAL.
          XVAR ARE PREDICTD,CUST,RVSALES,ENERGY,
          XSYS,EPAEXP.
/END
01    46.964  1646253.81691  5.98631352      0
30       .073         30.83751    .2433    0    0
{SYSTEM CARDS}
```

Figure 13.10   Using BMDP program 9R for utility data.

```
                                   ◊◊◊◊  SUBSETS WITH    1 VARIABLES  ◊◊◊◊
            ADJUSTED
 R-SQUARED  R-SQUARED        CP

  0.817397   0.810876     23.08   ENERGY

  0.645393   0.632728     69.31   CUST

  0.265525   0.239398    171.39   XSYS

  0.093593   0.061325    217.60   EPAEXP

  0.065821   0.032458    225.09   RVSALES


                                   ◊◊◊◊  SUBSETS WITH    2 VARIABLES  ◊◊◊◊
            ADJUSTED
 R-SQUARED  R-SQUARED        CP

  0.848538   0.837318     16.71   VARIABLE      COEFFICIENT   T-STATISTIC
                                  4 RVSALES       9.10847         2.36
                                  5 ENERGY        4.92118        11.81
                                    INTERCEPT   -35.4955

  0.845516   0.834073     17.52   VARIABLE      COEFFICIENT   T-STATISTIC
                                  3 CUST        0.0000293913       2.22
                                  5 ENERGY        3.88666         5.91
                                    INTERCEPT    -2.49607

  0.839993   0.828141     19.01   ENERGY      EPAEXP

  0.822398   0.809243     23.74   ENERGY      XSYS


                                   ◊◊◊◊  SUBSETS WITH    3 VARIABLES  ◊◊◊◊
            ADJUSTED
 R-SQUARED  R-SQUARED        CP

  0.902981   0.891787      4.08   VARIABLE      COEFFICIENT   T-STATISTIC
                                  3 CUST        0.0000481937       4.11
                                  5 ENERGY        3.97360         7.48
                                  7 EPAEXP      -0.00263167       -3.92
                                    INTERCEPT    -3.54013

  0.876960   0.862763     11.07   VARIABLE      COEFFICIENT   T-STATISTIC
                                  4 RVSALES       9.97295         2.79
                                  5 ENERGY        5.43207        12.47
                                  7 EPAEXP      -0.00169767       -2.45
                                    INTERCEPT   -39.3384

  0.858214   0.841854     16.11   VARIABLE      COEFFICIENT   T-STATISTIC
                                  3 CUST        0.0000193443       1.33
                                  4 RVSALES       6.52570         1.53
                                  5 ENERGY        4.20744         6.23
                                    INTERCEPT   -26.0942

  0.850116   0.832822     18.29   RVSALES     ENERGY      XSYS

  0.848137   0.830615     18.82   CUST        ENERGY      XSYS

  0.843406   0.825337     20.09   ENERGY      XSYS        EPAEXP


                                   ◊◊◊◊  SUBSETS WITH    4 VARIABLES  ◊◊◊◊
            ADJUSTED
 R-SQUARED  R-SQUARED        CP

  0.910681   0.896390      4.01   VARIABLE      COEFFICIENT   T-STATISTIC
                                  3 CUST        0.0000395917       3.07
                                  4 RVSALES       5.11009         1.47
                                  5 ENERGY        4.22140         7.72
                                  7 EPAEXP      -0.00252887       -3.83
                                    INTERCEPT   -21.9783

  0.903256   0.887778      6.00   VARIABLE      COEFFICIENT   T-STATISTIC
                                  3 CUST        0.0000476526       3.93
                                  5 ENERGY        4.06203         6.40
                                  6 XSYS        -0.000725960      -0.27
                                  7 EPAEXP      -0.00260498       -3.77
                                    INTERCEPT    -3.36457

  0.877465   0.857859     12.94   VARIABLE      COEFFICIENT   T-STATISTIC
                                  4 RVSALES       9.74854         2.64
                                  5 ENERGY        5.53048        10.26
                                  6 XSYS        -0.000988663      -0.32
                                  7 EPAEXP      -0.00167422       -2.36
                                    INTERCEPT   -38.2917

  0.859480   0.836997     17.77   VARIABLE      COEFFICIENT   T-STATISTIC
                                  3 CUST        0.0000190466       1.29
                                  4 RVSALES       6.23054         1.42
                                  5 ENERGY        4.38466         5.62
                                  6 XSYS        -0.00155841      -0.47
                                    INTERCEPT   -24.6726

  0.744778   0.703942     48.60   CUST        RVSALES     XSYS        EPAEXP


                                   ◊◊◊◊  SUBSETS WITH    5 VARIABLES  ◊◊◊◊
            ADJUSTED
 R-SQUARED  R-SQUARED        CP

  0.910739   0.892107      6.00   VARIABLE      COEFFICIENT   T-STATISTIC
                                  3 CUST        0.0000394838       2.99
                                  4 RVSALES       5.06982         1.42
                                  5 ENERGY        4.24817         6.68
                                  6 XSYS        -0.000235828      -0.09
                                  7 EPAEXP      -0.00252101       -3.71
                                    INTERCEPT   -21.7760
```

Figure 13.11 Partial BMDP program 9R output for utility data.

(named RESIDUAL), and the standardized residuals (named STRESIDL) are available for plotting either as dependent or independent variables. These variables can be plotted, along with other variables, by specifying their name in the XVAR = or YVAR = sentence of the /PLOT paragraph.

Figure 13.10 illustrates the program using P-9R that has been written to provide model selection for the utility data, while Figure 13.11 represents partial output.

PROBLEMS

13.1. In Problem 10.3 we selected a random sample of 33 corporations from the Fortune 500 and provided data concerning net income, sales, and assets for the year 1979. Information was also available concerning stockholders' equity, the number of employees, earnings per share, and total return to investors. The results were as shown on page 391.

    (a) In this problem, we would like to predict net income based on four independent variables: sales, assets, stockholders' equity, and number of employees. Use a computer package (such as SPSS, SAS, or BMDP) to select the appropriate variables that should be included in the model. Be sure to

        (1) Determine $R_{p^*}^2$ for each model, and plot $R_{p^*}^2$ versus p^*. What model would you select using this approach?

        (2) Determine C_{p^*} for each model, and plot C_{p^*} versus p^*. What model would you select using this approach?

        (3) Use

           (i) stepwise regression

           (ii) forward selection

         and

           (iii) backward elimination

        to select the variables to be included in the model. Indicate the model that you would select using *each* of these approaches.

        (4) Evaluate the aptness of the model selected using each of these approaches.

        (5) Compare the results of the various model selection procedures. Which model would you ultimately select as the "best" model? Why?

        (6) Interpret the meaning of the regression coefficients in the selected model.

    (b) In this problem, we would like to predict *total return to investors* (%) based on six independent variables: net income, sales, assets, stockholders' equity, number of employees, and earnings per share. Use a computer package (such as SPSS, SAS, or BMDP) to do parts (1)–(6) as in part (a).

*13.2. A sample of 35 companies was randomly chosen from *Dun's Review* of "240 selected companies." We wish to develop a model to predict stock price (Y) based on the variables displayed in Table P13.2.

    (a) In this problem, we would like to predict the stock price based on the independent variables X_1–X_7. Use a computer package (such as SPSS, SAS, or BMDP) to select the appropriate variables that should be included in the model. Be sure to

Corporation	Stockholders' Equity ($ millions)	Number of Employees (000)	Earnings per Share	Total Return to Investors (%)
Mobil	10,513	213.5	9.46	65.91
DuPont	5,312	134.2	6.42	2.68
Caterpillar Tractor	3,065	89.3	5.69	−4.51
National Steel	1,414	38.8	6.56	.34
Eaton	936	57.8	5.89	17.56
American Cyanamid	1,347	44.2	3.52	40.28
Boise Cascade	1,178	35.7	6.52	32.00
Burroughs	2,136	56.5	7.45	10.03
Central Soya	266	9.3	2.23	15.92
Interco	684	43.0	6.37	14.71
National Distillers	914	15.7	4.04	66.25
Emhart	415	37.5	4.38	−9.02
U.S. Gypsum	625	20.5	7.59	29.00
Staley Mfg.	258	4.2	1.86	94.04
Timken	706	23.8	9.14	9.04
Smith Kline	713	18.1	3.85	40.31
Pitney-Bowes	357	25.2	4.02	42.24
Baker Int.	560	19.0	3.25	70.27
Witco Chemical	237	6.5	4.83	53.26
Champion Spark Plug	399	16.0	1.49	13.30
Joy Mfg.	395	12.7	3.82	16.42
AM Int.	228	20.0	1.38	−20.78
Hoover	228	21.5	3.13	29.19
Outboard Marine	277	12.8	2.00	2.84
Morton-Norwich	319	10.4	3.40	18.93
Sybron	239	15.3	2.05	−.61
Bemis	165	10.1	5.86	44.51
Con Agra	95	6.4	3.55	26.26
Dan River	183	15.5	3.90	60.72
Avery Int.	154	7.1	2.35	23.72
United Refining	70	2.2	7.15	110.05
Consolidated Papers	251	5.5	10.29	9.06
Dorsey	69	6.8	3.14	27.00

SOURCE: "The *Fortune* Directory of the 500 Largest Industrial Corporations," *Fortune*, May 5, 1980, pp. 276–299. © 1980 Time Inc. All rights reserved.

(1) Determine R_p^2. for each model, and plot R_p^2. versus p^*. What model would you select using this approach?

(2) Determine C_{p^*} for each model, and plot C_{p^*} versus p^*. What model would you select using this approach?

(3) Use
   (i) stepwise regression,
   (ii) forward selection,
   and
   (iii) backward elimination

TABLE P13.2 Sample of 35 Companies Selected from *Dun's Review*

Y: Stock Price
X_1: Dividend
X_2: Yield
X_3: Earnings per Share ($)
X_4: Sales in millions of dollars
X_5: Income in millions of dollars
X_6: ROS (Return on Sales)
X_7: ROE (Return on Equity)
X_8: Exchange Traded (1 = New York; 2 = American or Over the Counter)

Company	Y	X_1	X_2	X_3	X_4	X_5	X_6	X_7	X_8
Cross (A.T.)	40.00	.170	3.9	3.61	95.1	14.6	15.4	29.9	2
McDonough	41.25	.150	3.4	5.39	450.6	21.6	4.8	14.9	1
Brunswick	14.50	.100	6.0	2.39	1257.3	51.4	4.1	10.4	1
Leaseway Trn.	33.50	.170	3.9	3.62	937.2	42.9	4.6	20.6	1
Com. Clr. House	31.38	.130	3.2	2.11	215.8	19.6	9.1	65.7	2
Mallinckrodt	45.00	.175	2.7	3.40	392.5	31.9	8.1	13.9	2
Empl. Casualty	39.50	.210	3.0	5.55	178.8	17.2	9.6	23.6	2
Lib. Natl. Life Ins.	17.68	.225	6.8	2.67	409.3	50.8	12.4	20.1	2
Ohio Casualty	36.75	.330	4.8	7.11	809.2	82.5	10.2	23.9	2
Wstn Cas. & Surety	38.88	.350	4.7	7.70	330.8	30.8	9.3	18.2	2
Kaneb Services	31.88	.159	2.6	2.17	401.9	48.8	12.0	25.8	1
Crane	38.50	.320	4.1	5.39	1573.2	55.0	3.5	14.7	1
Am. Bnkrs. Life Asr.	11.00	.090	4.0	2.22	131.0	10.2	7.8	31.2	2
Toronto-Dom. Bank	30.50	.340	4.5	2.80	2739.0	106.4	3.9	14.3	2
Kennametal	32.88	.190	2.3	3.06	325.9	36.7	11.3	19.7	1
Huyck Corp.	22.25	.210	3.2	1.58	143.0	9.0	6.3	14.0	1
Std. Brands Paint	29.75	.180	2.4	2.70	182.7	14.3	7.8	16.2	1
Nevada Power	19.78	.610	11.7	3.37	175.1	17.1	9.8	13.1	1
Heinz (H.J.)	44.75	.630	5.0	6.24	2924.8	142.9	4.9	16.4	1
Nashua Corp.	28.13	.450	5.3	5.75	608.4	26.7	4.4	19.0	1
HB Fuller	12.38	.120	3.2	1.75	258.7	7.9	3.1	14.0	2
Diebold Inc.	44.75	.260	1.7	3.19	305.2	18.2	6.0	15.3	1
Kellogg	19.18	.450	6.9	2.13	1846.6	162.6	8.8	24.6	1
Caterpillar	56.00	.800	4.2	5.69	7613.2	491.6	6.5	16.0	1
Ryl. Bank Can.	52.75	.860	4.7	7.40	4215.5	270.7	6.4	20.9	2
Banco de Ponce	16.50	.451	7.3	4.96	107.7	7.7	7.1	13.3	2
Fla. P&L	26.50	1.020	10.0	4.22	1933.9	204.7	10.6	11.1	1
Moore Products	29.00	.310	2.7	2.84	48.3	5.7	11.8	19.8	2
Meyer (Fred)	38.50	.220	1.4	3.29	1060.2	22.4	2.1	14.3	2
Eagle-Picher	18.88	.380	4.7	3.03	590.0	30.7	5.2	15.3	1
Ga.-Pacific	26.38	.520	4.6	3.12	5207.0	327.0	6.3	18.2	1
Ctl. Tel. & Ut.	23.25	.880	8.6	3.34	750.5	83.1	11.1	15.1	1
Gnl. Shale	13.50	.470	7.3	2.21	61.8	5.5	8.9	13.5	2
MT-Dak Util.	22.88	.880	7.9	2.76	173.3	17.7	10.2	10.7	1
So. Union	45.25	.980	4.2	4.78	724.0	34.1	4.7	19.6	1

SOURCE: Reprinted with the special permission of *Dun's Review* (December 1980), Copyright 1980, Dun & Bradstreet Publications Corporation.

to select the variables to be included in the model. Indicate the model that you would select using each of these approaches.

(4) Evaluate the aptness of the model selected using *each* of these approaches.

(5) Compare the results of the various model selection procedures. Which model would you ultimately select at the "best" model? Why?

(6) Interpret the meaning of the regression coefficients in the selected model.

(b) In this problem, use X_8 (exchange traded) as a *dummy* variable (see Section 12.5) in addition to $X_1 - X_7$. Do parts (1)–(6) of part (a) with this variable included among those to be considered for selection. Compare the results of parts (a) and (b).

13.3. Table P13.3 represents information concerning the quantity and value of farms for each state. The variables measured were

Number: number of farms (thousands)

Size: average farm size (acres)

Value: average value per acre (thousands of dollars)

Big: number of farms with 2000+ acres

Small: number of farms with 10–49 acres

Total: total farm acreage (thousands)

(a) In this problem, we would like to predict total farm acreage (in thousands) based on the number of farms (in thousands), the number of large farms (greater than 2000 acres), and the number of small farms (between 10 and 49 acres). Use a computer package (such as SPSS, SAS, or BMDP) to select the appropriate variables that should be included in the model. Be sure to

(1) Determine $R_{p^*}^2$ for each model, and plot $R_{p^*}^2$ versus p^*. What model would would you select using this approach?

(2) Determine C_{p^*} for each model, and plot C_{p^*} versus p^*. What model would you select using this approach?

(3) Use
   (i) stepwise regression,
   (ii) forward selection,
   and
   (iii) backward elimination
   to select the variables to be included in the model. Indicate the model that you would select using *each* of these approaches.

(4) Evaluate the aptness of the model selected using each of these approaches.

(5) Compare the results of the various model selection procedures. Which model would you ultimately select as the "best" model? Why?

(6) Interpret the meaning of the regression coefficients in the selected model.

(b) In this problem, we would like to develop a model to predict the average value per acre (thousands of dollars). The following variables should be considered for selection: number of farms, average farm size, number of large farms (2000+ acres), number of small farms (between 10 and 49 acres), and total state acreage. Use a computer package (such as SPSS, SAS, or BMDP) to do parts (1)–(6) as in part (a).

TABLE P13.3 Farm Acreage by State

State	Number	Size	Value	Big	Small	Total
Alabama	56.7	209	364	643	15,475	11,853
Alaska	.3	5612	42	37	40	1,633
Arizona	5.8	6539	111	877	1,253	37,944
Arkansas	51.0	287	419	787	8,537	14,642
California	67.7	493	653	2,813	24,162	33,386
Colorado	25.5	1408	188	4,061	2,800	35,902
Connecticut	3.4	129	1525	6	888	440
Delaware	3.4	185	971	17	783	631
Florida	32.4	407	685	1,006	10,940	13,199
Georgia	54.9	253	474	723	11,118	13,878
Hawaii	3.0	702	485	70	812	2,119
Idaho	23.7	603	339	1,246	4,251	14,274
Illinois	111.0	262	846	229	13,420	29,095
Indiana	87.9	191	720	125	17,325	16,785
Iowa	126.1	262	719	150	9,841	33,045
Kansas	79.2	605	296	3,819	5,127	47,946
Kentucky	102.1	141	427	159	20,945	14,432
Louisiana	33.2	275	512	594	9,255	9,133
Maine	6.4	237	341	28	728	1,524
Maryland	15.2	174	1060	57	3,399	2,634
Massachusetts	4.5	134	961	9	1,185	602
Michigan	64.9	169	553	62	12,233	10,832
Minnesota	98.5	280	429	443	6,846	27,605
Mississippi	53.6	267	379	938	10,351	14,301
Missouri	115.7	258	396	509	13,919	29,801
Montana	23.3	2665	112	7,411	1,550	62,158
Nebraska	67.6	683	282	3,719	3,384	46,172
Nevada	2.1	5209	85	342	317	10,814
New Hampshire	2.4	210	564	5	360	506
New Jersey	7.4	130	1807	18	2,215	961
New Mexico	11.3	4170	78	2,610	1,694	47,046
New York	43.7	215	510	67	5,093	9,410
North Carolina	91.3	123	590	237	30,195	11,244
North Dakota	42.7	992	195	3,789	773	42,397
Ohio	92.2	170	706	76	16,163	15,668
Oklahoma	69.7	475	302	2,191	6,001	33,083
Oregon	26.8	682	250	1,731	8,292	18,241
Pennsylvania	53.2	154	734	45	8,363	8,186
Rhode Island	.6	102	1500	2	177	61
South Carolina	29.3	211	467	312	7,677	6,177
South Dakota	42.8	1074	145	4,444	1,549	45,978
Tennessee	93.7	140	467	233	24,759	13,103
Texas	174.1	771	243	10,038	20,951	134,185
Utah	12.2	871	188	822	2,948	10,610
Vermont	5.9	282	462	14	402	1,667
Virginia	52.7	184	558	264	11,521	9,678
Washington	29.4	567	350	1,657	8,739	16,662
West Virginia	16.9	207	300	60	2,160	3,497
Wisconsin	89.5	197	434	118	8,176	17,625
Wyoming	8.0	4274	80	2,607	403	34,272

SOURCE: Statistical Abstract of the United States, 1979

13.4. Using the real estate data base (see Appendix A), select a random sample of 50 houses. Develop a multiple regression model to predict the selling price by selecting a set of independent variables from among time period (TIME), age of house (AGE), lot size (LSQF), heating area (DSQF), presence or absence of central heating (HT), air conditioning (AIR), fireplace (FIREPL), brick exterior (EXTER), rating of built-in appliances (BLT), and type of garage (CAR).   ■

13.5. Using the real estate data base (see Appendix A), select a random sample of 50 houses. Develop a multiple regression model to predict the assessed value by selecting a set of independent variables from among age of house (AGE), heating area (DSQF), number of bathrooms (BATH), presence or absence of a fireplace (FIREPL), central air conditioning (AIR), and type of garage (CAR).   ■

13.6. Reviewing concepts:

   (a) In evaluating all possible regression models, what are the differences between C_{p^*} and $R_{p^*}^2$?

   (b) Compare and contrast the methods of stepwise regression, forward selection, and backward elimination for developing multiple regression models.

14

VIOLATING THE ASSUMPTIONS
AND OTHER PROBLEMS
IN REGRESSION ANALYSIS

14.1 INTRODUCTION

We may recall from Sections 8.5 and 8.11 that the appropriateness of the regression analysis was dependent on the validity of several assumptions pertaining to the fitted regression model. In particular, the least-squares estimates for the parameter values as well as the various confidence interval statements and tests of hypotheses required the assumptions that the residuals $\epsilon_i = Y_i - \hat{Y}_i$ (where $i = 1, 2, \ldots, n$) be independent and identically distributed according to a normal distribution with mean 0 and variance $\sigma^2_{Y|X}$. It is essential to realize that proceeding with a regression analysis as if these assumptions are true when in reality they are not may result in gross data misinterpretations and faulty conclusions.

As previously discussed in Sections 8.11 and 10.8, *residual analysis* is a powerful tool for evaluating the aptness of the regression assumptions. In Section 14.2 the value of utilizing residual analysis will be further amplified by focusing on a well-known example presented by Anscombe (1973). In the remainder of the chapter, we describe the consequences of violations in the regression assumptions that may be uncovered through residual analysis as well as introduce the problem of multicollinearity. Remedies that may be available for corrective action will be suggested. Thus, this chapter serves as a capstone to the topic of regression models by highlighting certain issues that should be considered during any analysis.

14.2 THE IMPORTANCE OF RESIDUAL ANALYSIS: THE ANSCOMBE (1973) DATA

The residual scatter plot is a fundamental tool of invaluable assistance in uncovering violations against the assumptions underlying the regression model. Such plots, for example, usually are the first clear signal a researcher will receive indicating violations in the assumptions of homoscedasticity and independent errors. Before pursuing this, however, we shall discuss what has become a classical pedagogical piece of statistical literature dealing with the kind of information only such plots are capable of uncovering.

TABLE 14.1   Four Sets of Artificial Data

Set A		Set B		Set C		Set D	
X_i	Y_i	X_i	Y_i	X_i	Y_i	X_i	Y_i
10	8.04	10	9.14	10	7.46	8	6.58
14	9.96	14	8.10	14	8.84	8	5.76
5	5.68	5	4.74	5	5.73	8	7.71
8	6.95	8	8.14	8	6.77	8	8.84
9	8.81	9	8.77	9	7.11	8	8.47
12	10.84	12	9.13	12	8.15	8	7.04
4	4.26	4	3.10	4	5.39	8	5.25
7	4.82	7	7.26	7	6.42	19	12.50
11	8.33	11	9.26	11	7.81	8	5.56
13	7.58	13	8.74	13	12.74	8	7.91
6	7.24	6	6.13	6	6.08	8	6.89

SOURCE: F. J. Anscombe, "Graphs in Statistical Analysis," *American Statistician*, Vol. 27 (1973), pp. 17–21.

Anscombe (1973) showed that for the four data sets given in Table 14.1 the following results would be obtained:

$$\hat{Y}_i = 3.0 + .5X_i$$

$$S_{Y|X} = 1.236$$

$$S_{b_1} = .118$$

$$r^2 = .667$$

$$\text{SSR} = \text{Explained Variation} = \sum_{i=1}^{n} (\hat{Y}_i - \bar{Y})^2 = 27.50$$

$$\text{SSE} = \text{Unexplained Variation} = \sum_{i=1}^{n} (Y_i - \hat{Y}_i)^2 = 13.75$$

$$\text{SST} = \text{Total Variation} = \sum_{i=1}^{n} (Y_i - \bar{Y})^2 = 41.25$$

Thus with respect to the pertinent statistics associated with a simple linear regression, the four data sets are identical. Had the researcher stopped his/her analysis at this point, valuable information in the data would be lost.

Table 14.2 gives the standardized residuals $\epsilon_i/S_{Y|X}$ for each of the data sets.

TABLE 14.2 Standardized Residuals

	Data Set A	Data Set B	Data Set C	Data Set D	
X_i	$\epsilon_i/S_{Y|X}$	$\epsilon_i/S_{Y|X}$	$\epsilon_i/S_{Y|X}$	X_i	$\epsilon_i/S_{Y|X}$
4	−.599	−1.536	.314	8	−.340
5	.145	−.614	.185	8	−1.003
6	1.002	.105	.064	8	.574
7	−1.359	.614	−.065	8	1.489
8	−.041	.922	−.186	8	1.189
9	1.059	1.027	−.315	8	.032
10	.032	.922	−.437	8	−1.416
11	−.138	.614	−.558	19	.000
12	1.487	.105	−.687	8	−1.165
13	−1.554	−.614	2.622	8	.736
14	−.033	−1.536	−.937	8	−.089

Source: F. J. Anscombe, "Graphs in Statistical Analysis," *American Statistician*, Vol. 27 (1973), pp. 17–21.

When the standardized residuals are plotted against $\hat{Y}$, we see how different the data sets are. Panels A, B, C, and D of Figure 14.1 graphically depict, for each data set, a plot of the standardized residuals against the fitted values $\hat{Y}$. While the plot for data set A does not show any obvious anomalies, this is not the case for data sets B, C, and D. The parabolic form of the residual plot for B probably indicates that the basic simple linear regression model should be augmented to include a second-order term. Hence for this data set we might try fitting the model

$$Y_i = \beta_0 + \beta_1 X_{1_i} + \beta_2 X_{1_i}^2 + \epsilon_i \tag{11.1}$$

as developed in Section 11.1. The plot for data set C clearly depicts what may very well be an *outlying* observation. If this is the case, the researcher may deem it appropriate to remove the outlier and re-estimate the basic model. The result of this exercise would probably be a relationship much different from what was originally uncovered. Similarly, the plot for data set D would be evaluated cautiously because the fitted model is so dependent on the outcome of a single response ($X_8 = 19$ and $Y_8 = 12.50$).

It is interesting and instructive to note that had we constructed the residual plots by using the independent variable as the X axis (instead of the estimated values $\hat{Y}$) the same conclusions would prevail.

In summary, residual scatter plots as a class of statistical methodology are

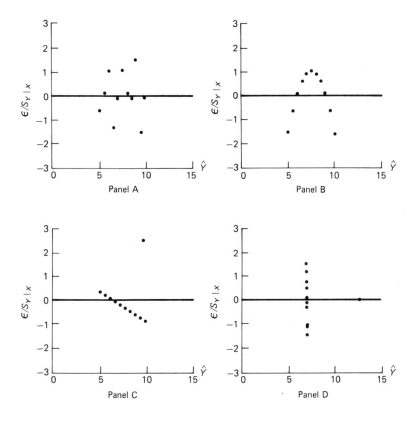

Figure 14.1 Plots of $\hat{Y}_i$ versus standardized residuals. Source: F. J. Anscombe, "Graphs in Statistical Analysis," *American Statistician*, Vol. 27 (1973), pp. 17–21.

often of vital importance to a complete regression analysis. The information they impart is so basic to a credible analysis that such scatter plots should always accompany the usual statistics of a regression study.

14.3 USING WEIGHTED LEAST SQUARES FOR VIOLATIONS OF HOMOSCEDASTICITY

In *ordinary least-squares* (OLS) estimation we assumed that the residuals were independent and identically distributed. Implicit in this assumption is that the residuals have a common variance $\sigma_{Y|X}^2$. Applying OLS procedures in multiple regression situations where the variance of the response variable fluctuates (as the level of at least one of the explanatory variables changes) will not impact at all upon the unbiasedness of the estimated regression coefficients; however, it could seriously affect the precision of such estimates as measured through their variances. Hence, the problem is important, and the researcher must attempt to examine this homogeneity of variance assumption.

As previously mentioned (see Section 8.11.2), heteroscedasticity is usually uncovered by plotting the residuals ϵ_i against an explanatory variable. A typical situation is depicted in Figure 14.2.

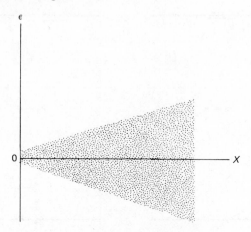

Figure 14.2   Example of heteroscedasticity.

14.3.1 Development of Weighted Least Squares

Here it can be argued that at least for the simple linear model

$$Y_i = \beta_0 + \beta_1 X_i + \epsilon_i \tag{8.2}$$

the variance of ϵ_i is a function of the level of X_i; that is, $\text{Var}(\epsilon_i) = kX_i^2$, where k is some positive constant. In a multiple regression model the researcher will usually make several *residual by explanatory variable* scatter plots—one for each explanatory variable. Once the "suspected variable" is uncovered, a simple transformation of the data often permits the application of OLS.

The procedure we have just described is an extension of OLS and is known as *weighted least squares* (WLS). It is designed to treat the case where the variance of the residuals may be a function of the levels of one or more explanatory variables. In general, the variance of the residuals may be expressed by

$$\text{Var}(\epsilon_i) = W_{ii}\sigma_{Y|X}^2 \tag{14.1}$$

where W_{ii} is assigned to the ith value of an explanatory variable ($i = 1, 2, \ldots, n$).*

To illustrate, suppose the initial model being used is

$$Y_i = \beta_0 + \beta_1 X_{1_i} + \cdots + \beta_p X_{p_i} + \epsilon_i \tag{14.2}$$

and that through the p possible residual scatter plots it is determined that for

---

*Note, of course, that when the W_{ii} are all the same the homoscedasticity assumption is said to hold.

the second explanatory variable $\text{Var}(\epsilon_i) = kX_{2_i}^2$, suggesting $W_{ii} = X_{2_i}^2$ for this model. Assuming no other obvious patterns in any of the other residual scatter plots, we attempt to find the appropriate set of *weights* ω_{ii} which, when applied to the variables in the regression model, will remove the effects of heteroscedasticity associated with the second explanatory variable. The appropriate weights for this example are

$$\omega_{ii} = \frac{1}{X_{2_i}^2} = \frac{1}{W_{ii}}$$

From the calculus, the following set of $p + 1$ *normal equations* are developed*:

I. $\quad \sum_{i=1}^{n} \omega_{ii} Y_i = b_0 \sum_{i=1}^{n} \omega_{ii} + b_1 \sum_{i=1}^{n} \omega_{ii} X_{1_i} + \cdots + b_p \sum_{i=1}^{n} \omega_{ii} X_{p_i}$

II. $\quad \sum_{i=1}^{n} \omega_{ii} X_{1_i} Y_i = b_0 \sum_{i=1}^{n} \omega_{ii} X_{1_i} + b_1 \sum_{i=1}^{n} \omega_{ii} X_{1_i}^2 + \cdots + b_p \sum_{i=1}^{n} \omega_{ii} X_{1_i} X_{p_i}$

$$
\begin{array}{ccccc}
\cdot & \cdot & \cdot & \cdot & \cdot \\
\cdot & \cdot & \cdot & \cdot & \cdot \quad (14.3) \\
\cdot & \cdot & \cdot & \cdot & \cdot \\
\end{array}
$$

$p + 1$. $\quad \sum_{i=1}^{n} \omega_{ii} X_{p_i} Y_i = b_0 \sum_{i=1}^{n} \omega_{ii} X_{p_i} + b_1 \sum_{i=1}^{n} \omega_{ii} X_{1_i} X_{p_i} + \cdots + b_p \sum_{i=1}^{n} \omega_{ii} X_{p_i}^2$

When solved simultaneously, these normal equations yield the $p + 1$ desired estimated regression coefficients $b_0, b_1, \ldots, b_p$.

14.3.2 Weighted Least Squares: A Matrix Approach

Using the matrix notation presented in Sections 9.8 and 10.11, the basic multiple regression model was given by

$$\mathbf{Y} = \mathbf{X}\boldsymbol{\beta} + \boldsymbol{\epsilon} \tag{9.24}$$

and the fitted model was stated as

$$\hat{\mathbf{Y}} = \mathbf{X}\hat{\boldsymbol{\beta}} \tag{9.28}$$

By using (14.1), the variance of the residuals can now be expressed in matrix form as

$$\text{Var}(\boldsymbol{\epsilon}) = \mathbf{W}\sigma_{Y|X}^2 \tag{14.4}$$

where $\mathbf{W}$ is an $n \times n$ diagonal matrix containing the elements W_{ii}:

---

*In using OLS, the least-squares principle states that the quantity $L = \sum_{i=1}^{n} [Y_i - (b_0 + b_1 X_{1_i} + \cdots + b_p X_{p_i})]^2$ be a *minimum total* with respect to $b_0, b_1, \ldots, b_p$ (see the footnote on p. 206). On the other hand, with weighted least squares we want the quantity $L = \sum_{i=1}^{n} \omega_{ii}[Y_i - (b_0 + b_1 X_{1_i} + \cdots + b_p X_{p_i})]^2$ to be a minimum total with respect to $b_0, b_1, \ldots, b_p$. To achieve this, we set the *partial derivatives* equal to zero (that is, $\partial L/\partial b_0 = 0$, $\partial L/\partial b_1 = 0, \ldots, \partial L/\partial b_p = 0$) and obtain the set of $p + 1$ normal equations presented in (14.3), which yield the desired regression coefficients.

$$
\mathbf{W} =
\begin{bmatrix}
W_{11} & 0 & 0 & \cdots & 0 \\
0 & W_{22} & 0 & \cdots & 0 \\
0 & 0 & W_{33} & \cdots & 0 \\
\cdot & \cdot & \cdot & & \cdot \\
\cdot & \cdot & \cdot & & \cdot \\
\cdot & \cdot & \cdot & & \cdot \\
0 & 0 & 0 & \cdots & W_{nn}
\end{bmatrix}
\tag{14.5}
$$

Note again that when $\mathbf{W} = \mathbf{I}$, the identity matrix, we are back to the (original) assumption of homoscedasticity.

Choosing the appropriate set of elements W_{ii}, the estimated regression coefficients under weighted least squares are given by

$$
\hat{\boldsymbol{\beta}} = (\mathbf{X}'\mathbf{W}^{-1}\mathbf{X})^{-1}\mathbf{X}'\mathbf{W}^{-1}\mathbf{Y}
\tag{14.6}
$$

Note again that when $\mathbf{W} = \mathbf{I}$, (14.6) reduces to precisely the estimates given by OLS (see Sections 9.8.1 and 10.11.1).

Confidence interval estimates and tests of hypotheses as developed and discussed in previous chapters can readily be implemented under WLS. Toward this end, had we assumed that the residuals under OLS were normally distributed with mean 0 and variance $W_{ii}\sigma_{Y|X}^2$, then it would follow that under WLS the residuals are normally distributed with mean 0 and variance $\sigma_{Y|X}^2$. Hence all the inferential arguments previously developed can be replicated for the transformed (WLS) model.

14.3.3 Implementation of Weighted Least Squares

While the mathematical construction of the WLS process is straightforward, its implementation with data can be difficult. The problem is primarily associated with a determination of the appropriate set of elements W_{ii} to comprise the $\mathbf{W}$ matrix. One approach to the problem which has gained wide acceptance is to employ a two-stage procedure whereby the original data are used to estimate the elements, and then these elements are used to generate the estimates for the regression coefficients given in (14.6). While the idea of such an adaptive procedure is sensible, there are two problems with its use: The procedure requires replicated measurements on each explanatory variable, and inferential arguments become more complex due to the random nature of the elements which make up $\mathbf{W}$.

14.3.4 An Example

To illustrate how we may employ the method of WLS, let us consider the following multiple regression problem containing two explanatory variables: A sample of 10 interactive terminals is selected from a large educational computer facility, and information is obtained pertaining to the age of a terminal in years (X_1), the median number of hours per day a terminal is used (X_2), and the number of service calls per annum required for terminal maintenance (Y). The data are presented in Table 14.3. By applying the OLS methods of Chapter 10, the fol-

TABLE 14.3   Maintenance Study

No. of Service Calls Y	Age X_1	Usage X_2
3	1	10
4	2	7
5	3	10
6	4	8
7	5	10
10	5	10
5	4	8
6	3	10
3	2	7
5	1	10

lowing least-squares regression model is fitted:

$$\hat{Y}_i = -2.745 + 1.044 X_{1_i} + .557 X_{2_i}$$

In addition, the following pertinent summary statistics are obtained:

$$S_{Y|X_{12}} = 1.15 \qquad S_{b_0} = 2.664 \qquad S_{b_1} = .259 \qquad S_{b_2} = .290$$

$$r_{Y.12}^2 = .7577 \qquad r_{Y.12} = .87$$

$$\text{SSR} = \text{Explained Variation} = \sum_{i=1}^{n} (\hat{Y}_i - \bar{Y})^2 = 29.095$$

$$\text{SSE} = \text{Unexplained Variation} = \sum_{i=1}^{n} (Y_i - \hat{Y}_i)^2 = 9.305$$

$$\text{SST} = \text{Total Variation} = \sum_{i=1}^{n} (Y_i - \bar{Y})^2 = 38.400$$

By using the fitted regression model, the set of standardized residuals are obtained and displayed in Table 14.4. Figures 14.3 and 14.4, respectively, depict

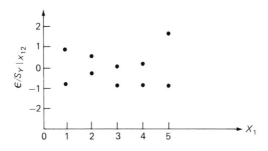

Figure 14.3   Plot of standardized residuals versus age (X_1).

TABLE 14.4 Obtaining the Standardized Residuals

Y_i	$\hat{Y}_i$	$\epsilon_i = Y_i - \hat{Y}_i$	$\epsilon_i / S_{Y|X_{12}}$
3	3.869	−.869	−.76
4	3.242	.758	.66
5	5.957	−.957	−.83
6	5.887	.113	.10
7	8.045	−1.045	−.91
10	8.045	1.955	1.70
5	5.887	−.887	−.77
6	5.957	.043	.04
3	3.242	−.242	−.21
5	3.869	1.131	.98

SOURCE: Table 14.3.

the residual scatter plots for each explanatory variable. While there does not appear to be any visual pattern to the plot of $\epsilon_i / S_{Y|X_{12}}$ versus X_1 (Figure 14.3), it is clear from Figure 14.4 that the (absolute) standardized residuals increase with increasing levels of X_2—indicating a violation in homoscedasticity (for that variable). Therefore, using the weights $\omega_{ii} = 1/X_{2_i}^2$, we develop the following normal equations:

$$\text{I.} \quad \sum_{i=1}^{n} \frac{Y_i}{X_{2_i}^2} = b_0 \sum_{i=1}^{n} \frac{1}{X_{2_i}^2} + b_1 \sum_{i=1}^{n} \frac{X_{1_i}}{X_{2_i}^2} + b_2 \sum_{i=1}^{n} \frac{1}{X_{2_i}}$$

$$\text{II.} \quad \sum_{i=1}^{n} \frac{X_{1_i} Y_i}{X_{2_i}^2} = b_0 \sum_{i=1}^{n} \frac{X_{1_i}}{X_{2_i}^2} + b_1 \sum_{i=1}^{n} \frac{X_{1_i}^2}{X_{2_i}^2} + b_2 \sum_{i=1}^{n} \frac{X_{1_i}}{X_{2_i}}$$

$$\text{III.} \quad \sum_{i=1}^{n} \frac{Y_i}{X_{2_i}} = b_0 \sum_{i=1}^{n} \frac{1}{X_{2_i}} + b_1 \sum_{i=1}^{n} \frac{X_{1_i}}{X_{2_i}} + b_2 n$$

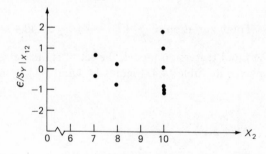

Figure 14.4 Plot of standardized residuals versus usage (X_2).

Based on the data in Table 14.3, the necessary computations are presented in Table 14.5. By solving the normal equations simultaneously, the following least-squares regression model is obtained:

$$\hat{Y}_i = -2.990 + 1.028 X_{1_i} + .591 X_{2_i}$$

of service calls per annum required for terminal maintenance (Y). The data are presented in Table 14.3. By applying the OLS methods of Chapter 10, the fol-

TABLE 14.3   Maintenance Study

No. of Service Calls Y	Age X_1	Usage X_2
3	1	10
4	2	7
5	3	10
6	4	8
7	5	10
10	5	10
5	4	8
6	3	10
3	2	7
5	1	10

lowing least-squares regression model is fitted:

$$\hat{Y}_i = -2.745 + 1.044X_{1_i} + .557X_{2_i}$$

In addition, the following pertinent summary statistics are obtained:

$$S_{Y|X_{12}} = 1.15 \qquad S_{b_0} = 2.664 \qquad S_{b_1} = .259 \qquad S_{b_2} = .290$$

$$r_{Y.12}^2 = .7577 \qquad r_{Y.12} = .87$$

$$\text{SSR} = \text{Explained Variation} = \sum_{i=1}^{n} (\hat{Y}_i - \bar{Y})^2 = 29.095$$

$$\text{SSE} = \text{Unexplained Variation} = \sum_{i=1}^{n} (Y_i - \hat{Y}_i)^2 = 9.305$$

$$\text{SST} = \text{Total Variation} = \sum_{i=1}^{n} (Y_i - \bar{Y})^2 = 38.400$$

By using the fitted regression model, the set of standardized residuals are obtained and displayed in Table 14.4. Figures 14.3 and 14.4, respectively, depict

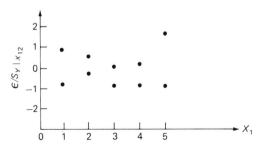

Figure 14.3   Plot of standardized residuals versus age (X_1).

TABLE 14.4 Obtaining the Standardized Residuals

Y_i	$\hat{Y}_i$	$\epsilon_i = Y_i - \hat{Y}_i$	$\epsilon_i/S_{Y|X_{12}}$
3	3.869	−.869	−.76
4	3.242	.758	.66
5	5.957	−.957	−.83
6	5.887	.113	.10
7	8.045	−1.045	−.91
10	8.045	1.955	1.70
5	5.887	−.887	−.77
6	5.957	.043	.04
3	3.242	−.242	−.21
5	3.869	1.131	.98

SOURCE: Table 14.3.

the residual scatter plots for each explanatory variable. While there does not appear to be any visual pattern to the plot of $\epsilon_i/S_{Y|X_{12}}$ versus X_1 (Figure 14.3), it is clear from Figure 14.4 that the (absolute) standardized residuals increase with increasing levels of X_2—indicating a violation in homoscedasticity (for that variable). Therefore, using the weights $\omega_{ii} = 1/X_{2_i}^2$, we develop the following normal equations:

$$\text{I.} \quad \sum_{i=1}^{n} \frac{Y_i}{X_{2_i}^2} = b_0 \sum_{i=1}^{n} \frac{1}{X_{2_i}^2} + b_1 \sum_{i=1}^{n} \frac{X_{1_i}}{X_{2_i}^2} + b_2 \sum_{i=1}^{n} \frac{1}{X_{2_i}}$$

$$\text{II.} \quad \sum_{i=1}^{n} \frac{X_{1_i}Y_i}{X_{2_i}^2} = b_0 \sum_{i=1}^{n} \frac{X_{1_i}}{X_{2_i}^2} + b_1 \sum_{i=1}^{n} \frac{X_{1_i}^2}{X_{2_i}^2} + b_2 \sum_{i=1}^{n} \frac{X_{1_i}}{X_{2_i}}$$

$$\text{III.} \quad \sum_{i=1}^{n} \frac{Y_i}{X_{2_i}} = b_0 \sum_{i=1}^{n} \frac{1}{X_{2_i}} + b_1 \sum_{i=1}^{n} \frac{X_{1_i}}{X_{2_i}} + b_2 n$$

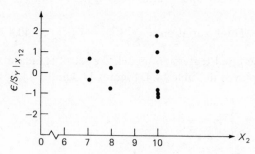

Figure 14.4 Plot of standardized residuals versus usage (X_2).

Based on the data in Table 14.3, the necessary computations are presented in Table 14.5. By solving the normal equations simultaneously, the following least-squares regression model is obtained:

$$\hat{Y}_i = -2.990 + 1.028X_{1_i} + .591X_{2_i}$$

TABLE 14.5   Necessary Computations for WLS Example

$\dfrac{Y_i}{X_{2_i}^2}$	$\dfrac{1}{X_{2_i}^2}$	$\dfrac{X_{1_i}}{X_{2_i}^2}$	$\dfrac{1}{X_{2_i}}$	$\dfrac{X_{1_i}}{X_{2_i}}$	$\dfrac{X_{1_i}^2}{X_{2_i}^2}$	$\dfrac{X_{1_i}Y_i}{X_{2_i}^2}$	$\dfrac{Y_i}{X_{2_i}}$	$\dfrac{Y_i^2}{X_{2_i}^2}$
.030	.010	.010	.100	.100	.010	.030	.300	.090
.082	.020	.041	.143	.286	.082	.163	.571	.327
.050	.010	.030	.100	.300	.090	.150	.500	.250
.094	.016	.063	.125	.500	.250	.375	.750	.563
.070	.010	.050	.100	.500	.250	.350	.700	.490
.100	.010	.050	.100	.500	.250	.500	1.000	1.000
.078	.016	.063	.125	.500	.250	.313	.625	.391
.060	.010	.030	.100	.300	.090	.180	.600	.360
.061	.020	.041	.143	.286	.082	.122	.429	.184
.050	.010	.010	.100	.100	.010	.050	.500	.250
.675	.132	.388	1.136	3.372	1.364	2.233	5.975	3.905

SOURCE: Table 14.3.

Of course, had we used a matrix approach (where the elements W_{ii} comprising the $\mathbf{W}$ matrix are given by $W_{ii} = X_{2_i}^2$), then for our data

$$\mathbf{Y} = \begin{bmatrix} 3 \\ 4 \\ \cdot \\ \cdot \\ \cdot \\ 5 \end{bmatrix} \qquad \mathbf{X} = \begin{bmatrix} 1 & 1 & 10 \\ 1 & 2 & 7 \\ \cdot & \cdot & \cdot \\ \cdot & \cdot & \cdot \\ \cdot & \cdot & \cdot \\ 1 & 1 & 10 \end{bmatrix} \qquad \mathbf{W}^{-1} = \begin{bmatrix} .010 & 0 & \cdots & 0 \\ 0 & .020 & \cdots & 0 \\ \cdot & \cdot & \cdot & \cdot \\ \cdot & \cdot & \cdot & \cdot \\ 0 & 0 & \cdots & .010 \end{bmatrix}$$

and the regression coefficients would have been obtained from

$$\hat{\boldsymbol{\beta}} = (\mathbf{X}'\mathbf{W}^{-1}\mathbf{X})^{-1}\mathbf{X}'\mathbf{W}^{-1}\mathbf{Y}$$

$$\hat{\boldsymbol{\beta}} = \begin{bmatrix} \sum_{i=1}^{n} \dfrac{1}{X_{2_i}^2} & \sum_{i=1}^{n} \dfrac{X_{1_i}}{X_{2_i}^2} & \sum_{i=1}^{n} \dfrac{1}{X_{2_i}} \\ \sum_{i=1}^{n} \dfrac{X_{1_i}}{X_{2_i}^2} & \sum_{i=1}^{n} \dfrac{X_{1_i}^2}{X_{2_i}^2} & \sum_{i=1}^{n} \dfrac{X_{1_i}}{X_{2_i}} \\ \sum_{i=1}^{n} \dfrac{1}{X_{2_i}} & \sum_{i=1}^{n} \dfrac{X_{1_i}}{X_{2_i}} & n \end{bmatrix}^{-1} \begin{bmatrix} \sum_{i=1}^{n} \dfrac{Y_i}{X_{2_i}^2} \\ \sum_{i=1}^{n} \dfrac{X_{1_i}Y_i}{X_{2_i}^2} \\ \sum_{i=1}^{n} \dfrac{Y_i}{X_{2_i}} \end{bmatrix} = \begin{bmatrix} -2.990 \\ 1.028 \\ .591 \end{bmatrix} \begin{matrix} \leftarrow b_0 \\ \leftarrow b_1 \\ \leftarrow b_2 \end{matrix}$$

The variance $(S_{Y|X_{12}}^2)$ is given by

$$S_{Y|X_{12}}^2 = \frac{1}{n-3}(\mathbf{Y}'\mathbf{W}^{-1}\mathbf{Y} - \hat{\boldsymbol{\beta}}'\mathbf{X}'\mathbf{W}^{-1}\mathbf{Y})$$

$$= \frac{\sum_{i=1}^{n}(Y_i^2/X_{2_i}^2) - b_0\sum_{i=1}^{n}(Y_i/X_{2_i}^2) - b_1\sum_{i=1}^{n}(X_{1_i}Y_i/X_{2_i}^2) - b_2\sum_{i=1}^{n}(Y_i/X_{2_i})}{n-3} \qquad (14.7)$$

Moreover, the variance of the regression coefficients are obtained from

$$\mathbf{S}_b = S_{Y|X_{12}}^2(\mathbf{X}'\mathbf{W}^{-1}\mathbf{X})^{-1} \qquad (14.8)$$

The following pertinent summary statistics are then computed:

$$S_{Y|X_{12}} = .11632 \qquad S_{b_0} = 2.182 \qquad S_{b_1} = .249 \qquad S_{b_2} = .249$$

Note that the standard errors of the estimated parameters are more precise under WLS than under OLS because of the violation in the homogeneity of variance assumption.

14.3.5 Computer Packages and Weighted Least Squares

Each of the three computer packages have control statements that enable weights to be applied to each observation. These weights can then be utilized as part of a regression program to perform WLS instead of OLS.

14.4 TREATING THE PROBLEM OF CORRELATED ERRORS

One of the assumptions of the basic regression model is the independence of the residual components. In particular, if ϵ_i and ϵ_j are the errors associated with the ith and jth observations, then they are assumed to be uncorrelated. When there is correlation in the errors, the basic model misses valuable information since there are no terms which can exploit the association.

Autocorrelation is a type of correlation that occurs when observations have a natural sequential order (such as hourly temperature readings or daily prices of particular stocks). Its occurrence is primarily attributable to the fact that adjacent residuals tend to be similar in the sense that large positive deviations are followed by other positive errors, and large negative differences are followed by other negative errors. When autocorrelated errors are present and no provision is made for this in the model, serious problems can arise in the analysis. The most compelling of these problems are the following:

1. While OLS estimates will still be unbiased, they will lose the favorable property of having minimum variance.
2. The estimate of $\sigma^2_{Y|X}$ and the various standard errors for the regression coefficients may be overly optimistic in the sense that their values will reflect smaller errors than is actually the case.
3. All inferential statements, that is, confidence intervals and hypothesis tests, will no longer be valid.

14.4.1 Detection of Autocorrelation

Plotting the residuals against time.   Perhaps the most straightforward approach for detecting autocorrelation for data collected over specified time periods is through the construction of graphical plots of residuals against time. Typically, when positive autocorrelation is present, a cluster of residuals of the same sign will occur—followed by another cluster of a different sign, etc.

Application.   To illustrate this phenomenon, let us consider the data set obtained by McNamara and Browne (1980). Table 14.6 presents quarterly data from March 1976 to March 1980 on the price of gold (response variable) and the price of petroleum (explanatory variable). The purpose here is to examine the relationship between the prices of gold and "black gold" over this period of time, and a regression equation was developed so that the former could be predicted based on a knowledge of the latter. For these data the fitted model is

$$\text{Gold} = -151.63 + 36.24 \text{ (Petroleum)}$$

TABLE 14.6   Study of U.S. Petroleum and Gold Prices

Month	Petroleum Price ($/barrel)	Gold Price ($/ounce)
March 1976	7.79	133.1
June 1976	7.99	126.2
September 1976	8.39	114.7
December 1976	8.55	134.4
March 1977	8.45	148.6
June 1977	8.44	140.8
September 1977	8.63	150.1
December 1977	8.75	161.1
March 1978	8.80	184.1
June 1978	9.05	184.1
September 1978	9.12	212.4
December 1978	9.27	208.1
March 1979	9.83	242.4
June 1979	11.70	279.4
September 1979	14.57	357.2
December 1979	17.03	459.0
March 1980	19.35	553.6

SOURCE: A. M. McNamara and J. J. Browne, "An Interesting Correlation—United States Oil and Gold Prices," *The New York Statistician*, Vol. 32, No. 2 (1980), p. 2.

The r^2 for this model is .974, and both regression coefficients are highly significant. On the surface it would appear that the analysis is complete and that a good-fitting model has been found. However, a plot of the residuals against time as shown in Figure 14.5 strongly suggests the presence of autocorrelation in the residuals. We can clearly identify three distinct clusters of residuals, each cluster having the same sign, while contiguous clusters have the opposite sign.

The Durbin-Watson procedure.   In addition to graphical displays, such autocorrelated errors can also be detected using the Durbin-Watson procedure. This procedure is based on the assumption that the residuals follow a first-order autoregressive model; that is, $\epsilon_i = \rho_a \epsilon_{i-1} + \delta_i$, where the δ_i are

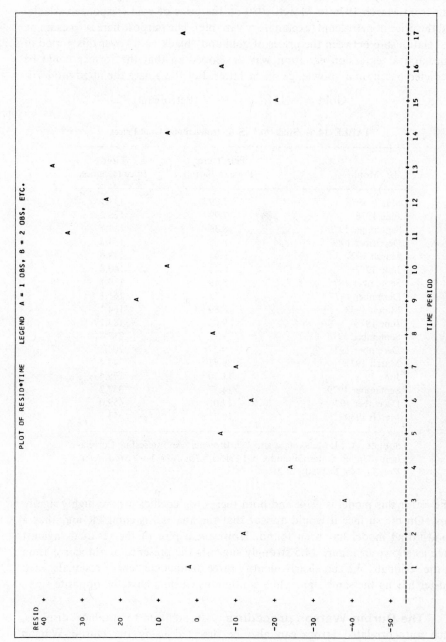

Figure 14.5 SAS plot of residuals versus time.

independently and identically distributed as $\mathcal{N}(0,\sigma^2_{Y|X})$. Absence of autocorrelation occurs when $\rho_a = 0$.

The basis of this procedure is the statistic

$$d = \frac{\sum_{i=2}^{n} (\epsilon_i - \epsilon_{i-1})^2}{\sum_{i=1}^{n} \epsilon_i^2} \tag{14.9}$$

The Durbin-Watson statistic d was devised to test the hypothesis $H_0: \rho_a = 0$ versus $H_1: \rho_a > 0$. Note that the alternative is for positive autocorrelation. Rarely are tests for negative autocorrelation performed; indeed, situations where negative autocorrelation occurs are uncommon [for further discussion, see Ostrom (1978)].

Published tables for critical values of the Durbin-Watson statistic are available (see Appendix B, Table B.9). For a given significance level, the table is entered according to the number of explanatory variables (p) and the sample size. Upper and lower critical values d_u and d_l are given, and the formal test is based on the following decision rules:

If $d < d_l$, reject H_0.

If $d > d_u$, do not reject H_0.

If $d_l \leq d \leq d_u$, the test is inconclusive.

To have a more complete conceptual understanding of what the Durbin-Watson procedure accomplishes, we should take a closer look at the composition of the d statistic presented in (14.9). If we expand the numerator $\sum_{i=2}^{n} (\epsilon_i - \epsilon_{i-1})^2$, we obtain $\sum_{i=2}^{n} \epsilon_i^2 + \sum_{i=2}^{n} \epsilon_{i-1}^2 - 2 \sum_{i=2}^{n} \epsilon_i \epsilon_{i-1}$. Dividing each of these terms by $\sum_{i=1}^{n} \epsilon_i^2$, the denominator in (14.9), gives us (approximately)

$$d \cong 2 - \frac{2 \sum_{i=2}^{n} \epsilon_i \epsilon_{i-1}}{\sum_{i=1}^{n} \epsilon_i^2} \tag{14.10}$$

Since the term $\sum_{i=2}^{n} \epsilon_i \epsilon_{i-1}$ will be close to $\sum_{i=1}^{n} \epsilon_i^2$ when successive residuals are positively (auto)correlated, the Durbin-Watson statistic d will approach zero in these situations (i.e., by using Table B.9 in Appendix B, if $d < d_l$, then the null hypothesis $H_0: \rho_a = 0$ is rejected). On the other hand, if the residuals are not correlated, $\sum_{i=2}^{n} \epsilon_i \epsilon_{i-1} \cong 0$, and the value of d will be close to 2. Indeed, there exists an approximate relationship between d and the magnitude of the *autocorrelation parameter* ρ_a given by

$$d \cong 2(1 - \rho_a) \tag{14.11}$$

From this approximation we see that d is close to 2 when $\rho_a = 0$ and near to zero when $\rho_a = 1$. The closer d is to 2, the more firm the evidence is of no positive autocorrelation (i.e., from Table B.9 in Appendix B, if $d > d_u$, then $H_0: \rho_a = 0$

is not rejected). More formally, however, the actual magnitude of the autocorrelation parameter ρ_a can be estimated from

$$\hat{\rho}_a = \frac{\sum_{i=2}^{n} \epsilon_i \epsilon_{i-1}}{\sum_{i=2}^{n} \epsilon_{i-1}^2} \tag{14.12}$$

As we shall see in the following section, such an estimate becomes important when we wish to remove the deleterious effects of autocorrelation once residual analysis or the Durbin-Watson procedure has made known its presence.

14.4.2 Removing Autocorrelation: The Cochrane-Orcutt Method

Cochrane and Orcutt (1949) suggested a procedure for the removal of autocorrelation. Assuming that the residuals follow a first-order autoregressive model, i.e., $\epsilon_i = \rho_a \epsilon_{i-1} + \delta_i$, they proposed transforming Y_i to $Y_i^* = Y_i - \rho_a Y_{i-1}$, X_i to $X_i^* = X_i - \rho_a X_{i-1}$, β_0 to $\beta_0^* = \beta_0 - \rho_a \beta_0$, and ϵ_i to $\delta_i = \epsilon_i - \rho_a \epsilon_{i-1}$ in the model

$$Y_i = \beta_0 + \beta_1 X_i + \epsilon_i \tag{14.13}$$

The effect of this transformation on (14.13) leads to the model

$$Y_i^* = \beta_0^* + \beta_1^* X_i^* + \delta_i \tag{14.14}$$

whose residuals δ_i satisfy the assumptions of the standard linear model.

The Cochrane-Orcutt method is an iterative procedure which works as follows: Using (14.13), we first estimate β_0 and β_1 by methods of OLS. We then calculate the residuals and obtain an estimate of the autocorrelation parameter ρ_a from (14.12). Using the variables $Y_i^* = Y_i - \hat{\rho}_a Y_{i-1}$ and $X_i^* = X_i - \hat{\rho}_a X_{i-1}$, we then fit the model given in (14.14) [the estimates of the parameters in the original equations are $\hat{\beta}_0 = \hat{\beta}_0^*/(1 - \hat{\rho}_a)$ and $\hat{\beta}_1 = \hat{\beta}_1^*$]. If the residuals in the second fitted model show no autocorrelation, we terminate the process; if not, we repeat the process again, but instead of using the OLS estimates, we use $\hat{\beta}_0^*/(1 - \hat{\rho}_a)$ and $\hat{\beta}_1^*$ as starting values.

To demonstrate the use of the Durbin-Watson and Cochrane-Orcutt procedures, we refer to the data set from McNamara and Browne (1980) presented in Table 14.6. The computations for d and $\hat{\rho}_a$ are displayed in Table 14.7. From this we note that the value of the Durbin-Watson statistic is .7595 and the estimate of the autocorrelation parameter $\hat{\rho}_a$ is .6201. Since d is significant, we go through the Cochrane-Orcutt procedure (see Table 14.8) using the transformed variables $Y_i^* = Y_i - .6201 Y_{i-1}$ and $X_i^* = X_i - .6201 X_{i-1}$. The refitted equation yields a d value equal to 1.7559, and hence the hypothesis $\rho_a = 0$ is not rejected at the .05 level. We are satisfied that the autocorrelation has been removed.

Table 14.9 presents a summary of the regression estimates obtained for the gold-"black gold" data by use of ordinary least squares and by the Cochrane-

TABLE 14.7 Computing the Durbin-Watson Statistic and Estimating the Autocorrelation Parameter

Observation i	Petroleum X_i	Gold Y_i	Predicted Value $\hat{Y}_i$	Residual $\epsilon_i = Y_i - \hat{Y}_i$	ϵ_{i-1}	$\epsilon_i - \epsilon_{i-1}$	$(\epsilon_i - \epsilon_{i-1})^2$	ϵ_i^2	ϵ_{i-1}^2	$\epsilon_i\epsilon_{i-1}$
1	7.79	133.1	130.647	2.453	—	—	—	6.017209	—	—
2	7.99	126.2	137.894	-11.694	2.453	-14.147	200.137609	136.749636	6.017209	-28.685382
3	8.39	114.7	152.389	-37.689	-11.694	-25.995	675.740025	1420.460721	136.749636	440.735166
4	8.55	134.4	158.187	-23.787	-37.689	13.902	193.265604	565.821369	1420.460721	896.508243
5	8.45	148.6	154.563	-5.963	-23.787	17.824	317.694976	35.557369	565.821369	141.841881
6	8.44	140.8	154.201	-13.401	-5.963	-7.438	55.323844	179.586801	35.557369	79.910163
7	8.63	150.1	161.085	-10.985	-13.401	2.416	5.837056	120.670225	179.586801	147.209985
8	8.75	161.1	165.434	-4.334	-10.985	6.651	44.235801	18.783556	120.670225	47.608990
9	8.80	184.1	167.246	16.854	-4.334	21.188	448.931344	284.057316	18.783556	-73.045236
10	9.05	184.1	176.305	7.795	16.854	-9.059	82.065481	60.762025	284.057316	131.376930
11	9.12	212.4	178.841	33.559	7.795	25.764	663.783696	1126.206481	60.762025	261.592405
12	9.27	208.1	184.277	23.823	33.559	-9.736	94.789696	567.535329	1126.206481	799.476057
13	9.83	242.4	204.569	37.831	23.823	14.008	196.224064	1431.184561	567.535329	901.247913
14	11.70	279.4	272.330	7.070	37.831	-30.761	946.239121	49.984900	1431.184561	267.465170
15	14.57	357.2	376.328	-19.128	7.070	-26.198	686.335204	365.880384	49.984900	-135.234960
16	17.03	459.0	465.469	-6.469	-19.128	12.659	160.250281	41.847961	365.880384	123.739032
17	19.35	553.6	549.537	4.063	-6.469	10.532	110.923024	16.507969	41.847961	-26.283547
							4881.776826	6427.613812	6411.105843	3975.462810

Summary statistics:

$b_0 = -151.632$ $S_{b_0} = 16.542$

$b_1 = 36.236$ $S_{b_1} = 1.525$

$r^2 = .974$ $S_{Y|X} = 20.700$

$n = 17$

$$d = \frac{4881.776826}{6427.613812} = .7595 \qquad \hat{\rho}_a = \frac{3975.462810}{6411.105843} = .6201$$

From Table B.9 in Appendix B, for $\alpha = .05$, $n = 17$, and $p = 1$, we observe $d_l = 1.13$. Since $d < d_l$, reject H_0: $\rho_a = 0$ and conclude that $\rho_a > 0$.

SOURCE: Table 14.6.

TABLE 14.8 Removing the Autocorrelation by the Cochrane-Orcutt
Procedure ($\hat{\rho}_a = .6201$)

Time Period i	$X_i^* = X_i - \hat{\rho}_a X_{i-1}$	$Y_i^* = Y_i - \hat{\rho}_a Y_{i-1}$	$\hat{Y}_i^*$	$\delta_i = Y_i^* - \hat{Y}_i^*$
1	—	—	—	—
2	3.15942	43.665	57.822	−14.157
3	3.43540	36.443	67.619	−31.176
4	3.34736	63.275	64.494	−1.219
5	3.14815	65.259	57.421	7.838
6	3.20016	48.653	59.268	−10.615
7	3.39636	62.790	66.233	−3.443
8	3.39854	68.023	66.311	1.712
9	3.37413	84.202	65.444	18.758
10	3.59312	69.940	73.219	−3.279
11	3.50810	98.240	70.200	28.040
12	3.61469	76.391	73.984	2.407
13	4.08167	113.357	90.563	22.794
14	5.60442	129.088	144.622	−15.534
15	7.31483	183.944	205.344	−21.400
16	7.99514	237.500	229.496	8.004
17	8.78970	268.974	257.703	11.271

Summary statistics:

$$\hat{\beta}_0^* = -54.3415 \qquad S_{\beta_0^*} = 10.9472$$
$$\hat{\beta}_1^* = 35.5012 \qquad S_{\beta_1^*} = 2.2806$$
$$r^2 = .945 \qquad S_{Y|X} = 16.7488$$
$$n = 16$$
$$\sum_{i=3}^{17} (\delta_i - \delta_{i-1})^2 = 6895.940281$$
$$\sum_{i=2}^{17} \delta_i^2 = 3927.296703$$
$$d = 1.7559$$

From Table B.9 in Appendix B, for $\alpha = .05$, $n = 16$, and $p = 1$, we observe $d_u = 1.37$.
Since $d > d_u$, do not reject H_0: $\rho_a = 0$. Thus,

$$\hat{\beta}_0 = \hat{\beta}_0^*/(1 - \hat{\rho}_a) = -54.3415/(1 - .6201) = -143.0415$$
$$\hat{\beta}_1 = \hat{\beta}_1^* = 35.5012$$

SOURCE: Tables 14.6 and 14.7.

Orcutt method as well as by SAS PROC AUTOREG [see SAS (1979) and
Figures 14.6 and 14.7]. It should be noted that although the three methods all
led to highly significant, good-fitting regression models for the gold-"black
gold" data, this will not always be the case. While major discrepancies between
the Cochrane-Orcutt and the SAS AUTOREG iterative estimation procedures
should not be expected, their generated results (see Table 14.9) are far more
appropriate than those obtained from ordinary least-squares procedures whenever significant autocorrelation is present.
  It is interesting to point out that the standard error of the slope parameter

TABLE 14.7 Computing the Durbin-Watson Statistic and Estimating the Autocorrelation Parameter

Observation i	Petroleum X_i	Gold Y_i	Predicted Value $\hat{Y}_i$	Residual $\epsilon_i = Y_i - \hat{Y}_i$	ϵ_{i-1}	$\epsilon_i - \epsilon_{i-1}$	$(\epsilon_i - \epsilon_{i-1})^2$	ϵ_i^2	ϵ_{i-1}^2	$\epsilon_i \epsilon_{i-1}$
1	7.79	133.1	130.647	2.453	—	—	—	6.017209	—	—
2	7.99	126.2	137.894	−11.694	2.453	−14.147	200.137609	136.749636	6.017209	−28.685382
3	8.39	114.7	152.389	−37.689	−11.694	−25.995	675.740025	1420.460721	136.749636	440.735166
4	8.55	134.4	158.187	−23.787	−37.689	13.902	193.265604	565.821369	1420.460721	896.508243
5	8.45	148.6	154.563	−5.963	−23.787	17.824	317.694976	35.557369	565.821369	141.841881
6	8.44	140.8	154.201	−13.401	−5.963	−7.438	55.323844	179.586801	35.557369	79.910163
7	8.63	150.1	161.085	−10.985	−13.401	2.416	5.837056	120.670225	179.586801	147.209985
8	8.75	161.1	165.434	−4.334	−10.985	6.651	44.235801	18.783556	120.670225	47.608990
9	8.80	184.1	167.246	16.854	−4.334	21.188	448.931344	284.057316	18.783556	−73.045236
10	9.05	184.1	176.305	7.795	16.854	−9.059	82.065481	60.762025	284.057316	131.376930
11	9.12	212.4	178.841	33.559	7.795	25.764	663.783696	1126.206481	60.762025	261.592405
12	9.27	208.1	184.277	23.823	33.559	−9.736	94.789696	567.535329	1126.206481	799.476057
13	9.83	242.4	204.569	37.831	23.823	14.008	196.224064	1431.184561	567.535329	901.247913
14	11.70	279.4	272.330	7.070	37.831	−30.761	946.239121	49.984900	1431.184561	267.465170
15	14.57	357.2	376.328	−19.128	7.070	−26.198	686.335204	365.880384	49.984900	−135.234960
16	17.03	459.0	465.469	−6.469	−19.128	12.659	160.250281	41.847961	365.880384	123.739032
17	19.35	553.6	549.537	4.063	−6.469	10.532	110.923024	16.507969	41.847961	−26.283547
							4881.776826	6427.613812	6411.105843	3975.462810

Summary statistics:

$b_0 = -151.632 \qquad S_{b_0} = 16.542$

$b_1 = 36.236 \qquad S_{b_1} = 1.525$

$r^2 = .974 \qquad S_{Y|x} = 20.700$

$n = 17$

$$d = \frac{4881.776826}{6427.613812} = .7595 \qquad \hat{\rho}_a = \frac{3975.462810}{6411.105843} = .6201$$

From Table B.9 in Appendix B, for $\alpha = .05$, $n = 17$, and $p = 1$, we observe $d_1 = 1.13$. Since $d < d_1$, reject H_0: $\rho_a = 0$ and conclude that $\rho_a > 0$.

SOURCE: Table 14.6.

Time Period i	$X_i^* = X_i - \hat{\rho}_a X_{i-1}$	$Y_i^* = Y_i - \hat{\rho}_a Y_{i-1}$	$\hat{Y}_i^*$	$\delta_i = Y_i^* - \hat{Y}_i^*$
1	—	—	—	—
2	3.15942	43.665	57.822	−14.157
3	3.43540	36.443	67.619	−31.176
4	3.34736	63.275	64.494	−1.219
5	3.14815	65.259	57.421	7.838
6	3.20016	48.653	59.268	−10.615
7	3.39636	62.790	66.233	−3.443
8	3.39854	68.023	66.311	1.712
9	3.37413	84.202	65.444	18.758
10	3.59312	69.940	73.219	−3.279
11	3.50810	98.240	70.200	28.040
12	3.61469	76.391	73.984	2.407
13	4.08167	113.357	90.563	22.794
14	5.60442	129.088	144.622	−15.534
15	7.31483	183.944	205.344	−21.400
16	7.99514	237.500	229.496	8.004
17	8.78970	268.974	257.703	11.271

Summary statistics:

$$\hat{\beta}_0^* = -54.3415 \qquad S_{\hat{\beta}_0^*} = 10.9472$$
$$\hat{\beta}_1^* = 35.5012 \qquad S_{\hat{\beta}_1^*} = 2.2806$$
$$r^2 = .945 \qquad S_{Y|X} = 16.7488$$
$$n = 16$$
$$\sum_{i=3}^{17} (\delta_i - \delta_{i-1})^2 = 6895.940281$$
$$\sum_{i=2}^{17} \delta_i^2 = 3927.296703$$
$$d = 1.7559$$

From Table B.9 in Appendix B, for $\alpha = .05$, $n = 16$, and $p = 1$, we observe $d_u = 1.37$. Since $d > d_u$, do not reject H_0: $\rho_a = 0$. Thus,

$$\hat{\beta}_0 = \hat{\beta}_0^*/(1 - \hat{\rho}_a) = -54.3415/(1 - .6201) = -143.0415$$
$$\hat{\beta}_1 = \hat{\beta}_1^* = 35.5012$$

SOURCE: Tables 14.6 and 14.7.

Orcutt method as well as by SAS PROC AUTOREG [see SAS (1979) and Figures 14.6 and 14.7]. It should be noted that although the three methods all led to highly significant, good-fitting regression models for the gold-"black gold" data, this will not always be the case. While major discrepancies between the Cochrane-Orcutt and the SAS AUTOREG iterative estimation procedures should not be expected, their generated results (see Table 14.9) are far more appropriate than those obtained from ordinary least-squares procedures whenever significant autocorrelation is present.

It is interesting to point out that the standard error of the slope parameter

TABLE 14.9 Comparing the Regression Estimates

Parameter Being Estimated	Method		
	Ordinary Least Squares	Cochrane-Orcutt	Iterative Estimation (SAS)
β_0	−151.632	−143.042	−143.326
β_1	36.236	35.501	35.519
Standard Error for β_1	1.525	2.281	2.064
ρ_a	— — — —	.620	.619
Coefficient of Determination	.974	.945	.952

SOURCE: Tables 14.7 and 14.8 and Figure 14.7.

in each of the refitted models is about 1.5 times as large as the standard error for the same parameter in the OLS fitted model. This, then, is a good example of how the researcher can be lulled into believing that estimates are efficient when indeed they only appear so because of the failure to account for correlated errors.

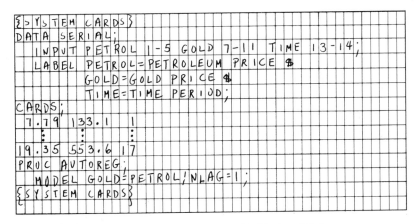

Figure 14.6 SAS program for gold-"black gold" data.

14.4.3 Using SAS PROC AUTOREG for Autocorrelated Regression

As mentioned in Section 14.4.2, SAS AUTOREG is an iterative estimation procedure which yields results similar to the Cochrane-Orcutt method for evaluating autocorrelated data in regression. The format for this procedure step is

PROCbAUTOREG;

$$\text{MODELb} \left\{ \begin{array}{l} \text{dependent} \\ \text{variable} \end{array} \right\} = \left\{ \begin{array}{l} \text{list of} \\ \text{independent} \\ \text{variables} \end{array} \right\} / \text{NLAG} = \{\text{value}\};$$

The "NLAG =" option of the model statement indicates the order of the autoregressive process. If it is not specified, an ordinary least-squares analysis will be performed. Figure 14.6 presents the complete SAS program that has been written to perform an autocorrelated regression analysis for the gold-"black gold" data. Figure 14.7 represents partial output for these data.

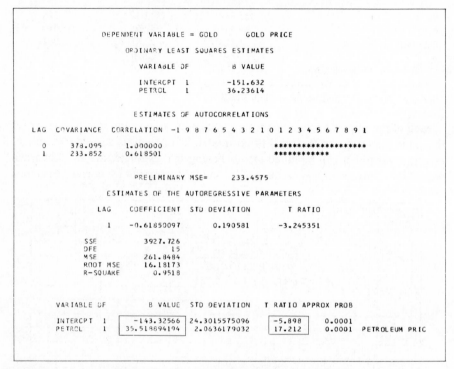

Figure 14.7   SAS output using PROC AUTOREG.

14.5 THE PROBLEM OF MULTICOLLINEARITY

Interpretation of a multiple regression analysis is best served when the explanatory variables comprising the model are themselves uncorrelated. When strong interrelationships exist, it is difficult if not impossible to assess the unique effects individual explanatory variables have upon the response variable. Nonorthogonality in the explanatory variables can result in such highly unstable regression coefficients that their values will be subject to dramatic alterations as a result of additions or deletions of variables or small changes in data points.

Multicollinearity is the term usually applied to the condition of strong interrelationships between the explanatory variables. It is not a problem of misspecification but more a problem of the data itself, and hence extreme caution should be exercised in interpreting the meaning of all estimates generated.

There are a number of signals by which we can red-flag multicollinearity. The most obvious indicator is the correlation matrix calculated from the set of explanatory variables. Large correlations point to strong linear associations, implying that certain variables may be surrogates for others with little or no effect as predictors themselves.

If we are convinced that the model has been appropriately specified, then two other indicators of multicollinearity may be helpful: (1) if the signs of certain regression coefficients are the reverse of what we would expect or (2) if important regression coefficients turn out to have large standard errors. Either of these two conditions should concern the researcher and lead to a serious investigation into questions of multicollinearity.

Identifying those linear combinations of the β parameters which can be accurately estimated is particularly problematic in the presence of multicollinearity. While in general we cannot eliminate the problem, there is an approach whereby the researcher works with a model induced through transforming the original variable set into a new and uncorrelated set. This approach, called *principal component analysis**, is a powerful technique—both in the detection of multicollinearity and as a vehicle in identifying those linear combinations of the regression coefficients that can be accurately estimated. For an in-depth discussion of principal components regression, see Goldstein and Dillon (1983).

PROBLEMS

***14.1.** Friedman and Meiselman (1963) examined the relationship between consumer expenditure (response variable) and the stock of money (explanatory variable). Both variables were measured in billions of current dollars for the United States. Table P14.1 presents quarterly data from 1952 to 1956:
  (a) Using ordinary least squares, fit a simple linear regression equation to these data.
  (b) Discuss your results in terms of:
    (1) $\hat{\beta}_1$
    (2) r^2
    (3) the significance of the fitted model.
  (c) Determine whether or not the assumption of independence of residuals necessary for ordinary least-squares analysis appears to have been violated
    (1) by plotting the residuals against time
    (2) by performing the Durbin-Watson test
    Discuss your findings.
  (d) If appropriate, remove the autocorrelation by using:
    (1) the Cochrane-Orcutt method
    (2) the SAS AUTOREG iterative estimation procedure.

*The method of principal components as a dimensionality reduction procedure is discussed in depth in Chapter 15.

TABLE P14.1   Consumer Expenditure and Money Stock

Year	Quarter	Consumer Expenditure	Money Stock	Year	Quarter	Consumer Expenditure	Money Stock
1952	1	214.6	159.3	1954	3	238.7	173.9
	2	217.7	161.2		4	243.2	176.1
	3	219.6	162.8	1955	1	249.4	178.0
	4	227.2	164.6		2	254.3	179.1
1953	1	230.9	165.9		3	260.9	180.2
	2	233.3	167.9		4	263.3	181.2
	3	234.1	168.3	1956	1	265.6	181.6
	4	232.3	169.7		2	268.2	182.5
1954	1	233.7	170.5		3	270.4	183.3
	2	236.5	171.6		4	275.6	184.3

SOURCE: M. Friedman and D. Meiselman, "The Relative Stability of Monetary Velocity and the Investment Multiplier in the United States, 1897–1958," in *Commission on Money and Credit Stabilization Policies*, Prentice-Hall, Englewood Cliffs, N.J., 1963, p. 266.

    (e) Set up a summary table similar to Table 14.9. Compare the results based on the three fitted models, and state your conclusions.

14.2. Table P14.2 presents monthly data for the 34-month period November 1977–August 1980 on the total number of industrial and commercial failures (response variable) and the end-of-month discount rate (explanatory variable) of the New York Federal Reserve Bank.

    (a) Using ordinary least squares, fit a simple linear regression equation to these data.

    (b) Discuss your results in terms of:

        (1) $\hat{\beta}_1$

        (2) r^2

        (3) the significance of the fitted model.

    (c) Determine whether or not the assumption of independence of residuals necessary for ordinary least-squares analysis appears to have been violated:

        (1) by plotting the residuals against time

        (2) by performing the Durbin-Watson test

    Discuss your findings.

    (d) If appropriate, remove the autocorrelation by using:

        (1) the Cochrane-Orcutt method

        (2) the SAS AUTOREG iterative estimation procedure

    (e) Set up a summary table similar to Table 14.9. Compare the results based on the three fitted models, and state your conclusions.

14.3. Reviewing concepts:

    (a) Why is residual analysis such an important tool for the researcher?

    (b) What is meant by *weighted least-squares* methods as opposed to OLS methods?

    (c) What is the purpose of the Durbin-Watson procedure?

    (d) What is meant by multicollinearity?

TABLE P14.2   Discount Rate and Business Failures

Month-Year	Discount Rate (%)	No. of Failures	Month-Year	Discount Rate (%)	No. of Failures
11-77	6.00	621	4-79	9.50	734
12-77	6.00	517	5-79	9.50	708
1-78	6.37	504	6-79	9.50	602
2-78	6.50	559	7-79	9.69	565
3-78	6.50	666	8-79	10.24	736
4-78	6.50	594	9-79	10.70	505
5-78	6.84	583	10-79	11.77	767
6-78	7.00	519	11-79	12.00	519
7-78	7.23	459	12-79	12.00	509
8-78	7.43	675	1-80	12.00	729
9-78	7.83	458	2-80	12.52	677
10-78	8.26	511	3-80	13.00	925
11-78	9.50	556	4-80	13.00	1068
12-78	9.50	535	5-80	12.94	975
1-79	9.50	642	6-80	11.40	1094
2-79	9.50	545	7-80	10.87	1141
3-79	9.50	732	8-80	10.00	1009

SOURCE: Data are taken from Series S-6 and S-15, *Survey of Current Business* and *Business Statistics Supplement*, U.S. Department of Commerce.

Part III

MULTIVARIATE METHODS

In the next few chapters we begin a discussion that is different from the material discussed thus far. Heretofore all the topics we have presented fall under a class of problems whereby the data analyzed result from measurements on only one attribute or response variable Y.* Interestingly, though, perhaps the more common situation in practice is for data to consist of measurements on a set of different variables. For example, students applying for admission to college present a profile of themselves by way of high school average, SAT scores, standing in class, etc. Thus each student is represented by a vector of measurements, the components of which are the individual variables associated with his or her high school experience.

In a more general sense, multivariate problems are those that are concerned with the analysis of n points in p-space, i.e., where each of n individuals or experimental units has associated with it a p-dimensional set of responses. A distinguishing feature which we exploit in the ensuing discussion is that while one may choose to consider the p-dimensional observations from person to person as being statistically independent, the observed components within each vector will usually be associated in some sense.

*In multiple regression, for example, there is but one response variable Y and a set of *fixed* explanatory variables.

Gnanadesikan (1977) points out that many researchers question the usefulness of multivariate methods as a response to difficulties and frustrations in implementing certain procedures in practice which far exceed those encountered in applying univariate methods. In particular he notes the following:

1. Because of the nature of the data, it is often difficult to know what one really wants to do. While this can also be said for univariate problems, the issues are compounded in the multivariate case.
2. Since researchers usually have many attributes or variables which can potentially be measured, there is no obvious natural value of p. Since the sample size is usually constrained by a host of considerations, the relative magnitude of p and n becomes crucial in determining what analyses or insights can be made.
3. Even with high-speed computers, many multivariate techniques are severely limited due to considerations relating to the magnitude of p and n.
4. Graphical representation of multivariate data is limited, although there have been recent strides attempting to fill the hiatus.
5. There is no unique linear ordering for variables in p-space. Some researchers find this to be very unsettling.

In writing a treatise on multivariate analysis, one would probably partition the work so that classes of similar problems are considered jointly. While ours is far from a complete treatment of multivariate methods, we have chosen to look at three basic problem areas: (1) procedures for dimensionality reduction, (2) some specific multivariate models, and (3) multidimensional classification.

It is our purpose in this text to focus primarily on the concepts and applications of these multivariate methods so that the researcher would be able to: (1) recognize a particular application situation; (2) conduct an appropriate multivariate analysis by accessing and utilizing a computer package; and (3) interpret the resulting computer output. Therefore, in our presentation only the essential formulas and mathematical expressions are shown since the theoretical aspects of multivariate analysis are more mathematically complex than that of our previous material. For those researchers wishing to focus on these aspects see Goldstein and Dillon (1983).

15

DIMENSIONALITY REDUCTION:
PRINCIPAL COMPONENTS
AND FACTOR ANALYSIS

15.1 INTRODUCTION

It is more the norm rather than the exception that social research deals with studies where measurements are multidimensional. Frequently the number of measurements for each subject is so large that analysis becomes difficult if not bewildering. The motivation for reducing the dimension when analyzing multiresponse data is a balance between attainment of parsimony for understanding and interpretation and the retention of sufficient information for adequate analysis. Using some decision mechanism, the researcher usually will achieve a reduction in the dimension of a problem. Our purpose in this chapter is to discuss two widely used statistical procedures—principal components and factor analysis—which frequently lend great assistance in this reduction process.

Under what conditions should we consider the process of dimensionality reduction? Survey questionnaires frequently contain more information than is needed for the task; indeed, usually so many items are included that redundancies exist. In such cases we might want to screen questions as a first step to further analysis. We often attempt to stabilize the scales in survey instruments through the formation of *linear combinations* of the measurements. In general, these are fewer in number than the original measurements and will exhibit more stable statistical properties. With large numbers of measurements taken, the researcher probably would attempt to prespecify a *space* which might be used as the basis of discrimination or classification. While there are other more subtle reasons for reducing the dimension of a problem, those given here are probably the most common.

To appreciate how principal components analysis and factor analysis assist in the achievement of dimensionality reduction, we must observe their application in real problems.

15.1.1 An Illustrative Example

Before we embark upon a technical discussion of principal components and factor analysis (as representatives of a class of dimensionality reduction techniques) it is helpful to examine briefly the results and interpretation of a typical analysis. While there are many differences between the two techniques and different reasons for their use, the example to be discussed is the result of applying a procedure which leads to a common solution for both principal components and factor analysis. In general, the two techniques will not yield identical results (as we shall establish more formally later); however, we have chosen this approach so that the reader may gain an appreciation of the types of problems for which the techniques are applicable.

The Brief Psychiatric Rating Scale (BPRS) consists of an inventory of 18 psychological variables:

$X_1 =$ somatic concern $X_{10} =$ hostility

$X_2 =$ anxiety $X_{11} =$ suspiciousness

$X_3 =$ emotional withdrawal $X_{12} =$ hallucinatory behavior

$X_4 =$ conceptual disorganization $X_{13} =$ motor retardation

$X_5 =$ guilt feelings $X_{14} =$ uncooperativeness

$X_6 =$ tension $X_{15} =$ unusual thought content

$X_7 =$ mannerisms and posturing $X_{16} =$ blunted effect

$X_8 =$ grandiosity $X_{17} =$ excitement

$X_9 =$ depressive mood $X_{18} =$ disorientation

Each variable is evaluated on a seven-point scale:

1 = not present   4 = moderate   7 = extremely severe

2 = very mild   5 = moderately severe

3 = mild   6 = severe

The data used in this example consist of the ratings given to 145 diagnosed schizophrenic patients. Typically, in both linear reduction techniques, the data are converted into either a covariance matrix or a correlation matrix. In the case at hand an 18×18 correlation matrix was computed from the 145 responses on each of the 18 variables from the BPRS. One's objectives in such analyses are commonly multifaceted; however, we usually wish to determine whether there are uncorrelated variables (fewer in number than 18), each of which is defined as a different *composite* of the original set in such a manner that it (1) accounts for a "large" portion of total variability in the original set of 18 variables and (2) can assist in explaining the relationships among the 18 variables. Toward this end it was decided that 5 composite variables were sufficient to satisfy both conditions. Typically, the output generated from an appropriate computer package for these new composite variables will appear as in Figure 15.1.

FIGURE 15.1   Partial Computer Output for the Psychiatric Example

	Composite 1	Composite 2	Composite 3	Composite 4	Composite 5
X_1	**0.53803**	−0.00786	0.11905	0.10645	0.20675
X_2	**0.74394**	−0.03681	0.04760	0.45337	0.01423
X_3	−0.03396	**0.66087**	−0.23770	0.05728	0.42124
X_4	0.04314	**0.79422**	0.26829	−0.01305	−0.06369
X_5	**0.82071**	0.07886	0.02227	−0.12288	−0.12866
X_6	**0.64648**	0.15473	0.07617	0.46208	−0.03229
X_7	−0.09527	0.51292	**0.54977**	−0.01348	0.23092
X_8	0.05559	0.05416	**0.67634**	0.00238	−0.28711
X_9	**0.76241**	−0.24816	0.00256	0.12489	0.33988
X_{10}	0.03973	0.07726	0.13092	**0.86673**	0.06997
X_{11}	0.38924	−0.20444	0.31819	**0.67928**	0.14071
X_{12}	0.01742	0.00918	**0.78609**	0.30599	0.14241
X_{13}	0.24681	0.12263	0.05897	−0.09197	**0.80381**
X_{14}	0.12622	**0.63562**	0.10493	0.34679	0.13873
X_{15}	0.28545	0.15623	**0.71540**	0.32141	0.08358
X_{16}	−0.00426	**0.58258**	−0.03375	0.11129	0.45281
X_{17}	0.19801	0.19482	0.16503	**0.73096**	−0.29448
X_{18}	−0.10561	**0.76806**	0.07886	−0.08614	−0.17208

15.1.2 Interpreting the Results

The first composite variable, which we may designate as $Y_{(1)}$, can be expressed mathematically as

$$Y_{(1)} = .53803X_1 + .74394X_2 - .03396X_3 + \cdots - .10561X_{18}$$

The coefficients (i.e., the *loadings*) here represent the correlation of $Y_{(1)}$ with the respective original variables; thus, .74394 can be interpreted as the correlation of $Y_{(1)}$ with the variable anxiety (X_2). Typically we interpret the composite variables on the basis of those variables having strong loading patterns. Using this approach, the physician who performed the analysis interpreted the composite $Y_{(1)}$ as a "patient moods" variable in that somatic concern (X_1), anxiety (X_2), guilt feelings (X_5), tension (X_6), and depressive mood (X_9) all have similar loading profiles within $Y_{(1)}$.

The second composite $Y_{(2)}$ was interpreted as a "withdrawal" variable with emotional withdrawal (X_3), conceptual disorganization (X_4), uncooperativeness (X_{14}), blunted effect (X_{16}), and disorientation (X_{18}) being the important defining variables. $Y_{(3)}$ was interpreted to be "thought disturbance," defined by a composite of mannerisms and posturing (X_7), grandiosity (X_8), hallucinatory behavior (X_{12}), and unusual thought content (X_{15}). The composite $Y_{(4)}$ was interpreted as a "belligerence" variable, with hostility (X_{10}), suspiciousness (X_{11}), and excitement (X_{17}) contributing most to its definition. Last, $Y_{(5)}$ was primarily defined in terms of "motor retardation" (X_{13}) since its loading .80381 dominated the composite.

In this example, the five composite variables accounted for 65% of the total variability in the original set of 18 variables. In general, there are as many composite variables as original variables. In the problem just described, however, any additional composite included would not contribute measurably to additional total variation.

15.2 PRINCIPAL COMPONENTS

15.2.1 Introduction

Suppose we have n subjects' responses to a questionnaire containing p items. A basic purpose of *principal components* is to account for the total variation among these n subjects in p-dimensional space by forming a new set of orthogonal and uncorrelated *composite* variates. As we shall see, each member of the new set of variates is a linear combination of the original set of measurements. The linear combinations will be generated in such a manner that *each successive composite variate will account for a smaller portion of total variation.* Hence the first composite (i.e., *principal component*) will have the largest variance, the second will have a variance smaller than the first but larger than the third, and so on. In general, the number of new composite variables that will be needed to account adequately for the total variation is less than p.

15.2.2 Development

To fix ideas, suppose that for a particular subject the observed responses to the p items on the questionnaire are represented by $\mathbf{X}' = [X_1, X_2, \ldots, X_p]$. For all subjects let the population variance-covariance† matrix of $\mathbf{X}$ be given by

---

†In practice, principal components evolve not from using the true variance-covariance matrix $\hat{\mathbf{\Sigma}}$ but from its sample estimates

$$\hat{\mathbf{\Sigma}} = \frac{1}{n} \sum_{l=1}^{n} [\mathbf{X}_l - \bar{\mathbf{X}}][\mathbf{X}_l - \bar{\mathbf{X}}]' = (S_{x_{ij}}); \ i, j = 1, \ldots, p$$

where $\mathbf{X}'_l$ is the vector of observed responses to the p items for the lth subject and $\bar{\mathbf{X}}$ is the mean vector for the p items for all n subjects in the sample. Nonetheless, precisely the same methods and interpretations prevail as if the researcher were using $\mathbf{\Sigma}$.

$$\mathbf{\Sigma} = \begin{bmatrix} \sigma_{11} & \sigma_{12} & \cdots & \sigma_{1p} \\ \sigma_{21} & \sigma_{22} & \cdots & \sigma_{2p} \\ \cdot & \cdot & & \cdot \\ \cdot & \cdot & & \cdot \\ \cdot & \cdot & & \cdot \\ \sigma_{p1} & \sigma_{p2} & \cdots & \sigma_{pp} \end{bmatrix} \tag{15.1}$$

The elements along the main diagonal of $\mathbf{\Sigma}$ are, respectively, the true population variances of each of the p original variables. The elements off the main diagonal are the true population covariances (i.e., the manner in which pairs of original variables relate to each other—positively, negatively, or zero).

To find the first principal component $Y_{(1)}$, we seek a vector of coefficients $\mathbf{a}' = [a_1, a_2, \ldots, a_p]$ such that the variance of $\mathbf{a}'\mathbf{X}$ is a maximum over the class of all linear combinations $\mathbf{X}$ subject to the constraint $\mathbf{a}'\mathbf{a} = 1$. The reason for requiring that the coefficients be normalized (that is, $\mathbf{a}'\mathbf{a} = 1$) is that otherwise the variance of $\mathbf{a}'\mathbf{X}$ would increase by making the coordinates of $\mathbf{a}$ arbitrarily large. It can be shown that the set of coefficients (i.e., the vector) $\mathbf{a}$ defining the first principal component is that which corresponds to the largest *eigenvalue*[†] λ_1 of $\mathbf{\Sigma}$. In words, an eigenvalue is merely the *variance of a particular principal component*. Thus, the largest eigenvalue is the variance of the first principal component $Y_{(1)}$. Moreover, the set of coefficients defining this first principal component is called the eigenvector $\mathbf{a}_{(1)}$. The first principal component is given by $Y_{(1)} = \mathbf{a}'_{(1)}\mathbf{X}$.

The second principal component is obtained by finding a second normalized vector (set of coefficients) $\mathbf{a}_{(2)}$ orthogonal to $\mathbf{a}_{(1)}$ such that $Y_{(2)} = \mathbf{a}'_{(2)}\mathbf{X}$ has the second largest variance (i.e., eigenvalue) among all vectors satisfying the constraints $\mathbf{a}'_{(2)}\mathbf{a}_{(2)} = 1$, $\mathbf{a}'_{(1)}\mathbf{a}_{(2)} = 0$. The process continues until all p sets of coefficients (eigenvectors) are generated in such a manner that each is normalized and is orthogonal to the sets of coefficients generated for the other principal components.

The coefficients defining a given principal component have an interesting interpretation. Within the jth component the contribution of variable X_i is given by a_{ij}, where $Y_{(j)} = a_{1j}X_1 + a_{2j}X_2 + \cdots + a_{pj}X_p$. The magnitude and sign of a_{ij} gives the strength and direction of the relationship between X_i and $Y_{(j)}$. It can be shown that the covariance of X_i with the component $Y_{(j)}$ is $\sqrt{\lambda_j}\, a_{ij}$, where λ_j is the eigenvalue whose associated set of coefficients (eigenvector) defines the jth component $Y_{(j)}$. The set of coefficients or *component weights* a_{ij} can be converted into *component loadings* by dividing $\sqrt{\lambda_j}\, a_{ij}$ by the standard deviation of X_i. Component loadings represent the correlation between each variable X_i and the component $Y_{(j)}$. Note that if all the variables defining $\mathbf{X}$

[†]In general, the eigenvalues of a $p \times p$ matrix $\mathbf{B}$ are the solutions to the determinantal equation $|\mathbf{B} - \lambda\mathbf{I}| = 0$. Here the matrix $\mathbf{B}$ is the variance-covariance matrix $\mathbf{\Sigma}$. For further discussion see Hadley (1961), or Noble and Daniel (1977), or Rao (1973).

are first standardized, then $\boldsymbol{\Sigma}$ is a correlation matrix, and since all the variances will then be equal to 1, the scaled weights $\sqrt{\lambda_j}\, a_{ij}/S_{X_i} = \sqrt{\lambda_j} a_{ij}$ represent the correlation between X_i and $Y_{(j)}$. This was the case with the psychiatric example discussed in Sections 15.1.1 and 15.1.2.

15.2.3 Application

With a view toward explaining the dynamics of population change, Press and Wilson (1978) reported data on the 50 states dealing with (1) per capita income (in thousands of dollars), (2) birth rate (percent), and (3) death rate (percent). A principal components analysis was applied using the variance-covariance matrix for the first three variables as the basic input. The variance-covariance matrix for income, birth, and death is given in Table 15.1. Table 15.2

TABLE 15.1   Variance-Covariance Matrix for Demographic Data

Variable	Income	Birth	Death
Income	.3250	−.0405	−.0402
Birth	−.0405	.0595	−.0089
Death	−.0402	−.0089	.0611

Source: S. J. Press and S. Wilson, "Choosing Between Logistic Regression and Discriminant Analysis," JASA, Vol. 73 (1978), pp. 699–705.

TABLE 15.2   Principal Components for Demographic Data

Dimension	Component Weights		
	1	2	3
Income per Capita	.9816	.0053	.1960
Birth Rate	−.1386	−.6891	.7100
Death Rate	−.1387	.7271	.6750
Variance (eigenvalue)	.336	.069	.040
Cumulative Proportion of Total Variance	.755	.910	1.000

Source: S. J. Press and S. Wilson, "Choosing Between Logistic Regression and Discriminant Analysis," JASA, Vol. 73 (1978), pp. 699–705.

gives the coefficient weights for the three principal components along with each component variance (i.e., the eigenvalues of $\boldsymbol{\Sigma}$) and the cumulative proportion of total variance. Hence, for example, we see from Table 15.2 that the first principal component is

$$Y_{(1)} = +.9816 \,(\text{income}) - .1386 \,(\text{birth rate}) - .1387 \,(\text{death rate})$$

and $Y_{(1)}$ accounts for about 76% of the total variance. Since the components of $Y_{(1)}$ are dominated by the income weighting coefficient, the first principal component is essentially an income variable. Moreover, component 2 can be interpreted as a bipolar dimension depicting a contrast between birth and death rates.

Table 15.3 converts the component weights into component loadings by the scaling discussed in Section 15.2.2. From this table we see that the birth rate

TABLE 15.3   Principal Components for Demographic Data

	Component Loadings		
Dimension	1	2	3
Income per Capita	.99	.00	.07
Birth Rate	−.33	−.74	.58
Death Rate	−.32	.77	.55

and death rate are about equally correlated with component 3. To demonstrate the calculations used to generate Table 15.3, we note that to obtain the component loading .99 we multiply the component weight .9816 from Table 15.2 by $\sqrt{\lambda_1} = \sqrt{.336} = .5796$ and divide by the standard deviation of the income per capita variable $\sqrt{.3250} = .5700$. The other values are similarly computed. Thus the composite

$$Y_{(1)} = +.99(\text{income}) - .33(\text{birth rate}) - .32(\text{death rate})$$

has a correlation of $+.99$ with income, $-.33$ with birth rate, and $-.32$ with death rate since the loadings represent the correlation of the composite with each variable.

Our development of principal components and its illustration in the preceding example depended on a variance-covariance matrix as the basic input data. In almost all practical cases, however (including the problem discussed), the measurements comprising the responses vary in their units. Thus, while we can measure height in either inches or feet, the variance associated with the latter unit will be much smaller than that associated with the former. Because of the potential for mixing variables with widely different scales of measurement, it is difficult to justify that forming linear combinations in such cases is sensible from the viewpoint of clear interpretation. It is in situations like this that researchers first standardize their measurements before doing a component analysis, and hence the basic input is now a correlation matrix. The desirable property that the jth component generated explains the jth largest part of the total response variance is no longer as meaningful in this case. Furthermore, since principal components are not scale-invariant, there is no mathematical way in which to compare the two solutions. There is the additional problem using correlation input in dealing with intractable statistical inference (not the case with covariance inputs) when deciding upon how many components to maintain for analysis. All this being said, however, the great majority of principal components analysis is performed using standardized data.

To get an indication about how solutions would change between covariance and correlation starting matrices, the preceding example was redone by first converting the data to standardized units. In this case the correlation matrix is presented in Table 15.4.

TABLE 15.4   Correlation Matrix for Demographic Data

Variable	Income	Birth	Death
Income	1.00	−.29	−.28
Birth	−.29	1.00	−.15
Death	−.28	−.15	1.00

Table 15.5 presents both the component weights and component loadings. Using the weights, the first principal component is

$$Y_{(1)} = .76 \text{ (income)} - .46 \text{ (birth rate)} - .44 \text{ (death rate)}$$

Using the loadings, the composite becomes

$$Y_{(1)} = .88 \text{ (income)} - .53 \text{ (birth rate)} - .51 \text{ (death rate)}$$

TABLE 15.5   Component Weights and Loadings for Demographic Data

	Component Weights			Component Loadings		
Dimension	1	2	3	1	2	3
Income	.76	−.01	.64	.88	−.01	.46
Birth Rate	−.46	−.70	.55	−.53	−.75	.39
Death Rate	−.44	.72	.55	−.51	.77	.39

While the numbers have changed from the analysis based on covariance input, conclusions are essentially the same. However, when the units of measurement vary markedly, this generally will not happen. It is good advice to view principal components analysis more as an exploratory tool and as a basis for further analysis when data are first standardized. In either case care and good judgment are essential for the intelligent application of the technique.

15.2.4 How Many Components to Retain

One of our primary objectives in a principal components analysis is dimensionality reduction, and hence if we use all or most of the possible new variates, we are in a sense defeating our purpose. There is a rich class of statistical inference which can be helpful in determining how many components to generate.

Most of the available statistical tests are applicable only for large samples and deal with a determination of whether the generated eigenvalues (i.e., component variances) appear significantly different from zero. Moreover, many users of these tests do not find them particularly helpful, and hence more ad hoc procedures are commonplace.

When the variance-covariance matrix is the basic input, components are usually generated until some prespecified amount of total variation is accounted for. This, of course, is a very subjective and arbitrary *stopping rule*. It is, however, the most frequently utilized rule. In the case of correlation input the "root greater than 1" criterion is frequently employed. Originally proposed by Kaiser (1958), this criterion retains only those components whose eigenvalues are greater than 1. The rationale for this rule is that any component should account for more variance than any single variable in the standardized test score space. Both procedures are sensible and are recommended. There are also a number of graphical procedures which can be helpful in determining how many components to retain. For further discussion regarding these procedures see Goldstein and Dillon (1983).

15.3 FACTOR ANALYSIS

Perhaps more than any other branch of multivariate analysis, factor analysis has gained widespread acceptance in a great variety of disciplines. As with principal components, it is a linear reduction technique. However, factor analysis has more inherent structure since it assumes a specified model implying a reduced form of the input matrix. That is, the factor analytic model presumes the existence of a smaller set of factors that can reproduce exactly the correlation in the larger set of variables.

15.3.1 Development

The basic model in factor analysis is usually expressed by

$$\mathbf{X} = \mathbf{\Lambda f} + \mathbf{Y} \qquad (15.2)$$

where $\mathbf{X} = p$-dimensional vector of observed responses

$\mathbf{\Lambda} = p \times q$ matrix of unknown constants called *factor loadings*

$\mathbf{f} = q$-dimensional vector of unobservable variables called *common factors*

$\mathbf{Y} = p$-dimensional vector of unobservable variables called *unique factors*

We assume that the variance-covariance matrix of $\mathbf{Y}$ is a diagonal matrix $\mathbf{\Phi}$ with entries ϕ_i^2 and that all covariances between $\mathbf{Y}$ and $\mathbf{f}$ are zero. The basic model along with the associated assumptions imply that $\mathbf{\Sigma}_{XX}$, the variance-covariance matrix of $\mathbf{X}$, is expressible as

$$\mathbf{\Sigma}_{XX} = \mathbf{\Lambda \Sigma}_{ff}\mathbf{\Lambda}' + \mathbf{\Phi} \qquad (15.3)$$

Standardizing the vector of common factors and assuming that they are pairwise uncorrelated, one is led to

$$\mathbf{\Sigma}_{XX} = \mathbf{\Lambda\Lambda'} + \mathbf{\Phi} \tag{15.4}$$

To get a better feel for the basic model, let us examine (15.2) in somewhat more detail. Writing (15.2) in terms of its elements we have

$$
\begin{bmatrix} X_1 \\ X_2 \\ \cdot \\ \cdot \\ \cdot \\ X_p \end{bmatrix}
=
\begin{bmatrix}
l_{11} & l_{12} & \cdots & l_{1q} \\
l_{21} & l_{22} & \cdots & l_{2q} \\
\cdot & \cdot & & \cdot \\
\cdot & \cdot & & \cdot \\
\cdot & \cdot & & \cdot \\
l_{p1} & l_{p2} & \cdots & l_{pq}
\end{bmatrix}
\begin{bmatrix} f_1 \\ f_2 \\ \cdot \\ \cdot \\ \cdot \\ f_q \end{bmatrix}
+
\begin{bmatrix} Y_1 \\ Y_2 \\ \cdot \\ \cdot \\ \cdot \\ Y_p \end{bmatrix}
$$

or

$$X_i = \sum_{j=1}^{q} l_{ij} f_j + Y_i \tag{15.5}$$

This representation shows directly that, for the case of uncorrelated and standardized common factors, the common factor loading l_{ij} expresses the correlation between the jth factor and the variable X_i. If we do not assume, however, that the common factors are pairwise uncorrelated, then the same interpretation does not prevail.

A further implication of the basic model as shown in (15.2) is that for the case of uncorrelated and standardized common factors, the correlation r_{ij} between any two variables X_i and X_j is expressible in terms of factor loadings by

$$r_{ij} = \sum_{k=1}^{q} l_{ik} l_{jk} \tag{15.6}$$

Note that from (15.3)

$$\text{Var}(X_i) = \sigma_{ii} = \sum_{j=1}^{q} l_{ij}^2 + \phi_i^2 \tag{15.7}$$

and hence we can think of l_{ij}^2 as the contribution of the common factor f_j to the variance of X_i. The contribution of all the factors to the variance of X_i (that is, $\sum_{j=1}^{q} l_{ij}^2$) is called the *communality* of X_i; ϕ_i^2 is the *uniqueness* of X_i and measures the extent to which the common factors fail to account for the variance of X_i. The total contribution of f_j to the variances of all the variables is

$$v_j = \sum_{i=1}^{p} l_{ij}^2 \tag{15.8}$$

and hence the total contribution of all the common factors to the total variance of all the variables is the total communality

$$v = \sum_{j=1}^{q} v_j \tag{15.9}$$

To summarize,

$$\text{Total Variance} = \sum_{i=1}^{p} \sigma_{ii}$$

$$= \sum_{i=1}^{p} \sum_{j=1}^{q} l_{ij}^2 + \sum_{i=1}^{p} \phi_i^2 \qquad (15.10)$$

$$= v + \phi^2$$

$$= \text{Total Communality} + \text{Total Uniqueness}$$

15.3.2 The Reduced Correlation Matrix and Communalities

As was the case in principal components analysis, frequently the researcher will standardize the original variables before starting a factor analysis. Assuming standardized and uncorrelated common factors, the basic factorization is of the correlation matrix $\mathbf{R}$,

$$\mathbf{R} = \mathbf{\Lambda}\mathbf{\Lambda}' + \mathbf{\Phi} \qquad (15.11)$$

Note that this is identical to (15.4). Replacing each unit variance on the main diagonal of $\mathbf{R}$ by the respective communality for each variable results in a *reduced correlation matrix* $\mathbf{R}_*$. If the rank[†] of $\mathbf{R}_*$ is m, then the smallest number of linearly independent factors which will account for the correlation is m. Clearly the rank of $\mathbf{R}_*$ is directly affected by the values chosen for the communalities, and therefore the estimation process becomes important since our interest centers on how much the rank of $\mathbf{R}$ can be reduced through appropriate choices for $h_i^2 = \sum_{j=1}^{m} l_{ij}^2, i = 1, 2, \ldots, p$. Lowering the rank of the reduced correlation matrix leads to a diminution in the dimension of the factor space which is in line with trying to come upon a more parsimonious representation of the data.

There are a number of procedures which have been suggested for estimating communalities; none, however, are very satisfactory. Perhaps the simplest is to estimate the ith communality to be the highest observed positive correlation of X_i with the remaining $p - 1$ other observed variables. Another method uses *triads*, that is, the ith communality is taken to be $r_{ij}r_{ik}/r_{jk}$, where X_j and X_k are the two variables which correlate most highly with X_i. While there are a number of other procedures, the one which has gained greatest support is to estimate the ith communality by the squared multiple correlation (SMC) of X_i with the remaining other variables. It is known that $h_i^2 = \sum_{j=1}^{q} l_{ij}^2$ is bounded from below by the SMC, and with its interpretation (namely, measuring the proportion of the observed total variability in X_i that is explained by its regression on the $p - 1$ other variables—providing a measure of shared variance), it is the most reasonable method proposed thus far. Indeed, most computer packages use this procedure to generate communalities.

---

†The rank of a symmetric matrix is the number of nonzero eigenvalues that the matrix emits. For further discussion see Hadley (1961) or Noble and Daniel (1977).

15.3.3 Illustrating the Concepts of Factor Analysis

As an illustration of factor analysis we consider a correlation matrix on nine variables (Table 15.6). All variables were first standardized to have unit variance, and therefore the variance-covariance matrix is a correlation matrix. The first three factors generated are given in Table 15.7.

TABLE 15.6   Correlations Between Nine Variables

Variable	1	2	3	4	5	6	7	8	9
1	1.00								
2	.81	1.00							
3	.59	.67	1.00						
4	.58	.29	.19	1.00					
5	.53	.25	.16	.67	1.00				
6	.17	.13	.08	.23	.29	1.00			
7	.33	.39	.28	.17	.22	.70	1.00		
8	.22	.29	.21	.08	.12	.52	.59	1.00	
9	.40	.41	.29	.28	.33	.77	.83	.62	1.00

TABLE 15.7   Factor Matrix for First Three Factors

Variable	f_1	f_2	f_3	h_i^2
1	+.76	+.50	+.12	.84
2	+.69	+.41	+.49	.90
3	+.54	+.33	+.48	.63
4	+.54	+.48	−.43	.70
5	+.56	+.31	−.48	.64
6	+.63	−.49	−.29	.72
7	+.74	−.51	+.07	.82
8	+.56	−.43	+.13	.53
9	+.82	−.40	−.02	.83

$$\sum_{i=1}^{9} h_i^2 = 6.61$$

No attempt is being made to interpret the factors, but rather we concern ourselves with basic calculations. Toward this end, recall that the communality of a variable is the sum of the squared factor loadings over all factors. Thus for variable 1 we have

$$(.76)^2 + (.50)^2 + (.12)^2 = .84$$

which can be interpreted to mean that 84% of the variance in variable 1 is explained by the first three common factors. Since the variance of variable 1 has been standardized to unity, the uniqueness is merely the complement of the communality (communality + uniqueness = 1), and therefore the uniqueness for variable 1 is .16. We can conclude that 16% of the variance in variable 1 is not accounted for by the first three common factors. The total communality is $\sum_{i=1}^{9} h_i^2 = 6.61$, and since the total variance is 9, the total uniqueness is 2.39.

Reproducing the correlation between variables 4 and 5, r_{45}, for the case of uncorrelated and standardized common factors involves the calculation (15.6):

$$r_{45} = \sum_{k=1}^{3} l_{4k}l_{5k}$$
$$= .54(.56) + .48(.31) + (-.43)(-.48) = .6576$$

Note, however, from the correlation matrix given in Table 15.6 that $r_{45} = .67$; since the difference between .6576 and .67 is not marked, it can be said that the three factors satisfactorily reproduce the original correlation.

15.3.4 Rotation of Factor Solution

When we introduced the notion of factoring the variance-covariance matrix $\mathbf{\Sigma}_{xx}$ (or the correlation matrix), we did not indicate that alternative solutions were possible and may be equally as valid. In fact, if the matrix $\mathbf{\Lambda}$ of factor loadings is postmultiplied by any orthogonal matrix $\mathbf{A}$, then $\mathbf{\Sigma}_{xx}$ will be reproducible through $\mathbf{\Lambda A}$ as well as through $\mathbf{\Lambda}$. The matrix of factor loadings $\mathbf{\Lambda}$ represents a particular interpretation of the data, that is, the variance-covariance or correlation matrix in terms of a set of factors. The rotated matrix of factor loadings $\mathbf{\Lambda A}$ represents an alternative interpretation of the data which, in a mathematical sense, is equally as valid. The rotational process of factor analysis allows the researcher a degree of flexibility by presenting a multiplicity of views of the same data set in order to aid in interpretation.

Many procedures used to rotate the matrix of factor loadings do so in a manner to achieve *simple structure*, a criterion developed by Thurstone (1947). The major characteristics of simple structure are the following:

1. Any column of the factor loading matrix should have mostly small values, as close to zero as possible.
2. Any given row of the matrix of factor loadings should have nonzero entries in only a few columns.
3. Any two columns of the matrix of factor loadings should exhibit a different pattern of high and low loadings.

The idea of simple structure is not limited to orthogonal rotations but is equally reasonable for *oblique rotations*, that is, rotations which lead to nonorthogonal solutions.

Most computer software packages for factor analysis contain various rotational procedures. Factor loadings are automatically rotated to achieve certain criteria. Although options are given for oblique rotations, orthogonal rotations are without question the most frequently employed. Kaiser's (1958, 1959) *Varimax* method for factor rotation is probably the most popular of the computer-generated procedures. The Varimax method rotates factors so that the variance of the squared factor loadings for a given factor is made large. Most of the popular computer packages use Varimax rotation either with raw factor loadings or with *normalized* loadings, that is, by first dividing each variable loading by the square root of its communality. By scaling, all variables are given equal weight in the rotation. However, some authors have argued against such scaling, especially when communalities are very small.

Because of the computational complexities inherent in generating a matrix of factor loadings, today factor analyses are done almost exclusively by computer. Perhaps the most frequently employed methods are the *principal factor solution* and the *maximum likelihood solution*. Discussion of how each procedure operates is beyond the scope of this text. For an in-depth presentation, we recommend the reader to Harman (1967); however, at least for the beginner, a step-by-step discussion of the nuances of each algorithm is not necessary in order to access an available computer package.

15.3.5 Application

The following data set was discussed by Kendall (1975). To satisfy one of the conditions for employment in a large organization, 48 job applicants were interviewed and judged by a personnel officer on the following 15 variables:

1. Form of application letter
2. Appearance
3. Academic ability
4. Likeability
5. Self-confidence
6. Lucidity
7. Honesty
8. Salesmanship
9. Experience
10. Drive
11. Ambition
12. Grasp
13. Potential
14. Keenness to join
15. Suitability

Since a large number of variables were measured and a number of large correlations obtained, it was decided that a factor analysis would be helpful in uncovering a smaller number of dimensions that may have been used by the officer in making his determination.

Kendall employed a maximum likelihood solution to the correlation matrix presented in Table 15.8. Table 15.9 gives the results for the unrotated solution, and anything more than a cursory examination shows that the factors are difficult to interpret. However, when the solution given in Table 15.9 is rotated by the Varimax method (Table 15.10), the seven-factor solution makes good sense. The first factor loads heavily on variables 5, 6, 8, 10, 11, 12, and 13 and can be interpreted as a dimension reflecting an "extroverted personality" usually associated with a salesperson. Factor 2 loads most heavily with variables 4

TABLE 15.8   Correlation Matrix for the Applicant Data

Variable	1	2	3	4	5	6	7	8	9	10	11	12	13	14	15
1	1.00	.24	.04	.31	.09	.23	−.11	.27	.55	.35	.28	.34	.37	.47	.59
2		1.00	.12	.38	.43	.37	.35	.48	.14	.34	.55	.51	.51	.28	.38
3			1.00	.00	.00	.08	−.03	.05	.27	.09	.04	.20	.29	−.32	.14
4				1.00	.30	.48	.65	.35	.14	.39	.35	.50	.61	.69	.33
5					1.00	.81	.41	.82	.02	.70	.84	.72	.67	.48	.25
6						1.00	.36	.83	.15	.70	.76	.88	.78	.53	.42
7							1.00	.23	−.16	.28	.21	.39	.42	.45	.00
8								1.00	.23	.81	.86	.77	.73	.55	.55
9									1.00	.34	.20	.30	.35	.21	.69
10										1.00	.78	.71	.79	.61	.62
11											1.00	.78	.77	.55	.43
12												1.00	.88	.55	.53
13													1.00	.54	.57
14														1.00	.40
15															1.00

SOURCE: Reproduced by permission of the Publishers, Charles Griffin & Company Ltd., of London and High Wycombe, from Kendall, *Multivariate Analysis*, Second Edition (1980).

**TABLE 15.9   Maximum Likelihood Factor Solution of Applicant Data
with 7 Factors, Unrotated**

				Factor Loadings			
Variable	1	2	3	4	5	6	7
1	.090	−.134	−.338	.400	.411	−.001	.277
2	−.466	.171	.037	−.002	.517	−.194	.167
3	−.131	.466	.153	.143	−.031	.330	.316
4	.004	−.023	−.318	−.362	.657	.070	.307
5	−.093	.017	.434	−.092	.784	.019	−.213
6	.281	.212	.330	−.037	.875	.001	.000
7	−.133	.234	−.181	−.807	.494	.001	−.000
8	−.018	.055	.258	.207	.853	.019	−.180
9	−.043	.173	−.345	.522	.296	.085	.185
10	−.079	−.012	.058	.241	.817	.417	−.221
11	−.265	−.131	.411	.201	.839	−.000	−.001
12	.037	.202	.188	.025	.875	.077	.200
13	−.112	.188	.109	.061	.844	.324	.277
14	.098	−.462	−.336	−.116	.807	−.001	.000
15	−.056	.293	−.441	.577	.619	.001	−.000

SOURCE: Reproduced by permission of the Publishers, Charles Griffin & Company Ltd., of
London and High Wycombe, from Kendall, *Multivariate Analysis,* Second Edition (1980).

**TABLE 15.10   Maximum Likelihood Factor Solution of Applicant Data
with 7 Factors, Varimax Rotation**

				Factor Loadings			
Variable	1	2	3	4	5	6	7
1	.129	.074	.665	−.096	.017	−.042	.267
2	.329	.242	.182	.095	.611	−.013	−.006
3	.048	−.017	.097	.688	.043	.007	.008
4	.249	.759	.252	−.058	.090	−.096	.204
5	.882	.184	−.082	−.074	.190	.059	−.045
6	.907	.266	.136	.046	−.042	−.290	−.016
7	.199	.911	−.224	−.013	.174	−.094	−.204
8	.875	.082	.264	−.076	.140	.043	−.058
9	.073	−.027	.718	.158	.069	.036	.009
10	.780	.197	.386	.026	−.051	.398	−.023
11	.874	.036	.157	−.052	.382	.142	.205
12	.775	.346	.286	.172	.143	−.159	.111
13	.703	.409	.354	.329	.140	.070	.193
14	.432	.540	.381	−.540	−.013	.099	.275
15	.313	.079	.909	.049	.142	.027	−.214

SOURCE: Reproduced by permission of the Publishers, Charles Griffin & Company Ltd., of
London and High Wycombe, from Kendall, *Multivariate Analysis*, Second Edition (1980).

and 7 and may be defined as an "agreeable personality." Variables 1, 9, and 15 load heavily on factor 3 and may be viewed as an "experience" dimension. Factors 4 and 5 load heavily on only one variable each, namely "academic ability" (3) and "appearance" (2). Factors 6 and 7 do not seem to exhibit a clear signal which would be easily interpreted. Their inclusion in the analysis was decided on the basis that they accounted for a sufficient amount of total variability.

15.4 HOW DO PRINCIPAL COMPONENTS AND FACTOR ANALYSIS DIFFER?

There continues to be considerable confusion in many quarters about the differences between principal components and factor analysis. Recall that in principal components analysis we find linear combinations of the original variables such that the jth component generated has the jth largest variance. Even though a few components may account for a large portion of the total variance, all p components are needed to recover the correlations exactly. In contrast, the common factor model posits the existence of a number of factors smaller than the number of original variables which will reproduce the correlations exactly but which may not account for as much variance as does the same number of principal components. Finally, it is important to note that while in principal components analysis the factors are linear combinations of the observable variables, in common factor analysis the factors are linear combinations of only the common parts of the variables. Hence it is understandable that principal components is viewed as variance-oriented, whereas in common factor analysis the specific variance is expressed separately, and, as such, it is correlation- or covariance-oriented.

15.5 COMPUTER PACKAGES AND PRINCIPAL COMPONENTS AND FACTOR ANALYSIS: USE OF SPSS, SAS, AND BMDP

In this section we shall discuss how to use either SPSS, SAS, or BMDP to perform principal components and/or factor analysis. The use of these packages will be illustrated by referring to the job interview example (see Section 15.3.5).

15.5.1 Using the SPSS FACTOR Subprogram for Principal Components and Factor Analysis

The FACTOR subprogram of SPSS can be utilized for various types of principal components and factor analysis. The format of the FACTOR procedure statements is

1	16
FACTOR	VARIABLES = {list of variables}/
	TYPE = PA1/
	NFACTORS = {value}/
	MINEIGEN = {value}/
	ROTATE = VARIMAX/
OPTIONS	{list of options}
STATISTICS	{list of statistics}

The VARIABLES statement provides a list of the variables to be included in the factor analysis. Although several alternatives are available [see Nie et al. (1975)], the use of TYPE = PA1 provides for an initial principal components analysis. The NFACTORS statement indicates the number of factors to be extracted with a maximum equal to the number of variables. The MINEIGEN statement provides for the deletion of all factors with eigenvalues below a specified value. If no MINEIGEN value is specified, the default value is 1.0. However, if both the NFACTORS and MINEIGEN statements are included, the NFACTORS statement will take precedence. The ROTATE statement indicates the method of rotation of axes. Although several other methods are available (such as QUARTIMAX, EQUIMAX, or OBLIQUE), we shall utilize the VARIMAX rotation method (see Section 15.3.4).

Among the STATISTICS available are STATISTICS 1 (which provides means and standard deviations), STATISTICS 2 (which provides the correlation matrix), STATISTICS 4 (which indicates communalities, eigenvalues, and the proportion of total and common variance), STATISTICS 5 (which provides the initial unrotated factor matrix), STATISTICS 6 (which provides the rotated factor matrix), and STATISTICS 8 (which plots the unstated factors).

If matrix input (i.e., a correlation matrix) is utilized, as it was in the job interview example, OPTIONS 3 and 9 are needed. OPTIONS 3 informs SPSS that the input is in the matrix form, while OPTIONS 9 specifies that the order of variables on the correlation matrix is the same as on the VARIABLES LIST statement (which is used instead of the DATA LIST statement). These STATISTICS and OPTIONS statements are followed by a READ MATRIX statement which precedes the data (i.e., correlation) matrix. The format of the data matrix *must be* 8F10.7 (10 characters for each data value, 7 of which follow the decimal point, with 8 data values on each card).

Figure 15.2 represents the complete SPSS program that has been written to perform a principal components and factor analysis on the job interview data. Figure 15.3 represents partial output obtained from this program.

Figure 15.2 SPSS program for job interview example.

```
{SYSTEM CARDS}
RUN NAME        FACTOR ANALYSIS OF JOB INTERVIEWS
VARIABLE LIST   X1,X2,X3,X4,X5,X6,X7,X8,X9,X10,X11,X12,X13,X14,X15
INPUT FORMAT    FIXED(8F10.7/7F10.7)
INPUT MEDIUM    CARD
N OF CASES      48
VAR LABELS      X1, FORM OF APPLICATION LETTER/
                X2, APPEARANCE/
                X3, ACADEMIC ABILITY/
                X4, LIKEABILITY/
                X5, SELF CONFIDENCE/
                X6, LUCIDITY/
                X7, HONESTY/
                X8, SALESMANSHIP/
                X9, EXPERIENCE/
                X10, DRIVE/
                X11, AMBITION/
                X12, GRASP/
                X13, POTENTIAL/
                X14, KEENNESS TO JOIN/
                X15, SUITABILITY/
FACTOR          VARIABLES=X1 TO X15/
                TYPE=PA1/
                NFACTORS=7/
                MINEIGEN=0.5/
                ROTATE=VARIMAX/
OPTIONS         3, 9
STATISTICS      2, 4, 5, 6, 8
READ MATRIX
1.0000000  0.2400000  0.0400000  0.3100000  0.3000000  0.0900000  0.3300000  0.4700000 -0.1100000  0.2700000
0.5500000  0.3500000  0.2800000  0.3400000  0.3700000  0.2500000  0.5300000  0.5900000
     ...
0.5900000  0.3800000  0.1400000  0.3300000  0.2500000  0.4200000  0.0000000  0.0000000
0.6900000  0.6200000  0.4300000  0.5300000  0.5700000  0.4000000  1.0000000  0.5500000
FINISH
{SYSTEM CARDS}
```

FACTOR	EIGENVALUE	PCT OF VAR	CUM PCT
1	7.50395	50.0	50.0
2	2.06148	13.7	63.8
3	1.46768	9.8	73.6
4	1.20910	8.1	81.6
5	0.74143	4.9	86.6
6	0.48402	3.2	89.8
7	0.34408	2.3	92.1
8	0.31027	2.1	94.1
9	0.25965	1.7	95.9
10	0.20575	1.4	97.2
11	0.15093	1.0	98.3
12	0.09327	0.6	98.9
13	0.07628	0.5	99.4
14	0.05766	0.4	99.8
15	0.03441	0.2	100.0

FACTOR MATRIX USING PRINCIPAL FACTOR, NO ITERATIONS

	FACTOR 1	FACTOR 2	FACTOR 3	FACTOR 4	FACTOR 5	FACTOR 6	FACTOR 7
X1	0.44676	0.61880	0.37635	-0.12148	0.10168	0.42496	0.08504
X2	0.58285	-0.05019	-0.01995	0.28167	0.75188	-0.03325	0.00345
X3	0.10900	0.33907	-0.49450	0.71393	-0.18095	0.16113	0.18206
X4	0.61698	-0.18150	0.57968	0.35707	-0.09904	0.07837	-0.05714
X5	0.79807	-0.35611	-0.29930	-0.17939	0.00025	0.00377	0.06620
X6	0.86688	-0.18544	-0.18414	-0.06923	-0.17813	0.11744	-0.30132
X7	0.43330	-0.58195	0.36036	0.44570	-0.06052	-0.21591	0.06539
X8	0.88244	-0.05647	-0.24821	-0.22786	0.02960	-0.06262	0.00981
X9	0.36549	0.79438	0.09258	0.07431	-0.08999	-0.25962	-0.06758
X10	0.86261	0.06908	-0.09993	-0.16645	-0.17554	-0.17549	0.29665
X11	0.87185	-0.09840	-0.25565	-0.20948	0.13698	0.07573	0.12514
X12	0.90776	-0.03023	-0.13453	0.09726	-0.06359	0.10194	-0.24685
X13	0.91310	0.03250	-0.07327	0.21842	-0.10489	0.04666	-0.00366
X14	0.71033	-0.11478	0.55801	-0.23496	-0.10071	0.05911	0.14353
X15	0.64584	0.60374	0.10687	-0.02889	0.06413	-0.29308	-0.10537

VARIABLE	COMMUNALITY
X1	0.93708
X2	0.98841
X3	0.97293
X4	0.89636
X5	0.88989
X6	0.96088
X7	0.90948
X8	0.90031
X9	0.85878
X10	0.93618
X11	0.91921
X12	0.92786
X13	0.90108
X14	0.91856
X15	0.89497

VARIMAX ROTATED FACTOR MATRIX

	FACTOR 1	FACTOR 2	FACTOR 3	FACTOR 4	FACTOR 5	FACTOR 6	FACTOR 7
X1	0.12359	0.04204	0.42738	-0.00497	0.85336	0.09437	0.01521
X2	0.32636	0.21176	0.11729	0.05621	0.07715	0.90226	0.01101
X3	0.05396	-0.02816	0.13368	0.97451	-0.01201	0.03936	0.00014
X4	0.22106	0.85846	-0.13049	-0.01215	0.26494	0.09997	0.11479
X5	0.91144	0.15413	-0.08310	-0.04208	-0.05072	0.13904	-0.06943
X6	0.87938	0.25709	0.10119	0.01702	0.05912	-0.00285	0.32778
X7	0.20161	0.87606	-0.13423	0.00066	-0.22952	0.16057	-0.06982
X8	0.90070	0.07788	0.21967	-0.05953	0.05564	0.16142	-0.04510
X9	0.06497	-0.03039	0.88690	0.16270	0.20105	-0.01159	-0.00158
X10	0.79694	0.20942	0.35909	0.02333	0.10603	-0.02550	-0.34034
X11	0.89427	0.06033	-0.08680	-0.01813	0.16585	0.26018	-0.11304
X12	0.79690	0.30629	0.23598	0.14095	0.12915	0.14954	0.29053
X13	0.73031	0.40428	0.29019	0.26489	0.16244	0.14080	0.06079
X14	0.45932	0.56662	0.16988	-0.38607	0.42522	-0.03753	-0.16248
X15	0.33966	0.07614	0.84300	-0.01002	0.18417	0.17055	0.00713

Figure 15.3 Partial SPSS output for the job interview data.

15.5.2 Using SAS PROC FACTOR for Principal Components and Factor Analysis

The SAS FACTOR procedure performs various types of principal components and factor analyses. The format of the PROC FACTOR step is

PROCϸFACTORϸMETHOD = PRINϸNFACT = {value}
MINEIGEN = $\begin{cases} \text{positive integer} \\ \text{value} \end{cases}$ ϸPORTION = {value}
ROTATE = VARIMAXϸPLOTϸPREPLOT;
VARϸ{list of variables};

Although other options are available [see SAS (1979)], the METHOD = PRIN option provides for a principal components analysis (assuming the prior estimates of the communalities are all 1). The NFACT = option allows the user to specify the maximum number of factors for the model. The MINEIGEN = option provides for the deletion of all factors with eigenvalues below a specified value. The PORTION option limits the number of factors to a specified proportion of the total variation (sum of the eigenvalues). Although several other rotation options are available (such as EQUAMAX, QUAR-TIMAX, or PROMAX), setting ROTATE = VARIMAX will provide for Varimax rotation in the factor analysis (see Section 15.3.4). The PLOT option plots factor patterns after rotation, while the PREPLOT option plots factor patterns before rotation.

If matrix input (i.e., a correlation matrix) is utilized, as it was for the job interview example, the format for the DATA and INPUT statements becomes

DATA ϸ $\begin{cases} \text{name of} \\ \text{data set} \end{cases}$ ϸ (TYPE = CORR);
INPUT ϸ{list of variables}ϸ (format for each value);
__TYPE__ = 'CORR';
CARDS;
{data matrix}

The DATA statement indicates that the data are in the form of a correlation matrix. The INPUT statement names the variables to be included in the correlation matrix. The format for each of the data values is shown in parentheses. The __TYPE__ = 'CORR' statement indicates that the data which follow consist of the set of intercorrelations between the variables. The correlation matrix is in the form of a full matrix so that the main diagonal is included.

Figure 15.4 represents the complete SAS program that has been written to perform a principal components and factor analysis on the job interview data. Note that a set of IF THEN statements has been included after the __TYPE__ = 'CORR' statement in order to identify the set of variables provided in the correlation matrix. Figure 15.5 represents partial output obtained from this SAS program.

```
{SYSTEM CARDS}
DATA JOB(TYPE=CORR);
INPUT (X1 X2 X3 X4 X5 X6 X7 X8 X9 XA XB XC XD XE XF)(5.2);
TYPE='CORR';
IF _N_ = 1 THEN _NAME_='X1';
IF _N_ = 2 THEN _NAME_='X2';
IF _N_ = 3 THEN _NAME_='X3';
IF _N_ = 4 THEN _NAME_='X4';
IF _N_ = 5 THEN _NAME_='X5';
IF _N_ = 6 THEN _NAME_='X6';
IF _N_ = 7 THEN _NAME_='X7';
IF _N_ = 8 THEN _NAME_='X8';
IF _N_ = 9 THEN _NAME_='X9';
IF _N_ = 10 THEN _NAME_='XA';
IF _N_ = 11 THEN _NAME_='XB';
IF _N_ = 12 THEN _NAME_='XC';
IF _N_ = 13 THEN _NAME_='XD';
IF _N_ = 14 THEN _NAME_='XE';
IF _N_ = 15 THEN _NAME_='XF';
LABEL X1='FORM OF APPLICATION LETTER'
      X2='APPEARANCE'
      X3='ACADEMIC ABILITY'
      X4='LIKEABILITY'
      X5='SELF CONFIDENCE'
      X6='LUCIDITY'
      X7='HONESTY'
      X8='SALESMANSHIP'
      X9='EXPERIENCE'
      XA='DRIVE'
      XB='AMBITION'
      XC='GRASP'
      XD='POTENTIAL'
      XE='KEENNESS TO JOIN'
      XF='SUITABILITY';
CARDS;
1.00 0.24 0.04 0.31 0.09 0.23 -0.11 0.27 0.55 0.35 0.28 0.34 0.37 0.47 0.59
0.59 0.38 0.14 0.33 0.25 0.42 0.00 0.55 0.69 0.62 0.43 0.53 0.57 0.40 1.00
PROC FACTOR METHOD=PRIN NFACT=7 PORTION=.75 ROTATE=VARIMAX
SCORE PLOT PREPLOT DATA=JOB(TYPE=CORR);
VAR X1 X2 X3 X4 X5 X6 X7 X8 X9 XA XB XC XD XE XF;
{SYSTEM CARDS}
```

Figure 15.4 SAS program for job interview example.

442

PRINCIPAL AXIS

PRIOR ESTIMATES OF COMMUNALITY

X1	X2	X3	X4	X5	X6	X7	X8
1.000000	1.000000	1.000000	1.000000	1.000000	1.000000	1.000000	1.000000

X9	XA	XB	XC	XD	XE	XF
1.000000	1.000000	1.000000	1.000000	1.000000	1.000000	1.000000

	1	2	3	4	5	6	7	8
EIGENVALUES	7.503986	2.061498	1.467686	1.209097	0.741423	0.484018	0.344075	0.310272
PORTION	0.500	0.137	0.098	0.081	0.049	0.032	0.023	0.021
CUM PORTION	0.500	0.638	0.736	0.816	0.866	0.898	0.921	0.941

	9	10	11	12	13	14	15
EIGENVALUES	0.259652	0.205746	0.150932	0.093269	0.076283	0.057655	0.034407
PORTION	0.017	0.014	0.010	0.006	0.005	0.004	0.002
CUM PORTION	0.959	0.972	0.983	0.989	0.994	0.998	1.000

· 4 FACTORS WILL BE RETAINED.

FACTOR PATTERN

	FACTOR1	FACTOR2	FACTOR3	FACTOR4	
X1	0.44676	0.61880	0.37635	-0.12148	FORM OF APPLICATION LETTER
X2	0.58285	-0.05019	-0.01995	0.28166	APPEARANCE
X3	0.10900	0.33907	-0.49449	0.71391	ACADEMIC ABILITY
X4	0.61699	-0.18149	0.57967	0.35706	LIKEABILITY
X5	0.79807	-0.35610	-0.29930	-0.17939	SELF CONFIDENCE
X6	0.86688	-0.18543	-0.18414	-0.06923	LUCIDITY
X7	0.43330	-0.58195	0.36035	0.44569	HONESTY
X8	0.88244	-0.05647	-0.24821	-0.22786	SALESMANSHIP
X9	0.36549	0.79437	0.09258	0.07431	EXPERIENCE
XA	0.86261	0.06908	-0.09993	-0.16645	DRIVE
XB	0.87186	-0.09840	-0.25564	-0.20948	AMBITION
XC	0.90776	-0.03023	-0.13453	0.09726	GRASP
XD	0.91310	0.03250	-0.07327	0.21842	POTENTIAL
XE	0.71033	-0.11478	0.55800	-0.23495	KEENNESS TO JOIN
XF	0.64584	0.60373	0.10687	-0.02889	SUITABILITY

VARIMAX

ROTATED FACTOR PATTERN

	FACTOR1	FACTOR2	FACTOR3	FACTOR4	
X1	0.11447	0.83336	0.11063	-0.13808	FORM OF APPLICATION LETTER
X2	0.43964	0.14979	0.39417	0.22555	APPEARANCE
X3	0.06115	0.12744	0.00557	0.92792	ACADEMIC ABILITY
X4	0.21559	0.24667	0.87360	-0.08137	LIKEABILITY
X5	0.91896	-0.10368	0.16241	-0.06219	SELF CONFIDENCE
X6	0.86439	0.10195	0.25878	0.00642	LUCIDITY
X7	0.21715	-0.24607	0.86440	0.00341	HONESTY
X8	0.91799	0.20635	0.08773	-0.04938	SALESMANSHIP
X9	0.08530	0.84871	-0.05537	0.21919	EXPERIENCE
XA	0.79576	0.35407	0.15950	-0.05026	DRIVE
XB	0.91641	0.16268	0.10496	-0.04184	AMBITION
XC	0.80415	0.25872	0.34049	0.15153	GRASP
XD	0.73917	0.32885	0.42493	0.22980	POTENTIAL
XE	0.43597	0.36420	0.54105	-0.51862	KEENNESS TO JOIN
XF	0.37950	0.79807	0.07847	0.08221	SUITABILITY

ORTHOGONAL TRANSFORMATION MATRIX

	1	2	3	4
1	0.84091	0.36961	0.39521	0.00892
2	-0.20568	0.86427	-0.37657	0.26253
3	-0.43554	0.33672	0.62433	-0.55420
4	-0.24673	-0.05518	0.55876	0.78985

VARIANCE EXPLAINED BY EACH FACTOR

FACTOR1	FACTOR2	FACTOR3	FACTOR4
5.745474	2.735065	2.413961	1.347767

Figure 15.5 Partial SAS output for the job interview data.

443

15.5.3 Using BMDP Program 4M for Principal Components and Factor Analysis

The BMDP program 4M performs various types of principal components and factor analyses. The paragraphs of BMDP that are unique to P-4M are /FACTOR, /ROTATE, /PRINT, and /PLOT. The format of these paragraphs is

/FACTOR	FORM = $\begin{Bmatrix} \text{CORR} \\ \text{or} \\ \text{COVA} \end{Bmatrix}$.
	METHOD = PCA.
	NUMBER = {value}.
	CONSTANT = {value}.
/ROTATE	METHOD = VMAX.
/PRINT	CORR.
	COVA.
	PARTIAL.
	SHADE.
/PLOT	INITIAL = {value}.
	FINAL = {value}.

In the /FACTOR paragraph, the FORM sentence indicates the matrix to be factored, usually either a correlation (CORR) matrix or a covariance (COVA) matrix. Although several other options are available, the METHOD = PCA sentence provides for a principal components analysis. The NUMBER sentence specifies the maximum number of factors for the model. The CONSTANT sentence provides for the deletion of all factors with eigenvalues below a specified value.

The METHOD sentence of the /ROTATE paragraph indicates the method of factor rotation. Although several alternatives are available, VMAX is used for Varimax rotation, while NONE is utilized if only a principal components analysis is desired.

Although various options are available for the /PRINT paragraph, PARTIAL provides the partial correlations of the variables, and SHADE prints a correlation matrix in "shaded" form.

In the /PLOT paragraph, the INITIAL sentence specifies the number of unrotated factors to be plotted, while the FINAL sentence specifies the number of rotated factor loadings to be plotted.

Figure 15.6 represents the complete BMDP program that has been written to perform a principal components and factor analysis on the job interview data. Note that since the input data consists of a correlation matrix, a TYPE = CORR sentence has been included in the /INPUT paragraph. Figure 15.7 represents partial output obtained from this BMDP program.

```
//SYSTEM CARDS?
/PROBLEM    TITLE IS 'FACTOR ANALYSIS OF JOB INTERVIEWS'.
/INPUT      TYPE=CORR.
            VARIABLES ARE 15.
            FORMAT IS '(15F5.2)'.
/VARIABLE   NAMES ARE LETTER, APPEAR, ABILITY, LIKEABIL, SELFCONF,
            LUCIDITY, HONESTY, SALESMAN, EXPER, DRIVE, AMBITION,
            GRASP, POTENTL, KEENNESS, SUITABLE.
/FACTOR     FORM=CORR.
            METHOD=PCA.
            NUMBER=7.
            CONSTANT=0.5.
            METHOD=VMAX.
/ROTATE     PARTIAL.
/PRINT      SHADE.
/PLOT       INITIAL=4.
            FINAL=10.
/END
1.00  0.04  0.31  0.09  0.33  -0.11  0.27  0.55  0.35  0.28  0.34  0.37  0.47  0.59
0.59  0.38  0.33  0.14  0.35   0.42  0.22  0.55  0.69  0.62  0.43  0.53  0.57  0.40  1.00
//SYSTEM CARDS?
```

Figure 15.6   Using BMDP program 4M for job interview example.

FACTOR	VARIANCE EXPLAINED	CUMULATIVE PROPORTION OF TOTAL VARIANCE
1	7.503986	0.500266
2	2.061498	0.637699
3	1.467686	0.735545
4	1.209097	0.816151
5	0.741423	0.865579
6	0.484018	0.897847
7	0.344075	0.920786
8	0.310272	0.941470
9	0.259652	0.958781
10	0.205746	0.972497
11	0.150932	0.982559
12	0.093269	0.988777
13	0.076283	0.993863
14	0.057655	0.997706
15	0.034407	1.000000

THE VARIANCE EXPLAINED BY EACH FACTOR IS THE EIGENVALUE FOR THAT FACTOR.

TOTAL VARIANCE IS DEFINED AS THE SUM OF THE DIAGONAL ELEMENTS OF THE CORRELATION (COVARIANCE) MATRIX.

UNROTATED FACTOR LOADINGS (PATTERN)
FOR PRINCIPAL COMPONENTS

		FACTOR 1	FACTOR 2	FACTOR 3	FACTOR 4	FACTOR 5
LETTER	1	0.447	0.619	0.376	-0.121	0.102
APPEAR	2	0.583	-0.050	-0.020	0.282	0.752
ABILITY	3	0.109	0.339	-0.494	0.714	-0.181
LIKEABIL	4	0.617	-0.181	0.580	0.357	-0.099
SELFCONF	5	0.798	-0.356	-0.299	-0.179	0.000
LUCIDITY	6	0.867	-0.185	-0.184	-0.069	-0.178
HONESTY	7	0.433	-0.582	0.360	0.446	-0.061
SALESMAN	8	0.882	-0.056	-0.248	-0.228	0.030
EXPER	9	0.365	0.794	0.093	0.074	-0.090
DRIVE	10	0.863	0.069	-0.100	-0.166	-0.176
AMBITION	11	0.872	-0.098	-0.256	-0.209	0.137
GRASP	12	0.908	-0.030	-0.135	0.097	-0.064
POTENTL	13	0.913	0.032	-0.073	0.218	-0.105
KEENNESS	14	0.710	-0.115	0.558	-0.235	-0.101
SUITABLE	15	0.646	0.604	0.107	-0.029	0.064
VP		7.504	2.061	1.468	1.209	0.741

THE VP FOR EACH FACTOR IS THE SUM OF THE SQUARES OF THE ELEMENTS OF THE COLUMN OF THE FACTOR LOADING MATRIX CORRESPONDING TO THAT FACTOR. THE VP IS THE VARIANCE EXPLAINED BY THE FACTOR.

ROTATED FACTOR LOADINGS (PATTERN)

		FACTOR 1	FACTOR 2	FACTOR 3	FACTOR 4	FACTOR 5
LETTER	1	0.105	0.834	0.098	-0.150	0.105
APPEAR	2	0.324	0.148	0.214	0.056	0.901
ABILITY	3	0.062	0.120	-0.015	0.945	0.037
LIKEABIL	4	0.223	0.242	0.876	-0.044	0.100
SELFCONF	5	0.910	-0.107	0.143	-0.064	0.147
LUCIDITY	6	0.878	0.097	0.271	0.046	0.006
HONESTY	7	0.215	-0.251	0.852	0.026	0.153
SALESMAN	8	0.906	0.203	0.065	-0.057	0.162
EXPER	9	0.095	0.847	-0.047	0.239	-0.038
DRIVE	10	0.813	0.350	0.179	-0.009	-0.034
AMBITION	11	0.890	0.160	0.060	-0.073	0.268
GRASP	12	0.799	0.253	0.321	0.164	0.154
POTENTL	13	0.738	0.322	0.409	0.252	0.140
KEENNESS	14	0.456	0.364	0.573	-0.475	-0.034
SUITABLE	15	0.367	0.796	0.055	0.071	0.146
VP		5.624	2.714	2.285	1.315	1.045

Figure 15.7 Partial BMDP program 4M output for the job interview data.

446

PROBLEMS

15.1. Using the rotated solution given in Table 15.10,
   (a) Calculate the communality and uniqueness for each of the 15 variables.
   (b) Calculate the total communality. What percentage of the total variance do the common factors represent?
   (c) For the variables "appearance" and "honesty," compare their correlation (see Table 15.8) to the value of the correlation reproduced by the common factors.

*15.2. Table P15.2 presents the correlation matrix for five variables taken from a sample of nations around the world. The variables are

X_1: number of daily newspapers
X_2: radios per 1000 population
X_3: televisions per 1000 population
X_4: theater seats per 1000 population
X_5: literacy rate in percent

TABLE P15.2 Correlation Matrix

Variable	X_1	X_2	X_3	X_4	X_5
X_1	1.000	−0.212	−0.311	−0.126	−0.277
X_2	−0.212	1.000	0.715	0.562	0.528
X_3	−0.311	0.715	1.000	0.561	0.750
X_4	−0.126	0.562	0.561	1.000	0.517
X_5	−0.277	0.528	0.750	0.517	1.000

Use a computer package (such as SPSS, SAS, or BMDP) to perform a
   (a) Principal components analysis. Interpret the results.
   (b) Factor analysis using Varimax rotation. Interpret the results.

15.3. Table P15.3 presents seven characteristics of the 10 largest insurance companies in the United States in 1979. The characteristics are

X_1: assets (millions of dollars)
X_2: premium and annuity receipts (millions of dollars)
X_3: net investment income (millions of dollars)
X_4: net gain from operations (millions of dollars)
X_5: life insurance in force (millions of dollars)
X_6: increase in life insurance in force (millions of dollars)
X_7: number of employees

Use a computer package (such as SPSS, SAS, or BMDP) to perform a
   (a) Principal components analysis of the characteristics. Interpret the results.
   (b) Factor analysis of the characteristics using Varimax rotation. Interpret the results.

TABLE P15.3 Characteristics of the 10 Largest Life Insurance Companies

Company	X_1	X_2	X_3	X_4	X_5	X_6	X_7
Prudential	54,734	8008	3313	278.4	367,284	36,920	61,942
Metropolitan	44,968	5935	3089	233.9	323,589	36,055	51,000
Equitable Life	30,839	4577	1809	77.0	183,491	17,785	25,470
Aetna Life	18,549	4729	1228	194.4	127,619	15,354	17,756
New York Life	18,479	2599	1235	151.4	111,892	10,550	19,889
John Hancock	17,319	2171	1151	144.4	125,594	12,778	20,400
Conn. General	12,241	2435	842	137.2	73,574	9,390	9,346
Travelers	11,817	3642	801	159.9	95,653	7,899	39,488
Northwestern Mutual	10,554	1247	703	64.2	51,667	6,297	6,410
Mass. Mutual	8,341	1222	542	56.9	44,527	4,316	9,243

SOURCE: Peter D. Petre, "The Fortune Directory of the Largest Non-Industrial Companies," *Fortune* (July 14, 1980), pp. 146–159. © 1980 Time Inc. All rights reserved.

15.4. A large commercial banking institution recently collected data concerning various duties of those employees who maintained the job title of "aide." Each individual rated the 12 tasks on a five-point scale according to the importance that the task be carried out properly. The 12 tasks were

X_1: typing on electric typewriter

X_2: typing on manual typewriter

X_3: typing handwritten copy

X_4: typing from printed or typed copy

X_5: typing statistical reports

X_6: typing standard forms

X_7: typing letters or memos

X_8: correcting spelling and vocabulary errors

X_9: answering phones and taking messages

X_{10}: responding to customer inquiries

X_{11}: transferring calls to supervisors

X_{12}: placing internal calls to obtain information

The responses of 200 aides were obtained, and a correlation matrix (see Table P15.4) was developed. Use a computer package (such as SPSS, SAS, or BMDP) to perform a

(a) Principal components analysis of the tasks. Interpret the results.

(b) Factor analysis of the tasks using Varimax rotation. Interpret the results.

15.5. Reviewing concepts:

(a) What is the purpose of dimensionality reduction?

(b) What are the similarities and differences between principal components analysis and factor analysis?

(c) What is the purpose of a rotation of a factor solution?

TABLE P15.4   Correlation Matrix of 12 Tasks

Tasks	X_1	X_2	X_3	X_4	X_5	X_6	X_7	X_8	X_9	X_{10}	X_{11}	X_{12}
X_1	1.000	−.110	.485	.359	.272	.257	.432	.253	.283	.252	.361	.121
X_2		1.000	.179	.232	.238	.263	.196	.215	.113	.248	.156	.260
X_3			1.000	.444	.384	.382	.474	.370	.225	.229	.300	.094
X_4				1.000	.278	.385	.302	.348	.048	.098	.138	.075
X_5					1.000	.398	.439	.320	.135	.251	.250	.137
X_6						1.000	.266	.187	.199	.290	.192	.212
X_7							1.000	.302	.375	.401	.309	.217
X_8								1.000	.316	.382	.304	.270
X_9									1.000	.535	.629	.409
X_{10}										1.000	.575	.541
X_{11}											1.000	.391
X_{12}												1.000

16

MULTIDIMENSIONAL SCALING
AND CLUSTER ANALYSIS

16.1 INTRODUCTION AND OVERVIEW

Thus far in this text we have primarily been concerned with either evaluating the effect of a set of factors upon a response variable or predicting the value of a response variable based on a set of explanatory variables. In Chapter 15 we investigated two multivariate linear reduction procedures (principal components and factor analysis) which attempt to extract a smaller set of *factors* from a larger set of variables.

In this chapter we shall expand our discussion of multivariate methods to consider multidimensional scaling and cluster analysis. Like principal components and factor analysis, both of these techniques can be categorized as reduction procedures whereby either a measure of *similarity* or *distance* between objects is utilized as the basis for forming subgroups (i.e., clusters) or determining the dimensions (i.e., factors) that separate the objects on a geometric map. Multidimensional scaling and cluster analysis may also be classified as multivariate exploratory data procedures [see Gnanadesikan (1977)] in which a multidimensional graph of a set of objects is explored for its structure or subgroupings.

16.2 BASIC CONCEPTS OF MULTIDIMENSIONAL SCALING

Researchers in a wide variety of disciplines have often been concerned with investigating the perceived similarity among a set of objects. Thus, for example, psychologists have studied the perception of Morse code signals [see Kruskal and Wish (1978)], market researchers have studied the perceived similarity of product brands [see Green and Rao (1972)], while political scientists have examined the perceived similarity of various political candidates and parties [see Rommey et al. (1972)]. The objective of any multidimensional scaling analysis is to develop a map or configuration that locates the objects according to a measure of similarity that has been computed for all pairs of objects. As an aid toward understanding how this can be accomplished mathematically, we note the following: Figure 16.1 consists of a map of the actual location of eight

Figure 16.1   Geographic location of eight European cities.

European cities—Berlin, London, Madrid, Moscow, Paris, Rome, Stockholm, and Warsaw. From this we could determine the actual airline mileage between any two cities by simply measuring the distance according to the scale provided on the map. The entire set of these mileage distances is presented in matrix form in Table 16.1. When a map showing the location of the cities is available, it is, of course, quite simple to obtain such a table of distances. Now, however, imagine the opposite problem; that is, the set of distances between all pairs of cities are provided and we want to construct the map showing the location of the cities. Such a problem, which is a key aspect of multidimensional scaling, is not nearly as simple as obtaining the distances from the map. In our example, since we are

TABLE 16.1   Airline Distance in Miles Between Eight European Cities

City	Berlin	London	Madrid	Moscow	Paris	Rome	Stockholm	Warsaw
Berlin	—							
London	583	—						
Madrid	1165	785	—					
Moscow	1006	1564	2147	—				
Paris	548	214	655	1554	—			
Rome	737	895	851	1483	690	—		
Stockholm	528	942	1653	716	1003	1245	—	
Warsaw	322	905	1427	721	852	820	494	—

scaling eight objects (i.e., cities) in two dimensions (east-west and north-south), we need to obtain eight pairs of geometric coordinates. Moreover, for $n = 8$ objects being compared there are a total of $n(n - 1)/2 = 28$ different distances between objects that need to be evaluated in locating these coordinates. Thus the purpose of multidimensional scaling is to locate the coordinates of all objects in a graphical configuration or map so that the distances between the objects on the configuration have a direct relationship to the actual distances in the original data. Typically, any multidimensional scaling algorithm utilized attempts to maximize the *goodness of fit* of the fitted distances with these actual distances. **Metric multidimensional scaling** attempts to maintain a linear (or other) functional relationship between the plotted objects and the actual distances. **Nonmetric multidimensional scaling** (see Section 16.3) attempts to maintain only the rank order of the distances rather than the distances themselves.

Once the type of scaling to be performed (metric versus nonmetric) has been determined, the measure of distance between objects must be chosen so that the coordinates of the objects can be initially located. The most commonly used distance function is the *Euclidean* distance. For two dimensions we have

$$d_{ij} = \sqrt{(X_{i1} - X_{j1})^2 + (X_{i2} - X_{j2})^2} \qquad (16.1)$$

where X_{i1} = coordinate of object i in dimension 1
X_{j1} = coordinate of object j in dimension 1
X_{i2} = coordinate of object i in dimension 2
X_{j2} = coordinate of object j in dimension 2
d_{ij} = distance between stimulus i and stimulus j

This Euclidean distance function can be generalized to r dimensions as

$$d_{ij} = \sqrt{\sum_{k=1}^{r} (X_{ik} - X_{jk})^2} \qquad (16.2)$$

where X_{ik} = coordinate of object i in dimension k
X_{jk} = coordinate of object j in dimension k

The actual computations for a multidimensional scaling configuration are invariably performed by a computer using complicated forms of numerical analysis. However, the computational process can be described in the following manner: (1) A random starting configuration of coordinates is selected at the initial stage of the process. (2) The set of distances fit for this configuration is compared to the set of distances present in the data. (3) Since the objective is to develop a configuration that best represents the distances present in the original data, an evaluation is performed to measure the goodness of fit of the configuration obtained to the original data to determine whether the fit would be improved if the location of any coordinates were changed. Since the initial starting configuration usually is randomly selected, the location of the coordinates almost always can be changed to provide a better fit to the data. Thus this *fitting* process usually continues iteratively through many steps. At each step a new configuration is obtained that is a better fit to the original data than the previous configuration. The fitting process terminates when no further improvement can be obtained by altering the location of the points in the last configuration that has been fit.

If we refer back to our example concerning the distance between the eight European cities, multidimensional scaling methods may be utilized to obtain a map of these cities based only on the distances provided in Table 16.1. The resulting metric multidimensional configuration is presented as Figure 16.2.

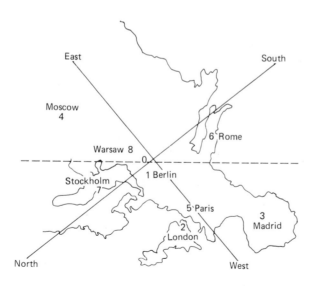

Figure 16.2 Configuration obtained from a metric multidimensional scaling analysis of the airline distance between eight European cities.

If we compare this multidimensional scaling map with the *actual* map of European cities depicted in Figure 16.1, we can observe several phenomena. Although the cities appear to be correctly positioned *relative* to each other, their location has been subjected to the following transformations:

1. Rotation about the origin
2. Translation of origin
3. Reflection (i.e., switching) of axes

Since the objective of multidimensional scaling is to maintain the distances between the cities *relative* to each other, the location of the coordinates of the set of cities is not unique. Thus in Figure 16.2 it appears that the north-south and east-west axes have been rotated away from the horizontal-vertical plane. In addition, the location of the origin is not uniquely determined and appears to have been somewhat shifted. Furthermore, not only have the axes been rotated and the origin shifted, but the axes themselves have been switched or *reflected*. For example, in Figure 16.2 we observe that both the north-south and east-west axes have been reflected about the origin. Thus Rome appears on the upper portion of the south-north axis, while Madrid appears in the lower portion of the east-west axis. Therefore, it is crucially important to view a multidimensional scaling configuration from several alternative perspectives when we attempt to interpret the factors or dimensions separating a set of objects.

These transformations have provided the set of objects (the eight European cities) with the property of *translation invariance* in which the difference between the objects is dependent only on the difference in the coordinates, not the coordinates themselves. Thus from this perspective the multidimensional scaling solution displayed as Figure 16.2 has provided an extremely accurate approximation to the actual map of the cities. In fact, for these data we have an almost perfect measure of goodness of fit. This measurement of goodness of fit, called *stress*, will be discussed in greater detail in Section 16.4.4.

16.3 NONMETRIC MULTIDIMENSIONAL SCALING

In metric multidimensional scaling the input distances are assumed to be ratio-scaled. However, *nonmetric* multidimensional scaling [see Shepard (1962)] assumes only an ordinal relationship among distances. That is, the criterion utilized in developing a multidimensional scaling configuration attempts to maintain only the rank order (i.e., *monotonicity*) of the distances. For example, if we refer to the airline mileages between the eight European cities (see Table 16.1), nonmetric multidimensional scaling would be concerned only with the rank order of these distances. Thus for the purposes of nonmetric multidimensional scaling, these mileages could be converted to a set of rank-order distances, with rank 1 given to the smallest distance and rank $n(n-1)/2$ given to the largest distance. The set of 28 rank-ordered distances for the eight European cities is presented as Table 16.2.

TABLE 16.2   Rank Order of the Distance Between Eight European Cities

City	Berlin	London	Madrid	Moscow	Paris	Rome	Stockholm	Warsaw
Berlin	—							
London	6	—						
Madrid	21	12	—					
Moscow	20	26	28	—				
Paris	5	1	7	25	—			
Rome	11	16	14	24	8	—		
Stockholm	4	18	27	9	19	22	—	
Warsaw	2	17	23	10	15	13	3	—

From Table 16.2 we note that the "closest" cities are London and Paris, the second closest cities are Berlin and Warsaw, while conversely the cities that are farthest apart are Moscow and Madrid. Figure 16.3 represents the configuration obtained from a nonmetric multidimensional scaling analysis of the airline distances of these eight European cities. The measure of the goodness of fit of this configuration to the actual data is based only on the degree to which the rank order of the actual distances has been preserved. Thus if we examine Figure 16.3, we can observe that London and Paris appear to be close to each other, as are Berlin and Warsaw, while Madrid and Moscow appear to be "far apart." In fact, except for the rotation and reflection of the axes, the nonmetric configuration seems remarkably similar to the metric configuration depicted in Figure 16.2.

16.4 MULTIDIMENSIONAL SCALING OF SIMILARITIES

16.4.1 Introduction

Thus far in our discussion of the basic principles of multidimensional scaling, we have referred to a model of the location of eight European cities. Although this model provided an excellent fit to the true location of the cities, the input data used for the analysis contained certain properties that are rarely found in any actual application. First, the distance between the cities was measured *without error*. Second, the actual location of the cities was available from a geographical map. Thus we were able to compare the configuration obtained from the multidimensional scaling model to the true map.

In reality, however, any actual set of data would not contain either of these properties. First, the measure of distance between the objects themselves is usually obtained only from a sample of respondents. Thus any measure of distance developed inherently contains *error* that represents the difference

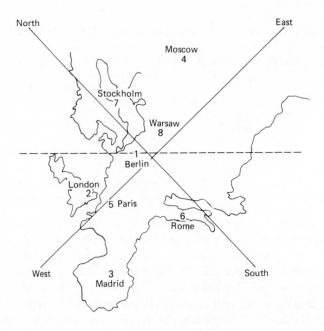

Figure 16.3 Configuration obtained from a nonmetric multidimensional scaling analysis of the airline distance between eight European cities.

between the *true* distance and the distance measured from the sample. In addition, the "distance" itself does not always represent a physical phenomenon such as miles, but often is only a *perceptual* distance. Second, the true mapping of the set of objects is rarely known (although various structures may sometimes be hypothesized). Thus any multidimensional scaling configuration needs to be carefully studied since the important dimensions or factors separating the objects must be interpreted *only from the configuration itself*. This is clearly in contrast to our example of the eight European cities wherein the dimensions were known to represent east-west and north-south geographical axes.

16.4.2 Alternative Data Collection Methods

We have observed that the basic input to a multidimensional scaling analysis consists of the set of "distances" that have been measured between all pairs of objects. Thus a key aspect of data collection concerns the manner in which this distance matrix is obtained. The distances themselves can represent either *similarities* or *dissimilarities* among the objects. Similarities occur when a "large" numerical value between object *i* and object *j* represents a "small" distance. For example, when working with psychophysical confusion matrices (such as Morse code signals), the measure of distance is based on the number of times that one signal is confused with another [see Kruskal and Wish (1978)]. The greater the frequency of the confusions, the more similar the two signals are considered. A second example of similarities occurs when the correlation matrix of a set of objects or variables is being analyzed (as in principal components or factor analysis). Since a high correlation between two variables represents a strong association, it would also represent a small distance between them. *Dissimilarities* occur when a small numerical value between object *i* and object *j* represents a small distance. Thus our discussion of the mapping of the European cities involved an example of dissimilarities because a small numerical value implied a small distance between any two cities.

Regardless of whether the distances measure similarities or dissimilarities, there exist two basic alternative procedures for data collection: *direct* methods and *derived* methods. Direct methods of data collection involve the evaluation of the similarity (or dissimilarity) of the set of objects without reference to any prespecified criteria. One example of this is the method of paired comparisons [see Green and Rao (1972)]. Let us suppose that there are five products that are to be compared (A, B, C, D, and E). The method of paired comparisons evaluates each *pair* of products and develops the rank order of the similarity for the set of pairs. With five products there would be [5(4)]/2 or 10 pairwise comparisons to be made (AB, AC, AD, AE, BC, BD, BE, CD, CE, and DE). These 10 pairs would be placed in their order of similarity such that the most similar pair is given the rank 1, the second most similar pair is given rank 2, and so on until the least similar pair is given the rank 10. The major advantage of this direct approach is that the paired ranks are developed without any constraint imposed by the researcher in the form of prespecified criteria. Unfortunately, however, when more than a limited number of products or objects are to be compared, this approach becomes extremely unwieldy due to the large number of comparisons that need to be made. For example, if 10 products are to be compared, 45 pairwise comparisons would be required; if 15 products are to be compared, 105 such pairwise evaluations would be needed. In such situations, an alternative data collection approach (i.e., *derived* methods) can be advantageous.

Derived methods of data collection usually involve the rating of each object on a set of scales that represent characteristics or attributes of the object being studied. Although any derived method based on a set of rating scales implicitly assumes that the set of scales has described all characteristics of the objects, these methods can be much less cumbersome than direct methods. The "distance" between any two objects can be developed by comparing their ratings on a set of scales. If the ratings are similar over the scales, the distance will be small. If the ratings of two objects are quite different over the set of scales, their distance measure will be large. Since large sets of ratings are usually involved, computer programs [see Rao (1970) or Westin and Dillon (1979)] are available to develop a distance matrix from a set of object ratings obtained from a group of respondents.

16.4.3 Application

Levine (1977) collected a set of data from a sample of 45 undergraduate students in order to study the perception of various sports. The following 13 sports were included:

1. Boxing (BX)
2. Basketball (BK)
3. Golf (G)
4. Swimming (SW)
5. Skiing (SK)
6. Baseball (BB)
7. Ping-Pong (PP)
8. Hockey (HK)
9. Handball (H)
10. Track and field (TF)
11. Bowling (BW)
12. Tennis (T)
13. Football (F)

In view of the large number of objects (i.e., sports), a derived method of data collection was utilized with six different seven-point bipolar rating scales. These scales were

Fast moving (1)	Slow moving (7)
Complicated rules (1)	Simple rules (7)
Team-oriented (1)	Individual (7)
Easy to play (1)	Hard to play (7)
Noncontact (1)	Contact (7)
Competition against opponent (1)	Competition against standard (7)

Each respondent rated each sport on each bipolar adjective. Once these ratings were compiled, a version of the DISTAN program [see Rao (1970)] was utilized to obtain the average distance matrix over all subjects and the average rating of each sport on each bipolar adjective. These results are presented as Tables 16.3 and 16.4.

Before using the average distance matrix to perform multidimensional scaling, it would be of value to briefly examine these tables. Table 16.3 indicates the "distance" between each pair of sports for an "average" respondent. The numerical values in Table 16.3 represent dissimilarities rather than similarities, since small differences in the ratings of two sports produce a small distance between them. Although the distances range from 2.20 to 5.15, it is more impor-

TABLE 16.3   Aggregate Derived Distance Matrix for the 13 Sports

Sport	BX	BK	G	SK	SW	BB	PP	HK	H	TF	BW	T
BK	3.85											
G	4.33	4.88										
SK	3.80	4.05	3.73									
SW	3.81	3.81	3.56	2.84								
BB	4.12	3.15	3.83	4.16	3.60							
PP	3.74	3.56	3.61	3.67	2.72	3.41						
HK	3.85	2.58	5.11	4.02	4.17	3.49	4.27					
H	3.41	3.24	3.92	3.25	2.80	3.34	2.58	3.52				
TF	3.81	3.36	3.88	3.20	2.84	3.37	3.06	3.72	2.75			
BW	4.07	4.23	2.72	3.75	2.89	3.32	2.87	4.58	3.13	3.26		
T	3.49	3.32	3.59	3.19	2.82	3.25	2.54	3.58	2.33	2.72	2.85	
F	3.86	2.51	5.15	4.38	4.41	3.43	4.35	2.20	3.68	3.84	4.67	3.69

SOURCE: D. M. Levine, "Nonmetric Multidimensional Scaling and Hierarchical Clustering: Procedures for the Investigation of the Perception of Sports," *Research Quarterly*, Vol. 48 (1977), pp. 341–348.

TABLE 16.4   Average Rating of Each Sport on Each Bipolar Adjective

Sport	(1) Fast Mvg. (7) Slow Mvg.	(1) Compl. (7) Simple	(1) Team (7) Indv.	(1) Easy to Play (7) Hard to Play	(1) Ncon. (7) Con.	(1) Comp Opp. (7) Comp Std.
Boxing	3.07	4.62	6.62	4.78	6.02	1.73
Basketball	1.84	3.78	1.56	3.82	4.89	2.27
Golf	6.13	4.49	6.58	3.84	1.82	4.11
Swimming	2.87	5.02	5.29	3.64	2.22	4.36
Skiing	2.13	4.60	5.96	5.22	2.51	4.71
Baseball	4.78	4.18	2.16	3.33	3.60	2.67
Ping-Pong	3.18	5.13	5.38	2.91	2.04	2.20
Hockey	1.71	3.22	1.82	5.04	5.96	2.49
Handball	2.53	4.67	4.78	3.71	2.78	2.31
Track & field	2.82	4.38	4.47	3.84	2.89	3.82
Bowling	5.07	5.16	5.40	3.11	1.60	3.73
Tennis	2.89	3.78	5.47	4.09	2.16	2.42
Football	2.42	2.76	1.44	5.00	6.47	2.33

SOURCE: D. M. Levine, "Nonmetric Multidimensional Scaling and Hierarchical Clustering: Procedures for the Investigation of the Perception of Sports," *Research Quarterly*, Vol. 48 (1977), pp. 341–348.

tant to note the differences in the distances between each pair of sports (for metric scaling) or the rank order of the paired distances (for nonmetric scaling). For example, we may observe that the smallest distance occurs between football and hockey, while the largest distance occurs between football and golf. If we examine Table 16.4, the basis for these results is clearly evident. Football and hockey are perceived as fast-moving, high-contact team sports that have complicated rules, are hard to play, and involve competition against an opponent. In contrast, football and golf are perceived quite differently. Football is considered fast moving, while golf is very slow moving; football is a team sport, while golf is an individual sport; football is a very high-contact sport, while golf is very much a noncontact sport.

Although some comparisons of the sports can be made from these tables, multidimensional scaling can reveal the overall structure of the group of sports. Nonmetric multidimensional scaling was utilized since it makes less rigid assumptions regarding the distances. The rotated two-dimensional graphical solution for the 13 sports is depicted in Figure 16.4.

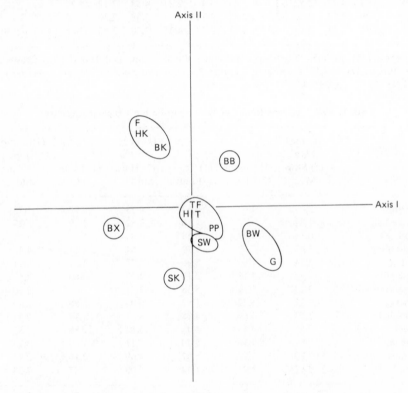

Figure 16.4 Rotated two-dimensional configuration for the similarity of the 13 sports. Source: D. M. Levine, "Nonmetric Multidimensional Scaling and Hierarchical Clustering: Procedures for the Investigation of the Perception of Sports," *Research Quarterly*, Vol. 48 (1977), pp. 341–348.

16.4.4 Interpreting the Results

When evaluating a multidimensional scaling solution there are two basic aspects that need to be considered:

1. Determination of the proper number of dimensions

2. Interpretation of the dimensions (or factors)

In evaluating the multidimensional scaling results, we need to consider both the number of dimensions and the goodness of fit for the particular configuration obtained. For a given set of data, an increase in the number of dimensions will improve the fit of the configuration to the actual data. However, we must realize that increasing the dimensionality complicates any analysis since the interpretation of more than two or three dimensions is usually quite difficult. Thus the objective in multidimensional scaling is to maximize the goodness of fit of a solution subject to minimizing the number of dimensions that need to be interpreted. Although several measures of goodness of fit have been suggested, one useful measure is called *stress* and is defined as

$$\text{Stress} = \sqrt{\frac{\sum_{i,j=1}^{m} (d_{ij} - \hat{d}_{ij})^2}{\sum_{i,j=1}^{m} (d_{ij} - \bar{d})^2}} \tag{16.3}$$

where d_{ij} = distance between a particular pair of objects (i, j)

$\hat{d}_{ij}$ = fitted regression value estimated from the original data for objects (i, j)

$\bar{d}$ = arithmetic mean of the distances

$m = n(n - 1)/2$, where n is the number of objects in the configuration

As the fit of the multidimensional scaling solution to the data improves, the stress statistic approaches 0. Thus, in practice, a set of data is usually scaled in a varying number of dimensions (ranging from one to four). Our sports data revealed a stress of .474 in one dimension, .196 in two dimensions, and .133 in three dimensions. In view of the relatively large difference in stress between one and two dimensions and the relatively small difference between the stress values for two and three dimensions, it would make sense to first examine the two-dimensional configuration presented as Figure 16.4. When evaluating a particular configuration, we need to (1) study the location of the objects relative to each other and (2) determine the factors or dimensions that appear to separate the various objects. From Figure 16.4 we may observe that there are several sub-groups or *clusters* of sports. Football, hockey, and basketball appear close to each other in the upper left-hand quadrant of the map; bowling and golf are clustered in the lower right-hand quadrant; track and field, handball, tennis, Ping-Pong, and swimming seem to be located at the center of the map. To interpret any factors, we need to determine the basis upon which the sports have been separated on the vertical and horizontal axes. If we examine the vertical axis (Axis II), we may observe that football, hockey, basketball, and baseball are at the top, while skiing and golf are at the bottom of this axis. By using Table 16.4, this axis can then be interpreted as a "team sport" factor or dimension—with a polarization of the team sports away from the nonteam sports. Referring to Table 16.4, each team sport has a low average rating on the team-individual rating scale, while each nonteam sport has a high average rating.

Now that the vertical axis has been interpreted, we need to interpret the factor relating to the horizontal axis (Axis I). Football, hockey, basketball, and boxing appear to be clearly separated from bowling and golf on this axis. To be able to evaluate this separation, we need to again refer to Table 16.4. Boxing, football, hockey, and basketball have been perceived as fast-moving contact sports, while bowling and golf are considered slow-moving noncontact sports. Thus we may conclude that the horizontal axis represents a "degree of action" factor combining both contact-noncontact and slow-moving/fast-moving characteristics.

Thus, to summarize, we have been able to interpret two factors that appear to separate the perception of the sports. These factors are "team versus individual" and degree of action which combines both contact-noncontact aspects and fast-moving/slow-moving characteristics.

16.5 BASIC CONCEPTS OF CLUSTER ANALYSIS

16.5.1 Introduction

In our discussion of multidimensional scaling, we have focused upon the geometric location of objects on a configuration in a specified number of dimensions. We have also attempted to interpret those dimensions that appear to be the basis on which the objects have been separated. Thus in our study of the perception of sports (Section 16.4) we were able to interpret the dimensions "team versus individual" and "degree of action" as the basis on which the various sports were being differentiated.

In contrast to multidimensional scaling, **cluster analysis** can be viewed as a *dimension-free classification procedure* that attempts to subdivide or partition a set of heterogeneous objects into relatively homogeneous groups. In other words, the objective of cluster analysis is to develop subgroupings such that objects within a particular subgroup are more like other objects within that subgroup than they are to objects in a different subgroup. Thus the outcome of cluster analysis is the development of a classification scheme that provides the sequence of groupings by which a set of objects is subdivided.

16.5.2 Basic Principles

During the past 10 years many alternative procedures have been developed to perform cluster analysis using various computer algorithms [see Anderberg (1973) and Hartigan (1975)]. Since the objective of any clustering procedure is to partition the objects into subgroups (or clusters) in a sequence of steps, most clustering algorithms begin by considering each object as its own distinct cluster. The first step in the clustering process involves placing two objects together in a single cluster according to the optimizing criteria of the algorithm while grouping each of the remaining objects separately. The second clustering step involves grouping the objects into either one cluster of three or two clusters of two (with each remaining object grouped separately). This clustering process continues sequentially until all objects have been merged into a single collectively exhaustive cluster.

Two principal issues determine the optimal formation of the clusters. The first consideration relates to the overall structure of the sequencing process. The most widely utilized approach involves a *hierarchical clustering* arrangement. Under this approach, once two objects are linked together at a particular stage, they are considered to be permanently merged and cannot be separated into different clusters later in the clustering process. Thus, clustering decisions at a particular step are conditioned upon the arrangement of objects at the previous step. One advantage of this hierarchical clustering approach is that the number of possible clustering choices at each step of the process continually diminishes.

The second issue concerns the measure of distance utilized in linking the objects together in clusters. Generally, there are three alternative approaches. The first approach, called *complete linkage*, bases the merger of two subsets of objects on the maximum distance between objects. This approach (also called the *farthest neighbor* or *diameter method*) seems to produce compact clusters of approximately equal size and appears to provide a more powerful representation of data structure [see Hubert and Baker (1976)] than the other two approaches.

The second approach is called *single linkage* (i.e., the *connectedness* or *nearest neighbor method*). This approach bases the merger of two subsets of objects on the minimum distance between objects. The single linkage approach often produces a single large chainlike cluster and several small clusters during its sequencing process.

The third approach, called *average linkage*, bases the merger of two subsets of objects on the average distance between objects. This third alternative can be viewed as a middle ground between the other two alternative linkage approaches.

16.5.3 Application

Now that the basic principles of cluster analysis have been developed, we can fix ideas by analyzing the sports data (see Section 16.4) using clustering methodology. As in multidimensional scaling, the basic input for cluster analysis usually consists of a distance matrix between all pairs of objects. Thus Table 16.3 was used as the input to the BMDP 1M cluster program. The tree diagram obtained (using the hierarchical complete linkage option) is presented as Figure 16.5.

The sequence in which the sports have been merged into clusters can be revealed by examining this tree diagram from left to right. The first pair of sports to cluster together is hockey and football, followed by handball and tennis. Subsequently, basketball merges with football and hockey, while Ping-Pong

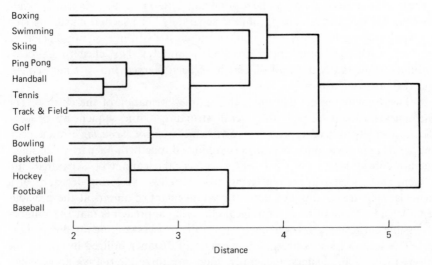

Figure 16.5 Hierarchical linkage of the sports from initial to final grouping as provided by the BMDP 1M computer program.

merges with handball and tennis. Next, golf and bowling merge into a single cluster and so on until all sports are grouped into a single cluster.

16.5.4 Interpreting the Results

The clustering results depicted in Figure 16.5 exhibit a clear consistency with those obtained from nonmetric multidimensional scaling. To facilitate this comparison, the clusters (obtained at the seven-cluster level) have already been superimposed onto the multidimensional scaling solution in Figure 16.4. In addition, it is particularly interesting to note the two-cluster solution that separates the team sports and nonteam sports. This outcome provides evidence to support the conjecture that the team sport characteristic is the principal dimension separating the various sports.

16.6 COMPUTER PACKAGES AND MULTIDIMENSIONAL SCALING AND CLUSTER ANALYSIS: USE OF SAS AND BMDP

16.6.1 Using SAS PROC ALSCAL for Multidimensional Scaling

The SAS ALSCAL procedure applies the *alternating least-squares scaling algorithm* [see Takane et al. (1977) and Schiffman et al. (1981)] to a variety of multidimensional scaling models including the basic model discussed in Sections 16.1–16.4. The format for using PROC ALSCAL is

$$\text{PROCþALSCALþLEVEL} = \left.\begin{array}{l}\text{RATIO or}\\ \text{INTERVAL}\end{array}\right\} \leftarrow \text{metric scaling}$$

$$\text{or}$$

$$\left.\begin{array}{l}\text{ORDINAL or}\\ \text{NOMINAL}\end{array}\right\} \leftarrow \text{nonmetric scaling}$$

PLOTþDIMENS = {value}þITER = {value};
VARþ{list of variables};

The LEVEL = option indicates whether metric (RATIO or INTERVAL) or nonmetric (ORDINAL or NOMINAL) scaling is being performed. The PLOT option provides a plot of the multidimensional scaling configuration. The DIMENS = option is utilized if other than two dimensions is requested. The ITER = option provides for additional iterations if more than 30 iterations are desired. The input data matrix typically consists of a symmetric half matrix of distances. If similarities are utilized instead of dissimilarities, the SIMILAR option may be invoked. For metric scaling, the DEGREE = option indicates the degree of the polynomial to be fitted to the data.

The VAR statement specifies the names of the variables to be used. If all variables are to be analyzed, this statement may be omitted.

Figure 16.6 represents the SAS program that has been written using PROC ALSCAL for the sports data. We may note that since the distance matrix serves as input, the INPUT statement provides the format for each distance and sport name. Figure 16.7 presents partial output for this SAS program.

16.6.2 Using the BMDP Program 1M for Cluster Analysis

Although there is a multitude of computer programs that perform cluster analysis, the BMDP package is particularly flexible since it can cluster either variables (objects) in program P-1M or cluster respondents (cases) in program P-KM. In addition, it can perform cluster analysis directly on a set of variables or accept a distance matrix as input.

The only paragraphs of BMDP that are unique to program P-1M are the /INPUT and /PROCEDURE paragraphs. The format of these paragraphs is

$$/\text{INPUTþVARIABLESþAREþ} \left\{\begin{array}{l}\text{number of}\\ \text{objects}\end{array}\right\} .$$

$$\text{TYPE} = \left\{\begin{array}{l}\text{DIST or}\\ \text{CORR or}\\ \text{SIMI or}\\ \text{COVA or}\\ \text{ANG}\end{array}\right\} .$$

$$\text{FORMATþISþ'(} \qquad \text{)'.}$$

$$/\text{PROCEDUREþLINKAGEþISþ} \left\{\begin{array}{l}\text{COMP or}\\ \text{SINGLE or}\\ \text{AVE}\end{array}\right\} .$$

```
$SYSTEM CARDS$
DATA SPORT;
INPUT(SPORT1-SPORT13 SPTNAME)(13*5.3 @67 $10.);
CARDS;
3.85                                                            BOXING
4.33 4.88                                                       BSKETBALL
3.80 4.05 3.73                                                  GOLF
3.81 3.81 3.56 2.84                                             SKIING
4.11 3.15 3.83 4.16 3.60                                        SWIMMING
3.74 3.56 3.61 3.67 4.41 3.72                                   BASEBALL
3.85 2.58 5.11 4.08 3.41 4.17 2.58                              PINGPONG
3.41 3.24 3.92 3.25 3.34 2.80 3.06 3.52                         HOCKEY
3.81 3.36 3.88 3.75 3.37 2.84 2.87 3.72 2.75 3.26               HANDBALL
4.07 4.33 2.72 3.19 3.32 2.81 2.54 4.58 3.13 3.26 2.72          TRKFIELD
3.49 3.32 3.59 3.38 3.25 2.82 3.58 3.33 2.72 2.85               BOWLING
3.86 2.51 5.15 3.19 3.43 4.20 2.20 3.68 3.84 4.67 3.69          TENNIS
                                                                FOOTBALL
PROC ALSCAL PLOT LEVEL=ORDINAL ITER=50;
$SYSTEM CARDS$
```

Figure 16.6 SAS PROC ALSCAL program for sports data.

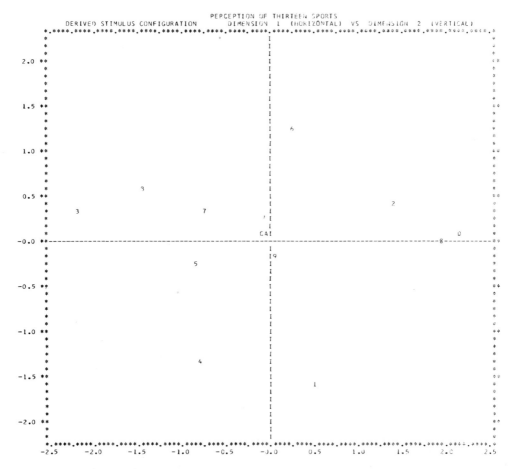

Figure 16.7 Partial output from SAS PROC ALSCAL for sports data.

The VARIABLES sentence specifies the number of objects being scaled. The TYPE sentence is included in the /INPUT paragraph only when a data matrix (not raw data) is provided. The TYPE of data could represent a correlation matrix (CORR), a similarities matrix (SIMI), a covariance matrix (COVA), an angular distance matrix (ANG), or a distance matrix (DIST). Unfortunately, when matrix input is involved, program P-1M accepts only a full-distance matrix rather than a lower half matrix such as displayed in Table 16.3. We should also note that the self-distances (which of course are zero) need to be entered as part of the matrix.

The /PROCEDURE paragraph defines the type of linkage used in the clustering procedure. The alternative methods are complete (COMP), single (SINGLE), or average (AVE). When the input consists of raw data, the /PROCEDURE paragraph must include a MEASURE = sentence that indicates the distance measure utilized. This measure can be the correlation (CORR), the

467

absolute value of the correlation (ABSCORR), the angular distance (ANG), or the absolute value of the angular distance (ABSANG).

Figure 16.8 represents the program that has been written using P-1M for the sports example. Figure 16.9 presents partial output of the BMDP program 1M for the sports data.

The sequence of clusters shown in Figure 16.9 utilizes the information concerning the distance or similarity when the cluster is formed. The smallest distance (2.20) occurred when hockey combined with football (sport 13). The next smallest distance was 2.33 between handball and tennis (sport 12). Then football and hockey combined with basketball (sport 2) to form a three-sport cluster at a distance of 2.58. This process continued until the entire clustering sequence was developed (see Figure 16.9).

16.6.3 Using BMDP Program KM to Cluster Respondents

In many applications the researcher wishes to cluster respondents instead of, or in addition to, a set of variables or objects. For example, in Problem 16.5 we would like to apply clustering methods to analyze the 10 largest insurance companies based on a set of seven quantitative variables. Since the companies (i.e., the "respondents") rather than the variables are to be clustered, we need to use program P-KM. If the variables were being clustered, then P-1M would be utilized.

The only unique portion of program P-KM concerns the /CLUSTER paragraph. The format of the /CLUSTER paragraph is

/CLUSTER♭NUMB $= \#_1, \#_2, \ldots$.   where each $\#$ represents the desired
                                                                   number of clusters

Another aspect concerns the use of alphabetic characters for naming the respondents (i.e., the insurance companies). Since an alphabetic variable is limited to an A4 format in BMDP, each insurance company (which has a 16-character label) is considered to consist of four "variables" (thus providing a format of 4A4 for each label).

Figure 16.10 presents the program that has been written using P-KM for the insurance company data.

16.6.4 Using SAS PROC CLUSTER for Cluster Analysis

The SAS CLUSTER procedure can be utilized to perform a hierarchical clustering analysis on a set of respondents. For example, in Problem 16.5 we would like to use clustering methods to analyze the 10 largest insurance companies based on a set of seven quantitative variables. The format for PROC CLUSTER is

PROC♭CLUSTER♭{options};
ID♭{variable name};
VAR♭{variable list};

Figure 16.8 Using BMDP program 1M for sports data.

```
{SYSTEM CARDS}
/PROBLEM  TITLE IS 'PERCEPTION OF THIRTEEN SPORTS'.
/INPUT    VARIABLES ARE 13.
          TYPE=DIST.
          FORMAT IS '(13F5.2)'.
/VARIABLE NAMES ARE BOXING,BSKTBALL,GOLF,SWIM,SKIING,BASEBALL,
          PINGPONG,HANDBALL,TRKFLD,BOWLING,TENNIS,FOOTBALL.
/PROCEDURE LINKAGE IS COMP.
/END
0.00 3.85 4.33 3.80 3.81 4.12 3.74 3.85 3.41 3.81 4.07 3.49 3.86
...  ...  ...  ...  ...  ...  ...  ...  ...  ...  ...  ...  ...
3.86 2.51 5.15 4.38 4.41 3.43 4.35 2.20 3.68 3.84 4.67 3.69 0.00
{SYSTEM CARDS}
```

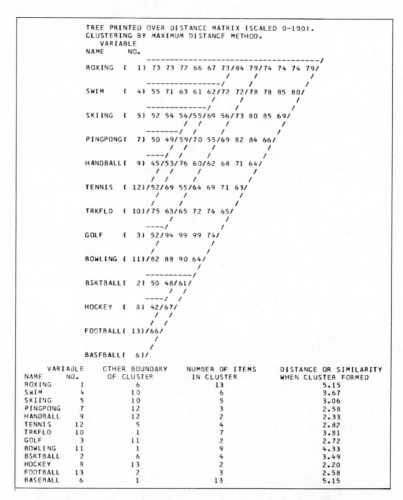

Figure 16.9 Partial BMDP program 1M output for sports data.

```
{SYSTEM CARDS}
/PROBLEM    TITLE IS 'INSURANCE COMPANY DATA'.
/INPUT      VARIABLES ARE 11.
     FORMAT IS '(4A4,F5.0,2F4.0,F5.1,F6.0,F5.0,F6.0)'.
/VARIABLE   NAMES ARE C01,C02,C03,C04,ASSETS,RECEIPTS,
            INCOME,NETGAIN,LIFEINS,CHANGINS,EMPLOY.
            LABELS ARE C01,C02,C03,C04.
/CLUST      3,4,5,6.
/END
PRUDENTIAL        54734800833132 78.4367284369 20061942
  :                :      :      :     :     :      :
MASS,MUTUAL       834111222 542 56.9 44527 4316  9243
{SYSTEM CARDS}
```

Figure 16.10 Using BMDP program KM for insurance data.

The options available as part of the PROC CLUSTER step include NOMAP (if no cluster map is desired) and N = {value} to indicate the number of clusters desired. The ID statement provides the values or categories of the identifying variable (i.e., the insurance companies) that is to be clustered. The VAR statement indicates the set of variables to be used in the cluster analysis.

Figure 16.11 represents the SAS program that has been written for the insurance company data. We may observe from Figure 16.11 that the symbol $ follows the variable name COMPANY in the INPUT statement since this variable represents an alphabetic label.

```
{SYSTEM CARDS}
DATA INSURE;
  INPUT COMPANY$ 1-16 ASSETS 17-21 RECEIPTS 22-25
    INCOME 26-29 NETGAIN 30-34 LIFEINS 35-40
    CHANGINS 41-45 EMPLOY 46-51;
  LABEL COMPANY=NAME OF INSURANCE COMPANY;
CARDS;
PRUDENTIAL            5473480083313278.43672843692006 1942
  :       :       :       :       :       :       :       :
MASS. MUTUAL         834 11222 542 56.9 44527 4316    9243
PROC PRINT;
PROC CLUSTER;
  ID COMPANY;
  VAR ASSETS RECEIPTS INCOME NETGAIN LIFEINS
    CHANGINS EMPLOY;
  TITLE CLUSTER ANALYSIS OF INSURANCE COMPANIES;
{SYSTEM CARDS}
```

Figure 16.11   SAS program for insurance data.

PROBLEMS

*16.1. (a) A market research firm has recently conducted a study concerning the perceived similarity and preference of different types of food:

1. Japanese	6. Mandarin (Chinese)
2. Cantonese (Chinese)	7. American
3. Szechuan (Chinese)	8. Spanish
4. French	9. Italian
5. Mexican	10. Greek

The average value (over 50 respondents) for each of the foods on the following three bipolar adjectives

Bland (1)   Spicy (7)
Light (1)   Heavy (7)
Low calories (1)   High calories (7)

was

	Rating Scale		
Food	Spicy/Bland	Heavy/Light	High/Low Calories
Japanese	2.8	3.2	3.4
Cantonese	2.6	5.3	5.4
Szechuan	6.6	3.6	3.0
French	3.5	4.5	5.1
Mexican	6.4	4.3	4.3
Mandarin	3.4	4.1	4.2
American	2.3	5.8	5.7
Spanish	4.7	5.4	4.9
Italian	4.6	6.0	6.2
Greek	5.3	4.7	6.0

The market researcher has utilized the multidimensional scaling model to investigate the similarity of perception of these 10 types of foods. The two-dimensional space of stimuli in the study revealed the following:

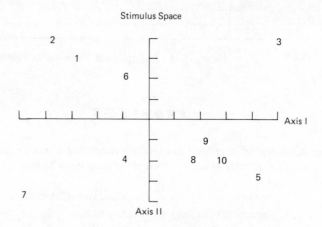

Stimulus Space

Analyze these data, and be sure to include the following:
(1) An interpretation of the two dimensions.
(2) An analysis of the similarities among the various types of foods.
(b) The similarities data of the average subject were analyzed using hierarchical clustering with the following results:

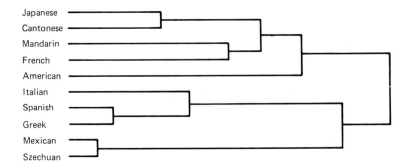

Analyze the results, and compare them to those obtained from multidimensional scaling [see part (a)].

16.2. A social science researcher has utilized the multidimensional scaling model to investigate the similarity of the perception of various academic fields of study:

1. Statistics	7. Psychology
2. Mathematics	8. Biology
3. Marketing	9. Accounting
4. Economics	10. History
5. Management	11. Computers
6. English literature	12. Art

The multidimensional scaling analysis revealed the following configuration for the 12 academic subjects based on a sample of 50 respondents:

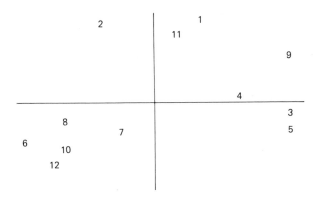

(a) Analyze these results, and be sure to include the following:
   (1) An interpretation of the two dimensions.
   (2) An analysis of the similarity of the academic fields.
(b) The similarities data were also analyzed using hierarchical clustering. The results were as follows:

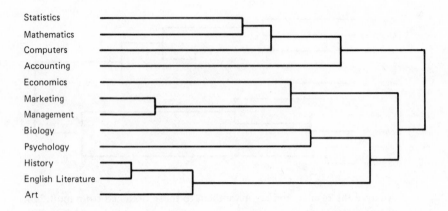

Analyze these results, and compare them to those obtained from the multi-dimensional scaling analysis in part (a) of this problem.

16.3. A study was conducted concerning the perceived similarity of 10 different soft drinks*:

1. Coca-Cola	6. Pepsi-Cola
2. Diet Pepsi	7. Tab
3. Seven-Up	8. Dr. Pepper
4. Welch's Grape Soda	9. Sunkist Orange Soda
5. C&C Cola	10. Sprite

The average value (for 58 respondents) for each of the soft drinks on the following seven bipolar adjectives

A	Flat (1)	Fizzy (7)
B	Expensive (1)	Inexpensive (7)
C	Hard to find (1)	Readily available (7)
D	Bland (1)	Flavorful (7)
E	Lack of aftertaste (1)	Presence of aftertaste (7)
F	Not sweet (1)	Sweet (7)
G	Does not satisfy thirst (1)	Satisfies thirst (7)

was

*The data are extracted from a study conducted by M. Mandelbaum and E. Reich as part of the requirements in a course taught by D. Levine.

16.4. Table 16.1 contained the airline distance (in miles) between eight European cities. Table P16.4 contains a matrix of distances in which random error (uniformly distributed between 0 and 200 miles) has been added to the actual distance.

TABLE P16.4   Airline "Pseudodistance" in Miles Between Eight European Cities

City	Berlin	London	Madrid	Moscow	Paris	Rome	Stockholm	Warsaw
Berlin	—							
London	621	—						
Madrid	1321	787	—					
Moscow	1037	1694	2345	—				
Paris	557	369	700	1560	—			
Rome	917	1011	951	1603	733	—		
Stockholm	720	1108	1748	893	1040	1294	—	
Warsaw	357	918	1536	730	1038	948	620	—

(a) Use this distance matrix to perform a *nonmetric* multidimensional scaling analysis of the location of the eight European cities.

(b) Do part (a) using metric multidimensional scaling.

(c) Compare the results obtained in parts (a) and (b) of the problem to the actual location of these cities depicted in Figure 16.1. What has been the effect on the multidimensional scaling map of adding randomly distributed error to the distances?

16.5. Referring to the data of Problem 15.3,

(a) Perform a hierarchical cluster analysis of the 10 largest insurance companies using BMDP program KM or SAS PROC CLUSTER (see Sections 16.6.3 and 16.6.4) with the complete linkage method.

(b) Perform a hierarchical cluster analysis of the *seven variables* (obtained from the insurance companies) using BMDP program 1M (see Section 16.6.2) with the complete linkage method.

16.6. Reviewing concepts:

(a) What is meant by multidimensional scaling?

(b) Describe the differences between metric and nonmetric multidimensional scaling.

(c) What is meant by cluster analysis?

(d) Compare and contrast multidimensional scaling and cluster analysis as data reduction methods.

Soft Drink	Rating Scale						
	A	B	C	D	E	F	
Coca-Cola	5.810	2.914	6.345	6.069	4.017	5.655	4.7
Diet Pepsi	4.810	3.190	5.707	3.293	5.345	3.672	3.1
Seven-Up	5.534	3.121	5.948	5.603	3.379	5.000	4.9
Welch's Grape Soda	3.655	3.655	4.052	4.207	4.897	5.241	3.4
C&C Cola	4.776	4.500	4.655	3.983	4.276	5.052	3.79
Pepsi-Cola	5.879	3.310	6.207	5.310	4.138	5.397	4.20
Tab	4.621	3.397	5.810	3.690	5.172	3.345	3.24
Dr. Pepper	4.862	3.466	5.121	4.655	4.810	4.724	4.034
Sunkist Orange Soda	4.483	3.828	4.466	5.379	4.293	5.172	4.603
Sprite	5.121	3.586	5.293	5.241	3.655	5.448	5.052

Table P16.3 represents the distance matrix among these 10 soft drinks for the average subject.

TABLE P16.3   Derived Distance Matrix for Soft Drinks for 58 Respondents

Soft Drink	Coke	Diet Pepsi	Seven-Up	Welch's	C&C Cola	Pepsi	Tab	Dr. Pepper	Sunkist
Diet Pepsi	3.96								
Seven-Up	2.56	3.78							
Welch's	4.43	4.23	4.08						
C&C Cola	3.99	3.84	3.80	3.68					
Pepsi	2.51	3.67	2.78	4.21	3.60				
Tab	4.06	2.75	3.90	4.41	3.85	3.74			
Dr. Pepper	3.42	3.51	3.26	3.59	3.30	3.14	3.36		
Sunkist	3.70	4.20	3.18	3.27	3.54	3.63	4.14	3.03	
Sprite	3.16	3.92	2.48	3.90	3.62	3.20	3.88	3.10	2.81

(a) Use a multidimensional scaling computer program (such as ALSCAL) to analyze these data.

(b) Use a computer program (such as BMDP program 1M) to perform a hierarchical cluster analysis of these soft drinks using the *complete linkage* option.

(c) Do part (b) of this problem using a *single linkage* option, and compare the results.

(d) (**Class Project**) Ask each student in the class to rate these 10 soft drinks on each of the seven bipolar adjectives. Use the DISTAN computer program [Rao (1970)] or the CANDI program [Westin and Dillon (1979)] to obtain the distance matrix for the average subject. For this distance matrix, perform multidimensional scaling and hierarchical cluster analysis, and compare the results to those obtained in parts (a)–(c).

CROSS-CLASSIFIED
CATEGORICAL DATA

17.1 INTRODUCTION

It is interesting to observe that with few exceptions data which are collected through the vehicle of questionnaires are in the form of frequency counts. That is, such data are obtained by merely classifying the observed responses into various distinct categories.* When the data are cross-classified (the purpose of which is usually to uncover interrelationships between the variables) the various classifications along with their constituent frequency counts are referred to as a *contingency table*. Examples of variables whose measurements would be defined as categorical include sex (male, female), preference for laundry detergent (brand A, brand B, brand C), age (under 18, 18–35, over 35), income level (low, middle, high), etc. When three or more categorical variables are cross-classified, the induced table is a multidimensional contingency table.

As we shall shortly observe, computational complexities in analyzing higher-order tables were responsible, in large measure, for the paucity of both research and application of techniques useful in dealing with such problems. Faced with analyzing higher-dimensional tables, researchers typically would collapse the dimensions in various ways so that sets of two-way tables would evolve. Analysis of a multivariate system was replaced through dependent considerations of only two variables at a time. Such an approach can be helpful in gaining insights to the data; however, this approach

*Such data are said to be measured on a nominal scale.

1. Compromises our ability to study higher-order associations between more than two variables at a time
2. Can disguise true associations since relationships observed from marginal tables frequently will be different from the same relationship in the presence of other variables
3. Precludes the ability to study simultaneously all pairwise associations because the several pairwise tests are not independent

While many models have been proposed to study contingency tables, the representation of the logarithm of theoretical frequencies by a linear model holds the greatest promise. Indeed, when we discuss the analysis of multiway tables, this will be the approach taken. Before we embark on the more involved multiway problem, we shall first discuss a number of important issues pertaining to two-way tables.

17.2 THE 2 × 2 TABLE

Table 17.1 represents what is typically called a 2×2 contingency table. The four cells are characterized by two variables A and B each assuming two levels, A_1, A_2 and B_1, B_2, respectively. The data are represented by the frequency counts x_{ij}, that is, the number of observations falling in the cell defined by row A_i and column B_j, $i = 1, 2$ and $j = 1, 2$. By using $+$ notation, the number of observations all having the common characteristic A_1 is denoted by x_{1+} and the number with A_2 by x_{2+}; the total number of observations is therefore $x_{++} = x_{1+} + x_{2+}$. By similar argument, we can also define the marginal totals $x_{+j}, j = 1, 2$.

TABLE 17.1   The 2 × 2 Table

Variable A	Variable B		
	B_1	B_2	Total
A_1	x_{11}	x_{12}	x_{1+}
A_2	x_{21}	x_{22}	x_{2+}
Total	x_{+1}	x_{+2}	x_{++}

For our purposes we shall consider two sampling schemes which can be employed to generate a table like that given by Table 17.1: (1) A fixed sample of size $n (= x_{++})$ is taken, and the counts x_{ij} are distributed among the four cells, or (2) independent samples of size $n_1 (= x_{1+})$ and $n_2 (= x_{2+})$ are taken from the row categories A_1 and A_2. Within each row, counts are distributed between B_1 and B_2. Sampling scheme 1 is sometimes referred to as *multinomial sampling*, while sampling scheme 2, characterized by fixed marginal totals, is known as *product multinomial sampling.* Let us for the moment consider sampling scheme 2.

17.2.1 Homogeneity of Proportions

Suppose we are interested in determining whether the counts are consistent with the hypothesis that the true proportion of individuals from group A_1 having characteristic B_1 is the same for the A_2 group. If we denote by P_1 the probability that an individual from A_1 has characteristic B_1 and by P_2 the probability that an individual from A_2 has characteristic B_1, then we are interested in testing the null hypothesis that $P_1 = P_2 = P$. Hence, if independent samples from A_1 and A_2 are available, then an approximate test of $P_1 = P_2$ uses the asymptotic normality of

$$Z \cong \frac{p_{s_1} - p_{s_2}}{\sqrt{\bar{p}(1 - \bar{p})[(1/n_1) + (1/n_2)]}} \tag{17.1}$$

where the two sample proportions are

$$p_{s_1} = \frac{x_{11}}{n_1}, \qquad p_{s_2} = \frac{x_{21}}{n_2}, \qquad \text{where } n_1 = x_{1+}, \quad n_2 = x_{2+}$$

and where

$$\bar{p} = \frac{x_{11} + x_{21}}{n_1 + n_2}$$

At a significance level of α, the null hypothesis is rejected if $Z \geq Z_{1-\frac{\alpha}{2}}$ or $Z \leq Z_{\frac{\alpha}{2}}$ (see Appendix B, Table B.1).

An alternative procedure to the preceding uses the Pearson χ^2 statistic [see Berenson and Levine (1983, Section 10.7)]:

$$\chi^2 = \sum_{\substack{\text{all} \\ \text{cells}}} \frac{(\text{observed} - \text{expected})^2}{\text{expected}} \tag{17.2}$$

where *observed* is the observed count x_{ij} and *expected* is the expected frequency $(\hat{m}_{ij})$ in cell (i,j). To obtain the expected frequencies, the value $n_1\bar{p}$ is used in cell $(1,1)$, while $n_1(1 - \bar{p})$ is used for cell $(1,2)$. Moreover, in cell $(2,1)$ the value $n_2\bar{p}$ is used, while in cell $(2,2)$ the value $n_2(1 - \bar{p})$ is used. The null hypothesis $(P_1 = P_2 = P)$ is rejected if the computed value in (17.2) exceeds $\chi^2_{1-\alpha;1}$, the α percentage point under a chi-square distribution with one degree of freedom (see Appendix B, Table B.2). To fix ideas, suppose we wanted to study the difference in attitude toward a governor's position on public transportation between urban and suburban registered voters. One sample from each group of voters was selected. The results are summarized in Table 17.2. From (17.1),

$$\bar{p} = \frac{672 + 588}{1600 + 1200} = \frac{1260}{2800} = .45$$

TABLE 17.2   **Attitude Toward the Governor's Position
of Two Samples of Voters**

| | Attitude | | |
Sample	Favor	Oppose	Total
1 (Urban)	672	928	1600
2 (Suburban)	588	612	1200
Total	1260	1540	2800

Thus

$$n_1 \bar{p} = 1600(.45) = 720 \qquad n_1(1 - \bar{p}) = 1600(.55) = 880$$
$$n_2 \bar{p} = 1200(.45) = 540 \qquad n_2(1 - \bar{p}) = 1200(.55) = 660$$

Therefore from (17.2),

$$\chi^2 = \frac{(672 - 720)^2}{720} + \frac{(928 - 880)^2}{880} + \frac{(588 - 540)^2}{540} + \frac{(612 - 660)^2}{660}$$
$$= 3.2 + 2.618 + 4.267 + 3.491$$
$$= 13.576$$

If the .05 level of significance were selected, since $\chi^2 = 13.576 > \chi^2_{.95;\,1} = 3.841$, the null hypothesis would be rejected, and we would conclude that there was substantial evidence of a difference between the two proportions.

17.2.2 Independence in 2 × 2 Tables

Another fundamental question that can be pondered is whether the two variables A and B are associated. Unlike the case discussed above, where the margins of one of the categories are fixed by design, here we shall require that the total sample $n = x_{++}$ is the only number that is fixed; cell counts are then distributed among the four cells according to a multinomial distribution* with parameter set $\{n, P_{11}, P_{12}, P_{21}, P_{22}\}$. Usually the presence of an association between two variables is uncovered through tests designed to assess the null hypothesis of independence. After a hypothesis of independence is rejected, we commonly employ a measure of association to determine the nature and degree of association.

If P_{ij} represents the *true probability* of an observation falling in cell (i,j), then the row variable will be independent of the column variable if

*The multinomial distribution is the multivariate extension of the binomial distribution to the case in which there are p parameters [see Goldstein and Dillon (1983)].

$$P_{ij} = P\{\text{row category} = i \text{ and column category} = j\}$$
$$= P\{\text{row category} = i\}P\{\text{column category} = j\}$$
$$= P_{i+}P_{+j}, \qquad i = 1, 2 \text{ and } j = 1, 2 \qquad (17.3)$$

Using properties of the multinomial distribution, we see that the expected number of observations in cell (i, j) is $m_{ij} = nP_{i+}P_{+j}$. The observed margins are commonly employed to estimate these expected cell counts by substituting x_{i+}/n for P_{i+} and x_{+j}/n for P_{+j}. Thus, $\hat{m}_{ij} = x_{i+}x_{+j}/n$. A test of independence can now be constructed using the Pearson statistic:

$$\chi^2 = \sum_{i,j} \frac{(x_{ij} - \hat{m}_{ij})^2}{\hat{m}_{ij}} = \sum_{i,j} \frac{[x_{ij} - (x_{i+}x_{+j})/n]^2}{(x_{i+}x_{+j})/n} \qquad (17.4)$$

The null hypothesis $H_0: P_{ij} = P_{i+}P_{+j}$ for all i, j will be rejected at the specified significance level α if $\chi^2 \geq \chi^2_{1-\alpha;\, 1}$.

An application of the Pearson χ^2 statistic is provided in the following example: A sample of 120 individuals were cross-classified according to whether they had ever been audited by the Internal Revenue Service and their type of employment—classified as white collar or blue collar. The results are summarized in Table 17.3.

TABLE 17.3   Cross-classification of Type of Employment and IRS Audit

Type of Employment	IRS Audit		Total
	Audited	Not Audited	
White Collar	55	20	75
Blue Collar	13	32	45
Total	68	52	120

For this example,

$H_0: P_{ij} = P_{i+}P_{+j}$,   there is no association between type of employment and IRS audit

$H_1: P_{ij} \neq P_{i+}P_{+j}$,   there is an association between type of employment and IRS audit

We can compute

$$\hat{m}_{11} = \frac{75(68)}{120} \qquad \hat{m}_{12} = \frac{75(52)}{120} \qquad \hat{m}_{21} = \frac{45(68)}{120} \qquad \hat{m}_{22} = \frac{45(52)}{120}$$
$$= 42.5 \qquad\qquad = 32.5 \qquad\qquad = 25.5 \qquad\qquad = 19.5$$

Thus from (17.4) we have

$$\chi^2 = \frac{(55 - 42.5)^2}{42.5} + \frac{(20 - 32.5)^2}{32.5} + \frac{(13 - 25.5)^2}{25.5} + \frac{(32 - 19.5)^2}{19.5}$$

$$= 3.676 + 4.808 + 6.127 + 8.013$$

$$= 22.624$$

If the .05 level of significance were selected, since $\chi^2 = 22.624 > \chi^2_{.95;\,1} = 3.841$, the null hypothesis would be rejected, and we would conclude that there was substantial evidence of an association between type of employment and the likelihood of an IRS audit.

Some authors advise that in a 2×2 contingency table the Pearson statistic be replaced by a *corrected* statistic proposed by Yates (1934):

$$\chi^2_* = \sum_{i,j} \frac{[|x_{ij} - (x_{i+}x_{+j}/n)| - \frac{1}{2}]^2}{x_{i+}x_{+j}/n} \tag{17.5}$$

Yates originally proposed this correction to make tail probabilities more closely in line with exact hypergeometric probabilities that can be computed when both row margins and column margins are fixed. Other authors advise against the general application of χ^2_*. To ensure that the statistic is asymptotically chi square, the researcher is better advised to use χ^2 since otherwise, as Grizzle (1967) and Conover (1974) have shown, the test using χ^2_* is too conservative in the sense that for a specified significance level the test statistic does not reject often enough.

17.2.3 Association in 2 × 2 Tables

As we previously stated, tests for independence are used to uncover whether a relationship exists between two variables, while measures of association assist the researcher in understanding the nature and extent of the relationship. In many ways studying association for discrete data is more complex than for continuous measurements. Unlike the bivariate normal distribution where the correlation coefficient gives us information relating to degree of association between two variables, association in a contingency table is a multidimensional phenomenon. For example, when two variables are cross-classified such that the first variable has I levels while the second has J levels, we shall require $(I - 1)(J - 1)$ functions of the parameters to completely specify association. Hence, in the 2×2 case one function of the parameters is sufficient.

There are a large number of measures of association for the 2×2 table; however, we shall restrict attention to only one. The reasons we focus only on the *cross-product ratio* are that many other measures of association are functions of it, and, moreover, it is a fundamental parameter for the models we shall discuss later in this chapter. The cross-product ratio or *odds ratio* is defined by

$$\theta = \frac{P_{11}P_{22}}{P_{12}P_{21}} \tag{17.6}$$

The range of values of θ is between 0 and ∞, its logarithm ranges between $-\infty$ and ∞. θ is symmetric in the sense that two values of θ, say θ_1 and θ_2 such that $\ln \theta_1 = -\ln \theta_2$, represent the same degree of association but in opposite directions. Perhaps the easiest interpretation of the odds ratio occurs when we consider fixed totals for rows or columns. Here the ratio $P_{11} \,|\, P_{12}$ is the odds of falling in the first column conditioned on being in the first row, and hence θ is the relative odds for the two rows.

Independence between rows and columns implies

$$\theta = \frac{P_{1+}P_{+1}P_{2+}P_{+2}}{P_{1+}P_{+2}P_{2+}P_{+1}} = 1 \qquad (17.7)$$

Further, θ is invariant under scale changes in rows and columns; that is, if $r_1 > 0, r_2 > 0, c_1 > 0, c_2 > 0$ are specified constants, then

$$\frac{(r_1 c_1 P_{11})(r_2 c_2 P_{22})}{(r_1 c_2 P_{12})(r_2 c_1 P_{21})} = \theta \qquad (17.8)$$

Since knowledge of θ requires knowing the true cell probabilities, a sample-based cross-product ratio may be defined in terms of the observed cell counts by

$$\hat{\theta} = \frac{x_{11} x_{22}}{x_{12} x_{21}} \qquad (17.9)$$

It can be shown that the asymptotic distribution of $\ln \hat{\theta}$ is normal with mean $\ln \theta$ and an approximate variance given by

$$\hat{\sigma}_\infty^2(\ln \hat{\theta}) = \frac{1}{x_{11} + x_{12} + x_{21} + x_{22}} \qquad (17.10)$$

The asymptotic normality of $\ln \hat{\theta}$ allows us to construct a $1 - \alpha$ confidence interval on $\ln \theta$:

$$\ln \hat{\theta} \pm Z_{1-\frac{\alpha}{2}} \hat{\sigma}_\infty(\ln \hat{\theta}) \qquad (17.11)$$

where $Z_{1-\frac{\alpha}{2}}$ is the $\left[1 - \dfrac{\alpha}{2}\right] \times 100$ percentile point from the standardized normal distribution (Appendix B, Table B.1).

If we refer back to Table 17.3, we can compute the cross-product ratio ($\hat{\theta}$) from (17.9) as

$$\hat{\theta} = \frac{55(32)}{20(13)} = 6.77$$

and

$$\ln \hat{\theta} = 1.91$$

The computed $\hat{\theta}$ points strongly to an association between the nature of employment and being audited by the IRS. This is clearly confirmed upon calculating a confidence interval for $\ln \theta$ at a confidence level of $1 - \alpha = .95$. From (17.11) we have

$$\ln \hat{\theta} \pm Z_{.975} \hat{\sigma}_{\infty}(\ln \hat{\theta})$$
$$1.91 \pm 1.96 \sqrt{.008}$$
$$1.91 \pm .179$$

so that

$$1.731 \leq \ln \theta \leq 2.089$$

In general, we would have sufficient evidence to conclude that the two categories are related if the confidence interval does not cover the number zero.

17.3 ASSOCIATION IN $I \times J$ TABLES

17.3.1 Introduction

Let us now consider measures of association when the row category A can assume I levels and the column category B can assume J levels. We shall discuss two very different approaches to the problem. One approach, due to Goodman and Kruskal (1954), uses the idea of *proportional reduction in error*, while the other, discussed in a paper by Light and Margolin (1971), uses a measure analogous to the *squared multiple correlation*.

17.3.2 The Goodman-Kruskal Lambda Measure

Goodman and Kruskal (1954) observed that if our goal is to predict the *column category B from the row category A* there are two situations under which prediction can take place: (1) A and B are independent and (2) A and B are functionally related. If, indeed, situation 1 is the case, then we cannot expect to do any better in predicting the B category from knowledge about A; otherwise there will be an improvement. They propose a measure which quantifies the improvement through a proportional reduction in error (PRE) measure defined as the relative improvement in predicting the B category when A is known as opposed to when the A category is unknown. In terms of the sample counts x_{ij}, $i = 1, 2, \ldots, I$ and $j = 1, 2, \ldots, J$, the measure which we denote by $\hat{\lambda}_{B|A}$ is defined by

$$\hat{\lambda}_{B|A} = \frac{\sum_{i=1}^{I} x_{im} - x_{+m}}{n - x_{+m}}, \qquad 0 \leq \hat{\lambda}_{B|A} \leq 1 \qquad (17.12)$$

where x_{im} is the largest frequency count in the ith row and x_{+m} is the largest column marginal total. The statistical inference needed to study the behavior of $\hat{\lambda}_{B|A}$ is assisted by the fact that asymptotically $\hat{\lambda}_{B|A}$ is normally distributed with mean

$$\lambda_{B|A} = \frac{\sum_{i=1}^{I} P_{im} - P_{+m}}{1 - P_{+m}} \qquad (17.13)$$

(where P_{im} is the largest value in the ith row and P_{+m} is the largest column marginal total from the table of true cell probabilities) and approximate variance

$$\hat{\sigma}^2_{\infty}(\hat{\lambda}_{B|A}) = \frac{\left(n - \sum\limits_{i=1}^{I} x_{im}\right)\left(\sum\limits_{i=1}^{I} x_{im} + x_{+m} - 2\overset{*}{\sum} x_{im}\right)}{(n - x_{+m})^3} \qquad (17.14)$$

where $\overset{*}{\sum} x_{im}$ is the largest of the I row frequency counts found in the column containing x_{+m}.

Application.   To assist in preparing an advertising strategy for a major foreign automobile company, a consulting team wished to determine the nature of the relationship (if one existed) between the age of a perspective buyer and the type of option package available that would be purchased with the car. A random sample of 150 individuals was taken and classified according to their age (under 30 or 30 and older) and the type of option package that would be purchased (A, B, or C). The results of this classification are presented in Table 17.4. It follows that

$$\hat{\lambda}_{B|A} = \frac{32 + 32 - 55}{150 - 55} = .095$$

TABLE 17.4   Cross-classifying Age and Option Package

	Option			
Age	A	B	C	Total
Under 30	20	18	32	70
30 or Older	25	32	23	80
Total	45	50	55	150

This shows that there is a proportional reduction in error, albeit small, in predicting the option package from knowledge of the age group of the buyer. A confidence interval for the true $\lambda_{B|A}$ (more accurately an approximate asymptotic confidence interval is, at level $1 - \alpha$, given by

$$\hat{\lambda}_{B|A} \pm Z_{1-\frac{\alpha}{2}} \hat{\sigma}_{\infty}(\hat{\lambda}_{B|A}) \qquad (17.15)$$

By using (17.14),

$$\hat{\sigma}^2_{\infty}(\hat{\lambda}_{B|A}) = \frac{[150 - (32 + 32)][(32 + 32) + 55 - 2(32)]}{(150 - 55)^3} = .0055$$

The 95 percent confidence interval is

$$.095 \pm 1.96\sqrt{.0055}$$

$$.095 \pm .146$$

or

$$-.051 \leq \lambda_{B|A} \leq .241$$

Since the confidence interval covers zero, we conclude that the computed $\hat{\lambda}_{B|A}$ is not significantly different from zero; there is no evidence of association.

17.3.3 The Light-Margolin Measure

As Light and Margolin (1971) point out, one big problem in using a measure of association based on the concept of total variation for categorical data is the tendency to think of variation as a measure of departure of a set of individual observations from their mean. Since for categorical data the mean is an undefined concept, alternative methods of looking at variation have to be found.* Using an *analysis of variance* approach, the total variation in the response variable (i.e., the total sum of squares) is given by

$$\text{SST} = \frac{n}{2} - \frac{1}{2n} \sum_{i=1}^{I} x_{i+}^2 \qquad (17.16)$$

while the within group sum of squares (SSW) is

*Noting that a sum of squares given by $\sum_i (x_i - \bar{x})^2$ may be written as

$$\text{SST} = \sum_i (x_i - \bar{x})^2 = \frac{1}{2n} \sum_{i=1}^{n} \sum_{j=1}^{n} (x_i - x_j)^2 = \frac{1}{2n} \sum_{i=1}^{n} \sum_{j=1}^{n} d_{ij}^2$$

where $d_{ij} = x_i - x_j$, Light and Margolin (1971) borrow an idea first proposed by Gini (1912) to define a measure of total variation. They argue as follows: Suppose the data are represented by the n responses given by $x_1, x_2, \ldots, x_n$ and that each x_i is defined purely by the cell of the table in which it falls. By defining the distance d_{ij} between response x_i and x_j as

$$d_{ij} = \begin{cases} 1 & \text{if } x_i \text{ and } x_j \text{ fall in the same cell} \\ 0 & \text{if not} \end{cases}$$

the variation for the categorical responses is expressed by:

$$\frac{1}{2n} \sum_{i=1}^{n} \sum_{j=1}^{n} d_{ij}^2 = \frac{1}{2n} \sum_{i=1}^{n} \sum_{j=1}^{n} d_{ij}$$

If the number of cells in the table is given by M and the n responses are distributed so that n_i responses are in cell i, where $\sum_{i=1}^{M} n_i = n$, then the variation of these responses is

$$\frac{1}{2n} \sum_{i \neq j} n_i n_j = \frac{1}{2n} \left(n^2 - \sum_{i=1}^{M} n_i^2 \right)$$

$$= \frac{n}{2} - \frac{1}{2n} \sum_{i=1}^{M} n_i^2$$

Following the structure of a one-way analysis of variance, we think of the M cells as evolving from J unordered experimental groups and I unordered response categories. Hence the number of responses in category i for group j, $i = 1, 2, \ldots, I$ and $j = 1, 2, \ldots, J$, is designated by x_{ij} and the total number of responses for group j by x_{+j}. The total number of responses in the ith category is x_{i+}, and therefore the total number of responses is expressed by

$$n = \sum_{j=1}^{J} x_{+j} = \sum_{i=1}^{I} x_{i+} = \sum_{i=1}^{I} \sum_{j=1}^{J} x_{ij}$$

Based on these, SST, SSW, and SSA are defined in (17.16) through (17.18) and the Light-Margolin measure of association is presented in (17.19).

$$\text{SSW} = \sum_{j=1}^{J} \left(\frac{x_{+j}}{2} - \frac{1}{2x_{+j}} \sum_{i=1}^{I} x_{ij}^2 \right)$$

$$= \frac{n}{2} - \frac{1}{2} \sum_{j=1}^{J} \frac{1}{x_{+j}} \sum_{i=1}^{I} x_{ij}^2 \tag{17.17}$$

The among groups sum of squares (SSA) is found by subtracting SSW from SST, and therefore

$$\text{SSA} = \frac{1}{2} \left(\sum_{j=1}^{J} \frac{1}{x_{+j}} \sum_{i=1}^{I} x_{ij}^2 \right) - \frac{1}{2n} \sum_{i=1}^{I} x_{i+}^2 \tag{17.18}$$

The measure of association proposed by Light and Margolin (1971) is interpretable in the same manner as the squared multiple correlation coefficient:

$$R^2 = \frac{\text{SSA}}{\text{SST}} = \frac{\left[\sum_{j=1}^{J} (1/x_{+j}) \sum_{i=1}^{I} x_{ij}^2 \right] - (1/n) \sum_{i=1}^{I} x_{i+}^2}{n - (1/n) \sum_{i=1}^{I} x_{i+}^2} \tag{17.19}$$

It can be shown that $R^2 = 0$ if $x_{ij}|x_{+j} = f_i$, $i = 1, 2, \ldots, I$ and $j = 1, 2, \ldots, J$; that is, if there is no association or no effect of group on the category. Further, $R^2 = 1$ if, for each j (where $j = 1, 2, \ldots, J$), there exists an i such that $P_{ij} = P_{+j}$, that is, if there is perfect predictability. Otherwise $0 < R^2 < 1$.

Application.   For the data given in our example we view *Option* as the response variable. In keeping with the preceding notation, we write the table (see Table 17.5) so that the response variable is the *row category*. By using (17.19), the R^2 for this data set is

$$R^2 = \frac{\begin{gathered} \{[(20)^2 + (18)^2 + (32)^2]/70\} + \{[(25)^2 + (32)^2 + (23)^2]/80\} \\ - \{[(45)^2 + (50)^2 + (55)^2]/150\} \end{gathered}}{150 - \{[(45)^2 + (50)^2 + (55)^2]/150\}} = .0187$$

TABLE 17.5   Re-expressing the Automobile Option Package Example

	Age		
Option	Under 30	30 or Older	Total
A	20	25	45
B	18	32	50
C	32	23	55
Total	70	80	150

SOURCE: Table 17.4.

Thus only 1.87% of the variation in option preference can be explained by variation in age.

17.4 LOG-LINEAR MODELS

17.4.1 Introduction and Development

Our discussion thus far has focused on two-way contingency tables with particular attention paid to issues of homogeneity of proportions, independence, and measures of association. It is, however, more the rule than the exception that research in the social sciences involves large multidimensional problems. Thus many variables are considered simultaneously, and questions of interdependence are usually of paramount importance. When many measurements are considered in tandem, it is often helpful to achieve a parsimonious description of the data by using an interpretable mathematical model. The degree to which the model is useful depends on how well it fits the data and how helpful it is in assisting in the interpretation of complicated relationships. Log-linear models are a class of mathematical models designed to assist in uncovering associations that exist when discrete variables are cross-classified to form a contingency table.

To start our discussion, suppose that variable A, having I levels, and variable B, having J levels, are cross-classified to form an $I \times J$ table. Let m_{ij} denote the expected frequency associated with cell (i,j) if a sample of n observations was distributed across the table. Consider the model

$$\ln m_{ij} = \mu + \mu_{A(i)} + \mu_{B(j)} + \mu_{AB(ij)} \tag{17.20}$$

with the constraints

$$\sum_i \mu_{A(i)} = \sum_j \mu_{B(j)} = \sum_i \mu_{AB(ij)} = \sum_j \mu_{AB(ij)} = 0$$

The parameters on the right-hand side of (17.20) are commonly referred to as μ terms. Their interpretation can, for some researchers, be clarified by using an analogy to the analysis of variance:

$$\mu = \sum_{i,j} \frac{\ln m_{ij}}{IJ} \tag{17.21}$$

$$\mu_{A(i)} = \sum_j \frac{\ln m_{ij}}{J} - \mu \tag{17.22}$$

$$\mu_{B(j)} = \sum_i \frac{\ln m_{ij}}{I} - \mu \tag{17.23}$$

$$\mu_{AB(ij)} = \ln m_{ij} - \mu_{A(i)} - \mu_{B(j)} + \mu \tag{17.24}$$

The constraints imposed on the interaction parameters $\mu_{AB(ij)}$, $i = 1, 2, \ldots,$ I and $j = 1, 2, \ldots, J$, show that $(I - 1)(J - 1)$ independent parameters are needed to fully account for association in an $I \times J$ table. Of course, in the special case where $I = J = 2$ only one parameter is needed. This we see immediately from the relations

$$\mu_{AB(11)} = -\mu_{AB(12)} = -\mu_{AB(21)} = \mu_{AB(22)} \tag{17.25}$$

based on the constraints on the model given by (17.20). Further, this common parameter is related to the cross-product ratio by

$$\mu_{AB(11)} = \frac{1}{4} \ln \frac{m_{11}m_{22}}{m_{12}m_{21}} \tag{17.26}$$

By following the lead given in (17.20), extensions of log-linear models for higher-order tables are straightforward. For example, for three-dimensional tables (i.e., three variables A, B, C with respective levels I, J, K) the complete model is expressed by

$$\ln m_{ijk} = \mu + \mu_{A(i)} + \mu_{B(j)} + \mu_{C(k)}$$
$$+ \mu_{AB(ij)} + \mu_{AC(ik)} + \mu_{BC(jk)}$$
$$+ \mu_{ABC(ijk)} \tag{17.27}$$

where all subscripted μ terms sum to zero over the range of any included index. Since one may view higher-order μ terms as measuring deviations from lower-order terms, it is not surprising that computing $\mu_{AB(ij)}$ necessitates first computing μ, $\mu_{A(i)}$, and $\mu_{B(j)}$; further, computing higher-order terms like $\mu_{ABC(ijk)}$ requires an evaluation of all lower-order relatives of $\mu_{ABC(ijk)}$. Because of these observations and other technical considerations, we shall restrict our discussion to log-linear models which satisfy the *hierarchy principal*:

1. For any two sets of indexes $\{\theta\}$ and $\{\theta'\}$ such that $\{\theta\} \subset \{\theta'\}$, $\mu_{\{\theta\}} = 0$ implies that $\mu_{\{\theta'\}} = 0$.
2. For any set of indexes $\{\theta''\}$, the fact that $\mu_{\{\theta''\}} \neq 0$ implies that all lower-order relatives of $\mu_{\{\theta''\}}$ are nonzero.

Thus, for example, the inclusion of $\mu_{AB(ij)}$ in a model means that $\mu_{A(i)}$ and $\mu_{B(j)}$ would also appear since they are lower-order relatives of $\mu_{AB(ij)}$. In a similar fashion, the inclusion of $\mu_{BC(jk)}$ in a model implies that $\mu_{B(j)}$ and $\mu_{C(k)}$ would also appear. On the other hand, if $\mu_{A(i)}$ or $\mu_{B(j)} = 0$, then $\mu_{AB(ij)}$ must be zero.

The model expressed in (17.20) represents a *saturated* model because it contains *all* the μ terms. Since researchers are interested in determining which terms can be eliminated without seriously altering the *goodness of fit* of the model, we shall focus on the set of possible *unsaturated* models. The specification of an *unsaturated* log-linear model is a statement in which μ terms are set equal to zero. For example, for all i, j, k, $\mu_{AB} = \mu_{AC} = \mu_{BC} = 0$ specifies a model of complete independence of the variables. That is, conditional on C, A and B are independent; conditional on B, A and C are independent; and conditional on A, B and C are independent.

We may also state the log-linear model equivalently in terms of which μ

terms remain rather than which terms are absent. Thus, this model could be alternatively written as [A][B][C]. Further, $\mu_{AC} = 0$ (i.e., the model $[AB][BC]$) means that conditional on B, A and C are independent. Of course, this also implies that the association of A with B and B with C is included in the model. Finally, we should note that the two other conditional independence models, namely $\mu_{AB} = 0$ (the model $[AC][BC]$) and $\mu_{BC} = 0$ (the model $[AB][AC]$), imply, respectively, the independence of A and B conditional on C and the independence of B and C conditional on A.

Even under the constraints imposed by the hierarchy principle, there are many unsaturated models that the researcher could consider. There are basically three conditions which we aspire to in deciding upon an appropriate model for a given set of data: parsimony, goodness of fit, and interpretability. By parsimony we mean a log-linear model with few parameters. All other things remaining equal, the model containing the fewest number of parameters to be estimated is the best model. The model, however, must also be able to provide estimated cell counts which in some well-defined sense come close to the observed cell counts. Last, the model must make sense to the researcher in terms of the physical process observed, an established theory, or other considerations.

Before we can adequately deal with these issues, however, we must discuss sampling processes and estimation procedures.

17.4.2 Sampling and Parameter Estimation

In the beginning of this chapter we discussed the sampling schemes used to generate 2×2 tables. For tests of homogeneity of proportions, independent samples were taken from the two margins of one of the variables, whereas in tests of independence, a fixed sample was taken, and the data were then distributed over the four cells. More generally, *product multinomial sampling* refers to a sampling scheme in which at least one of the variables is viewed as a design variable. In such a case we assume that the marginal totals over the levels of this particular variable are fixed. The observations which make up these fixed totals are then allocated among all the cells in the table. In *multinomial sampling* the total number of observations is the only thing that is fixed; cell counts and marginal totals then evolve according to the probability mechanism inherent in the data.

After the data are generated, interest centers upon procedures for estimating theoretical cell frequencies. Most statisticians interested in the mathematical basis of this subject believe that estimation using the *method of maximum likelihood* [see Bishop et al. (1975) and Fienberg (1980)] is the most sensible of those procedures available. In general, such estimates (1) are easy to compute, (2) are often simple functions of marginal totals, and (3) have favorable large sample properties. We now know, for example, that estimates of theoretical frequencies will be identical under both sampling schemes provided that in the case of product multinomial sampling the μ terms corresponding to fixed margins are included in the fitted model. In general, *closed form* expressions for the maximum likelihood estimates will not be available, and iterative procedures need to be used. One of the most common procedures (indeed the procedure which is employed in many computer packages used to generate maximum likelihood estimates for contingency table models) relies on the method of *iterative proportional fitting* [see Fienberg (1980)].

Before describing the nature of this numerical scheme, we shall state a result, attributed to Birch (1963), within the context of a specific model. Suppose in the three-variable problem we wish to fit the no second-order interaction model, i.e., $\mu_{ABC} = 0$. The maximum likelihood estimates for the cell frequencies under this model are functions of the margin totals x_{ij+}, x_{i+k}, and x_{+jk}. In general, the sufficient statistics will be the observed marginal totals corresponding to the highest-order μ terms in the model. Let us denote the set of maximum likelihood estimators by $\{\hat{m}_{ijk}\}$, and hence the maximum likelihood estimators for m_{ij+}, m_{i+k}, and m_{+jk} are given, respectively, by $\hat{m}_{ij+}$, $\hat{m}_{i+k}$, and $\hat{m}_{+jk}$. Birch showed that the estimates of m_{ij+}, m_{i+k}, and m_{+jk} must satisfy

$$\hat{m}_{ij+} = x_{ij+}$$

$$\hat{m}_{i+k} = x_{i+k}$$

$$\hat{m}_{+jk} = x_{+jk}$$

and that there is a unique set of elementary cell estimates that satisfies the conditions of the model and these marginal constraints.

The actual generation of these estimates by iterative proportional fitting has as its basis Birch's two results. Initial values are given by $\hat{m}_{ijk}^{(0)} = 1$. These starting values are adjusted through the first iteration by

$$\hat{m}_{ijk}^{(1)} = \hat{m}_{ijk}^{(0)} \frac{x_{ij+}}{\hat{m}_{ij+}^{(0)}}$$

$$\hat{m}_{ijk}^{(2)} = \hat{m}_{ijk}^{(1)} \frac{x_{i+k}}{\hat{m}_{i+k}^{(1)}}$$

$$\hat{m}_{ijk}^{(3)} = \hat{m}_{ijk}^{(2)} \frac{x_{+jk}}{\hat{m}_{+jk}^{(2)}}$$

The process continues until there is small difference between successive adjusted values, i.e., until

$$|\hat{m}_{ijk}^{(v)} - \hat{m}_{ijk}^{(v-1)}| < \epsilon \tag{17.28}$$

where ϵ is a prespecified constant. Convergence will always occur, usually after only a few iterations.

17.4.3 Assessing the Fit of a Postulated Model

After a particular unsaturated model is postulated and cell estimates are derived, it is necessary to have a measure of how well the observed frequencies and the estimated frequencies conform. The most familiar measure of fit is based on the Pearson statistic [see (17.2)]. To fix ideas, let us for simplicity use single subscripts in denoting the set of observed frequencies $\{x_i\}$ and the set of estimated or expected frequencies $\{\hat{m}_i\}$ under the model. The Pearson fit statistic

$$\chi^2 = \sum_i \frac{(x_i - \hat{m}_i)^2}{\hat{m}_i} \tag{17.29}$$

is known to be asymptotically chi square with degrees of freedom appropriate for the particular set $\{\hat{m}_i\}$. While this statistic has been a mainstay for many years, most researchers prefer an alternative statistic based on the *likelihood ratio* property. The basis for the preference rests primarily on certain partitioning properties that the latter possesses which is useful in model building. In general, these properties are not shared by the Pearson statistic.

The likelihood ratio statistic G^2 is given by

$$G^2 = 2 \sum_i x_i \ln\left(\frac{x_i}{\hat{m}_i}\right) \tag{17.30}$$

Like χ^2, G^2 is asymptotically chi square with degrees of freedom dependent on which unsaturated hierarchical model is fitted. In particular, the number of degrees of freedom will equal the number of cells ($I \times J \times K$) in the table minus the number of independent fitted parameters defined by the model [see Fienberg (1980) for a more detailed discussion].

Referring to the complete (saturated) model

$$\begin{aligned}
\ln m_{ijk} = \mu &+ \mu_{A(i)} + \mu_{B(j)} + \mu_{C(k)} \\
&+ \mu_{AB(ij)} + \mu_{AC(ik)} + \mu_{BC(jk)} \\
&+ \mu_{ABC(ijk)}, \quad \text{for all } i, j, k
\end{aligned} \tag{17.31}$$

each μ term has associated with it a specified number of parameters. This is summarized in Table 17.6.

TABLE 17.6 μ Terms and Number
of Fitted Parameters for a Three-
Dimensional Log-Linear Model

Term	Number of Fitted Parameters
μ	1
μ_A	$I - 1$
μ_B	$J - 1$
μ_C	$K - 1$
μ_{AB}	$(I - 1)(J - 1)$
μ_{AC}	$(I - 1)(K - 1)$
μ_{BC}	$(J - 1)(K - 1)$
μ_{ABC}	$(I - 1)(J - 1)(K - 1)$

We should realize that G^2 attains its minimum value of zero when, for all i, $x_i = \hat{m}_i$. Thus for the fully saturated model (which includes all μ terms and has zero degrees of freedom), $x_i = \hat{m}_i$ so that $G^2 = 0$; for any unsaturated model, $G^2 > 0$ (and the degrees of freedom also will be > 0). Since we are interested only in choosing among the unsaturated models, we need to determine the set of possible alternative hierarchical models that may be considered. If we examine the $I \times J \times K$ contingency table, there are eight alternative hierarchical log-linear models that may be considered. These models and their associated degrees of freedom are summarized in Table 17.7.

TABLE 17.7   Hierarchical Log-Linear Models for $I \times J \times K$ Contingency Table

Model	Deleted Terms	Degrees of Freedom
M_1: $[A][B][C]$	$\mu_{AB} = \mu_{AC} = \mu_{BC} = \mu_{ABC} = 0$	$IJK - I - J - K + 2$
M_2: $[AB][C]$	$\mu_{AC} = \mu_{BC} = \mu_{ABC} = 0$	$(K-1)(IJ-1)$
M_3: $[AC][B]$	$\mu_{AB} = \mu_{BC} = \mu_{ABC} = 0$	$(J-1)(IK-1)$
M_4: $[A][BC]$	$\mu_{AB} = \mu_{AC} = \mu_{ABC} = 0$	$(I-1)(JK-1)$
M_5: $[AB][AC]$	$\mu_{BC} = \mu_{ABC} = 0$	$I(J-1)(K-1)$
M_6: $[AB][BC]$	$\mu_{AC} = \mu_{ABC} = 0$	$J(I-1)(K-1)$
M_7: $[AC][BC]$	$\mu_{AB} = \mu_{ABC} = 0$	$K(I-1)(J-1)$
M_8: $[AB][AC][BC]$	$\mu_{ABC} = 0$	$(I-1)(J-1)(K-1)$

Automobile preference example.   To illustrate the ideas that we have been discussing, we may refer to Table 17.8, which reports data on 530 individuals cross-classified on the basis of (A) sex, (B) place of residence—urban or rural, and (C) preference for domestic or foreign cars. A complete listing of all unsaturated models fitted by using the BMDP-program 4F (see Section 17.6) along with their degrees of freedom and computed G^2 statistics is given in Table 17.9.

An examination of Table 17.9 reveals four good-fitting models (indicated by *) if an $\alpha = .05$ level of significance is utilized. Thus the question that must be answered concerns which model should be chosen as the "best-fitting" model.

We may recall that the G^2 statistic possessed certain partitioning properties that are useful for model selection. To demonstrate this, let us define

TABLE 17.8   Automobile Preference Data

		Preference (C)		
Sex (A)	Residence (B)	Domestic	Foreign	Total
Male	Urban	218	29	416
	Rural	155	14	
Female	Urban	65	16	114
	Rural	26	7	
Total		464	66	530

TABLE 17.9   Alternative Log-Linear Models
for the Automobile Preference Example

Model	Degrees of Freedom	G^2
M_1: $[A][B][C]$	4	13.87
M_2: $[AB][C]$	3	8.54
M_3: $[AC][B]$	3	6.68*
M_4: $[A][BC]$	3	12.57
M_5: $[AB][AC]$	2	1.35*
M_6: $[AB][BC]$	2	7.25
M_7: $[AC][BC]$	2	5.38*
M_8: $[AB][AC][BC]$	1	.60*

$$L(\mathbf{x}) = \sum_i x_i \ln x_i \tag{17.32}$$

and

$$L(\hat{\mathbf{m}}) = \sum_i x_i \ln \hat{m}_i \tag{17.33}$$

Note that

$$G^2 = 2[L(\mathbf{x}) - L(\hat{\mathbf{m}})] \tag{17.34}$$

Because the Pearson χ^2 statistic does not have this same partitioning property as G^2, we shall use only the likelihood ratio statistic G^2 in our analysis.

Suppose we have two models which we denote by [1] and [2]. Suppose further that all the μ terms in [2] are contained in model [1]. We say that [2] is *nested* within [1]. Let the parameter estimates for model [1] be denoted by $\{\hat{m}^{[1]}\}$. Then the G^2 statistic for model [2] can be written as

$$
\begin{aligned}
G^2([2]) &= -2[L(\hat{\mathbf{m}}^{[2]}) - L(\mathbf{x})] \\
&= 2[L(\hat{\mathbf{m}}^{[1]}) - L(\hat{\mathbf{m}}^{[2]})] + 2[L(\mathbf{x}) - L(\hat{\mathbf{m}}^{[1]})] \\
&= G^2([2]|[1]) + G^2([1])
\end{aligned}
\tag{17.35}
$$

The first term on the right-hand side of (17.35) is called the *conditional likelihood ratio* statistic for model [2] given model [1]. It therefore follows as expected that $G^2([2]) > G^2([1])$, and if asymptotically $G^2([1])$ is chi square with degrees of freedom ν_1, then $G^2([2]|[1])$ is also asymptotically chi square with degrees of freedom $\nu_2 - \nu_1$. Many of the stepwise procedures for model building use this fact in determining whether a given μ term should be added or deleted from the model.

Thus we may use this principle to compare the models which have acceptable fits. In our three-way table, it makes sense to begin with the most complicated unsaturated model and attempt to eliminate μ terms through a backward elimination process. Thus we may begin by comparing the $[AB][AC][BC]$ model M_8 to the $[AB][AC]$ model M_5 (since the G^2 for M_5 is substantially less than that of M_7). We may compute

$$G^2(M_5 \mid M_8) = G^2(M_5) - G^2(M_8)$$
$$= 1.35 - .60$$
$$= .75$$

Since $G^2(M_5 \mid M_8) = .75 < \chi^2_{.95;\,1} = 3.841$, we would conclude from a model-building perspective that there was no substantial difference in fit between the two models. In the interest of parsimony, we would choose the $[AB][AC]$ model M_5 and delete the μ_{BC} term. Continuing with this approach, since the $[AC][B]$ model M_3 is also an acceptable fitting model, we need to determine whether the $[AB]$ term can be deleted from the $[AB][AC]$ model.

Using (17.35), we have

$$G^2(M_3 \mid M_5) = G^2(M_3) - G^2(M_5)$$
$$= 6.68 - 1.35$$
$$= 5.33$$

Since $G^2(M_3 \mid M_5) = 5.33 > \chi^2_{.95;\,1} = 3.841$, we would conclude that there is a substantial difference between the $[AC][B]$ and $[AB][AC]$ models. Thus our model-building process would conclude by choosing the $[AB][AC]$ model as the best-fitting model. This model includes the association of sex and residence and sex and preference. Moreover, it states that given the sex of the individual, preference is independent of the place of residence.

Branch preference example.   Now that we have analyzed data contained in a three-way contingency table, to demonstrate some of the more involved ideas previously discussed, we can consider the results of a survey of 2409 individuals dealing with their preferences for a particular branch of a savings institution. Four variables were considered: variable A, whether the person is familiar with the branch; variable B, whether the person feels the branch is conveniently located; variable C, whether the person had previously patronized the branch; and variable D, whether the person strongly recommends the branch. Level 1 of each variable designates a positive response to the question, while level 2 designates a negative response. The full multinomial scheme of sampling was used. Table 17.10 gives the distribution of responses over the 16 cells of the contingency table.

To get an idea of how some unsaturated models fit the data, the following models were initially considered (appended to each model is the value of G^2 obtained from the BMDP program 4F, to be discussed in Section 17.6):

Model M_1: $[ABC][ABD][ACD][BCD]$ or $\mu_{ABCD} = 0$
$\quad\quad$ ($G^2 = 3.27$, 1 degree of freedom)
Model M_2: $[AB][AC][AD][BC][BD][CD]$ or $\mu_{ABC} = \mu_{ABD} = \mu_{ACD} = \mu_{BCD}$
$\quad\quad = 0$
$\quad\quad\quad\quad$ ($G^2 = 5.48$, 5 degrees of freedom)

TABLE 17.10   Branch Preference for Saving Institution

			Variables	
				A
C	*D*	*B*	1	2
1	1	1	423	187
		2	459	412
	2	1	49	47
		2	68	127
2	1	1	13	84
		2	17	407
	2	1	0	22
		2	3	91

Model M_3: $[A][B][C][D]$ or $\mu_{AB} = \mu_{AC} = \mu_{AD} = \mu_{BC} = \mu_{BD} = \mu_{CD} = 0$
($G^2 = 810.22$, 11 degrees of freedom)

The first model contains all first-order associations between each pair of variables as well as all second-order associations among three variables. The second model contains only the complete set of first-order associations, while the third model contains only the main effects and thus postulates that each variable is independent of the others. It is reasonable to expect that a good-fitting unsaturated model can be found between models M_2 and M_3. We can approach the problem essentially two ways: deleting some first-order interactions from model M_2 as we did in the automobile preference example *or* adding some first-order interactions to model M_3. The backward elimination procedure continues until the deletion of an additional term results in a poor-fitting model; the forward selection procedure continues until the addition of terms leads to an unnecessarily too saturated model. In our example we use the former.

To illustrate these ideas, we start with model M_2 and determine which first-order interaction μ term is best to delete. The process will evolve using a significance level of $\alpha = .05$. Table 17.11 gives the results of the backward elimination procedure obtained from the BMDP program 4F, where, at each stage, we look at the conditional G^2 statistics found by differencing the various postulated models from model M_2. We see that the first-order μ term that should be dropped is μ_{BD} ($G^2 = 1.77$, d.f. $= 1$), and hence our first pass leads us to consider model M_8, $[AB][AC][AD][BC][CD]$.

Starting with model M_8, we go through the same process to determine if any additional first-order terms can be deleted. Table 17.12 summarizes the backward elimination procedure starting with model M_8. Holding to a strict $\alpha =$

TABLE 17.11   **Results of Deleting Each Two-Factor Effect
from the Model [AB][AC][AD][BC][BD][CD]**

Model	Effect	Degrees of Freedom	G^2	Prob.
M_4: [AC][AD][BC][BD][CD]		6	62.06	.0000
	Difference due to AB	1	56.58	.0000
M_5: [AB][AD][BC][BD][CD]		6	541.72	.0000
	Difference due to AC	1	536.24	.0000
M_6: [AB][AC][BC][BD][CD]		6	40.87	.0000
	Difference due to AD	1	35.39	.0000
M_7: [AB][AC][AD][BD][CD]		6	36.00	.0000
	Difference due to BC	1	30.52	.0000
M_8: [AB][AC][AD][BC][CD]		6	7.25	.2984
	Difference due to BD	1	1.77	.1832
M_9: [AB][AC][AD][BC][BD]		6	9.20	.1626
	Difference due to CD	1	3.72	.0537

TABLE 17.12   **Results of Deleting Each Two-Factor Effect
from the Model [AB][AC][AD][BC][CD]**

Model	Effect	Degrees of Freedom	G^2	Prob.
M_{10}: [AC][AD][BC][CD]		7	67.45	.0000
	Difference due to AB	1	60.20	.0000
M_{11}: [AB][AD][BC][CD]		7	542.72	.0000
	Difference due to AC	1	535.47	.0000
M_{12}: [AB][AC][BC][CD]		7	46.26	.0000
	Difference due to AD	1	39.01	.0000
M_{13}: [AB][AC][AD][CD]		7	37.26	.0000
	Difference due to BC	1	30.01	.0000
M_{14}: [AB][AC][AD][BC]		7	10.46	.1639
	Difference due to CD	1	3.21	.0731

.05, we can delete the interaction term CD ($G^2 = 3.21$, 1 d.f.). In deference to achieving parsimony, it is worthwhile to observe that the model [AB][AC][AD] [BC] yields an acceptable fit. However, in the interest of interpretability, the data are better explained through the use of M_8. Thus from this perspective the model-building process need not evaluate any other models in which additional terms are deleted. The model chosen, [AB][AC][AD][BC][CD], includes the association of "familiarity" and "location," "familiarity" and "previously patronized," "familiarity" and "recommendation," "location" and "previously patronized," and "previously patronized" and "recommendation." Moreover, given these associations, "location" is independent of "recommendation."

For the model chosen (M_8) the fitted values ($\hat{m}_{ijkl}$) were then obtained. By using a different lens to observe how the model performs, the standardized residuals defined as

$$\text{Standardized Residual} = \frac{\text{Observed Count} - \text{Fitted Count}}{\sqrt{\text{Fitted Count}}}$$

$$= \frac{x_{ijkl} - \hat{m}_{ijkl}}{\sqrt{\hat{m}_{ijkl}}} \tag{17.36}$$

were obtained (see Table 17.13). Generally speaking, standardized residuals in absolute value in excess of 2.0 would be cause for some concern. This was not the case here.

TABLE 17.13   Standardized Residuals

			Variables	
			A	
C	*D*	*B*	1	2
1	1	1	.940	.574
		2	−.190	−.339
	2	1	−.875	−.711
		2	.840	.454
2	1	1	1.149	−.453
		2	.754	.206
	2	1	−.996	.381
		2	.590	−.165

17.5 LOGIT MODELS

In our previous discussion of log-linear models we made no distinction between the variables making up the cross-classification. In many cases interest centers upon the relationship of a set of design or explanatory variables on at least one response variable. If the sample sizes for each combination of the explanatory variables are fixed by design, then from our previous discussion we need to include the μ terms relating the interactions between such variables in the model to ensure that the estimated marginal totals conform to fixed margins. Usually we are not interested in investigating relationships between the explanatory variables but want to instead determine the nature of the effects they have upon the response variables. When there is one response variable assuming two levels, it is reasonable to model the behavior of the *log odds* of one level of the response to the other on the basis of the explanatory variables. After a suitable log-linear model is found, the secondary analysis is performed, and ultimately what we arrive at is a table of log odds effects from which we can better understand how changes in the combined levels of the explanatory variables affect the response.

To illustrate these ideas, suppose variables A, B, and C are cross-classified, with variable C (having two levels) being viewed as the response variable while A and B are the design variables. Let us suppose that the "best" parsimonious model is given by $[AB][AC][BC]$ or $\mu_{ABC} = 0$. The log-linear model is therefore

$$\ln m_{ijk} = \mu + \mu_{A(i)} + \mu_{B(j)} + \mu_{C(k)} + \mu_{AB(ij)}$$
$$+ \mu_{AC(ik)} + \mu_{BC(jk)} \tag{17.37}$$

The corresponding logit model is defined by

$$\text{logit}_{ij} = \ln\left(\frac{m_{ij1}}{m_{ij2}}\right) = [\mu - \mu] + [\mu_{A(i)} - \mu_{A(i)}] + [\mu_{B(j)} - \mu_{B(j)}]$$
$$+ [\mu_{C(1)} - \mu_{C(2)}] + [\mu_{AB(ij)} - \mu_{AB(ij)}]$$
$$+ [\mu_{AC(i1)} - \mu_{AC(i2)}] + [\mu_{BC(j1)} - \mu_{BC(j2)}]$$

Eliminating terms, we obtain

$$\text{logit}_{ij} = \ln\left(\frac{m_{ij1}}{m_{ij2}}\right) = [\mu_{C(1)} - \mu_{C(2)}] + [\mu_{AC(i1)} - \mu_{AC(i2)}] + [\mu_{BC(j1)} - \mu_{BC(j2)}]$$

From (17.25), since $\mu_{C(1)} = -\mu_{C(2)}$, $\mu_{AC(i1)} = -\mu_{AC(i2)}$, and $\mu_{BC(j1)} = -\mu_{BC(j2)}$,

$$\text{logit}_{ij} = \ln\left(\frac{m_{ij1}}{m_{ij2}}\right) = 2[\mu_{C(1)} + \mu_{AC(i1)} + \mu_{BC(j1)}]$$

and thus

$$\text{logit}_{ij} = \ln\left(\frac{m_{ij1}}{m_{ij2}}\right) = W + W^{A\bar{C}}_{(i)} + W^{B\bar{C}}_{(j)} \tag{17.38}$$

where $W = 2\mu_{C(1)}$, $W^{A\bar{C}}_{(i)} = 2\mu_{AC(i1)}$, and $W^{B\bar{C}}_{(j)} = 2\mu_{BC(j1)}$

and where the bar over the superscripted variable identifies that variable which the odds pertain to. The μ terms from which the W terms can be computed are available as part of the output that may be obtained from the BMDP-4F computer program (see Section 17.6).

To illustrate the logit model, let us return to the automobile preference example (Section 17.4.3). Assume that "preference for an automobile" is considered as the response variable, while "sex" and "area of residence" are the explanatory variables. In our discussion of that example it was argued that $[AB][AC]$ was the best parsimonious model, where variable $A = $ sex, variable $B = $ area of residence, and variable $C = $ preference.

The log-linear model for this configuration is

$$\ln m_{ijk} = \mu + \mu_{A(i)} + \mu_{B(j)} + \mu_{C(k)} + \mu_{AB(ij)} + \mu_{AC(ik)} \tag{17.39}$$

It follows that the corresponding logit model is

$$\ln\left(\frac{m_{ij1}}{m_{ij2}}\right) = [\mu_{C(1)} - \mu_{C(2)}] + [\mu_{AC(i1)} - \mu_{AC(i2)}]$$

$$= 2[\mu_{C(1)} + \mu_{AC(i1)}]$$

$$= W + W_{(i)}^{AC} \qquad (17.40)$$

From Figure 17.2 we may obtain the estimated effects for the $[AB][AC]$ model. These effects and the W terms needed for the corresponding logit model are summarized in Table 17.14.

TABLE 17.14   Estimated Effects and Logits Obtained from the $[AB][AC]$ Model for the Car Preference Data

	Car Preference		
Explanatory Variable	Domestic	Foreign	W Terms
$\mu_{C(1)}$ (constant)	.88	−.88	1.76
$\mu_{AC(i1)}$			
Male	.20	−.20	.40
Female	−.20	+.20	−.40

From Table 17.14 the fact that the W term is positive for males as compared to females indicates that males are more likely than females to prefer domestic instead of foreign cars. We may also determine from (17.40) the odds ratio for either males or females. Thus for males ($i = 1$)

$$\text{logit}_{1j} = \ln\left(\frac{\hat{m}_{1j1}}{\hat{m}_{1j2}}\right) = 1.76 + .40 = 2.16$$

while for females ($i = 2$)

$$\text{logit}_{2j} = \ln\left(\frac{\hat{m}_{2j1}}{\hat{m}_{2j2}}\right) = 1.76 - .40 = 1.36$$

We should again note here that since $\mu_{BC} = 0$ in this model, there is no effect on preference due to residence.

By using these results, since each logit is positive, the odds ratio for both males and females is better than 1 to 1. The odds ratio for each may be obtained from

$$\text{Odds Ratio} = e^{\text{logit}_{ij}} = \frac{\hat{m}_{ij1}}{\hat{m}_{ij2}} \qquad (17.41)$$

Thus

$$\text{Odds Ratio (Males)} = e^{2.16} = 8.671$$

and

$$\text{Odds Ratio (Females)} = e^{1.36} = 3.896$$

Alternatively, from (17.41) we could obtain the odds ratio from the ratio of fitted cell counts (see Figure 17.2) for domestic versus foreign car preference for a particular combination of sex and residence. Thus for males

$$\frac{\hat{m}_{111}}{\hat{m}_{112}} = \frac{221.47}{25.53} = 8.675$$

while for females

$$\frac{\hat{m}_{211}}{\hat{m}_{212}} = \frac{64.66}{16.34} = 3.957$$

We should note that any differences in the odds ratio between the two computational methods are due to rounding errors that may occur in computing the logits by the iterative proportional fitting algorithm used by the computer package.

This odds ratio can be converted to a proportion by using the transformation suggested by Berkson (1944) as

$$\hat{P}_{ij} = \frac{e^{\text{logit}_{ij}}}{1 + e^{\text{logit}_{ij}}} = \frac{\hat{m}_{ij1}/\hat{m}_{ij2}}{1 + (\hat{m}_{ij1}/\hat{m}_{ij2})} \qquad (17.42)$$

Thus for males

$$\hat{P}_{1j} = \frac{8.675}{1 + 8.675} = .897$$

while for females

$$\hat{P}_{2j} = \frac{3.957}{1 + 3.957} = .798$$

Therefore, we would estimate from the model that 89.7% of the males and 79.8% of the females would prefer domestic over foreign automobiles.

Had we selected the [AB][AC][BC] model as the best model, there would have been estimated effects due to μ_{BC} as well as μ_C and μ_{AC}, and different logits would be obtained for each level of variable B (residence) as well as each level of variable A (sex).

17.6 COMPUTER PACKAGES AND MULTIWAY CONTINGENCY TABLES: USE OF BMDP PROGRAM 4F

In this section we shall explain how to use the BMDP program 4F [see Dixon et al. (1981)] to develop log-linear models for multidimensional contingency tables. BMDP program 4F can be utilized either with raw data or when cell frequency counts are available. The only paragraphs of BMDP unique to P-4F are the /TABLE, /FIT, and /PRINT paragraphs. In addition, since qualitative (i.e., discrete) variables are involved, a /CATEGORY (instead of a /GROUP) paragraph is needed. The format for all these paragraphs is

/CATEGORY   NAMES (i) ARE $\begin{Bmatrix} \text{label of} \\ \text{category 1} \end{Bmatrix}$, $\begin{Bmatrix} \text{label of} \\ \text{category 2} \end{Bmatrix}$, $\cdots$.

CODES (i) ARE $\begin{Bmatrix} \text{code of} \\ \text{category 1} \end{Bmatrix}$, $\begin{Bmatrix} \text{code of} \\ \text{category 2} \end{Bmatrix}$, $\cdots$.

/TABLE   INDICES = variable 1, variable 2, ...

$$\text{SYMBOL} = \begin{Bmatrix} \text{symbol for} \\ \text{variable 1} \end{Bmatrix}, \begin{Bmatrix} \text{symbol for} \\ \text{variable 2} \end{Bmatrix}, \cdots .$$

COUNTþISþFREQ. ←— (required only if data consists
of cell frequencies)

/FIT
$$\begin{Bmatrix} \text{ALL.} \longleftarrow \text{(only for a 2 or 3 way table)} \\ \text{or} \\ \text{MODELþIS} \begin{Bmatrix} \text{terms (symbols) to be included} \\ \text{in the model} \end{Bmatrix} \end{Bmatrix}.$$

ASSOCIATION = {value}.

$$\begin{Bmatrix} \text{ADD} = \text{MULTIPLE.} \\ \text{and/or} \\ \text{DELETE} = \text{SIMPLE.} \end{Bmatrix}$$

/PRINT   EXPECTED.
                STANDARDIZED.
                LAMBDA.

The /CATEGORY paragraph provides labels for each qualitative variable to be analyzed. The value i in the NAMES sentence indicates the ordered number of the variable in the /VARIABLE paragraph. Thus for the automobile preference data of Section 17.4.3, SEX is variable 1, RESIDE (i.e., residence) is variable 2, and PREFER (i.e., preference) is variable 3.

The /TABLE paragraph of P-4F forms the multiway table desired. The INDICES sentence names the variables to be used in forming the multiway table. The SYMBOL sentence applies one-digit alphabetic symbols to each of the variables listed in the INDICES sentence. These symbols are used to describe models and label variables in the output. The COUNT IS FREQ sentence is required only if the input data consist of the *cell frequencies* and not the raw data. For such cases (see Figure 17.1) each "data" card or line consists of the level of each variable along with the frequency count for that cell. Moreover, the cell frequency must be named as a variable.

The /FIT paragraph is used to specify the model(s) to be investigated. If a three-way table is under consideration, the ALL sentence will fit all possible log-linear models to the data. The MODEL sentence names the symbols for the terms to be included in the model. Thus, referring to Figure 17.1, the symbols SR, SP, and RP would be used if we were fitting a log-linear model that included the association of sex and residence (SR), sex and preference (SP), and residence and preference (RP). The ASSOCIATION = sentence prints the tests that the partial and marginal associations up to the given interaction level are equal to zero. For example, if ASSOCIATION = 3, each interaction that has up to 3 factors is tested for its partial and marginal association. The ADD = MULTIPLE sentence adds, in turn, each additional simple or multiple effect. The DELETE = SIMPLE sentence deletes, in turn, each additional simple effect.

The /PRINT paragraph provides for additional printed output. The EXPECTED sentence provides for the output of expected cell frequencies. The STANDARDIZED sentence provides the output of standardized residuals. Finally, the LAMBDA sentence provides the estimated effects which are needed for the logit model.

Figure 17.1 represents the program that has been written using BMDP-4F for the automobile preference problem. Figure 17.2 represents annotated partial output obtained from BMDP program 4F.

```
{SYSTEM CARDS}
/PROBLEM    TITLE IS 'CAR PREFERENCE ANALYSIS'.
/INPUT      VARIABLES ARE 4.
            FORMAT IS '(3(F1.0,1X),F3.0)'.
            CASES ARE 8.
/VARIABLE   NAMES ARE SEX,RESIDE,PREFER,FREQ.
/CATEGORY   NAMES(1) ARE MALE,FEMALE.
            CODES(1) ARE 1,2.
            NAMES(2) ARE URBAN,RURAL.
            CODES(2) ARE 1,2.
            NAMES(3) ARE DOMESTIC,FOREIGN.
            CODES(3) ARE 1,2.
/TABLE      INDICES=SEX,RESIDE,PREFER.
            SYMBOL=S,R,P.
            COUNT IS FREQ.
/FIT        ALL.
            MODEL IS SR,SP,RP.
            MODEL IS SR,SP.
            ADD=SIMPLE.
            DELETE=SIMPLE.
/PRINT      EXPECTED.
            STANDARDIZED.
            LAMBDA.
/END
1 1 1 218
1 1 2 29
1 2 1 55
1 2 2 14
2 1 1 65
2 1 2 16
2 2 1 26
2 2 2 7
{SYSTEM CARDS}
```

Figure 17.1   Using BMDP program 4F for the automobile preference data.

```
     THE FOLLOWING TABLE IS ANALYZED.
PREFER   RESIDE   I SEX    (S)
  P        R      I  MALE    FEMALE
------------------------------------
DOMESTIC  URBAN   I   218      65
          RURAL   I   155      26
                  I
FOREIGN   URBAN   I    29      16
          RURAL   I    14       7
------------------------------------
```

```
     THE TOTAL FREQUENCY IS    530
     THE RESULTS OF FITTING ALL K-FACTOR MARGINALS.
     THIS IS A SIMULTANEOUS TEST THAT ALL K+1 AND HIGHER FACTOR INTERACTIONS ARE ZERO
```

K-FACTOR	D.F.	LR CHISQ	PROB.	PEARSON CHISQ	PROB.
0(MEAN)	7	563.31	0.0	644.22	0.0
1	4	13.87	0.0077	14.46	0.0060
2	1	0.59	0.4405	0.61	0.4357

```
     A SIMULTANEOUS TEST THAT ALL K-FACTOR INTERACTIONS ARE ZERO.
     THE ENTRIES ARE DIFFERENCES IN THE ABOVE TABLE.
```

K-FACTOR	D.F.	LR CHISQ	PROB.	PEARSON CHISQ	PROB.
1	3	549.45	0.0	629.75	0.0
2	3	13.27	0.0041	13.86	0.0031
3	1	0.59	0.4405	0.61	0.4357

```
     ALL MODELS ARE REQUESTED--
```

MODEL	DF	LIKELIHOOD-RATIO CHISQ	PROB.	PEARSON CHISQ	PROB.
S	6	380.44	0.0	350.61	0.0
R	6	533.07	0.0	608.44	0.0
P	6	226.98	0.0	210.39	0.0
S,R	5	350.20	0.0	311.66	0.0
S,P	5	44.11	0.0000	44.13	0.0000
R,P	5	196.74	0.0	182.97	0.0
S,R,P	4	13.87	0.0077	14.46	0.0060
SR	4	344.88	0.0	302.84	0.0
SP	4	36.92	0.0000	36.11	0.0000
RP	4	195.44	0.0	180.75	0.0
SR,P	3	8.54	0.0360	9.09	0.0281
SP,R	3	6.68	0.0829	6.52	0.0887
RP,S	3	12.57	0.0057	12.83	0.0053
SR,SP	2	1.35	0.5079	1.32	0.5156
RS,RP	2	7.25	0.0267	8.26	0.0161
PS,PR	2	5.38	0.0677	5.21	0.0737
SR,SP,RP	1	0.59	0.4405	0.61	0.4357

```
     THE FOLLOWING MODEL WAS FIT.
     NO.     MODEL
      2    SR,SP.

     THE FITTED VALUES
```

```
PREFER   RESIDE   I SEX     (S)
  P        R      I  MALE      FEMALE
------------------------------------
DOMESTIC  URBAN   I  221.469   64.658
          RURAL   I  151.531   26.342
                  I
FOREIGN   URBAN   I   25.531   16.342
          RURAL   I   17.469    6.658
------------------------------------
```

```
     STANDARDIZED RESIDUALS = (OBSERVED-FITTED)/SQRT(FITTED)
```

```
PREFER   RESIDE   I SEX     (S)
  P        R      I  MALE      FEMALE
------------------------------------
DOMESTIC  URBAN   I  -0.233     0.043
          RURAL   I   0.282    -0.067
                  I
FOREIGN   URBAN   I   0.686    -0.085
          RURAL   I  -0.830     0.133
------------------------------------
```

Figure 17.2 Partial BMDP program 4F output for the automobile preference data.

```
                    ESTIMATES OF THE LOG-LINEAR PARAMETERS (LAMBDA)
         SEX     (S)
           MALE      FEMALE
         ------------------------
           0.549    -0.549
         ------------------------


                    ESTIMATES OF THE LOG-LINEAR PARAMETERS (LAMBDA)
         RESIDE  (R)
           URBAN     RURAL
         ------------------------
           0.319    -0.319
         ------------------------


                    ESTIMATES OF THE LOG-LINEAR PARAMETERS (LAMBDA)
         PREFER  (P)
           DOMESTIC FOREIGN
         ------------------------
           0.884    -0.884
         ------------------------


                    ESTIMATES OF THE LOG-LINEAR PARAMETERS (LAMBDA)
         RESIDE  I  SEX    (S)
           R    I    MALE     FEMALE
         ------------------------------
         URBAN  I  -0.130    0.130
         RURAL  I   0.130   -0.130
         ------------------------------


                    ESTIMATES OF THE LOG-LINEAR PARAMETERS (LAMBDA)
         PREFER  I  SEX    (S)
           P    I    MALE     FEMALE
         ------------------------------
         DOMESTIC I   0.196   -0.196
         FOREIGN  I  -0.196    0.196
         ------------------------------
```

Figure 17.2 (Continued)

PROBLEMS

17.1. A random sample of 80 first-year MBA students at a large metropolitan university were cross-classified on the basis of their undergraduate liberal arts major, humanities versus social/natural sciences, and the orientation of their graduate program, quantitative versus qualitative:

TABLE P17.1   Graduate School Study

Undergraduate Major	MBA Program	
	Quantitative	Qualitative
Social/Natural	36	14
Humanities	12	18

Assess the claim by the Dean of the School of Business that the nature of the undergraduate program is independent of the elected MBA discipline ($\alpha = .05$).

17.2. A large corporation was interested in determining whether there exists an association between the commuting time of their employees and whether or not

stress-related problems were observed on the job. The study was restricted to
assembly line employees who work an 8-hour shift of mostly repetitive tasks.

TABLE P17.2 Corporate Study

	Stress	
Commuting Time	Observed	Not Observed
Under 15 min.	12	20
15 min. to 45 min.	20	33
Over 45 min.	22	9

(a) Use the Proportional Reduction in Error measure (PRE) in predicting stress
based on commuting time. What conclusions can you make?

(b) Using stress as the response category, use the Light-Margolin R^2 on the data.
Compare and contrast with the results in part (a). *Hint:* Transpose Table
P17.2 so that the response variable "stress" is the row category.

17.3. Data reported by Wagner et al. (1976) for the purpose of developing criteria for
appropriate medical care for children are given in Table P17.3. Physicians were
requested to respond to a questionnaire dealing with various health related issues
for children. Considering whether or not the physician responded to the questionnaire as the response variable, fit an appropriate log-linear model to the
data. Discuss and interpret your results.

TABLE P17.3 Health Issues

		Questionnaire Status	
Medical Specialty	Year of Graduation	Respondents	Nonrespondents
Infectious Disease	1940–1950	9	10
Pediatricians	1951–1960	23	14
	After 1960	8	2
General	1940–1950	85	58
Pediatricians	1951–1960	102	60
	After 1960	49	21
Family	1940–1950	42	51
Physicians	1951–1960	81	61
	After 1960	27	35

Source: Reprinted by permission of *The New England Journal of Medicine*; E. H.
Wagner, R. A. Greenberg, P. B. Imrey, C. A. Williams, S. H. Wolf, and M. A. Ibrahim,
"Influence of Training and Experience on Selecting Criteria to Evaluate Medical
Care," *The New England Journal of Medicine*, Vol. 294 (1976), pp. 871–876.

*17.4. Data originally presented in *The New England Journal of Medicine* (1978) for a
case-control study of estrogens and endometrial cancer are given in Table P17.4.

TABLE P17.4   Cancer Study

	Uterine Bleeding		No Bleeding	
	Case	Control	Case	Control
Estrogen Takers	43	18	1	5
Nonestrogen Takers	99	71	6	55

Source: Reprinted, by permission of *The New England Journal of Medicine:* R. I. Horwitz and A. R. Feinstein, "Alternative Analytic Methods for Case-Control Studies of Estrogens and Endometrial Cancer," *The New England Journal of Medicine*, Vol. 299, No. 20 (Nov. 16, 1978), p. 1091.

Using the case-control variable as the response, form a logit model, and interpret the results.

17.5. (a) For the data of Table P17.5, use a computer package to fit each possible log-linear model to analyze the association of central air conditioning, having a brick exterior, and having a fireplace. Determine the best log-linear model using an $\alpha = .05$ level of significance.

TABLE P17.5   Cross-classification of Central Air Conditioning, Brick Exterior, and Fireplace

Central Air Conditioning	Brick Exterior	Fireplace	
		Yes	No
Yes	Yes	180	76
	No	14	19
No	Yes	2	2
	No	6	36

(b) Assuming that you wish to predict the ownership of a fireplace based on the presence (or absence) of central air conditioning and a brick exterior, develop a logit model, and predict the probability of having a fireplace for each combination of central air conditioning and brick exterior.   ■

17.6. Reviewing concepts:
   (a) What is a log-linear model?
   (b) What are the main differences between the Pearson χ^2 statistic and the likelihood ratio statistic G^2 in assessing the fit of a postulated model?
   (c) What is a logit model?
   (d) Compare and contrast the log-linear versus logit models in multidimensional contingency table analysis.

18

DISCRIMINANT ANALYSIS

18.1 INTRODUCTION

Discriminant analysis as a body of methodology could be of assistance in the following problems:

1. Based on responses received from owners of foreign-made cars and owners of American-made cars, it is desired, on the basis of pertinent measurements, to profile a potential foreign car purchaser.
2. On the basis of a financial analysis taken in connection with a business loan application, a banker needs to determine whether the applicant is a good risk or has a potential for bankruptcy.
3. A psychologist administers a battery of tests to a patient. On the basis of information known about two mental illnesses, A and B, the psychologist needs to decide which of the two symptomatically similar diseases the patient has.
4. College admission officers may desire to set up an objective criterion for admitting a student to study. On the basis of high school average, SAT scores, ranking in class, etc., the college will classify the student as a good risk or a poor risk.

More generally, discriminant analysis is concerned with the following problem: Given that a population can be partitioned into k distinct groups $G_1, G_2, \ldots, G_k$, given that an observation $\mathbf{X} = [X_1, X_2, \ldots, X_p]'$ is known to belong to one of these groups, and given that it is unknown to which group it belongs, develop a rule for assigning $\mathbf{X}$ in such a manner that the chance of a misclassification is made as small as possible. In most instances sample-based assignment rules are employed and are based on *training samples* taken directly from $G_1, G_2, \ldots, G_k$ or sampled from the mixed population. In either case, fundamental in the deployment of a rule is the assumption that the training samples are correctly classified; if otherwise, serious problems will probably result in the analysis.

18.2 DISCRIMINATION BETWEEN TWO GROUPS

18.2.1 Development

The most common problem in discriminant analysis deals with situations where only two groups G_1 and G_2 define the partitions of the population. We shall assume that n observations have been sampled from the mixed population and that with sufficient effort we are confident that correct identification can be made. The n observations are therefore split so that n_i are from G_i, $i = 1, 2$, and $n_1 + n_2 = n$.

In 1936, Fisher proposed a method for handling this problem which, even today, is the most frequently employed discriminant rule. Denoting the true mean vector of G_i by $\boldsymbol{\mu}_i = [\mu_{i1}, \mu_{i2}, \ldots, \mu_{ip}]'$, $i = 1, 2$, and assuming that the two variance-covariance matrices $\boldsymbol{\Sigma}_1$ and $\boldsymbol{\Sigma}_2$ have a common value $\boldsymbol{\Sigma}$, Fisher proposed, as a criterion for goodness of an allocation rule, finding a linear combination of $\mathbf{X}$ (the point to be assigned to either G_1 or G_2) so that the ratio of the difference of the means of the compound in G_1 and G_2 to its common variance is maximized. Denoting the linear compound by $Z = \mathbf{aX}$, where $\mathbf{a} = [a_1, a_2, \ldots, a_p]$, we need to find $\mathbf{a}$ so that we maximize

$$\delta = \frac{(\mathbf{a}\boldsymbol{\mu}_1 - \mathbf{a}\boldsymbol{\mu}_2)^2}{\mathbf{a}\boldsymbol{\Sigma}\mathbf{a}'} \tag{18.1}$$

It can be shown that the vector of coefficients $\mathbf{a}$ is proportional to $(\boldsymbol{\mu}_1 - \boldsymbol{\mu}_2)'\boldsymbol{\Sigma}^{-1}$ and therefore that $\mathbf{a}$ is not unique. Indeed, the set of coefficients can be multiplied by any constant without altering the discrimination rule.

Since the parameters $\boldsymbol{\mu}_1, \boldsymbol{\mu}_2$, and $\boldsymbol{\Sigma}$ will rarely if ever be known, the discriminant coefficients $(\boldsymbol{\mu}_1 - \boldsymbol{\mu}_2)'\boldsymbol{\Sigma}^{-1}$ are replaced by the estimated coefficients†

$$(\bar{\mathbf{X}}_1 - \bar{\mathbf{X}}_2)'\mathbf{S}^{-1} \tag{18.2}$$

†Note that $\mathbf{S}$ is the *pooled* sample variance-covariance matrix. If the two unknown population variance-covariance matrices $\boldsymbol{\Sigma}_1$ and $\boldsymbol{\Sigma}_2$ are assumed to have common value $\boldsymbol{\Sigma}$, then the respective sample variance-covariance matrices $\mathbf{S}_1$ and $\mathbf{S}_2$ are combined as in (18.2) to estimate the true $\boldsymbol{\Sigma}$. In practice, the sample variance-covariance matrix usually employed is one which provides an unbiased estimate for $\boldsymbol{\Sigma}$. Here, $\mathbf{S}_1, \mathbf{S}_2$, and $\mathbf{S}$ each provide unbiased estimates for $\boldsymbol{\Sigma}$. In Chapter 15, however, recall that the sample variance-covariance matrix ($\hat{\boldsymbol{\Sigma}}$) employed in the study of principal components was the maximum likelihood estimate for $\boldsymbol{\Sigma}$.

where

$$\bar{\mathbf{X}}_i = \frac{\sum_{l=1}^{n_i} X_{il}}{n_i}, \qquad \bar{\mathbf{X}} = \frac{\sum_{i=1}^{2}\sum_{l=1}^{n_i} X_{il}}{n}$$

$$\mathbf{S} = \frac{\sum_{i=1}^{2} n_i(\bar{\mathbf{X}}_i - \bar{\mathbf{X}})(\bar{\mathbf{X}}_i - \bar{\mathbf{X}})'}{n - 2}$$

The assignment rule therefore becomes the following: Assign $\mathbf{X}$ to G_i if

$$|(\bar{\mathbf{X}}_1 - \bar{\mathbf{X}}_2)'\mathbf{S}^{-1}(\mathbf{X} - \bar{\mathbf{X}}_i)| \leq |(\bar{\mathbf{X}}_1 - \bar{\mathbf{X}}_2)'\mathbf{S}^{-1}(\mathbf{X} - \bar{\mathbf{X}}_j)| \qquad (18.3)$$

for $i \neq j$, $i = 1, 2$, and $j = 1, 2$. It is well known that if the subpopulations are characterized by multivariate normal distributions† such that observations in $G_i \sim \mathcal{N}_p(\boldsymbol{\mu}_i, \boldsymbol{\Sigma})$, then assuming equal prior probabilities of group membership, the rule given in (18.3) minimizes the probability of misclassification. An equivalent representation of (18.3) uses the midpoint

$$\bar{Y} = \frac{(\bar{\mathbf{X}}_1 - \bar{\mathbf{X}}_2)'\mathbf{S}^{-1}\bar{\mathbf{X}}_1 + (\bar{\mathbf{X}}_1 - \bar{\mathbf{X}}_2)'\mathbf{S}^{-1}\bar{\mathbf{X}}_2}{2}$$
$$= \tfrac{1}{2}(\bar{\mathbf{X}}_1 - \bar{\mathbf{X}}_2)'\mathbf{S}^{-1}(\bar{\mathbf{X}}_1 + \bar{\mathbf{X}}_2) \qquad (18.4)$$

as the point of separation. The mean values of the discriminant function for group i, obtained by applying the vector of weighting coefficients to the mean scores of the original variables for each group, that is, $\bar{Y}_i = (\bar{\mathbf{X}}_1 - \bar{\mathbf{X}}_2)'\mathbf{S}^{-1}\bar{\mathbf{X}}_i$, are commonly referred to as group *centroids*. Note that with unequal sample sizes the point of separation would be obtained as a weighted average of group centroids

$$\bar{Y}* = \frac{n_2\bar{Y}_1 + n_1\bar{Y}_2}{n_1 + n_2} \qquad (18.5)$$

Taking the difference in group centroids yields the statistic

$$D^2 = \bar{Y}_1 - \bar{Y}_2 = (\bar{\mathbf{X}}_1 - \bar{\mathbf{X}}_2)'\mathbf{S}^{-1}(\bar{\mathbf{X}}_1 - \bar{\mathbf{X}}_2) \qquad (18.6)$$

which is the sample estimate of the *Mahalanobis distance*. This D^2 statistic can be used to determine if the between group differences are statistically significant in the sense of mean separation. Large values would give us some comfort that future observations can, on the basis of the characteristics measured, be successfully classified. A statistical test can be constructed using the fact that

$$\frac{n_1 n_2}{(n_1 + n_2)}\left[\frac{(n_1 + n_2 - p - 1)}{(n_1 + n_2 - 2)p}\right]D^2 \sim F_{p, n_1 + n_2 - p - 1} \qquad (18.7)$$

To summarize, the discriminant function is a linear combination of the original measurements selected so that the squared difference between the two

†A random vector $\mathbf{X}$ is said to have a p-dimensional normal distribution if the density of $\mathbf{X}$ is given by

$$f(\mathbf{X}) = \frac{1}{(2\pi)^{p/2}|\boldsymbol{\Sigma}|^{1/2}}e^{\{-(1/2)(\mathbf{X}-\boldsymbol{\mu})'\boldsymbol{\Sigma}^{-1}(\mathbf{X}-\boldsymbol{\mu})\}}$$

where $E(\mathbf{X}) = \boldsymbol{\mu}$ and $\text{Var}(\mathbf{X}) = \boldsymbol{\Sigma}$.

compound means is maximal relative to the within groups variance. When the population is partitioned into two groups, a single function will suffice; however, in the case of several groups one weighted combination may distinguish certain of the groups but not others. For the multiple group problem a second or possibly a third composite may be required to distinguish between the groups. For an in-depth discussion of such problems, see Goldstein and Dillon (1983).

18.2.2 Application

To illustrate the construction of a discriminant function, we shall use part of a data set compiled by the Rhode Island Statewide Planning Agency. A purpose of the analysis was to assess key variables principally in the area of cost and waiting time in predicting modes of transportation. To fix ideas, we consider 732 trips, the destination of which was an individual's place of work. Of the total number of work trips, 366 were by automobile, while the remaining 366 were accomplished via bus transportation. Three variables were examined as possible determinants of an individual's mode of travel choice:

X_1: relative cash costs given by "cost by bus"/"cost by auto"

X_2: income of the individual traveler

X_3: expected waiting time at the trip origin

For purposes of analysis, all variables were measured in logarithmic form. Tables 18.1 and 18.2 present the sample means and pooled variance-covariance matrix.

TABLE 18.1   Sample Means (Logarithmic Form)

Variable	Automobile (G_1)	Bus (G_2)
X_1: Relative Cost	1.963	1.293
X_2: Income	1.270	1.793
X_3: Waiting Time	.646	.381

TABLE 18.2   Pooled Variance-Covariance Matrix

Variable	X_1: Relative Cost	X_2: Income	X_3: Waiting Time
X_1: Relative Cost	.1272	−.0516	−.0037
X_2: Income	−.0516	.1205	−.0021
X_3: Waiting Time	−.0037	−.0021	.3211

The sample-based vector of discriminant coefficients is given by

$$\hat{\mathbf{a}} = (\bar{\mathbf{X}}_1 - \bar{\mathbf{X}}_2)'\mathbf{S}^{-1}$$

$$= [(1.963 - 1.293)(1.270 - 1.793)(.646 - .381)]$$

$$\times \begin{bmatrix} .1272 & -.0516 & -.0037 \\ -.0516 & .1205 & -.0021 \\ -.0037 & -.0021 & .3211 \end{bmatrix}^{-1} \qquad (18.8)$$

$$= [4.281, -2.493, .8585]$$

and therefore the discriminant function is expressed by the linear compound

$$Y = 4.281X_1 - 2.493X_2 + .8585X_3 \qquad (18.9)$$

In almost all cases the vector of discriminant coefficients is standardized so that scale effects due to measurement differences in the variables are removed. The standardized coefficients are then interpreted as a measure of a variable's worthiness as a discriminator. This was not done in the present example since any scale differences are probably removed, or at least significantly dampened, by the logarithmic transform. Noting that $\bar{Y}_1 = 5.792$ and $\bar{Y}_2 = 1.392$, we see that high discriminant scores are associated with G_1, the automobile travel group; consequently, we conclude the greater the relative cost of bus to automobile travel (X_1) and the greater the waiting time (X_3), the more likely is the choice of an automobile as a travel mode to one's place of work. The negative sign for the coefficient of X_2 in (18.9) implies that higher-income persons tend to use public transportation to work.

The point of separation between G_1 and G_2 is the midpoint $(\bar{Y}_1 + \bar{Y}_2)/2 = (5.792 + 1.392)/2 = 3.592$. Thus the assignment rule is as follows: Assign to the automobile travel mode (G_1) if $Y > 3.592$ and to bus transit (G_2) if $Y < 3.592$. If the unlikely event $Y = 3.592$ occurs, assignment is made randomly.

One of the methods social scientists have traditionally used in assessing how well a discriminant rule functions is to form a *confusion matrix*. The matrix for the data at hand is shown in Table 18.3, and hence the percent of cases

TABLE 18.3   Confusion Matrix for Transportation Data

Actual Group	Predicted Group Membership	
	Group 1	Group 2
Group 1	281	85
Group 2	31	335

correctly classified is 84.15 (i.e., 616 individuals out of 732). While this method of assessment is commonplace, a little thought would indicate its overly optimistic nature. Since the data being classified are precisely the same data that are used in constructing the discriminant function, there is an optimistic bias built into the calculation. There are other methods which can be employed to get estimates of probabilities of correct classification which do not suffer, to the same degree, from large built-in bias. Perhaps the most simple procedure is to split the available training sample in half, with one set being used to generate the discriminant function while the other is used to assess its worth. In this case independent samples are being used for the separate procedures and probably would result in a more precise estimate for probabilities of correct classification. For an in-depth discussion of procedures available for nonerror rate estimation, see Goldstein and Dillon (1983).

Last, using (18.7) we may test whether the group differences are statistically significant. For these data, $D^2 = 4.40$ so that we have

$$F = \frac{366(366)}{(366 + 366)}\left[\frac{(366 + 366 - 3 - 1)}{(366 + 366 - 2)(3)}\right](4.40) = 267.66$$

With 3 and 728 degrees of freedom, this value is significant at any reasonable significance level, and hence we would conclude that the populations are sufficiently separated to make the exercise of discrimination worthwhile.

18.3 TESTING WHETHER A SUBSET OF THE MEASUREMENTS IS SUFFICIENT FOR DISCRIMINATION

In our discussion thus far we have assumed that all p components of the measurement vector are needed to adequately define an allocation rule. This usually will not be the case; indeed, frequently only a portion of the measurements is needed. In a sense the remaining variables are superfluous, and while their inclusion will as a rule not harm the analysis, the additional information they impart may be very small.

Rao (1973) has suggested a method [assuming $G_i \sim \mathcal{N}_p(\boldsymbol{\mu}_i, \boldsymbol{\Sigma})$] whereby variables can be screened for their worthiness as discriminants. Toward this end, suppose $\mathbf{X} = [X_1, X_2, \ldots, X_p]'$ is partitioned into $\mathbf{X}_1 = [X_1, X_2, \ldots, X_{p_1}]'$ and $\mathbf{X}_2 = [X_{p_1+1}, \ldots, X_p]'$. Let D_p^2 represent the Mahalanobis distance as defined in (18.6) on $\mathbf{X}$, while $D_{p_1}^2$ is defined only on the subset $\mathbf{X}_1$. Thus, the statistic

$$\left(\frac{n_1 + n_2 - p - 1}{p - p_1}\right)\left[\frac{C(D_p^2 - D_{p_1}^2)}{1 + CD_{p_1}^2}\right] \sim F_{p - p_1, n_1 + n_2 - p - 1} \qquad (18.10)$$

where $C = n_1 n_2/(n_1 + n_2)(n_1 + n_2 - 2)$.

An important special case occurs where $\mathbf{X}_2$ contains only one component, that is, $p_1 = p - 1$, in which case we wish to determine if a single specified variable has discriminating power. The statistic

$$\left(\frac{n_1 + n_2 - p - 1}{1}\right)\left[\frac{C(D_p^2 - D_{p-1}^2)}{1 + CD_{p-1}^2}\right] \qquad (18.11)$$

is distributed as F with 1 and $n_1 + n_2 - p - 1$ degrees of freedom or equivalently as t^2 with $n_1 + n_2 - p - 1$ degrees of freedom. Significant values for

(18.11) would lead the researcher to conclude that the measurement X_p is needed for discriminatory power.

To illustrate, we return to the transportation data. The sample-based Mahalanobis distance using all three variables is, from (18.6),

$$D_p^2 = (\bar{\mathbf{X}}_1 - \bar{\mathbf{X}}_2)'\mathbf{S}^{-1}(\bar{\mathbf{X}}_1 - \bar{\mathbf{X}}_2) = 4.400$$

Suppose we delete X_2, income, from consideration. The two subsets are now defined as $\mathbf{X}_1 = [X_1, X_3]'$ and $\mathbf{X}_2 = [X_2]'$. The Mahalanobis distance defined on $\mathbf{X}_1$ is

$$D_{p-1}^2 = [.670, .265] \begin{bmatrix} .1272 & -.0037 \\ -.0037 & .3211 \end{bmatrix}^{-1} \begin{bmatrix} .670 \\ .265 \end{bmatrix}$$

$$= [.670, .265] \begin{bmatrix} 7.87 & .091 \\ .091 & 3.11 \end{bmatrix} \begin{bmatrix} .670 \\ .265 \end{bmatrix} = 3.784$$

The statistic give in (18.11) becomes

$$\frac{728}{1} \left[\frac{.2507(4.400 - 3.784)}{1 + (.2507)(3.784)} \right] = 57.68$$

which at the .05 level is significant, and hence we conclude that the income variable cannot be deleted without losing significant discriminatory power.

18.4 THE NEED FOR ALTERNATIVE PROCEDURES

In our discussion of Fisher's procedure no distributional constraints were imposed on the analysis. Our only requirement was that the variance-covariance matrices within the two subpopulations be equal. Fisher's linear discriminant makes good intuitive sense and is optimal in the sense of minimizing the probability of misclassification in the case where $G_i \sim \mathcal{N}_p(\boldsymbol{\mu}_i, \boldsymbol{\Sigma})$, $i = 1, 2$. However, we know at least in a theoretical sense that we can do better when different characteristics are known about G_i.

It is beyond the scope of our treatment here to present a very general discussion of the discrimination problem. We can, however, state without proof a result which will serve as the basis for motivating a search for more general procedures. To be specific, suppose that the probability distribution within the subpopulations is such that the density function in G_i is $f_i(\mathbf{X})$. For example, in the case where the density in G_i is $\mathcal{N}_p(\boldsymbol{\mu}_i, \boldsymbol{\Sigma})$, then if $\mathbf{X} = [X_1, X_2, \ldots, X_p]'$,

$$f_i(\mathbf{X}) = (2\pi)^{-p/2} |\boldsymbol{\Sigma}|^{-1} \exp[-\tfrac{1}{2}(\mathbf{X} - \boldsymbol{\mu})'\boldsymbol{\Sigma}^{-1}(\mathbf{X} - \boldsymbol{\mu})] \qquad (18.12)$$

If the prior probabilities of membership in G_i are P_i, $i = 1, 2$, then it can be shown that an optimal partition of the sample space in the sense of minimizing the probability of misclassification is given by

$$\begin{aligned} \mathfrak{D}_1 &= \{\mathbf{X} : P_1 f_1(\mathbf{X}) > P_2 f_2(\mathbf{X})\} \\ \mathfrak{D}_2 &= \{\mathbf{X} : P_1 f_1(\mathbf{X}) < P_2 f_2(\mathbf{X})\} \end{aligned} \qquad (18.13)$$

Operationally, (18.13) means that an observation is to be assigned to G_i if $\mathbf{X} \in \mathfrak{D}_i$. In general, if $P_1 f_1(\mathbf{X}) = P_2 f_2(\mathbf{X})$, then $\mathbf{X}$ is randomly assigned.

The partition $\mathfrak{D} = (\mathfrak{D}_1, \mathfrak{D}_2)$ is of little practical interest since it assumes that the underlying densities are completely specified. Sample-based versions of $\mathfrak{D}$ come about essentially in two ways: The parametric form of $f_i(\cdot)$ is known, and sample estimates are used in place of unknown parameters, *or* nonparametric density estimates are employed. In either case, provided the estimation procedure is statistically consistent, the estimated partitions will evolve into a sequence of sample-based partitions which at least for large samples will provide good error rate probabilities. For example, Fisher's rule is basically that given in (18.13) with $f_i \sim \mathcal{N}_p(\mathbf{\mu}_i, \mathbf{\Sigma})$, with the sample estimates $\bar{\mathbf{X}}_i$ replacing $\mathbf{\mu}_i$ and with the *pooled* variance-covariance matrix S replacing $\mathbf{\Sigma}$.

Often researchers do not feel that assuming equal variance-covariance matrices is appropriate in given problems (even though tests for equality of $\mathbf{\Sigma}_1$ and $\mathbf{\Sigma}_2$ may not lead to rejection, they would prefer modeling as if $\mathbf{\Sigma}_1 \neq \mathbf{\Sigma}_2$). In the case where the densities are multivariate normal with parameters $\mathbf{\mu}_i$ and $\mathbf{\Sigma}_i$ in G_i, the optimal partition in the sense given above is, assuming equal prior probabilities,

$$\mathfrak{D}_1 = \{\mathbf{X}: \tfrac{1}{2} \ln |\mathbf{\Sigma}_1||\mathbf{\Sigma}_2|^{-1} - \tfrac{1}{2}(\mathbf{X} - \mathbf{\mu}_1)'\mathbf{\Sigma}_1^{-1}(\mathbf{X} - \mathbf{\mu}_1)$$
$$+ \tfrac{1}{2}(\mathbf{X} - \mathbf{\mu}_2)'\mathbf{\Sigma}_2^{-1}(\mathbf{X} - \mathbf{\mu}_2) > 0\} \tag{18.14}$$

This partitioning of the sample space is usually referred to as the *quadratic discriminant* function and was first attributed to Smith (1947). Some researchers ignore tests for equality of variance-covariance matrices, pretend that $\mathbf{\Sigma}_1 = \mathbf{\Sigma}_2$, and proceed to apply Fisher's linear discriminant function. Although research is limited in the sense that only small-scale Monte Carlo experiments have been conducted, the evidence indicates that the Fisher rule and the quadratic discriminant function will be more in disagreement as: the training samples decrease in size; the variance-covariance matrices get further apart; the mean vectors get closer together; and the number of variables increases. Since computer packages such as SPSS, SAS, and BMDP now contain options to test $\mathbf{\Sigma}_1 = \mathbf{\Sigma}_2$, it seems foolish not to do so before embarking on an analysis (see Section 18.6).

18.5 DISCRIMINANT ANALYSIS WITH CATEGORICAL DATA

18.5.1 Development

In the social and behavioral sciences, deviations from normality appear more the rule than the exception. Frequently measurement is so primitive that responses of the form "present-absent" or "clear-milky-cloudy" are the only variable indicators of that which is observed. It is in situations where measurements are not available on an interval or ratio scale (Section 1.3) that classification techniques designed for discrete data appear to be most compelling.

As stated above, when the underlying probability densities defining the two groups are known, an optimal classification procedure in the sense of minimizing misclassification probability can be developed. Clearly, if the multivariate data under consideration consist of coordinates which are discrete-valued, then procedures derived assuming, say, underlying multinomial structures would appear more sensible in capturing the available information rather than blatantly applying the procedures discussed thus far.

18.5.2 The Full Multinomial Rule

To focus on the problem, let us consider discrete random variables X_1, $X_2, \ldots, X_p$ each assuming at most a finite number of distinct values $s_1, s_2, \ldots, s_p$; hence, there are $\prod_{i=1}^{p} s_i$ possible types of measurements. We shall assume that the underlying densities within each group are multinomial. In each subpopulation G_i, each of the $s = \prod_{i=1}^{p} s_i$ points has a probability attached it; that is, G_i is characterized by the multinomial probability $\boldsymbol{\theta} = (\theta_{i1}, \theta_{i2}, \ldots, \theta_{is})$, where for any $j, \theta_{ij} > 0$ and $\sum_{i,j} \theta_{ij} = 1$.

When n observations are sampled from the combined population, the number $N_i(\mathbf{X})$ from G_i having characteristic measurement $\mathbf{X}$ and the total from G_i, $N_i = \sum N_i(\mathbf{X})$, are binomial random variables. Intuitive estimates of prior probabilities and class conditional densities yield simple estimates of the parameters which define the optimal partition $\mathfrak{D} = (\mathfrak{D}_1, \mathfrak{D}_2)$, viz.,

$$\hat{P}_i \hat{f}_i(\mathbf{X}) = \frac{N_i}{n} \frac{N_i(\mathbf{X})}{N_i} = \frac{N_i(\mathbf{X})}{n} \tag{18.15}$$

The sample-based version of $\mathfrak{D}$ is given by the following: Assign $\mathbf{X}$ to G_1 if $N_1(\mathbf{X}) > N_2(\mathbf{X})$. The event $N_1(\mathbf{X}) = N_2(\mathbf{X})$ can occur with positive probability; hence, if this occurs, $\mathbf{X}$ is randomly assigned. Estimating the *discriminant scores* $P_i f_i(\mathbf{X})$ in the manner given in (18.15) yields the discriminant usually referred to as the *full multinomial rule*. While the rule has good large-sample properties, its use with small or moderate sample sizes can be a problem. A few variables each containing only two or three levels result in a proliferation of measurement states—many of which contain no frequency counts—and hence no assignment outside of random allocation is possible.

In the preceding formulation of the full multinomial rule we assumed that the data were sampled from a combined or mixed population. Frequently two independent samples of size n_1 and n_2 are selected from G_1 and G_2, respectively. In the case of independent random samples, prior probabilities are not usually estimated but are specified. By letting $n_i(\mathbf{X})$ denote the number of individuals from G_i, $i = 1, 2$, with characteristic measurement $\mathbf{X}$, it follows under the case of equal prior probabilities that $\mathbf{X}$ is assigned to G_1 if $n_1(\mathbf{X})/n_1 > n_2(\mathbf{X})/n_2$ and that $\mathbf{X}$ is assigned to G_2 if $n_1(\mathbf{X})/n_1 < n_2(\mathbf{X})/n_2$. The point is randomly assigned if the ratios are equal.

18.5.3 Application

To illustrate the ideas expressed above, we consider a data set discussed by Goldstein and Dillon (1983). The study was undertaken to investigate the relationship between store choice and a number of product- and shopper-specific variables. The two population groups consist of shoppers who either purchased audio equipment from a specialty store (G_1) or from a full-line department store (G_2). Not all the variables measured will be discussed here; we shall, for illustrative purposes, concentrate only on the three dichotomous variables: (O) opinion leadership trait [(1) leader, (2) follower]; (S) respondents' sex [male (1), female (2)]; and (I) information-seeking activity [(1) yes, (2) no]. Independent samples of sizes 113 and 106 were taken from G_1 and G_2. Table 18.4 gives the eight possible states along with the observed frequency distributions.

TABLE 18.4 Frequency Distribution Across States

State	Observed Frequency (G_1)	Observed Frequency (G_2)
1 1 1	26	32
1 1 2	0	0
1 2 1	50	33
1 2 2	4	0
2 1 1	8	20
2 1 2	0	0
2 2 1	23	20
2 2 2	2	1
	113	106

Utilization of the full multinomial rule for the data of Table 18.4 results in the following classifications: When $\mathbf{X} = (1,1,1)$, assignment is made to G_2 since $\frac{26}{113} < \frac{32}{106}$. On the other hand, for state $(1,2,1)$, classification to G_1 occurs since $\frac{50}{113} > \frac{33}{106}$. Assignment of the remaining states proceeds in a similar manner with states $(1,2,2)$, $(2,2,1)$, and $(2,2,2)$ classified into G_1, while state $(2,1,1)$ is assigned to G_2. Last, points having states $(1,1,2)$ or $(2,1,2)$ are randomly assigned to G_1 and G_2 since in both cases the state counts are zero. The entire set of classifications is summarized in Table 18.5.

Table 18.6 presents the confusion matrix for the data of Table 18.5.

Since there are 34 respondents in group 1 and 54 respondents in group 2 that have been misclassified, the error rate is $(34 + 54)/(113 + 106) = 39.7\%$.

18.5.4 The Logit Rule

Goldstein and Dillon (1978) discuss a number of alternative approaches to estimating discriminant scores which do not suffer from the sparseness problem to the degree exhibited by the full multinomial rule. Of all the models

TABLE 18.5   Classifications Using the Full Multinomial Rule

State	G_1 (Specialty) Observed Frequency	G_2 (Department) Observed Frequency	Classification Rule	Number of Misclassifications
1 1 1	26	32	G_2	26
1 1 2	0	0	Random	0
1 2 1	50	33	G_1	33
1 2 2	4	0	G_1	0
2 1 1	8	20	G_2	8
2 1 2	0	0	Random	0
2 2 1	23	20	G_1	20
2 2 2	2	1	G_1	1

TABLE 18.6   Confusion Matrix for the Store Choice Data Using the Full Multinomial Rule

Actual Group	Predicted Group Membership	
	Group 1 (Specialty)	Group 2 (Department)
Group 1 (Specialty)	79	34
Group 2 (Department)	54	52

presented, the authors recommend expressing state frequencies using log-linear models and defining allocation rules in terms of differences in the logarithm of particular state frequencies. As we have seen in Chapter 17, such models allow us to: use *goodness of fit statistics* as a tool of model building; provide a "handle" on the sparseness issue; and incorporate information pertaining to orderings of the variables.

To demonstrate the procedure, let us consider a three-variable problem (X_1, X_2, X_3) which under multinomial sampling generates a table of dimension $2 \times J \times K$. We view the first variable as one identifying the two groups under consideration, that is, $X_1 = 1 \Leftrightarrow G_1$, $X_1 = 2 \Leftrightarrow G_2$. The logarithm of m_{ijk}, the theoretical frequency in cell (i,j,k), is expressed as

$$\ln m_{ijk} = \mu + \mu_{A(i)} + \mu_{B(j)} + \mu_{C(k)}$$
$$+ \mu_{AB(ij)} + \mu_{AC(ik)} + \mu_{BC(jk)} + \mu_{ABC(ijk)} \tag{18.16}$$

The parameters are assumed to satisfy the constraints of such models discussed in Chapter 17. Assuming equal prior probabilities, we again call on the optimal partition $\mathfrak{D} = (\mathfrak{D}_1, \mathfrak{D}_2)$ to define a sample-based *logit rule*: Assign a response characterized by $X_2 = j$, $X_3 = k$ to G_1 if

$$\ln\left(\frac{m_{1jk}}{m_{2jk}}\right) > 0 \tag{18.17}$$

If the right-hand side assumes the value zero, the observation is randomly assigned.

We can think of the frequency distribution given in Table 18.4 as a $2 \times 2 \times 2 \times 2$ contingency table where in all (i,j,k,l) we mean the state characterized by group G_i, opinion leadership at level j, sex at level k, and information seeking at level l. By using established procedures for model building, as discussed in Chapter 17, the best parsimonious model is given by

$$[GS] \qquad [GO] \qquad [GI]$$

or in terms of μ parameters,

$$\ln m_{ijkl} = \mu + \mu_{A(i)} + \mu_{B(j)} + \mu_{C(k)} + \mu_{D(l)} \tag{18.18}$$
$$+ \mu_{AB(ij)} + \mu_{AC(ik)} + \mu_{AD(il)}$$

where $A = G$, $B = S$, $C = O$, $D = I$. Note that since the group-defining variable is being considered as the response while the others are explanatory, it is, almost by definition, included in all the interaction terms and would be included in any higher-order terms if a less parsimonious model were deemed appropriate.

Since only first-order interactions are being used in the fitted model given in (18.18), the logit model which will define the induced discriminant employed will be only of first order; that is,

$$\ln\left(\frac{m_{1jkl}}{m_{2jkl}}\right) = [\mu_{A(1)} - \mu_{A(2)}] + [\mu_{AB(1j)} - \mu_{AB(2j)}]$$
$$+ [\mu_{AC(1k)} - \mu_{AC(2k)}] + [\mu_{AD(1l)} - \mu_{AD(2l)}] \tag{18.19}$$
$$= W + W^{AB}_{(j)} + W^{AC}_{(k)} + W^{AD}_{(l)}$$

Classification of any observation whose explanatory components are given by j, k, l into G_1 occurs if the estimates of the W parameters given in (18.19) are such that

$$\ln\left(\frac{m_{1jkl}}{m_{2jkl}}\right) = \hat{W} + \hat{W}^{AB}_{(j)} + \hat{W}^{AC}_{(k)} + \hat{W}^{AD}_{(l)} > 0$$

If < 0, the point is assigned to G_2; otherwise assignment is made randomly.

Classification could be made either from the W values (see Section 17.5) or from the natural logarithm of the ratio of expected state (cell) counts. Table 18.7 presents the expected state counts under the fitted model. From Table 18.7, since the logit associated with state $\mathbf{X} = (1,1,1)$ is $\ln(22.793/31.586) < 0$, it follows that responses in this state are classified into G_2. Similarly, when $\mathbf{X} = (1,2,1)$, classification is made into G_1 since $\ln(52.960/32.801) > 0$. The entire set of classifications for the eight states is summarized in Table 18.8.

From Tables 18.8 and 18.5, we observe that the logit and full multinomial rules resulted in identical classification in all states except $(1,1,2)$ and $(2,1,2)$. Responses in these two states had to be randomly allocated under the full multinomial procedure, while the logit rule resulted in a forced classification. Table 18.9 presents the confusion matrix for the data of Table 18.8.

TABLE 18.7   Expected State Counts Using Model
$[GS] [GO] [GI]$

State	Expected Count (G_1)[a]	Expected Count (G_2)[a]
1 1 1	22.793 (.202)	31.586 (.304)
1 1 2	1.278 (.011)	.301 (.003)
1 2 1	52.960 (.469)	32.801 (.304)
1 2 2	2.970 (.026)	.312 (.003)
2 1 1	9.402 (.083)	19.923 (.192)
2 1 2	.527 (.005)	.190 (.002)
2 2 1	21.846 (.193)	20.690 (.192)
2 2 2	1.224 (.011)	.298 (.002)

[a]Numbers given in parentheses represent the relative expected frequencies in each state.

TABLE 18.8   Classifications Using the Logit Rule

State	G_1 (Specialty) Observed Frequency	G_2 (Department) Observed Frequency	Classification Rule	Number of Misclassifications
1 1 1	26	32	G_2	26
1 1 2	0	0	G_1	0
1 2 1	50	33	G_1	33
1 2 2	4	0	G_1	0
2 1 1	8	20	G_2	8
2 1 2	0	0	G_1	0
2 2 1	23	20	G_1	20
2 2 2	2	1	G_1	1

TABLE 18.9   Confusion Matrix for Store Choice Data Using the Logit Rule

Actual Group	Predicted Group Membership	
	Group 1 (Specialty)	Group 2 (Department)
Group 1 (Specialty)	79	34
Group 2 (Department)	54	52

Since there are 34 respondents in group 1 and 54 respondents in group 2 who have been misclassified, the error rate is $(34 + 54)/(113 + 106) = 39.7\%$.

It is interesting to observe that the full multinomial rule, when applied to the store choice data set, also resulted in a 39.7% error rate. In terms of a log-linear representation, the full multinomial rule is equivalent to using a completely saturated design. It is noteworthy that the fitted model which uses only 8 parameters and the full model characterized by 15 independent parameters yield the same error rate probabilities. Last, unlike the full model, the unsaturated design eliminated all random zeros, a great help in defining the discriminant.

18.6 COMPUTER PACKAGES AND DISCRIMINANT ANALYSIS: USE OF SPSS, SAS, AND BMDP

In this section we shall discuss how to use SPSS, SAS, and BMDP for discriminant analysis. We shall explain how to use these computer packages by referring to the data of Problem 18.1. In this problem, we wish to perform a discriminant analysis of the success of students in an MBA program based on undergraduate grade point index GPI (X_1) and Graduate Management Aptitude Test Score (X_2).

For data entry purposes, "success" is entered in column 1, X_1 is entered in columns 3–6, and X_2 is entered in columns 8–10.

18.6.1 Using the SPSS DISCRIMINANT Procedure for Discriminant Analysis

The SPSS DISCRIMINANT subprogram can be used for either discriminant analysis or stepwise discriminant analysis [see Goldstein and Dillon (1983)]. The format of the DISCRIMINANT procedure statements is

```
1                       16
DISCRIMINANT            GROUPS = {variable} (min, max)/
                                 {name    }
                        VARIABLES = {list of variables}/
                        ANALYSIS = {variables to be included in} /
                                   {discriminant analysis      }
                        METHOD = DIRECT/
                        PRIORS = {EQUAL or      }/
                                 {SIZE or       }
                                 {list of values}
```

The GROUPS = statement indicates the response variable that is to be used to classify variables into groups. The "min" and "max" values indicate the range of the categories for the classifying variable. The VARIABLES = statement lists the full set of variables that are to be analyzed. The ANALYSIS = statement indicates the variables to be used in a particular discriminant analysis. The METHOD = DIRECT statement is used when a stepwise analysis is not being performed. Other methods (such as WILKS, MAHAL, and RAO) are available for stepwise discriminant analysis [see Nie et al. (1975), Hull and Nie (1981), and Goldstein and Dillon (1983)]. The PRIORS = statement assigns prior probabilities for classification purposes: EQUAL sets equal probabilities for each group, SIZE sets probabilities based on the sample sizes in each group, while a "list of values" can also be provided to assign probabilities to each group.

Among the options available, OPTIONS 5 provides a table of correct and incorrect classifications, OPTIONS 6 prints discriminant scores and classification results, OPTIONS 7 prints a plot of cases, OPTIONS 10 prints a territorial map, OPTIONS 11 provides unstandardized discriminant function coefficients, while OPTIONS 12 prints classification functions.

The statistics available are STATISTICS 1 (mean of each variable), STATISTICS 2 (standard deviation of each variable), STATISTICS 3 (a pooled within groups covariance matrix), STATISTICS 4 (a pooled within groups correlation matrix), STATISTICS 6 (a one-way ANOVA on each discriminating variable), and STATISTICS 7 (a test of the equality of group covariance matrices).

Figure 18.1 represents the complete SPSS program that has been written to perform a discriminant analysis on the data of Problem 18.1. Figure 18.2 represents partial output for these data.

```
{SYSTEM CARDS}
RUN NAME              DISCRIMINANT ANALYSIS OF MBAS
DATA LIST             FIXED(1)/1 SUCCESS 1,GPI 3-6,
                      SCORE 8-10
VAR LABELS            SUCCESS,SUCCESS IN MBA PROGRAM/
                      GPI,UNDERGRAD GRADE POINT INDEX/
                      SCORE,GMAT SCORE/
VALUE LABELS          SUCCESS(1)YES(2)NO/
READ INPUT DATA
1 2.75 688
:   :    :
2 3.57 536
END INPUT DATA
DISCRIMINANT          GROUPS=SUCCESS(1,2)/
                      VARIABLES=GPI,SCORE/
                      ANALYSIS=GPI SCORE/
                      METHOD=DIRECT/
                      PRIORS=SIZE/
OPTIONS               5,6,7,10,11,12
STATISTICS            1,2,3,4,6,7
FINISH
{SYSTEM CARDS}
```

Figure 18.1   SPSS program for the data of Problem 18.1.

18.6.2 Using SAS PROC DISCRIM for Discriminant Analysis

The SAS DISCRIM procedure can be used to perform a discriminant analysis. The format of the PROC DISCRIM step is

PROCьDISCRIMь{options};

```
                        NUMBER OF CASES
          SUCCESS    UNWEIGHTED     WEIGHTED  LABEL

             1          20           20.0     YES
             2          10           10.0     NO

          TOTAL         30           30.0

       GROUP MEANS

          SUCCESS      GPI           SCORE

             1         3.43950       659.00000
             2         3.33600       575.20000

          TOTAL        3.40500       631.06667

       GROUP STANDARD DEVIATIONS

          SUCCESS      GPI           SCORE

             1         0.35806       43.25567
             2         0.24559       29.73886

          TOTAL        0.32431       55.80936

POOLED WITHIN-GROUPS COVARIANCE MATRIX WITH       28 DEGREES OF FREEDOM

            GPI           SCORE

GPI        0.1063834
SCORE      1.611354      1553.914


POOLED WITHIN-GROUPS CORRELATION MATRIX
            GPI           SCORE

GPI        1.00000
SCORE      0.12533       1.00000

CORRELATIONS WHICH CANNOT BE COMPUTED ARE PRINTED AS 99.0.

WILKS' LAMBDA (U-STATISTIC) AND UNIVARIATE F-RATIO
WITH   1 AND        28 DEGREES OF FREEDOM

   VARIABLE   WILKS' LAMBDA         F          SIGNIFICANCE
   --------   -------------   -------------    ------------

   GPI         0.97659          0.6713           0.4195
   SCORE       0.48170          30.13            0.0000

                  CLASSIFICATION FUNCTION COEFFICIENTS
                  (FISHER'S LINEAR DISCRIMINANT FUNCTIONS)

                  SUCCESS =        1                2
                                 YES              NO

                  GPI        26.32103          26.16248
                  SCORE      0.3967963         0.3430324
                  (CONSTANT) -176.4154         -143.3937

  TEST OF EQUALITY OF GROUP COVARIANCE MATRICES USING BOX'S M

     THE RANKS AND NATURAL LOGARITHMS OF DETERMINANTS PRINTED ARE THOSE
     OF THE GROUP COVARIANCE MATRICES.

        GROUP LABEL                   RANK   LOG DETERMINANT

           1 YES                        2       5.390093
           2 NO                         2       3.476753
        POOLED WITHIN-GROUPS
        COVARIANCE MATRIX              2       5.091995

     BOX'S M      APPROXIMATE F   DEGREES OF FREEDOM   SIGNIFICANCE
     8.8733         2.6831           3,      7384.8      0.0453

     CLASSIFICATION RESULTS -

                             NO. OF    PREDICTED GROUP MEMBERSHIP
            ACTUAL GROUP     CASES        1            2
     --------------------    ------    --------    --------

     GROUP       1             20          19           1
     YES                                 95.0%        5.0%

     GROUP       2             10           2           8
     NO                                  20.0%       80.0%

     PERCENT OF "GROUPED" CASES CORRECTLY CLASSIFIED:  90.00%
```

Figure 18.2 Partial output obtained from the SPSS subprogram DISCRIMINANT for the data of Problem 18.1.

Among the options available are

$$\begin{Bmatrix} \text{SIMPLE} & \text{POOL} = \{\text{YES or NO or TEST}\} & \text{SLPOOL} = .\underline{\hspace{1cm}} \\ \text{WCOV} & \text{WCORR} \quad \text{PCOV} \quad \text{LIST} \end{Bmatrix};$$

$$\text{CLASS}\beta \begin{Bmatrix} \text{name of} \\ \text{classifying variable} \end{Bmatrix};$$

$$\text{VAR}\beta \begin{Bmatrix} \text{variable} \\ \text{list} \end{Bmatrix};$$

$$\text{ID}\beta \begin{Bmatrix} \text{identifying} \\ \text{variable} \end{Bmatrix};$$

$$\text{PRIORS}\beta \begin{Bmatrix} \text{PROP or} \\ \begin{Bmatrix} \text{category} \\ \text{name 1} \end{Bmatrix} = \underline{\hspace{1cm}} \begin{Bmatrix} \text{category} \\ \text{name 2} \end{Bmatrix} = \underline{\hspace{1cm}} \cdots \end{Bmatrix};$$

The use of various options on the PROC DISCRIM step can produce different types of output. The SIMPLE (or S) option provides simple descriptive statistics for each variable. The POOL = option determines whether or not the within group covariance matrices are to be pooled. Use of POOL = TEST provides a statistical test for equality of covariance matrices and pools them unless the test statistic is significant (at the level specified in the SLPOOL = option). WCOV provides within group covariance matrices, WCORR provides within group correlation matrices, while PCOV produces the pooled covariance matrix. Finally, the LIST option prints classification results for each observation.

The CLASS statement (which must accompany PROC DISCRIM) indicates the variable to be used to classify groups. The VAR statement provides the variables to be included in the discriminant analysis. The ID statement is needed only to provide the value of the identifying (classification) variable for the LIST option. The PRIORS statement assigns prior probabilities for classification purposes. Equal classification probabilities will be assumed if this step is omitted. PRIORS PROP sets probabilities proportional to sample size. If other than equal or proportional probabilities are desired, the prior probability can be set for each category level.

Figure 18.3 represents the complete SAS program that has been written to perform a discriminant analysis for the data of Problem 18.1. Figure 18.4 represents partial output for these data. We may observe from Figure 18.4 that since the covariance matrices were tested and found unequal, the discriminant function obtained is based on the separate group covariance matrices.

18.6.3 Using BMDP Program 7M for Discriminant Analysis

The BMDP program 7M can be used for either discriminant analysis or stepwise discriminant analysis [see Goldstein and Dillon (1983)]. The paragraphs unique to a discriminant analysis are /VARIABLE, /GROUP, and /PRINT.

```
{SYSTEM CARDS}
DATA MBA;
    INPUT SUCCESS 1 GPI 3-6 SCORE 8-10;
LABEL SUCCESS=SUCCESS IN MBA PROGRAM
      GPI=UNDERGRAD GRADE POINT INDEX
      SCORE=GMAT SCORE;
CARDS;
1 2.75 688
:   :   :
:   :   :
2 3.57 536
PROC FORMAT;
    VALUE SUCCESSF 1=YES 2=NO;
PROC DISCRIM SIMPLE POOL=TEST SLPOOL=.05
WCOV WCORR PCOV LIST;
CLASS SUCCESS;
    VAR GPI SCORE;
ID SUCCESS; PRIORS PROP;
FORMAT SUCCESS SUCCESSF.;
{SYSTEM CARDS}
```

Figure 18.3 SAS program for the data of Problem 18.1.

The format of these paragraphs is

/VARIABLE	NAMESɃAREɃvariable$_1$, variable$_2$, ...
	GROUPINGɃIS {name of classifying variable}.
/GROUP	PRIORɃ=Ƀ {prior prob. of category 1}, {prior prob. of category 2}, ...
/PRINT	WITHIN.

A GROUPING sentence which identifies the classification variable *must* be included as part of the /VARIABLE paragraph. CODES and NAMES sentences should be specified as part of the /GROUP paragraph in order to label the classification groups. If the groups have unequal prior probabilities, then a PRIOR = sentence can be used to assign probabilities to each group. The /PRINT paragraph can be used to add or delete information from the output. The WITHIN sentence provides the within group covariance matrix, while the NO STEP sentence causes the deletion of the output of intermediate steps in the discriminant analysis.

Figure 18.5 represents the program that has been written for the data of Problem 18.1 using BMDP-7M. Figure 18.6 represents partial output obtained for these data (see p. 528).

```
                    DISCRIMINANT ANALYSIS     SIMPLE STATISTICS

                             SUCCESS = YES
                                                                                    STANDARD
VARIABLE           N                SUM                MEAN              VARIANCE    DEVIATION

GPI               20          68.79000000          3.43950000          0.12820500   0.35805726
SCORE             20       13180.00000000        659.00000000       1871.05263158  43.25566589
-----------------------------------------------------------------------------------------------

                             SUCCESS = NO

GPI               10          33.36000000          3.33600000          0.06031556   0.24559225
SCORE             10        5752.00000000        575.20000000        884.40000000  29.73886346

                    DISCRIMINANT ANALYSIS     WITHIN COVARIANCE MATRICES

                        SUCCESS = YES          DF =    19

                        VARIABLE               GPI                SCORE

                        GPI               0.12820500            4.54473684
                        SCORE             4.54473684         1871.05263158

                        ---------------------------------------------------

                        SUCCESS = NO           DF =     9

                        VARIABLE               GPI                SCORE

                        GPI               0.06031556           -4.58133333
                        SCORE            -4.58133333          884.40000000

                    DISCRIMINANT ANALYSIS     POOLED COVARIANCE MATRIX    DF =    28

                        VARIABLE               GPI                SCORE

                        GPI               0.10638339            1.61135714
                        SCORE             1.61135714         1553.91428571

            DISCRIMINANT ANALYSIS     WITHIN COVARIANCE MATRIX INFORMATION

            SUCCESS            COVARIANCE        NATURAL LOG OF DETERMINANT
                               MATRIX RANK       OF THE COVARIANCE MATRIX

            YES                    2                 5.39009253

            NO                     2                 3.47675195

            POOLED                 2                 5.09199530


TEST CHI-SQUARE VALUE =              9.26580455    WITH       3 DF    PROB > CHI-SQ = 0.0260

SINCE THE CHI-SQUARE VALUE IS SIGNIFICANT AT THE 0.0500 LEVEL, THE WITHIN COVARIANCE MATRICES WILL BE USED
IN THE DISCRIMINANT FUNCTION.

REFERENCE: KENDALL,M.G. AND A.STUART  THE ADVANCED THEORY OF STATISTICS VOL.3 P266 & 282.

        DISCRIMINANT ANALYSIS     PAIRWISE SQUARED GENERALIZED DISTANCES BETWEEN GROUPS
```

$$D^2(I|J) = (\bar{X}_I - \bar{X}_J)' \, COV_J^{-1} \, (\bar{X}_I - \bar{X}_J) + LN |COV_J| - 2 LN PRIOR_J$$

```
                        GENERALIZED SQUARED DISTANCE TO SUCCESS

            FROM SUCCESS                    YES                 NO

                        YES             6.20102275          21.51435233
                        NO             10.03965647           5.67397652

            NUMBER OF OBSERVATIONS AND PERCENTS CLASSIFIED INTO SUCCESS:
FROM
SUCCESS     YES      NO     TOTAL

NO           2        8       10
          20.00    80.00   100.00

YES         19        1       20
          95.00     5.00   100.00

TOTAL       21        9       30
PERCENT   70.00    30.00   100.00

PRIORS    0.6667  0.3333
```

Figure 18.4 Partial output obtained from SAS DISCRIM for the data of Problem 18.1.

```
{SYSTEM CARDS}
/INPUT    VARIABLES ARE 3.
          FORMAT IS '(F1.0,F5.2,F4.0)'.
/VARIABLE MAMES ARE SUCCESS, GPI,SCORE.
          GROUPING IS SUCCESS.
/GROUP    CODES(1) ARE 1,2.
          NAMES(1) ARE YES,NO.
          PRIOR=.667,.333.
/PRINT    WITHIN.
/END
1 2.75 688
: : :
2 3.57 536
{SYSTEM CARDS}
```

Figure 18.5 Using BMDP program 7M for the data of Problem 18.1.

18.7 MULTIVARIATE METHODS: A BRIEF SUMMARIZATION

The development of high-speed computer technology coupled with the wide availability of statistical software has in large measure resulted in a rebirth of interest in multivariate analysis. Researchers, especially in the social and behavioral sciences, are prolific users of multivariate techniques, and there is every reason to believe that both the frequency and scope of this use will expand.

While our treatment of multivariate statistical methods in this text is far from complete, we did highlight some of the important areas of the subject. It is important, however, that the serious reader seek out further sources in each subject area discussed. A fairly complete discussion of problems in discriminant analysis can be found in Lachenbruch (1975) and Goldstein and Dillon (1978). A number of excellent books in the area of multidimensional frequency data are available; in particular, Bishop et al. (1975), Fienberg (1980), and Haberman (1980). Kruskal and Wish (1978) or Schiffman et al. (1981) is a good place to start for a more in-depth discussion of multidimensional scaling; their bibliography is extensive and up-to-date. The most thorough and extensively illustrated source for clustering algorithms can be found in Hartigan (1975). Principal components and factor analysis, the most classical of the techniques discussed, have not seen recent treatment in the textbook literature; however, Harman (1967) is the classic treatment.

There are a number of important areas in multivariate analysis which we did not discuss. Most notable are graphical procedures used for uncovering internal dependencies and the detection of outliers; Gnanadesikan (1977) is an excellent source. A discussion of directional data in a multivariate context can be found in Mardia et al. (1979). An up-to-date general multivariate text written

```
  MEANS

      GROUP =   YES           NC            ALL GPS.
VARIABLE
  2 GPI          3.43950       3.33600       3.40500
  3 SCORE      659.00000     575.19995     631.06665

COUNTS             2C.           1C.           30.

   STANCARD DEVIATICNS

      GROUP =   YES           NC            ALL GPS.
VARIABLE
  2 GPI          0.35806       0.24559       C.32616
  3 SCCRE       43.25565      29.73885      39.41965

   COEFFICIENTS CF VARIATICN

      GROUP =   YES           NC            ALL GPS.
VARIABLE
  2 GPI          0.10410       0.C7362       0.09579
  3 SCCRE        0.06564       0.05170       C.C6247

WITHIN CCVARIANCE MATRIX

                    GPI           SCCRE
                     2             3

GPI        2      0.10638
SCORE      3      1.61135    1553.51429

*********************************************************************************************
STEP NUMBER   0

  VARIABLE        F TO FORCE  TCLERANCE  *   VARIABLE        F TC FORCE   TCLERANCE
                  REMCVE LEVEL           *                   ENTER LEVEL
                  DF=  1   29            *               DF=  1   28
                                         *   2 GPI         0.671    1    1.000000
                                         *   3 SCCRE      30.128    1    1.030000
*********************************************************************************************

STEP NUMEER   1
VARIABLE ENTEREC   3 SCORE

  VARIABLE        F TC FORCE  TCLERANCE  *   VARIAELE        F TC FORCE   TCLERANCE
                  REMCVE LEVEL           *                   ENTER LEVEL
                  DF=  1   28            *               DF=  1   27
  3 SCORE        30.128    1   1.C00000  *   2 CPI         0.0C8    1    C.984293

U-STATISTIC CR WILKS' LAMBDA    C.4816959    CEGREES CF FREECCM    1    1        28
APPRCXIMATE F-STATISTIC         30.128       CEGREES CF FREEDCM    1.00   28.00

F - MATRIX        DEGREES UF FREEDCM =    1    28

             YES
NC          30.13
CLASSIFICATICN FUNCTICNS

      GRCUP =   YES           NC
VARIABLE
  3 SCCRE         0.42405       0.37016

CGNSTAN1        -140.14262    -107.55817

CLASSIFICATICN MATRIX

GROUP     PERCENT   NUMBER OF CASES CLASSIFIED INTC GRCUP -
          CCRRECT
                      YES     NC
 YES       95.0       19       1
 NC        80.0        2       8

 TCTAL     9C.0       21       9

JACKKNIFEC CLASSIFICATIUN

GROUP     PERCENT   NUMBER OF CASES CLASSIFIED INTC GRCUP -
          CCRRECT
                      YES     NU
 YES       95.0       19       1
 NC        80.0        2       8

 TCTAL     90.0       21       9
```

Figure 18.6   Partial output obtained from BMDP-7M for the data of Problem 18.1.

for the social scientist and amply illustrated with real data is Goldstein and Dillon (1983).

PROBLEMS

*18.1. The Director of Graduate Studies at a large college of business would like to predict the success of students in an MBA program. Two explanatory variables, the undergraduate grade point index (GPI) and the Graduate Management Aptitude Test (GMAT) scores, were available for a random sample of 20 "successful" students and 10 "unsuccessful" students. The results are displayed in Table P18.1.

TABLE P18.1   Comparison of MBA Students

"Successful" MBA Students		"Unsuccessful" MBA Students	
GPI	GMAT	GPI	GMAT
2.75	688	2.93	617
2.81	647	3.05	557
3.03	652	3.11	599
3.10	608	3.24	616
3.06	680	3.36	594
3.17	639	3.41	567
3.24	632	3.45	542
3.41	639	3.60	551
3.37	619	3.64	573
3.46	665	3.57	536
3.57	694		
3.62	641		
3.66	594		
3.69	678		
3.70	624		
3.78	654		
3.84	718		
3.77	692		
3.79	632		
3.97	784		

(a) Use a computer package (such as SPSS, SAS, or BMDP) to perform a discriminant analysis of the two groups.

(b) Using the obtained discriminant function, how should the Director of Graduate Studies classify an applicant who achieved a grade point index of 3.33 and who scored 560 on the GMAT?

18.2. The marketing department of a large nationally franchised lawn service company would like to study the characteristics that differentiate homeowners who either have or do not have a lawn service. A random sample of 30 homeowners was selected—15 with a lawn service and 15 without a lawn service. Information was collected concerning family income (thousands of dollars), lawn size (thousands of square feet), attitude toward outdoor recreation, number of teenagers, and age. The results are displayed in Table P18.2.

TABLE P18.2   Lawn Service Study

Annual Family Income ($000)	Lawn Size (000 ft²)	Attitude Score	Number of Teenagers	Age
	Homeowners with Lawn Service			
38.2	6.9	5	1	43
49.4	8.3	5	1	39
29.7	10.8	4	2	40
46.7	10.1	3	1	55
47.6	10.3	5	1	49
42.1	6.8	5	2	53
56.1	7.2	3	0	51
29.8	3.3	6	3	48
23.7	4.7	3	2	41
35.3	5.7	6	0	45
26.3	10.9	4	2	43
56.4	8.3	4	2	62
45.3	7.8	3	3	52
34.4	6.3	3	0	34
37.8	7.2	6	1	45
	Homeowners without Lawn Service			
16.9	3.0	5	2	38
17.4	4.3	5	1	45
43.0	1.9	8	2	47
23.1	4.5	6	0	37
24.9	1.7	7	1	39
18.9	3.2	5	2	37
26.8	4.6	6	1	45
22.7	7.9	5	1	46
23.6	5.6	6	3	37
29.3	6.0	6	2	39
18.6	4.5	5	2	47
36.4	9.1	7	3	36
21.7	4.2	7	1	38
27.1	9.4	6	2	44
32.6	2.3	5	0	32

   (a) Use a computer package (such as SPSS, SAS, or BMDP) to perform a discriminant analysis of the two groups of homeowners.
   (b) Using the obtained discriminant function, how should the marketing department classify a 40-year-old homeowner with a family income of $28,000, a lawn size of 6500 square feet, an attitude score of 5, and two teenagers?

*18.3. The data in Table P18.3 are a modified version of the data originally collected by Reis and Smith (1963) and given in Goldstein and Dillon (1978). They result from an experiment in which 1008 people were given two brands of detergent, L and M, and subsequently asked questions regarding four variables: X_1, water softness—soft (1) or hard (2); X_2, previous use of brand M—yes (1) or no (2); X_3, water temperature—high (1) or low (2); and the grouping variable brand preference—L or M. Use the full multinomial rule and the logit rule as discriminants. Compare the error rates for the two procedures.

TABLE P18.3   Observed Frequency Distributions for Data on Detergent Preference

	Brand Preference	
$X_1\ X_2\ X_3$	L (observed)	M (observed)
1 1 1	19	29
1 1 2	57	49
1 2 1	29	27
1 2 2	63	53
2 1 1	24	43
2 1 2	37	52
2 2 1	46	30
2 2 2	68	42
	339	325

SOURCE: Extracted from T. N. Reis and H. T. Smith, "The Use of Chi-Square for Preference Testing in Multidimensional Problems," *Chemical Engineering Progress*, Vol. 59, pp. 39–43, 1963 [in Goldstein, M., and W. R. Dillon, *Discrete Discriminant Analysis* (New York: Wiley, 1978)].

18.4. We wish to develop a model that can be used in classifying houses that have or do not have fireplaces (see Appendix A for the real estate data base).

    (a) Select a random sample of 30 houses, and use the variables age (AGE), lot size (LSQF), heating area (DSQF), number of bathrooms (BATH), and assessed value (AV) to develop a discriminant function for fireplace ownership. Determine the number of houses of each type in the sample that have been correctly classified.

    (b) Select a second sample of 30 *different* houses, and apply the discriminant function developed in part (a) to classify these houses.

    (c) Explain the difference in the results in parts (a) and (b) of this problem. ■

18.5. Referring to the data of Problem 17.5 (see Table P17.5), consider the homeowners as two subgroups: those who have or do not have a fireplace. Use the full multinomial rule and the logit rule as discriminants. Compare the error rates of the two procedures. ■

18.6. Reviewing concepts:

    (a) What is meant by a discriminant function?

    (b) In what ways are discriminant analysis essentially different from multiple regression analysis?

    (c) In what ways does the logit rule differ from the full multinomial rule when classifying responses?

Appendix A

REAL ESTATE DATA BASE

The data included in this appendix pertain to all single-family houses that were sold ($N = 335$) in a small city in the southwestern United States during a 20-month period in the late 1970s. Information was collected relating to each of the following variables:

Columns 1–2: TIME–time in months from beginning of period under study

Columns 4–8: AGE–age of the house in years

Columns 10–14: LSQF–lot area in sq ft according to assessor's records

Columns 16–19: DSQF–heating area of the dwelling in sq ft

Columns 21–23: BATH–number of bathrooms

Column 25: HT–presence (C) or absence (S) of central heating

Column 27: FIREPL–presence (Y) or absence (N) of a fireplace

Column 29: POOL–presence (Y) or absence (N) of a swimming pool

Column 31: AIR–presence (C) or absence (N) of central air conditioning

Column 33: BLT–grade based on presence or absence of a set of built-in features such as a dishwasher, range, washer-dryer, etc., categorized as poor (P), average (A), and good (G)

Column 35: CAR–type of garage or parking facility categorized as none (N), carport (P), single (S), and double (D)

Column 37: EXTER–presence (B) or absence (F) of a brick exterior

Columns 39–43: AV–assessed value during 1976–1977 reappraisal

Columns 45–49: XPRICE–sales price in dollars

Columns 51–55: ASKP–asking price in dollars

Column 57: LOCAT–location according to geographic boundary

REAL ESTATE DATA BASE

OBS	TIME	AGE	LSQF	DSQF	BATH	HT	FIREPL	POOL	AIR	BLT	CAR	EXTER	AV	XPRICE	ASKP	LOCAT
1	7	21.00	6750	938	1.0	S	Z	Z	Z	P	P	F	11980	15000	15500	LOC1
2	10	25.00	8190	1000	1.0	S	Z	Z	Z	A	S	F	12930	25850	26500	LOC1
3	11	27.00	8710	1008	1.0	S	Z	Z	Z	A	S	F	15550	26000	27000	LOC1
4	18	28.00	7800	952	1.0	U	Y	Z	U	A	N	B	13050	27200	29900	LOC1
5	11	25.00	13837	1994	2.0	U	Z	Z	Z	P	P	F	30250	40220	40220	LOC1
6	5	24.83	6325	713	2.0	S	Z	Z	Z	P	P	F	8280	12000	12000	LOC2
7	5	25.03	9515	1152	2.0	U	Z	Z	Z	P	S	F	15420	21000	21000	LOC2
8	10	25.00	9500	1000	1.5	C	Y	Z	U	G	S	B	10140	19000	22900	LOC2
9	13	14.75	14941	1736	1.0	D	Z	Z	C	A	Z	F	26060	36500	38950	LOC2
10	3	21.00	17500	1325	1.0	S	Y	Z	Z	P	D	B	20410	29900	30500	LOC2
11	13	27.00	9750	1490	2.0	U	Z	Z	U	G	Z	B	27630	54000	31000	LOC2
12	3	20.00	15000	1997	2.0	C	Y	Z	C	A	N	B	16620	35000	63500	LOC2
13	13	25.00	17290	1410	2.0	C	Z	Z	C	P	Z	B	19770	37900	35500	LOC2
14	11	25.08	12580	1706	1.0	U	Z	Z	C	G	D	F	23070	43000	37900	LOC2
15	9	23.00	54000	1594	1.0	S	Z	Z	C	A	O	F	31850	29500	45900	LOC2
16	8	31.00	10920	1332	1.0	C	Z	Z	U	P	S	F	15030	25900	30000	LOC2
17	11	28.50	10620	1180	2.0	U	Z	Z	C	A	O	B	11550	26000	25900	LOC2
18	17	31.00	10560	950	2.0	C	Z	Z	Z	P	S	F	12530	24000	27000	LOC2
19	12	7.25	11250	809	2.0	C	Z	Z	U	A	N	F	12740	34000	25900	LOC2
20	13	20.50	16479	1571	1.0	C	Z	Z	C	A	S	B	19390	33000	35500	LOC2
21	11	24.00	15000	1323	2.0	C	Z	Z	C	G	S	F	19230	32200	36500	LOC2
22	19	28.00	11250	1358	1.0	U	Z	Z	Z	G	S	F	15640	33500	34000	LOC2
23	12	21.58	14976	1376	1.5	C	Z	Z	U	P	P	F	15590	41750	34000	LOC2
24	9	27.00	15834	1761	1.0	C	Z	Z	Z	P	S	F	23860	33500	42500	LOC2
25	7	27.00	14520	1600	1.0	S	Z	Z	Z	P	O	F	21940	19800	34500	LOC2
26	1	26.00	12925	907	1.0	C	Z	Z	C	P	Z	F	15460	19000	22000	LOC2
27	6	30.00	9000	1000	1.0	U	Z	Z	Z	P	O	F	16280	19343	20000	LOC2
28	2	23.24	9100	671	1.0	C	Z	Z	U	A	O	F	10800	18900	19000	LOC2
29	17	24.24	7800	735	1.0	C	Z	Z	Z	A	S	F	10560	24000	19500	LOC2
30	17	26.00	6875	1092	1.0	C	Z	Z	Z	A	O	F	10560	26850	24500	LOC2
31	8	23.58	7332	858	1.0	C	Z	Z	C	P	Z	F	11810	25500	27900	LOC2
32	5	23.58	12000	1100	2.0	S	Z	Z	U	P	N	F	13770	24500	26900	LOC2
33	10	32.00	10800	960	1.0	C	Z	Z	Z	P	S	F	15620	29700	27900	LOC2
34	1	17.67	8400	1984	2.0	S	Z	Z	Z	P	S	F	20530	19500	32500	LOC2
35	12	27.00	8125	907	1.0	C	Z	Z	C	P	S	F	13770	18500	26750	LOC2
36	5	22.00	7700	998	1.0	S	Z	Z	Z	P	Z	F	13160	24500	24900	LOC2
37	4	21.00	7995	1054	1.5	C	Z	Z	U	A	S	F	14620	22000	19900	LOC2
38	9	7.50	6300	1235	1.0	S	Z	Z	Z	A	N	F	15540	23750	25000	LOC2
39	9	27.00	6500	994	1.5	C	Z	Z	C	A	Z	F	14590	13000	23000	LOC2
40	3	27.00	9620	558	1.5	C	Z	Z	C	A	P	B	7950	19000	24900	LOC2
41	13	0.00	5000	816	1.0	C	Z	Z	C	G	S	F	10180	28850	14500	LOC2
42	13	0.00	5000	1120	1.5	C	Z	Z	C	A	Z	F	16910	27850	19500	LOC2
43	13	0.00	5000	1008	1.5	C	Z	Z	C	A	Z	B	13490	29950	28850	LOC2
44	6	3.00	5330	1232	1.5	C	Z	Z	C	A	Z	F	15700	25750	28850	LOC2
45	6	3.00	5030	1008	1.5	S	Z	Z	C	A	Z	F	15680	26750	29950	LOC2
46	6	0.00	5000	1120	1.5	C	Z	Z	C	A	Z	F	16720	27750	25750	LOC2
47	7	3.00	5330	1008	1.5	C	Z	Z	C	A	Z	F	15710	31500	26750	LOC2
48	6	25.92	14344	1426	2.0	C	Z	Z	C	A	N	F	20840	32900	25750	LOC2
49	7	21.42	19096	2012	2.0	C	Z	Z	Z	G	S	F	34910	37000	32900	LOC2
50	10	27.00	9500	1312	1.0	S	Y	Z	Z	A	O	B	22250	32000	52800	LOC3
51	11	24.50	9630	1240	1.5	C	Y	Z	Z	A	P	F	17120	33000	37000	LOC2
52	11	21.00	15000	1415	2.0	C	Y	Z	U	A	S	F	24310	34500	33500	LOC2
53	13	11.67	14471	1391	1.5	S	Z	Z	U	A	D	B	23490	33500	33500	LOC3
54	9	8.33	11310	1493	2.0	C	Z	Z	C	A	D	B	23940	34500	35175	LOC3

REAL ESTATE DATA BASE

OBS	TIME	AGE	LSQF	DSQF	BATH	HT	FIREPL	POOL	AIR	BLT	CAR	EXTER	AV	XPRICE	ASKP	LOCAT
55	12	1.33	11310	1406	2.0	C	N	Z	C	A	O	B	22360	33300	34900	LOC3
56	17	10.33	9750	1530	2.0	C	N	Z	C	G	O	B	23020	43500	44900	LOC3
57	12	8.00	10140	1543	2.0	C	N	Z	C	A	O	B	23610	39500	42200	LOC3
58	14	9.00	10140	1399	2.0	C	N	Z	C	G	Z	B	24550	39000	41950	LOC3
59	10	8.17	10140	1270	2.0	C	N	Z	C	A	O	B	20730	34200	35200	LOC3
60	9	8.33	13900	1427	2.0	C	N	Z	C	A	O	B	23340	40500	41900	LOC3
61	5	12.25	10140	1367	2.0	C	N	Z	C	G	O	B	22190	38900	39900	LOC3
62	7	13.83	10140	1278	1.5	C	N	Z	C	A	O	B	21360	33500	35000	LOC3
63	10	9.17	13900	1549	2.0	C	N	Z	C	A	O	B	25640	39500	40500	LOC3
64	11	11.08	10140	1588	2.5	C	Y	Z	C	G	O	B	24030	38650	39300	LOC3
65	13	13.50	20973	2100	2.0	C	N	Z	C	A	O	B	26230	57500	52000	LOC3
66	14	2.00	10464	1521	2.0	C	N	Z	C	G	O	B	22780	36500	38800	LOC3
67	13	12.33	10430	1497	2.0	C	N	Z	C	G	O	B	21280	42500	43500	LOC3
68	12	13.75	14135	2261	2.0	C	Y	Z	C	G	P	B	33340	57500	60000	LOC3
69	9	10.63	11050	1886	2.0	C	N	Z	C	G	O	B	27360	45400	47900	LOC3
70	12	7.17	10400	1585	2.0	C	N	Z	C	G	O	B	26830	41000	43700	LOC3
71	7	7.00	11856	1573	2.0	C	Y	Z	C	G	O	B	24460	32500	34500	LOC3
72	5	3.00	10400	1696	2.0	C	N	Z	C	G	O	B	30150	42000	45900	LOC3
73	12	8.42	10400	2345	2.0	C	Y	Z	C	G	O	B	34260	42500	51000	LOC3
74	11	5.50	15750	2487	3.0	C	Y	Z	C	G	O	B	40080	63900	63000	LOC3
75	11	10.08	11050	1736	2.0	C	Y	Z	C	G	O	B	27100	48500	49000	LOC3
76	5	7.25	10660	1825	2.0	C	Y	Z	C	G	O	B	28310	41500	41900	LOC3
77	9	1.67	18430	1550	2.0	C	Y	Z	C	G	O	B	27960	42000	42700	LOC3
78	7	0.00	7475	1708	2.0	C	Y	Z	C	G	O	B	27070	41500	42000	LOC3
79	5	1.67	8635	1776	2.0	C	Y	Z	C	G	O	B	28900	40275	42500	LOC3
80	7	0.00	7475	1387	2.0	C	Y	Z	C	G	O	B	22930	35500	39500	LOC3
81	1	0.00	11475	1529	2.0	C	N	Z	C	G	O	B	28870	35500	37900	LOC3
82	12	1.00	23618	1482	2.0	C	N	Y	C	G	O	B	25260	42900	42900	LOC3
83	7	2.92	8395	1436	2.0	C	N	Z	C	G	S	B	23280	39100	39100	LOC3
84	5	3.03	9240	1228	2.5	C	N	Z	C	G	O	B	20920	35000	37900	LOC3
85	19	3.83	19000	2621	2.0	C	Y	Z	C	G	D	B	44210	81000	86000	LOC3
86	3	1.92	19000	1960	2.0	C	N	Z	C	G	P	B	32540	57000	59500	LOC3
87	14	2.92	19000	1999	2.0	C	Y	Z	C	G	P	B	32540	63900	64900	LOC3
88	12	3.42	22800	2224	2.0	C	N	Z	C	G	O	B	34360	56700	58900	LOC3
89	3	1.75	15400	1364	2.0	C	Y	Z	C	A	O	B	37460	57900	59900	LOC4
90	12	0.00	9380	1280	2.0	C	N	Z	C	G	D	B	23720	39950	39950	LOC4
91	17	0.00	8820	1151	1.0	C	N	Z	C	A	Z	B	22540	35000	38900	LOC4
92	12	0.00	8820	1264	2.0	S	N	Z	C	A	O	F	22420	38000	38000	LOC4
93	11	0.00	13969	1513	2.0	S	Y	Z	C	P	O	F	27310	37650	37650	LOC4
94	1	0.00	11900	1326	1.0	C	Y	Z	N	P	S	B	16970	43750	43750	LOC4
95	1	0.00	9988	1542	1.0	C	N	Z	C	P	O	F	21300	27900	27900	LOC4
96	13	21.50	9988	1054	1.0	S	Y	Z	N	A	P	F	13190	31000	32000	LOC4
97	7	27.00	9100	978	1.0	S	N	Z	C	A	O	F	23210	25000	25500	LOC4
98	14	29.00	13320	1588	2.0	C	N	Z	C	A	S	B	16690	41800	43900	LOC4
99	18	24.00	11700	978	1.0	S	Y	Z	N	P	O	F	13980	27000	28000	LOC4
100	8	30.00	9000	1382	1.0	C	N	Z	C	P	N	L	13210	22000	22900	LOC4
101	12	27.00	12090	1163	1.5	C	Y	Z	C	P	Z	F	24650	25000	30000	LOC4
102	12	21.42	13500	1736	2.0	C	Y	Z	C	G	O	L	20530	39200	39900	LOC4
103	8	18.75	11440	1440	1.5	C	N	Z	C	G	O	B	28510	37000	37000	LOC4
104	18	16.42	12907	1950	2.0	C	Y	Z	C	G	P	B	24500	42500	43900	LOC4
105	17	12.58	10810	1539	2.0	C	Y	Z	C	G	P	B	20420	42500	42500	LOC4
106	11	32.00	15000	1200	1.0	C	N	Z	N	P	D	F	22020	29000	29900	LOC4
107	10	30.00	14640	2350	2.0	C	Y	Z	C	G	O	L	22020	50000	52000	LOC4
108	15	30.00	13300	1400	2.0	S	Y	Z	N	P	S	F	22600	25000	25000	LOC4

REAL ESTATE DATA BASE

OBS	TIME	AGE	LSQF	DSQF	BATH	HT	FIREPL	POOL	AIR	BLT	CAR	EXTER	AV	XPRICE	ASKP	LOCAT
109	6	16.00	14967	1551	2.0	C	Y	Z	C	A	D	B	25820	39000	39900	LOC4
110	15	11.33	9500	1960	2.0	C	N	Z	C	A	D	B	28580	45000	46900	LOC4
111	4	7.17	9520	1815	2.0	C	N	Z	C	P	D	B	26200	42000	44500	LOC4
112	13	8.83	9525	1858	2.0	C	Y	Z	C	G	D	B	29060	53500	54500	LOC4
113	5	5.42	8750	1397	1.0	S	N	Z	C	G	O	F	23830	36300	34900	LOC4
114	5	29.00	7800	792	1.0	S	Z	Z	Z	P	N	F	8570	11000	11000	LOC4
115	9	27.00	7860	1122	1.0	S	Z	Z	Z	P	Z	F	11050	18850	19950	LOC4
116	9	32.00	6000	879	1.0	S	Z	Z	Z	P	Z	F	8290	15100	18500	LOC4
117	15	29.00	8910	845	1.0	S	Z	Z	Z	P	N	F	10660	18000	24900	LOC4
118	2	28.00	14700	1020	1.0	S	Z	Z	Z	P	Z	F	11360	18000	18000	LOC4
119	17	30.00	9000	1032	1.0	S	Z	Z	Z	P	Z	F	11790	27800	29500	LOC4
120	4	32.00	11605	710	1.0	S	Z	Z	Z	P	N	F	11490	18900	19900	LOC4
121	2	31.00	14420	945	1.0	U	Z	Z	Z	P	N	F	9930	11900	11900	LOC4
122	15	32.00	10875	1277	1.0	C	Z	Z	U	P	P	F	10340	18040	19900	LOC4
123	13	33.00	15000	1149	1.0	C	N	Z	C	P	S	B	15150	28750	28900	LOC5
124	1	27.00	11250	1444	1.5	C	Y	Z	C	P	O	B	13210	20000	22500	LOC5
125	17	0.00	8400	1408	2.0	C	N	Z	C	G	N	B	25820	38300	37900	LOC5
126	8	1.75	17118	1320	2.0	C	Z	Z	Z	A	Z	F	22350	38500	39500	LOC5
127	8	0.00	8400	1295	2.0	C	Y	Z	Z	G	Z	B	22880	38500	38000	LOC5
128	9	1.17	8400	1764	2.0	C	N	Z	Z	G	N	B	24330	36790	39500	LOC5
129	8	0.67	8400	1429	2.0	C	Y	Z	Z	G	Z	B	28720	39500	41900	LOC5
130	10	0.00	8750	1502	2.0	C	N	Z	C	G	D	B	24720	41900	41900	LOC5
131	10	0.00	8400	1545	2.0	C	Y	Z	C	G	O	B	26550	44500	45000	LOC5
132	15	1.00	8400	1396	2.0	C	Y	Z	C	G	P	B	24650	45000	45000	LOC5
133	10	0.00	9000	1462	2.0	C	N	Z	C	G	O	F	25070	39500	39500	LOC5
134	6	0.00	9375	1462	2.0	C	N	Y	C	G	P	B	25190	38000	39300	LOC5
135	5	0.00	9375	1426	2.0	C	N	Z	C	G	P	B	25280	38000	39300	LOC5
136	7	0.00	9263	1500	2.0	C	Y	Z	C	G	O	B	25490	43600	43200	LOC5
137	11	0.67	8531	1426	2.0	C	Y	Z	C	G	N	B	28940	41000	41900	LOC5
138	17	2.00	11200	1455	2.0	C	Y	Z	C	G	Z	B	26640	43750	44500	LOC5
139	5	0.00	9000	1433	2.0	C	Y	Z	C	G	O	B	25860	37000	38500	LOC5
140	6	0.00	8977	1796	2.0	C	Y	Z	C	G	O	B	26050	38500	38500	LOC5
141	1	0.00	16800	1711	2.0	C	Y	Z	C	A	O	B	30930	45000	45000	LOC5
142	7	0.00	9100	1770	2.0	C	Y	Z	C	G	P	B	29550	43500	43500	LOC5
143	2	0.00	9100	1717	2.0	C	Y	Z	C	G	O	B	29840	47000	47500	LOC5
144	13	0.00	17500	1704	2.0	C	Y	Z	C	G	P	B	30450	47000	47500	LOC5
145	7	0.00	11859	2184	2.5	C	Y	Z	C	A	O	B	30160	48300	48300	LOC5
146	4	5.42	12320	1708	2.0	C	Y	Z	C	G	N	B	41130	68500	73000	LOC5
147	8	11.50	15000	2344	2.5	C	Y	Z	C	A	O	B	27440	42900	42900	LOC5
148	12	11.67	14000	1643	2.0	C	Y	Z	C	G	N	F	38810	62000	64500	LOC5
149	5	12.33	11700	1422	1.5	C	N	Z	C	G	N	B	25550	38000	39900	LOC5
150	11	12.33	11700	2073	2.0	C	Z	Z	C	G	S	B	22940	31500	31500	LOC5
151	12	6.25	14100	1580	2.0	C	N	Z	C	A	P	B	25800	45220	45220	LOC5
152	10	14.42	12000	1447	2.0	C	Y	Z	C	G	P	F	25180	36000	36350	LOC5
153	5	8.33	12000	1631	2.0	C	Y	Z	C	A	S	B	25670	38000	38900	LOC5
154	6	20.25	30300	2150	2.5	C	Y	Z	C	G	N	B	25220	37500	37500	LOC5
155	11	23.17	29000	1325	2.0	C	Z	Y	C	G	D	L	32260	39000	39000	LOC5
156	6	20.42	21509	2079	2.5	C	Z	Z	C	G	S	B	22270	30500	30500	LOC5
157	11	6.25	15370	3137	3.0	C	Z	Z	C	A	P	B	24660	48500	51500	LOC5
158	15	11.25	30918	1539	2.0	C	Y	Z	C	A	S	L	45770	68500	68500	LOC5
159	12	9.42	9075	1843	2.5	C	Y	Z	C	A	P	B	22370	42900	42900	LOC5
160	12	8.58	16200	1843	3.0	C	Y	Z	C	G	D	B	32490	45000	47900	LOC5
161	12	7.00	13980	1510	2.0	C	Z	Z	C	G	D	B	24920	40500	41900	LOC5
162	6	13.33	12000	1446	2.0	C	Z	Z	C	G	D	B	24320	39000	39000	LOC5

OBS	TIME	AGE	LSQF	DSQF	BATH	HT	FIREPL	POOL	AIR	BLT	CAR	EXTER	AV	XPRICE	ASKP	LOCAT
163	7	11.75	12567	1337	2.0	C	Y	N	C	A	D	B	23400	34555	33900	LOC5
164	18	2.53	17400	1738	2.0	C	N	Z	C	P	D	F	25270	40500	42000	LOC5
165	8	17.92	12900	1675	2.0	C	Y	Z	N	P	N	B	22180	35000	35000	LOC5
166	17	18.92	12900	1664	2.0	C	Z	Z	C	G	D	B	25450	42350	42900	LOC5
167	17	11.67	13900	1759	2.0	C	Z	Z	C	G	S	B	26130	42500	43500	LOC5
168	16	5.92	6150	1025	1.0	C	Z	Z	C	P	D	B	11660	18000	19000	LOC5
169	7	6.42	8645	1366	2.0	C	Z	Z	C	G	D	B	21840	35500	42000	LOC5
170	5	6.58	13827	1228	2.0	C	Z	Z	C	C	D	B	23570	34000	34500	LOC5
171	14	7.00	13629	1289	2.5	C	Z	Z	C	A	D	B	23200	36500	37400	LOC5
172	17	3.25	9977	1740	2.0	C	Z	Z	C	A	S	B	20900	35000	35000	LOC5
173	9	4.25	8775	1396	2.0	C	Z	Z	C	A	D	B	22430	41500	41500	LOC5
174	9	11.00	10000	1501	2.0	C	Z	Z	C	A	D	B	25270	38000	39900	LOC5
175	3	13.33	10625	1649	2.0	C	Z	Y	C	A	D	B	24800	39000	39900	LOC5
176	11	17.58	10000	1160	2.0	C	Z	Z	C	G	D	B	18230	33000	34000	LOC5
177	5	0.00	11050	1754	2.0	C	Z	Z	C	G	P	B	30810	51000	51000	LOC5
178	18	12.33	10125	1544	2.0	C	Z	Z	C	G	D	B	24490	42000	42750	LOC5
179	7	7.42	12250	1934	2.0	C	Y	Z	C	G	D	B	29110	46000	47000	LOC5
180	11	11.75	11875	1907	2.0	C	Z	Z	C	A	D	B	29400	47500	48400	LOC5
181	11	4.00	10400	1754	2.0	C	Y	Z	C	G	D	B	30530	48700	49900	LOC5
182	11	0.00	11072	1408	2.0	C	Y	Z	C	G	D	B	23090	43200	43600	LOC5
183	12	2.92	10300	1703	2.5	C	Y	Z	C	A	P	B	30100	51300	51300	LOC6
184	13	4.50	11600	2106	2.0	C	Y	Z	C	G	D	B	35610	63900	63900	LOC6
185	16	3.33	8088	1506	2.0	C	Y	Z	C	G	D	B	27320	48500	48500	LOC6
186	12	2.75	15214	1885	2.0	C	Y	Z	C	G	D	B	33770	55750	57000	LOC6
187	7	0.00	13500	1741	2.0	C	Y	Z	C	G	D	B	32500	49900	49900	LOC6
188	7	0.00	13348	1977	2.0	C	Y	Z	C	G	D	B	34710	47900	48900	LOC6
189	9	3.58	17516	1882	2.0	C	Y	Y	C	A	D	B	39160	65500	65500	LOC6
190	4	3.25	12555	1704	2.0	C	Y	Z	C	G	P	B	26940	45000	45000	LOC6
191	11	6.08	12600	2086	2.0	C	Y	Z	C	G	D	B	36940	55750	64900	LOC6
192	11	5.50	30030	1674	2.0	C	Y	Z	C	G	D	B	34210	56000	59500	LOC6
193	2	4.42	23475	2234	2.5	C	Y	Z	C	G	D	B	36310	65500	69000	LOC6
194	3	1.75	9766	1447	2.0	C	N	Z	C	A	D	F	26650	39000	40500	LOC6
195	2	3.00	16612	2031	2.0	C	Y	Z	C	G	D	B	34780	50800	50800	LOC6
196	13	2.08	10010	1894	2.0	C	Y	Z	C	G	D	B	31430	53000	55900	LOC6
197	6	0.00	9870	1814	2.0	C	Y	Y	C	A	D	B	29890	46000	49500	LOC6
198	3	1.67	9927	1694	2.0	C	Y	Z	C	G	D	B	28490	44500	45500	LOC6
199	4	0.00	8925	1506	2.0	C	Y	Z	C	G	D	B	29980	47000	47500	LOC6
200	13	0.00	8925	1589	2.0	C	Y	Z	C	G	D	B	27950	47392	48300	LOC6
201	5	1.42	11775	1738	2.0	C	Y	Z	C	A	D	B	34620	49900	49900	LOC6
202	6	3.50	13202	1624	3.0	C	Y	Z	C	G	D	B	28810	47200	47500	LOC6
203	1	0.75	14240	1607	2.0	C	Y	Z	C	G	D	B	30540	48550	47900	LOC6
204	13	1.17	8101	1627	2.0	C	Y	Z	C	A	D	B	29840	48000	48000	LOC6
205	7	2.00	12613	1927	2.0	C	Y	Z	C	G	D	B	35910	62000	63900	LOC6
206	2	2.42	11900	1679	3.0	C	Y	Z	C	A	N	B	30130	46900	47900	LOC6
207	14	3.17	2734	2734	3.0	C	Y	Z	C	G	D	B	48220	78500	82900	LOC6
208	9	2.92	9600	1611	2.0	C	Y	Z	C	G	D	B	28800	45450	46500	LOC6
209	6	3.17	7740	1634	2.0	C	Y	Z	C	G	D	B	29310	44500	46900	LOC6
210	1	3.08	9300	1825	2.0	C	Y	Z	C	G	D	B	34650	54000	55000	LOC6
211	6	3.08	10388	1551	2.0	C	Y	Z	C	A	D	B	28860	44000	44900	LOC6
212	13	2.42	10067	1644	2.0	C	Y	Z	C	G	D	B	28690	47500	49500	LOC6
213	7	2.42	11652	1450	2.0	C	N	Z	C	A	S	B	30250	36800	37900	LOC6
214	5	2.83	15045	1807	2.0	C	Z	Z	C	G	N	B	33320	45200	45900	LOC6
215	5	0.00	7680	1423	2.0	C	Y	Z	C	A	D	B	25790	39000	40500	LOC6
216	3	1.75	11900	1570	2.0	C	Y	Z	C	G	D	B	29130	40900	41900	LOC7

REAL ESTATE DATA BASE

OBS	TIME	AGE	LSQF	DSQF	BATH	HT	FIREPL	POOL	AIR	BLT	CAR	EXTER	AV	XPRICE	ASKP	LOCAT
217	9	1.67	10308	1476	2.0	C	Y	Z	U	A	D	B	27980	44075	44900	LOC7
218	12	1.17	10248	1692	2.0	C	Y	Z	U	G	D	B	29900	50400	52900	LOC7
219	12	1.75	10537	1590	2.0	C	Y	Z	U	A	D	B	28450	46000	46800	LOC7
220	11	2.75	10440	1578	2.0	C	Y	Z	U	G	D	B	27390	42000	42000	LOC7
221	2	1.75	11256	1648	2.0	C	Y	Z	U	G	D	B	29730	41500	41500	LOC7
222	12	2.17	11477	1477	2.0	C	Y	Z	U	A	D	B	26470	44900	44900	LOC7
223	9	2.17	8250	1494	2.0	C	Y	Z	U	G	D	B	28820	46900	46900	LOC7
224	12	2.75	9150	1504	2.0	C	Y	Z	U	G	D	B	29170	47400	48400	LOC7
225	3	1.75	13179	1564	2.0	C	Y	Z	U	G	D	B	27380	45750	45750	LOC7
226	18	3.75	10400	1844	2.0	C	Y	Z	U	G	D	B	37060	56000	57500	LOC7
227	5	1.00	15400	1804	2.0	C	Y	Z	U	G	D	B	35820	52500	52500	LOC7
228	7	0.00	13100	1731	2.0	C	Y	Z	U	G	D	B	34350	54000	54900	LOC7
229	9	0.00	13029	1731	2.0	C	Y	Z	U	G	D	B	33100	55750	55750	LOC7
230	6	0.00	11500	1981	2.0	C	Y	Z	U	G	D	F	40610	52500	52500	LOC7
231	18	0.00	19500	2156	2.0	C	Y	Z	U	G	D	B	25530	41000	42900	LOC7
232	11	2.75	10450	1460	2.0	C	Y	Z	U	G	D	B	25270	42000	44000	LOC7
233	13	2.75	10482	1479	2.0	C	Y	Z	U	G	D	B	25960	42000	42500	LOC7
234	10	2.75	9000	1455	2.0	C	Y	Z	U	G	D	B	27200	41600	44250	LOC7
235	7	1.75	10800	1542	2.0	C	Y	Z	U	G	D	B	27290	37000	37900	LOC7
236	12	2.67	11300	1328	2.5	C	Y	Z	U	G	D	B	40930	69000	73000	LOC7
237	11	1.75	17559	2204	2.0	C	N	Z	U	G	D	B	36280	56500	56500	LOC7
238	4	0.00	11288	1915	2.0	C	Y	Z	U	G	D	B	30320	50900	50900	LOC7
239	12	0.00	11610	1741	2.0	C	Y	Z	U	G	D	B	29480	45900	45900	LOC7
240	9	0.00	10125	1586	2.0	C	Y	Z	U	G	D	B	28190	46900	46900	LOC7
241	13	0.00	10125	1494	2.0	C	Y	Z	U	A	D	B	33800	52000	53100	LOC7
242	9	0.00	11429	1708	2.0	C	Y	Z	U	A	D	B	29240	48900	48900	LOC7
243	6	0.00	11561	1599	2.0	C	Y	Z	U	G	D	B	31250	48900	48900	LOC7
244	8	0.00	10500	1484	2.0	C	Y	Z	U	G	D	B	35240	56000	56000	LOC7
245	16	0.00	12000	1941	2.0	C	Y	Z	U	G	D	B	43420	61350	61350	LOC7
246	13	0.83	13500	2207	2.0	C	Y	Z	U	G	D	B	33980	58000	58000	LOC7
247	13	0.00	11250	1762	2.0	C	Y	Z	U	G	D	B	36600	56000	57000	LOC7
248	6	0.00	9375	1908	2.0	C	Y	Z	U	G	D	B	30130	48500	49900	LOC7
249	8	0.00	9375	1792	2.0	C	Y	Z	U	G	D	B	30190	45800	45800	LOC7
250	13	0.00	9375	1632	2.0	C	Y	Z	U	G	D	B	31540	48500	48500	LOC7
251	11	0.00	9317	1740	2.0	C	Y	Z	U	G	D	B	41110	69900	69900	LOC7
252	6	0.00	9375	2362	2.0	C	Y	Z	U	G	D	B	33410	50000	50000	LOC7
253	10	0.00	9375	1951	2.0	C	Y	Z	U	A	D	B	28170	44000	42500	LOC7
254	6	0.00	9375	1432	2.0	C	Y	Z	U	G	D	B	31540	48900	48900	LOC7
255	4	0.00	9375	1521	2.0	C	Y	Z	U	G	D	B	31450	49600	49600	LOC7
256	11	0.00	9375	1716	2.0	C	Y	Z	U	G	D	B	34460	48750	48750	LOC7
257	17	0.00	9816	1778	2.0	C	Y	Z	U	G	D	B	35450	55000	55900	LOC7
258	12	0.00	7375	1880	2.0	C	Y	Z	U	G	D	F	27210	43900	43900	LOC7
259	10	0.00	11200	1396	2.0	C	Y	Z	U	G	D	B	30880	49900	48900	LOC7
260	11	0.00	9375	1549	2.0	C	Y	Z	U	G	D	B	33060	52000	52200	LOC7
261	11	0.00	9375	1678	2.0	C	Y	Z	U	A	D	B	30080	49900	49900	LOC7
262	17	0.00	9375	1626	2.0	C	Y	Z	U	G	D	B	31150	48900	49900	LOC7
263	12	0.00	9375	1741	2.0	C	Y	Z	U	G	D	B	32800	47000	47500	LOC7
264	10	0.00	11250	1894	2.0	C	Y	Z	U	G	D	B	30240	51900	51900	LOC7
265	10	0.00	11250	1543	2.0	C	Y	Z	U	G	D	B	35880	59500	59500	LOC7
266	7	0.00	9375	1764	2.0	C	Y	Z	U	A	D	B	37220	58500	59500	LOC7
267	7	0.00	9375	2054	2.0	C	Y	Z	U	G	D	B	31860	49900	49900	LOC7
268	7	0.00	9375	1581	2.0	C	Y	Z	U	G	D	B	28270	42800	42800	LOC7
269	10	0.00	12500	1498	2.0	C	Y	Z	U	G	D	B	26800	39000	43500	LOC7
270	5	0.00	12500	1388	2.0	C	Y	Z	U	G	D	B	28200	42750	43900	LOC7

REAL ESTATE DATA BASE

OBS	TIME	AGE	LSQF	DSQF	BATH	HT	FIREPL	POOL	AIR	BLT	CAR	EXTER	AV	XPRICE	ASKP	LOCAT
271	11	0.75	1122	1370	2.0	U	Y	Z	U	G	O	B	26040	44000	46900	LOC7
272	5	0.00	11232	1483	2.0	U	Y	Z	U	G	O	B	31880	46900	46900	LOC7
273	7	0.00	12010	1602	2.0	U	Y	Z	U	G	O	B	32130	46900	46900	LOC7
274	7	0.00	12675	1530	2.0	U	Y	Z	U	G	O	B	28720	46600	46600	LOC7
275	4	0.00	15015	2049	2.0	U	Y	Z	U	G	O	B	39180	56350	56350	LOC7
276	4	0.00	14490	1896	2.0	U	Y	Z	U	G	O	B	36660	54500	54500	LOC7
277	3	0.00	13333	1610	2.0	U	Y	Z	U	G	O	B	31610	47900	47900	LOC7
278	3	0.00	13122	1616	2.0	U	Y	Z	U	A	O	B	35020	49900	49900	LOC7
279	13	0.00	11950	1584	2.0	U	Y	Z	U	G	O	B	28640	46800	46800	LOC7
280	12	0.00	12065	1482	2.0	U	Y	Z	U	A	O	B	28110	46800	46800	LOC7
281	12	0.00	12033	1584	2.0	U	Y	Z	U	A	O	B	28640	46500	46500	LOC7
282	12	0.00	12000	1649	2.0	U	Y	Z	U	G	O	B	35900	50400	50400	LOC7
283	12	0.00	12300	1734	2.0	U	Y	Z	U	G	O	B	33840	50200	49300	LOC7
284	13	0.00	1491	1491	2.0	U	Y	Z	U	G	O	B	29740	46900	46900	LOC7
285	13	0.08	8174	1608	2.0	U	Y	Z	U	G	O	B	29760	51000	52500	LOC7
286	1	0.00	11025	1025	2.0	U	Y	Z	U	G	O	B	29260	46500	46500	LOC7
287	5	0.00	10530	1562	2.0	U	Y	Z	U	G	O	F	42610	64400	62900	LOC7
288	6	0.00	9598	2352	2.5	U	Y	Z	U	G	O	B	42610	73000	75000	LOC7
289	12	0.83	9598	2352	2.5	U	Y	Z	U	A	O	F	48750	75000	66600	LOC7
290	1	0.00	13320	2233	2.0	U	Y	Z	U	G	O	B	36310	60600	59100	LOC7
291	14	0.00	11250	1894	2.0	U	Y	Z	U	A	O	F	37080	61500	62500	LOC7
292	14	0.00	11250	1914	2.5	U	Y	Z	U	G	O	F	36080	75000	77450	LOC7
293	11	0.00	11250	2569	2.5	U	Y	Z	U	G	O	L	38320	58800	61500	LOC7
294	14	0.58	11250	2000	2.0	U	Y	Z	U	G	O	L	35980	58800	58800	LOC7
295	5	0.00	11250	1770	2.0	U	Y	Z	U	G	O	L	39830	62730	61180	LOC7
296	18	0.00	11250	1967	2.0	U	Y	Z	U	G	O	F	35550	76000	76250	LOC7
297	6	0.00	11475	1936	2.5	U	Y	Z	U	G	O	B	52680	59500	59900	LOC7
298	12	0.00	11737	2580	2.5	U	Y	Z	U	G	O	F	37230	52900	52900	LOC7
299	3	0.00	10530	1906	2.0	U	Y	Z	U	G	O	B	34470	43750	45750	LOC7
300	7	0.00	9600	1950	2.0	U	Y	Z	U	G	O	B	30590	45900	45900	LOC7
301	12	1.75	12920	1612	2.0	U	Y	Z	U	G	O	B	28400	45290	46900	LOC7
302	12	0.00	13558	1389	2.0	U	Y	Z	U	G	O	B	29390	43508	45250	LOC7
303	13	0.00	10760	1492	2.0	U	Y	Z	U	G	O	B	21270	47000	49600	LOC7
304	18	0.58	16000	1460	2.0	U	Y	Z	U	G	O	B	28510	50300	50300	LOC7
305	11	0.00	11355	1561	2.0	U	Y	Z	U	G	O	B	30310	47300	47300	LOC7
306	12	0.00	10117	1687	2.0	U	Y	Z	U	G	O	B	29850	48600	48600	LOC7
307	12	0.00	11594	1594	2.0	U	Y	Z	U	A	O	B	30220	47900	47900	LOC7
308	12	0.00	7825	1620	2.0	U	Y	Z	U	G	O	B	27900	46330	46800	LOC7
309	18	0.00	9057	1470	2.0	U	Y	Z	U	G	O	B	30920	45900	45900	LOC7
310	16	0.00	9160	1550	2.0	U	Y	Z	U	G	O	B	28640	47900	46900	LOC7
311	16	0.00	11175	1439	2.0	U	Y	Z	U	G	O	F	28610	45900	45900	LOC7
312	17	0.00	10886	1522	2.0	U	Y	Z	U	G	O	B	34450	45600	58900	LOC7
313	12	0.00	9432	1926	2.0	U	Y	Z	U	G	O	B	28070	46750	47900	LOC7
314	17	0.00	9000	1441	2.0	U	Y	Z	U	G	O	B	29070	51700	45900	LOC7
315	12	0.00	9000	1543	2.0	U	Y	Z	U	G	O	B	26430	53500	45900	LOC7
316	11	0.00	12600	1392	2.0	U	Y	Z	U	G	O	B	28810	49670	45600	LOC7
317	17	0.00	11500	1461	2.0	U	Y	Z	U	G	O	B	31880	50600	47000	LOC7
318	13	0.00	11319	1691	2.0	U	Y	Z	U	A	O	B	.	53500	51500	LOC7
319	10	0.00	10830	1776	2.0	U	Y	Z	U	A	O	B	29280	48200	53500	LOC7
320	12	0.00	12233	1565	2.0	U	Y	Z	U	A	O	B	30370	49900	49870	LOC7
321	12	0.00	12827	1654	2.0	U	Y	Z	U	G	O	B	31000	50600	50600	LOC7
322	13	0.00	11381	1711	2.0	U	Y	Z	U	G	O	B	29220	53500	53500	LOC7
323	13	0.00	10696	1640	2.0	U	Y	Z	U	A	O	B	30960	48200	48200	LOC7
324	16	0.00	10509	1737	2.0	U	Y	Z	U	A	O	B		49900	49900	LOC7

REAL ESTATE DATA BASE

OBS	TIME	AGE	LSQF	DSQF	BATH	HT	FIREPL	POOL	AIR	BLT	CAR	EXTER	AV	XPRICE	ASKP	LOCAT
325	18	0	11199	1385	2	C	Y	N	C	G	D	B	29330	46900	46900	LOC7
326	13	0	7930	1505	2	C	Y	N	C	G	D	B	25780	39900	39900	LOC7
327	14	0	9150	1426	2	C	Y	N	C	A	D	B	25260	42500	42500	LOC7
328	17	0	9125	1288	2	C	Y	N	C	G	P	B	24420	36900	36900	LOC7
329	18	0	7930	1369	2	C	Y	N	C	G	P	B	23140	40900	40900	LOC7
330	4	0	7800	1144	1	C	N	N	C	A	P	B	20770	34200	34200	LOC7
331	4	0	6900	1087	2	C	N	N	C	A	D	B	21650	36900	39800	LOC7
332	14	0	6890	1288	2	C	Y	N	C	A	D	B	24090	40500	40500	LOC7
333	14	0	13769	1333	2	C	Y	N	C	G	D	B	26520	44500	44500	LOC7
334	12	0	9636	1315	2	C	Y	N	C	G	D	B	25240	41600	41500	LOC7
335.	12	0	10920	1315	2	C	Y	N	C	G	D	B	25620	41800	41800	LOC7

Appendix B

TABLES

TABLE B.1   The Standardized Normal Distribution[a]

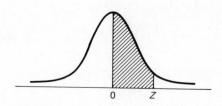

Z	.00	.01	.02	.03	.04	.05	.06	.07	.08	.09
.0	.0000	.0040	.0080	.0120	.0160	.0199	.0239	.0279	.0319	.0359
.1	.0398	.0438	.0478	.0517	.0557	.0596	.0636	.0675	.0714	.0753
.2	.0793	.0832	.0871	.0910	.0948	.0987	.1026	.1064	.1103	.1141
.3	.1179	.1217	.1255	.1293	.1331	.1368	.1406	.1443	.1480	.1517
.4	.1554	.1591	.1628	.1664	.1700	.1736	.1772	.1808	.1844	.1879
.5	.1915	.1950	.1985	.2019	.2054	.2088	.2123	.2157	.2190	.2224
.6	.2257	.2291	.2324	.2357	.2389	.2422	.2454	.2486	.2518	.2549
.7	.2580	.2612	.2642	.2673	.2704	.2734	.2764	.2794	.2823	.2852
.8	.2881	.2910	.2939	.2967	.2995	.3023	.3051	.3078	.3106	.3133
.9	.3159	.3186	.3212	.3238	.3264	.3289	.3315	.3340	.3365	.3389
1.0	.3413	.3438	.3461	.3485	.3508	.3531	.3554	.3577	.3599	.3621
1.1	.3643	.3665	.3686	.3708	.3729	.3749	.3770	.3790	.3810	.3830
1.2	.3849	.3869	.3888	.3907	.3925	.3944	.3962	.3980	.3997	.4015
1.3	.4032	.4049	.4066	.4082	.4099	.4115	.4131	.4147	.4162	.4177

Z	.00	.01	.02	.03	.04	.05	.06	.07	.08	.09
1.4	.4192	.4207	.4222	.4236	.4251	.4265	.4279	.4292	.4306	.4319
1.5	.4332	.4345	.4357	.4370	.4382	.4394	.4406	.4418	.4429	.4441
1.6	.4452	.4463	.4474	.4484	.4495	.4505	.4515	.4525	.4535	.4545
1.7	.4554	.4564	.4573	.4582	.4591	.4599	.4608	.4616	.4625	.4633
1.8	.4641	.4649	.4656	.4664	.4671	.4678	.4686	.4693	.4699	.4706
1.9	.4713	.4719	.4726	.4732	.4738	.4744	.4750	.4756	.4761	.4767
2.0	.4772	.4778	.4783	.4788	.4793	.4798	.4803	.4808	.4812	.4817
2.1	.4821	.4826	.4830	.4834	.4838	.4842	.4846	.4850	.4854	.4857
2.2	.4861	.4864	.4868	.4871	.4875	.4878	.4881	.4884	.4887	.4890
2.3	.4893	.4896	.4898	.4901	.4904	.4906	.4909	.4911	.4913	.4916
2.4	.4918	.4920	.4922	.4925	.4927	.4929	.4931	.4932	.4934	.4936
2.5	.4938	.4940	.4941	.4943	.4945	.4946	.4948	.4949	.4951	.4952
2.6	.4953	.4955	.4956	.4957	.4959	.4960	.4961	.4962	.4963	.4964
2.7	.4965	.4966	.4967	.4968	.4969	.4970	.4971	.4972	.4973	.4974
2.8	.4974	.4975	.4976	.4977	.4977	.4978	.4979	.4979	.4980	.4981
2.9	.4981	.4982	.4982	.4983	.4984	.4984	.4985	.4985	.4986	.4986
3.0	.49865	.49869	.49874	.49878	.49882	.49886	.49889	.49893	.49897	.49900
3.1	.49903	.49906	.49910	.49913	.49916	.49918	.49921	.49924	.49926	.49929
3.2	.49931	.49934	.49936	.49938	.49940	.49942	.49944	.49946	.49948	.49950
3.3	.49952	.49953	.49955	.49957	.49958	.49960	.49961	.49962	.49964	.49965
3.4	.49966	.49968	.49969	.49970	.49971	.49972	.49973	.49974	.49975	.49976
3.5	.49977	.49978	.49978	.49979	.49980	.49981	.49981	.49982	.49983	.49983
3.6	.49984	.49985	.49985	.49986	.49986	.49987	.49987	.49988	.49988	.49989
3.7	.49989	.49990	.49990	.49990	.49991	.49991	.49992	.49992	.49992	.49992
3.8	.49993	.49993	.49993	.49994	.49994	.49994	.49994	.49995	.49995	.49995
3.9	.49995	.49995	.49996	.49996	.49996	.49996	.49996	.49996	.49997	.49997

[a]Entry represents area under the standardized normal distribution from the mean to Z.

TABLE B.2 Critical Values of χ^2 [a]

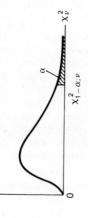

Upper-Tail Areas

Degrees of Freedom, ν	.99	.98	.95	.90	.80	.70	.50	.30	.20	.10	.05	.02	.01	.001
1	$.0^3157$	$.0^3628$	$.0^3393$	.0158	.0642	.148	.455	1.074	1.642	2.706	3.841	5.412	6.635	10.827
2	.0201	.0404	.103	.211	.446	.713	1.386	2.408	3.219	4.605	5.991	7.824	9.210	13.815
3	.115	.185	.352	.584	1.005	1.424	2.366	3.665	4.642	6.251	7.815	9.837	11.345	16.268
4	.297	.429	.711	1.064	1.649	2.195	3.357	4.878	5.989	7.779	9.488	11.668	13.277	18.465
5	.554	.752	1.145	1.610	2.343	3.000	4.351	6.064	7.289	9.236	11.070	13.388	15.086	20.517
6	.872	1.134	1.635	2.204	3.070	3.828	5.348	7.231	8.558	10.645	12.592	15.033	16.812	22.457
7	1.239	1.564	2.167	2.833	3.822	4.671	6.346	8.383	9.803	12.017	14.067	16.622	18.475	24.322
8	1.646	2.032	2.733	3.490	4.594	5.527	7.344	9.524	11.030	13.362	15.507	18.168	20.090	26.125
9	2.088	2.532	3.325	4.168	5.380	6.393	8.343	10.656	12.242	14.684	16.919	19.679	21.666	27.877
10	2.558	3.059	3.940	4.865	6.179	7.267	9.342	11.781	13.442	15.987	18.307	21.161	23.209	29.588
11	3.053	3.609	4.575	5.578	6.989	8.148	10.341	12.899	14.631	17.275	19.675	22.618	24.725	31.264
12	3.571	4.178	5.226	6.304	7.807	9.034	11.340	14.011	15.812	18.549	21.026	24.054	26.217	32.909
13	4.107	4.765	5.892	7.042	8.634	9.926	12.340	15.119	16.985	19.812	22.362	25.472	27.688	34.528
14	4.660	5.368	6.571	7.790	9.467	10.821	13.339	16.222	18.151	21.064	23.685	26.873	29.141	36.123
15	5.229	5.985	7.261	8.547	10.307	11.721	14.339	17.322	19.311	22.307	24.996	28.259	30.578	37.697

16	5.812	6.614	7.962	9.312	11.152	12.624	15.338	18.418	20.465	23.542	26.296	29.633	32.000	39.252
17	6.408	7.255	8.672	10.085	12.002	13.531	16.338	19.511	21.615	24.769	27.587	30.995	33.409	40.790
18	7.015	7.906	9.390	10.865	12.857	14.440	17.338	20.601	22.760	25.989	28.869	32.346	34.805	42.312
19	7.633	8.567	10.117	11.651	13.716	15.352	18.338	21.689	23.900	27.204	30.144	33.687	36.191	43.820
20	8.260	9.237	10.851	12.443	14.578	16.266	19.337	22.775	25.038	28.412	31.410	35.020	37.566	45.315
21	8.897	9.915	11.591	13.240	15.445	17.182	20.337	23.858	26.171	29.615	32.671	36.343	38.932	46.797
22	9.542	10.600	12.338	14.041	16.314	18.101	21.337	24.939	27.301	30.813	33.924	37.659	40.289	48.268
23	10.196	11.293	13.091	14.848	17.187	19.021	22.337	26.018	28.429	32.007	35.172	38.968	41.638	49.728
24	10.856	11.992	13.848	15.659	18.062	19.943	23.337	27.096	29.553	33.196	36.415	40.270	42.980	51.179
25	11.524	12.697	14.611	16.473	18.940	20.867	24.337	28.172	30.675	34.382	37.652	41.566	44.314	52.620
26	12.198	13.409	15.379	17.292	19.820	21.792	25.336	29.246	31.795	35.563	38.885	42.856	45.642	54.052
27	12.879	14.125	16.151	18.114	20.703	22.719	26.336	30.319	32.912	36.741	40.113	44.140	46.963	55.476
28	13.565	14.847	16.928	18.939	21.588	23.647	27.336	31.391	34.027	37.916	41.337	45.419	48.278	56.893
29	14.256	15.574	17.708	19.768	22.475	24.577	28.336	32.461	35.139	39.087	42.557	46.693	49.588	58.302
30	14.953	16.306	18.493	20.599	23.364	25.508	29.336	33.530	36.250	40.256	43.773	47.962	50.892	59.703

[a]For larger values of v, the expression $\sqrt{2\chi^2} - \sqrt{2v - 1}$ may be used as a normal deviate with unit variance, remembering that the probability for χ^2 corresponds with that of a single tail of the normal curve.

SOURCE: Table B.2 is taken from Table IV of R. A. Fisher and F. Yates, *Statistical Tables for Biological, Agricultural and Medical Research*, published by Longman Group, Ltd., London (previously published by Oliver & Boyd Ltd., Edinburgh) and by permission of the authors and publishers.

TABLE B.3 Critical Values of t

Degrees of Freedom, ν	Upper-Tail Areas												
	.45	.40	.35	.30	.25	.20	.15	.10	.05	.025	.01	.005	.0005
1	.158	.325	.510	.727	1.000	1.376	1.963	3.078	6.314	12.706	31.821	63.657	636.619
2	.142	.289	.445	.617	.816	1.061	1.386	1.886	2.920	4.303	6.965	9.925	31.598
3	.137	.277	.424	.584	.765	.978	1.250	1.638	2.353	3.182	4.541	5.841	12.941
4	.134	.271	.414	.569	.741	.941	1.190	1.533	2.132	2.776	3.747	4.604	8.610
5	.132	.267	.408	.559	.727	.920	1.156	1.476	2.015	2.571	3.365	4.032	6.859
6	.131	.265	.404	.553	.718	.906	1.134	1.440	1.943	2.447	3.143	3.707	5.959
7	.130	.263	.402	.549	.711	.896	1.119	1.415	1.895	2.365	2.998	3.499	5.405
8	.130	.262	.399	.546	.706	.889	1.108	1.397	1.860	2.306	2.896	3.355	5.041
9	.129	.261	.398	.543	.703	.883	1.100	1.383	1.833	2.262	2.821	3.250	4.781
10	.129	.260	.397	.542	.700	.879	1.093	1.372	1.812	2.228	2.764	3.169	4.587
11	.129	.260	.396	.540	.697	.876	1.088	1.363	1.796	2.201	2.718	3.106	4.437
12	.128	.259	.395	.539	.695	.873	1.083	1.356	1.782	2.179	2.681	3.055	4.318
13	.128	.259	.394	.538	.694	.870	1.079	1.350	1.771	2.160	2.650	3.012	4.221
14	.128	.258	.393	.537	.692	.868	1.076	1.345	1.761	2.145	2.624	2.977	4.140
15	.128	.258	.393	.536	.691	.866	1.074	1.341	1.753	2.131	2.602	2.947	4.073

16	.128	.258	.392	.535	.690	.865	1.071	1.337	1.746	2.120	2.583	2.921	4.015
17	.128	.257	.392	.534	.689	.863	1.069	1.333	1.740	2.110	2.567	2.898	3.965
18	.127	.257	.392	.534	.688	.862	1.067	1.330	1.734	2.101	2.552	2.878	3.922
19	.127	.257	.391	.533	.688	.861	1.066	1.328	1.729	2.093	2.539	2.861	3.883
20	.127	.257	.391	.533	.687	.860	1.064	1.325	1.725	2.086	2.528	2.845	3.850
21	.127	.257	.391	.532	.686	.859	1.063	1.323	1.721	2.080	2.518	2.831	3.819
22	.127	.256	.390	.532	.686	.858	1.061	1.321	1.717	2.074	2.508	2.819	3.792
23	.127	.256	.390	.532	.685	.858	1.060	1.319	1.714	2.069	2.500	2.807	3.767
24	.127	.256	.390	.531	.685	.857	1.059	1.318	1.711	2.064	2.492	2.797	3.745
25	.127	.256	.390	.531	.684	.856	1.058	1.316	1.708	2.060	2.485	2.787	3.725
26	.127	.256	.390	.531	.684	.856	1.056	1.315	1.706	2.056	2.479	2.779	3.707
27	.127	.256	.389	.531	.684	.855	1.057	1.314	1.703	2.052	2.473	2.771	3.690
28	.127	.256	.389	.530	.683	.855	1.056	1.313	1.701	2.048	2.467	2.763	3.674
29	.127	.256	.389	.530	.683	.854	1.055	1.311	1.699	2.045	2.462	2.756	3.659
30	.127	.256	.389	.530	.683	.854	1.055	1.310	1.697	2.042	2.457	2.750	3.646
40	.126	.255	.388	.529	.681	.851	1.050	1.303	1.684	2.021	2.423	2.704	3.551
60	.126	.254	.387	.527	.679	.848	1.046	1.296	1.671	2.000	2.390	2.660	3.460
120	.126	.254	.386	.526	.677	.845	1.041	1.289	1.658	1.980	2.358	2.617	3.373
∞	.126	.253	.385	.524	.674	.842	1.036	1.282	1.645	1.960	2.326	2.576	3.291

SOURCE: Table B.3 is taken from Table III of R. A. Fisher and F. Yates, *Statistical Tables for Biological, Agricultural and Medical Research*, published by Longman Group, Ltd., London (previously published by Oliver & Boyd Ltd., Edinburgh) and by permission of the authors and publishers.

TABLE B.4   Critical Values of F[a]

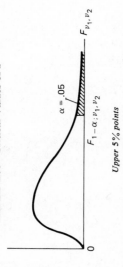

$\alpha = .05$

$F_{1-\alpha,\,\nu_1,\,\nu_2}$

$F_{\nu_1,\,\nu_2}$

Upper 5% points

ν_1 ν_2	1	2	3	4	5	6	7	8	9	10	12	15	20	24	30	40	60	120	∞
1	161.4	199.5	215.7	224.6	230.2	234.0	236.8	238.9	240.5	241.9	243.9	245.9	248.0	249.1	250.2	251.1	252.2	253.3	254.3
2	18.51	19.00	19.16	19.25	19.30	19.33	19.35	19.37	19.38	19.40	19.41	19.43	19.45	19.45	19.46	19.47	19.48	19.49	19.50
3	10.13	9.55	9.28	9.12	9.01	8.94	8.89	8.85	8.81	8.79	8.74	8.70	8.66	8.64	8.62	8.59	8.57	8.55	8.53
4	7.71	6.94	6.59	6.39	6.26	6.16	6.09	6.04	6.00	5.96	5.91	5.86	5.80	5.77	5.75	5.72	5.69	5.66	5.63
5	6.61	5.79	5.41	5.19	5.05	4.95	4.88	4.82	4.77	4.74	4.68	4.62	4.56	4.53	4.50	4.46	4.43	4.40	4.36
6	5.99	5.14	4.76	4.53	4.39	4.28	4.21	4.15	4.10	4.06	4.00	3.94	3.87	3.84	3.81	3.77	3.74	3.70	3.67
7	5.59	4.74	4.35	4.12	3.97	3.87	3.79	3.73	3.68	3.64	3.57	3.51	3.44	3.41	3.38	3.34	3.30	3.27	3.23
8	5.32	4.46	4.07	3.84	3.69	3.58	3.50	3.44	3.39	3.35	3.28	3.22	3.15	3.12	3.08	3.04	3.01	2.97	2.93
9	5.12	4.26	3.86	3.63	3.48	3.37	3.29	3.23	3.18	3.14	3.07	3.01	2.94	2.90	2.86	2.83	2.79	2.75	2.71
10	4.96	4.10	3.71	3.48	3.33	3.22	3.14	3.07	3.02	2.98	2.91	2.85	2.77	2.74	2.70	2.66	2.62	2.58	2.54
11	4.84	3.98	3.59	3.36	3.20	3.09	3.01	2.95	2.90	2.85	2.79	2.72	2.65	2.61	2.57	2.53	2.49	2.45	2.40
12	4.75	3.89	3.49	3.26	3.11	3.00	2.91	2.85	2.80	2.75	2.69	2.62	2.54	2.51	2.47	2.43	2.38	2.34	2.30
13	4.67	3.81	3.41	3.18	3.03	2.92	2.83	2.77	2.71	2.67	2.60	2.53	2.46	2.42	2.38	2.34	2.30	2.25	2.21
14	4.60	3.74	3.34	3.11	2.96	2.85	2.76	2.70	2.65	2.60	2.53	2.46	2.39	2.35	2.31	2.27	2.22	2.18	2.13
15	4.54	3.68	3.29	3.06	2.90	2.79	2.71	2.64	2.59	2.54	2.48	2.40	2.33	2.29	2.25	2.20	2.16	2.11	2.07
16	4.49	3.63	3.24	3.01	2.85	2.74	2.66	2.59	2.54	2.49	2.42	2.35	2.28	2.24	2.19	2.15	2.11	2.06	2.01
17	4.45	3.59	3.20	2.96	2.81	2.70	2.61	2.55	2.49	2.45	2.38	2.31	2.23	2.19	2.15	2.10	2.06	2.01	1.96
18	4.41	3.55	3.16	2.93	2.77	2.66	2.58	2.51	2.46	2.41	2.34	2.27	2.19	2.15	2.11	2.06	2.02	1.97	1.92
19	4.38	3.52	3.13	2.90	2.74	2.63	2.54	2.48	2.42	2.38	2.31	2.23	2.16	2.11	2.07	2.03	1.98	1.93	1.88
20	4.35	3.49	3.10	2.87	2.71	2.60	2.51	2.45	2.39	2.35	2.28	2.20	2.12	2.08	2.04	1.99	1.95	1.90	1.84
21	4.32	3.47	3.07	2.84	2.68	2.57	2.49	2.42	2.37	2.32	2.25	2.18	2.10	2.05	2.01	1.96	1.92	1.87	1.81
22	4.30	3.44	3.05	2.82	2.66	2.55	2.46	2.40	2.34	2.30	2.23	2.15	2.07	2.03	1.98	1.94	1.89	1.84	1.78
23	4.28	3.42	3.03	2.80	2.64	2.53	2.44	2.37	2.32	2.27	2.20	2.13	2.05	2.01	1.96	1.91	1.86	1.81	1.76
24	4.26	3.40	3.01	2.78	2.62	2.51	2.42	2.36	2.30	2.25	2.18	2.11	2.03	1.98	1.94	1.89	1.84	1.79	1.73
25	4.24	3.39	2.99	2.76	2.60	2.49	2.40	2.34	2.28	2.24	2.16	2.09	2.01	1.96	1.92	1.87	1.82	1.77	1.71
26	4.23	3.37	2.98	2.74	2.59	2.47	2.39	2.32	2.27	2.22	2.15	2.07	1.99	1.95	1.90	1.85	1.80	1.75	1.69
27	4.21	3.35	2.96	2.73	2.57	2.46	2.37	2.31	2.25	2.20	2.13	2.06	1.97	1.93	1.88	1.84	1.79	1.73	1.67
28	4.20	3.34	2.95	2.71	2.56	2.45	2.36	2.29	2.24	2.19	2.12	2.04	1.96	1.91	1.87	1.82	1.77	1.71	1.65
29	4.18	3.33	2.93	2.70	2.55	2.43	2.35	2.28	2.22	2.18	2.10	2.03	1.94	1.90	1.85	1.81	1.75	1.70	1.64
30	4.17	3.32	2.92	2.69	2.53	2.42	2.33	2.27	2.21	2.16	2.09	2.01	1.93	1.89	1.84	1.79	1.74	1.68	1.62
40	4.08	3.23	2.84	2.61	2.45	2.34	2.25	2.18	2.12	2.08	2.00	1.92	1.84	1.79	1.74	1.69	1.64	1.58	1.51
60	4.00	3.15	2.76	2.53	2.37	2.25	2.17	2.10	2.04	1.99	1.92	1.84	1.75	1.70	1.65	1.59	1.53	1.47	1.39
120	3.92	3.07	2.68	2.45	2.29	2.17	2.09	2.02	1.96	1.91	1.83	1.75	1.66	1.61	1.55	1.50	1.43	1.35	1.25
∞	3.84	3.00	2.60	2.37	2.21	2.10	2.01	1.94	1.88	1.83	1.75	1.67	1.57	1.52	1.46	1.39	1.32	1.22	1.00

TABLE B.4 (Continued)

$\alpha = .025$

F_{ν_1, ν_2}

$F_{1-\alpha\,;\,\nu_1,\,\nu_2}$

Upper 2.5% points

ν_2 \ ν_1	1	2	3	4	5	6	7	8	9	10	12	15	20	24	30	40	60	120	∞
1	647.8	799.5	864.2	899.6	921.8	937.1	948.2	956.7	963.3	968.6	976.7	984.9	993.1	997.2	1001	1006	1010	1014	1018
2	38.51	39.00	39.17	39.25	39.30	39.33	39.36	39.37	39.39	39.40	39.41	39.43	39.45	39.46	39.46	39.47	39.48	39.49	39.50
3	17.44	16.04	15.44	15.10	14.88	14.73	14.62	14.54	14.47	14.42	14.34	14.25	14.17	14.12	14.08	14.04	13.99	13.95	13.90
4	12.22	10.65	9.98	9.60	9.36	9.20	9.07	8.98	8.90	8.84	8.75	8.66	8.56	8.51	8.46	8.41	8.36	8.31	8.26
5	10.01	8.43	7.76	7.39	7.15	6.98	6.85	6.76	6.68	6.62	6.52	6.43	6.33	6.28	6.23	6.18	6.12	6.07	6.02
6	8.81	7.26	6.60	6.23	5.99	5.82	5.70	5.60	5.52	5.46	5.37	5.27	5.17	5.12	5.07	5.01	4.96	4.90	4.85
7	8.07	6.54	5.89	5.52	5.29	5.12	4.99	4.90	4.82	4.76	4.67	4.57	4.47	4.42	4.36	4.31	4.25	4.20	4.14
8	7.57	6.06	5.42	5.05	4.82	4.65	4.53	4.43	4.36	4.30	4.20	4.10	4.00	3.95	3.89	3.84	3.78	3.73	3.67
9	7.21	5.71	5.08	4.72	4.48	4.32	4.20	4.10	4.03	3.96	3.87	3.77	3.67	3.61	3.56	3.51	3.45	3.39	3.33
10	6.94	5.46	4.83	4.47	4.24	4.07	3.95	3.85	3.78	3.72	3.62	3.52	3.42	3.37	3.31	3.26	3.20	3.14	3.08
11	6.72	5.26	4.63	4.28	4.04	3.88	3.76	3.66	3.59	3.53	3.43	3.33	3.23	3.17	3.12	3.06	3.00	2.94	2.88
12	6.55	5.10	4.47	4.12	3.89	3.73	3.61	3.51	3.44	3.37	3.28	3.18	3.07	3.02	2.96	2.91	2.85	2.79	2.72
13	6.41	4.97	4.35	4.00	3.77	3.60	3.48	3.39	3.31	3.25	3.15	3.05	2.95	2.89	2.84	2.78	2.72	2.66	2.60
14	6.30	4.86	4.24	3.89	3.66	3.50	3.38	3.29	3.21	3.15	3.05	2.95	2.84	2.79	2.73	2.67	2.61	2.55	2.49
15	6.20	4.77	4.15	3.80	3.58	3.41	3.29	3.20	3.12	3.06	2.96	2.86	2.76	2.70	2.64	2.59	2.52	2.46	2.40
16	6.12	4.69	4.08	3.73	3.50	3.34	3.22	3.12	3.05	2.99	2.89	2.79	2.68	2.63	2.57	2.51	2.45	2.38	2.32
17	6.04	4.62	4.01	3.66	3.44	3.28	3.16	3.06	2.98	2.92	2.82	2.72	2.62	2.56	2.50	2.44	2.38	2.32	2.25
18	5.98	4.56	3.95	3.61	3.38	3.22	3.10	3.01	2.93	2.87	2.77	2.67	2.56	2.50	2.44	2.38	2.32	2.26	2.19
19	5.92	4.51	3.90	3.56	3.33	3.17	3.05	2.96	2.88	2.82	2.72	2.62	2.51	2.45	2.39	2.33	2.27	2.20	2.13
20	5.87	4.46	3.86	3.51	3.29	3.13	3.01	2.91	2.84	2.77	2.68	2.57	2.46	2.41	2.35	2.29	2.22	2.16	2.09
21	5.83	4.42	3.82	3.48	3.25	3.09	2.97	2.87	2.80	2.73	2.64	2.53	2.42	2.37	2.31	2.25	2.18	2.11	2.04
22	5.79	4.38	3.78	3.44	3.22	3.05	2.93	2.84	2.76	2.70	2.60	2.50	2.39	2.33	2.27	2.21	2.14	2.08	2.00
23	5.75	4.35	3.75	3.41	3.18	3.02	2.90	2.81	2.73	2.67	2.57	2.47	2.36	2.30	2.24	2.18	2.11	2.04	1.97
24	5.72	4.32	3.72	3.38	3.15	2.99	2.87	2.78	2.70	2.64	2.54	2.44	2.33	2.27	2.21	2.15	2.08	2.01	1.94
25	5.69	4.29	3.69	3.35	3.13	2.97	2.85	2.75	2.68	2.61	2.51	2.41	2.30	2.24	2.18	2.12	2.05	1.98	1.91
26	5.66	4.27	3.67	3.33	3.10	2.94	2.82	2.73	2.65	2.59	2.49	2.39	2.28	2.22	2.16	2.09	2.03	1.95	1.88
27	5.63	4.24	3.65	3.31	3.08	2.92	2.80	2.71	2.63	2.57	2.47	2.36	2.25	2.19	2.13	2.07	2.00	1.93	1.85
28	5.61	4.22	3.63	3.29	3.06	2.90	2.78	2.69	2.61	2.55	2.45	2.34	2.23	2.17	2.11	2.05	1.98	1.91	1.83
29	5.59	4.20	3.61	3.27	3.04	2.88	2.76	2.67	2.59	2.53	2.43	2.32	2.21	2.15	2.09	2.03	1.96	1.89	1.81
30	5.57	4.18	3.59	3.25	3.03	2.87	2.75	2.65	2.57	2.51	2.41	2.31	2.20	2.14	2.07	2.01	1.94	1.87	1.79
40	5.42	4.05	3.46	3.13	2.90	2.74	2.62	2.53	2.45	2.39	2.29	2.18	2.07	2.01	1.94	1.88	1.80	1.72	1.64
60	5.29	3.93	3.34	3.01	2.79	2.63	2.51	2.41	2.33	2.27	2.17	2.06	1.94	1.88	1.82	1.74	1.67	1.58	1.48
120	5.15	3.80	3.23	2.89	2.67	2.52	2.39	2.30	2.22	2.16	2.05	1.94	1.82	1.76	1.69	1.61	1.53	1.43	1.31
∞	5.02	3.69	3.12	2.79	2.57	2.41	2.29	2.19	2.11	2.05	1.94	1.83	1.71	1.64	1.57	1.48	1.39	1.27	1.00

TABLE B.4 (Continued)

α = .01

F_{ν_1, ν_2}

$F_{1-\alpha;\,\nu_1,\,\nu_2}$

Upper 1% points

ν_2 \\ ν_1	1	2	3	4	5	6	7	8	9	10	12	15	20	24	30	40	60	120	∞
1	4052	4999.5	5403	5625	5764	5859	5928	5982	6022	6056	6106	6157	6209	6235	6261	6287	6313	6339	6366
2	98.50	99.00	99.17	99.25	99.30	99.33	99.36	99.37	99.39	99.40	99.42	99.43	99.45	99.46	99.47	99.47	99.48	99.49	99.50
3	34.12	30.82	29.46	28.71	28.24	27.91	27.67	27.49	27.35	27.23	27.05	26.87	26.69	26.60	26.50	26.41	26.32	26.22	26.13
4	21.20	18.00	16.69	15.98	15.52	15.21	14.98	14.80	14.66	14.55	14.37	14.20	14.02	13.93	13.84	13.75	13.65	13.56	13.46
5	16.26	13.27	12.06	11.39	10.97	10.67	10.46	10.29	10.16	10.05	9.89	9.72	9.55	9.47	9.38	9.29	9.20	9.11	9.02
6	13.75	10.92	9.78	9.15	8.75	8.47	8.26	8.10	7.98	7.87	7.72	7.56	7.40	7.31	7.23	7.14	7.06	6.97	6.88
7	12.25	9.55	8.45	7.85	7.46	7.19	6.99	6.84	6.72	6.62	6.47	6.31	6.16	6.07	5.99	5.91	5.82	5.74	5.65
8	11.26	8.65	7.59	7.01	6.63	6.37	6.18	6.03	5.91	5.81	5.67	5.52	5.36	5.28	5.20	5.12	5.03	4.95	4.86
9	10.56	8.02	6.99	6.42	6.06	5.80	5.61	5.47	5.35	5.26	5.11	4.96	4.81	4.73	4.65	4.57	4.48	4.40	4.31
10	10.04	7.56	6.55	5.99	5.64	5.39	5.20	5.06	4.94	4.85	4.71	4.56	4.41	4.33	4.25	4.17	4.08	4.00	3.91
11	9.65	7.21	6.22	5.67	5.32	5.07	4.89	4.74	4.63	4.54	4.40	4.25	4.10	4.02	3.94	3.86	3.78	3.69	3.60
12	9.33	6.93	5.95	5.41	5.06	4.82	4.64	4.50	4.39	4.30	4.16	4.01	3.86	3.78	3.70	3.62	3.54	3.45	3.36
13	9.07	6.70	5.74	5.21	4.86	4.62	4.44	4.30	4.19	4.10	3.96	3.82	3.66	3.59	3.51	3.43	3.34	3.25	3.17
14	8.86	6.51	5.56	5.04	4.69	4.46	4.28	4.14	4.03	3.94	3.80	3.66	3.51	3.43	3.35	3.27	3.18	3.09	3.00
15	8.68	6.36	5.42	4.89	4.56	4.32	4.14	4.00	3.89	3.80	3.67	3.52	3.37	3.29	3.21	3.13	3.05	2.96	2.87
16	8.53	6.23	5.29	4.77	4.44	4.20	4.03	3.89	3.78	3.69	3.55	3.41	3.26	3.18	3.10	3.02	2.93	2.84	2.75
17	8.40	6.11	5.18	4.67	4.34	4.10	3.93	3.79	3.68	3.59	3.46	3.31	3.16	3.08	3.00	2.92	2.83	2.75	2.65
18	8.29	6.01	5.09	4.58	4.25	4.01	3.84	3.71	3.60	3.51	3.37	3.23	3.08	3.00	2.92	2.84	2.75	2.66	2.57
19	8.18	5.93	5.01	4.50	4.17	3.94	3.77	3.63	3.52	3.43	3.30	3.15	3.00	2.92	2.84	2.76	2.67	2.58	2.49
20	8.10	5.85	4.94	4.43	4.10	3.87	3.70	3.56	3.46	3.37	3.23	3.09	2.94	2.86	2.78	2.69	2.61	2.52	2.42
21	8.02	5.78	4.87	4.37	4.04	3.81	3.64	3.51	3.40	3.31	3.17	3.03	2.88	2.80	2.72	2.64	2.55	2.46	2.36
22	7.95	5.72	4.82	4.31	3.99	3.76	3.59	3.45	3.35	3.26	3.12	2.98	2.83	2.75	2.67	2.58	2.50	2.40	2.31
23	7.88	5.66	4.76	4.26	3.94	3.71	3.54	3.41	3.30	3.21	3.07	2.93	2.78	2.70	2.62	2.54	2.45	2.35	2.26
24	7.82	5.61	4.72	4.22	3.90	3.67	3.50	3.36	3.26	3.17	3.03	2.89	2.74	2.66	2.58	2.49	2.40	2.31	2.21
25	7.77	5.57	4.68	4.18	3.85	3.63	3.46	3.32	3.22	3.13	2.99	2.85	2.70	2.62	2.54	2.45	2.36	2.27	2.17
26	7.72	5.53	4.64	4.14	3.82	3.59	3.42	3.29	3.18	3.09	2.96	2.81	2.66	2.58	2.50	2.42	2.33	2.23	2.13
27	7.68	5.49	4.60	4.11	3.78	3.56	3.39	3.26	3.15	3.06	2.93	2.78	2.63	2.55	2.47	2.38	2.29	2.20	2.10
28	7.64	5.45	4.57	4.07	3.75	3.53	3.36	3.23	3.12	3.03	2.90	2.75	2.60	2.52	2.44	2.35	2.26	2.17	2.06
29	7.60	5.42	4.54	4.04	3.73	3.50	3.33	3.20	3.09	3.00	2.87	2.73	2.57	2.49	2.41	2.33	2.23	2.14	2.03
30	7.56	5.39	4.51	4.02	3.70	3.47	3.30	3.17	3.07	2.98	2.84	2.70	2.55	2.47	2.39	2.30	2.21	2.11	2.01
40	7.31	5.18	4.31	3.83	3.51	3.29	3.12	2.99	2.89	2.80	2.66	2.52	2.37	2.29	2.20	2.11	2.02	1.92	1.80
60	7.08	4.98	4.13	3.65	3.34	3.12	2.95	2.82	2.72	2.63	2.50	2.35	2.20	2.12	2.03	1.94	1.84	1.73	1.60
120	6.85	4.79	3.95	3.48	3.17	2.96	2.79	2.66	2.56	2.47	2.34	2.19	2.03	1.95	1.86	1.76	1.66	1.53	1.38
∞	6.63	4.61	3.78	3.32	3.02	2.80	2.64	2.51	2.41	2.32	2.18	2.04	1.88	1.79	1.70	1.59	1.47	1.32	1.00

$^aF = S_1^2/S_2^2$, where S_1^2 and S_2^2 are independent mean squares estimating a common variance σ^2 and based on ν_1 and ν_2 degrees of freedom, respectively.

SOURCE: Reprinted from E. S. Pearson and H. O. Hartley, eds., Table 18 of *Biometrika Tables for Statisticians, Vol. I*, 3rd ed., 1966, by permission of the *Biometrika* Trustees, London.

TABLE B.5 Critical Values of Hartley's F_{max} Test[a]

Upper 5% points ($\alpha = .05$)

v \ c	2	3	4	5	6	7	8	9	10	11	12
2	39.0	87.5	142	202	266	333	403	475	550	626	704
3	15.4	27.8	39.2	50.7	62.0	72.9	83.5	93.9	104	114	124
4	9.60	15.5	20.6	25.2	29.5	33.6	37.5	41.1	44.6	48.0	51.4
5	7.15	10.8	13.7	16.3	18.7	20.8	22.9	24.7	26.5	28.2	29.9
6	5.82	8.38	10.4	12.1	13.7	15.0	16.3	17.5	18.6	19.7	20.7
7	4.99	6.94	8.44	9.70	10.8	11.8	12.7	13.5	14.3	15.1	15.8
8	4.43	6.00	7.18	8.12	9.03	9.78	10.5	11.1	11.7	12.2	12.7
9	4.03	5.34	6.31	7.11	7.80	8.41	8.95	9.45	9.91	10.3	10.7
10	3.72	4.85	5.67	6.34	6.92	7.42	7.87	8.28	8.66	9.01	9.34
12	3.28	4.16	4.79	5.30	5.72	6.09	6.42	6.72	7.00	7.25	7.48
15	2.86	3.54	4.01	4.37	4.68	4.95	5.19	5.40	5.59	5.77	5.93
20	2.46	2.95	3.29	3.54	3.76	3.94	4.10	4.24	4.37	4.49	4.59
30	2.07	2.40	2.61	2.78	2.91	3.02	3.12	3.21	3.29	3.36	3.39
60	1.67	1.85	1.96	2.04	2.11	2.17	2.22	2.26	2.30	2.33	2.36
∞	1.00	1.00	1.00	1.00	1.00	1.00	1.00	1.00	1.00	1.00	1.00

Upper 1% points ($\alpha = .01$)

v \ c	2	3	4	5	6	7	8	9	10	11	12
2	199	448	729	1036	1362	1705	2063	2432	2813	3204	3605
3	47.5	85	120	151	184	21(6)	24(9)	28(1)	31(0)	33(7)	36(1)
4	23.2	37	49	59	69	79	89	97	106	113	120
5	14.9	22	28	33	38	42	46	50	54	57	60
6	11.1	15.5	19.1	22	25	27	30	32	34	36	37
7	8.89	12.1	14.5	16.5	18.4	20	22	23	24	26	27
8	7.50	9.9	11.7	13.2	14.5	15.8	16.9	17.9	18.9	19.8	21
9	6.54	8.5	9.9	11.1	12.1	13.1	13.9	14.7	15.3	16.0	16.8
10	5.85	7.4	8.6	9.6	10.4	11.1	11.8	12.4	12.9	13.4	13.9
12	4.91	6.1	6.9	7.6	8.2	8.7	9.1	9.5	9.9	10.2	10.6
15	4.07	4.9	5.5	6.0	6.4	6.7	7.1	7.3	7.5	7.8	8.0
20	3.32	3.8	4.3	4.6	4.9	5.1	5.3	5.5	5.6	5.8	5.9
30	2.63	3.0	3.3	3.4	3.6	3.7	3.8	3.9	4.0	4.1	4.2
60	1.96	2.2	2.3	2.4	2.4	2.5	2.5	2.6	2.6	2.7	2.7
∞	1.00	1.0	1.0	1.0	1.0	1.0	1.0	1.0	1.0	1.0	1.0

$^a F_{\max} = S^2_{\text{largest}} / S^2_{\text{smallest}} \sim F_{\max(1-\alpha; c, v)}$. S^2_{largest} is the largest and S^2_{smallest} the smallest in a set of c independent mean squares, each based on v degrees of freedom.

SOURCE: Reprinted from E. S. Pearson and H. O. Hartley, eds., Table 31 of *Biometrika Tables for Statisticians, Vol. 1*, 3rd ed., 1966, by permission of the *Biometrika* Trustees, London.

TABLE B.6 Critical Values[a] of the Studentized Range Q

Upper 5% points ($\alpha = .05$)

v \ η	2	3	4	5	6	7	8	9	10	11	12	13	14	15	16	17	18	19	20
1	18.0	27.0	32.8	37.1	40.4	43.1	45.4	47.4	49.1	50.6	52.0	53.2	54.3	55.4	56.3	57.2	58.0	58.8	59.6
2	6.09	8.3	9.8	10.9	11.7	12.4	13.0	13.5	14.0	14.4	14.7	15.1	15.4	15.7	15.9	16.1	16.4	16.6	16.8
3	4.50	5.91	6.82	7.50	8.04	8.48	8.85	9.18	9.46	9.72	9.95	10.15	10.35	10.52	10.69	10.84	10.98	11.11	11.24
4	3.93	5.04	5.76	6.29	6.71	7.05	7.35	7.60	7.83	8.03	8.21	8.37	8.52	8.66	8.79	8.91	9.03	9.13	9.23
5	3.64	4.60	5.22	5.67	6.03	6.33	6.58	6.80	6.99	7.17	7.32	7.47	7.60	7.72	7.83	7.93	8.03	8.12	8.21
6	3.46	4.34	4.90	5.31	5.63	5.89	6.12	6.32	6.49	6.65	6.79	6.92	7.03	7.14	7.24	7.34	7.43	7.51	7.59
7	3.34	4.16	4.68	5.06	5.36	5.61	5.82	6.00	6.16	6.30	6.43	6.55	6.66	6.76	6.85	6.94	7.02	7.09	7.17
8	3.26	4.04	4.53	4.89	5.17	5.40	5.60	5.77	5.92	6.05	6.18	6.29	6.39	6.48	6.57	6.65	6.73	6.80	6.87
9	3.20	3.95	4.42	4.76	5.02	5.24	5.43	5.60	5.74	5.87	5.98	6.09	6.19	6.28	6.36	6.44	6.51	6.58	6.64
10	3.15	3.88	4.33	4.65	4.91	5.12	5.30	5.46	5.60	5.72	5.83	5.93	6.03	6.11	6.20	6.27	6.34	6.40	6.47
11	3.11	3.82	4.26	4.57	4.82	5.03	5.20	5.35	5.49	5.61	5.71	5.81	5.90	5.99	6.06	6.14	6.20	6.26	6.33
12	3.08	3.77	4.20	4.51	4.75	4.95	5.12	5.27	5.40	5.51	5.62	5.71	5.80	5.88	5.95	6.03	6.09	6.15	6.21
13	3.06	3.73	4.15	4.45	4.69	4.88	5.05	5.19	5.32	5.43	5.53	5.63	5.71	5.79	5.86	5.93	6.00	6.05	6.11
14	3.03	3.70	4.11	4.41	4.64	4.83	4.99	5.13	5.25	5.36	5.46	5.55	5.64	5.72	5.79	5.85	5.92	5.97	6.03
15	3.01	3.67	4.08	4.37	4.60	4.78	4.94	5.08	5.20	5.31	5.40	5.49	5.58	5.65	5.72	5.79	5.85	5.90	5.96
16	3.00	3.65	4.05	4.33	4.56	4.74	4.90	5.03	5.15	5.26	5.35	5.44	5.52	5.59	5.66	5.72	5.79	5.84	5.90
17	2.98	3.63	4.02	4.30	4.52	4.71	4.86	4.99	5.11	5.21	5.31	5.39	5.47	5.55	5.61	5.68	5.74	5.79	5.84
18	2.97	3.61	4.00	4.28	4.49	4.67	4.82	4.96	5.07	5.17	5.27	5.35	5.43	5.50	5.57	5.63	5.69	5.74	5.79
19	2.96	3.59	3.98	4.25	4.47	4.65	4.79	4.92	5.04	5.14	5.23	5.32	5.39	5.46	5.53	5.59	5.65	5.70	5.75
20	2.95	3.58	3.96	4.23	4.45	4.62	4.77	4.90	5.01	5.11	5.20	5.28	5.36	5.43	5.49	5.55	5.61	5.66	5.71
24	2.92	3.53	3.90	4.17	4.37	4.54	4.68	4.81	4.92	5.01	5.10	5.18	5.25	5.32	5.38	5.44	5.50	5.54	5.59
30	2.89	3.49	3.84	4.10	4.30	4.46	4.60	4.72	4.83	4.92	5.00	5.08	5.15	5.21	5.27	5.33	5.38	5.43	5.48
40	2.86	3.44	3.79	4.04	4.23	4.39	4.52	4.63	4.74	4.82	4.91	4.98	5.05	5.11	5.16	5.22	5.27	5.31	5.36
60	2.83	3.40	3.74	3.98	4.16	4.31	4.44	4.55	4.65	4.73	4.81	4.88	4.94	5.00	5.06	5.11	5.16	5.20	5.24
120	2.80	3.36	3.69	3.92	4.10	4.24	4.36	4.48	4.56	4.64	4.72	4.78	4.84	4.90	4.95	5.00	5.05	5.09	5.13
∞	2.77	3.31	3.63	3.86	4.03	4.17	4.29	4.39	4.47	4.55	4.62	4.68	4.74	4.80	4.85	4.89	4.93	4.97	5.01

Upper 1% points ($\alpha = .01$)

v \ η	2	3	4	5	6	7	8	9	10	11	12	13	14	15	16	17	18	19	20
1	90.0	135	164	186	202	216	227	237	246	253	260	266	272	277	282	286	290	294	298
2	14.0	19.0	22.3	24.7	26.6	28.2	29.5	30.7	31.7	32.6	33.4	34.1	34.8	35.4	36.0	36.5	37.0	37.5	37.9
3	8.26	10.6	12.2	13.3	14.2	15.0	15.6	16.2	16.7	17.1	17.5	17.9	18.2	18.5	18.8	19.1	19.3	19.5	19.8
4	6.51	8.12	9.17	9.96	10.6	11.1	11.5	11.9	12.3	12.6	12.8	13.1	13.3	13.5	13.7	13.9	14.1	14.2	14.4
5	5.70	6.97	7.80	8.42	8.91	9.32	9.67	9.97	10.24	10.48	10.70	10.89	11.08	11.24	11.40	11.55	11.68	11.81	11.93
6	5.24	6.33	7.03	7.56	7.97	8.32	8.61	8.87	9.10	9.30	9.49	9.65	9.81	9.95	10.08	10.21	10.32	10.43	10.54
7	4.95	5.92	6.54	7.01	7.37	7.68	7.94	8.17	8.37	8.55	8.71	8.86	9.00	9.12	9.24	9.35	9.46	9.55	9.65
8	4.74	5.63	6.20	6.63	6.96	7.24	7.47	7.68	7.87	8.03	8.18	8.31	8.44	8.55	8.66	8.76	8.85	8.94	9.03
9	4.60	5.43	5.96	6.35	6.66	6.91	7.13	7.32	7.49	7.65	7.78	7.91	8.03	8.13	8.23	8.32	8.41	8.49	8.57
10	4.48	5.27	5.77	6.14	6.43	6.67	6.87	7.05	7.21	7.36	7.48	7.60	7.71	7.81	7.91	7.99	8.07	8.15	8.22
11	4.39	5.14	5.62	5.97	6.25	6.48	6.67	6.84	6.99	7.13	7.25	7.36	7.46	7.56	7.65	7.73	7.81	7.88	7.95
12	4.32	5.04	5.50	5.84	6.10	6.32	6.51	6.67	6.81	6.94	7.06	7.17	7.26	7.36	7.44	7.52	7.59	7.66	7.73
13	4.26	4.96	5.40	5.73	5.98	6.19	6.37	6.53	6.67	6.79	6.90	7.01	7.10	7.19	7.27	7.34	7.42	7.48	7.55
14	4.21	4.89	5.32	5.63	5.88	6.08	6.26	6.41	6.54	6.66	6.77	6.87	6.96	7.05	7.12	7.20	7.27	7.33	7.39
15	4.17	4.83	5.25	5.56	5.80	5.99	6.16	6.31	6.44	6.55	6.66	6.76	6.84	6.93	7.00	7.07	7.14	7.20	7.26
16	4.13	4.78	5.19	5.49	5.72	5.92	6.08	6.22	6.35	6.46	6.56	6.66	6.74	6.82	6.90	6.97	7.03	7.09	7.15
17	4.10	4.74	5.14	5.43	5.66	5.85	6.01	6.15	6.27	6.38	6.48	6.57	6.66	6.73	6.80	6.87	6.94	7.00	7.05
18	4.07	4.70	5.09	5.38	5.60	5.79	5.94	6.08	6.20	6.31	6.41	6.50	6.58	6.65	6.72	6.79	6.85	6.91	6.96
19	4.05	4.67	5.05	5.33	5.55	5.73	5.89	6.02	6.14	6.25	6.34	6.43	6.51	6.58	6.65	6.72	6.78	6.84	6.89
20	4.02	4.64	5.02	5.29	5.51	5.69	5.84	5.97	6.09	6.19	6.29	6.37	6.45	6.52	6.59	6.65	6.71	6.76	6.82
24	3.96	4.54	4.91	5.17	5.37	5.54	5.69	5.81	5.92	6.02	6.11	6.19	6.26	6.33	6.39	6.45	6.51	6.56	6.61
30	3.89	4.45	4.80	5.05	5.24	5.40	5.54	5.65	5.76	5.85	5.93	6.01	6.08	6.14	6.20	6.26	6.31	6.36	6.41
40	3.82	4.37	4.70	4.93	5.11	5.27	5.39	5.50	5.60	5.69	5.77	5.84	5.90	5.96	6.02	6.07	6.12	6.17	6.21
60	3.76	4.28	4.60	4.82	4.99	5.13	5.25	5.36	5.45	5.53	5.60	5.67	5.73	5.79	5.84	5.89	5.93	5.98	6.02
120	3.70	4.20	4.50	4.71	4.87	5.01	5.12	5.21	5.30	5.38	5.44	5.51	5.56	5.61	5.66	5.71	5.75	5.79	5.83
∞	3.64	4.12	4.40	4.60	4.76	4.88	4.99	5.08	5.16	5.23	5.29	5.35	5.40	5.45	5.49	5.54	5.57	5.61	5.65

^{a} Range$/S_Y \sim Q_{1-\alpha;\,\eta,\,v}$. η is the size of the sample from which the range is obtained, and v is the number of degrees of freedom of S_Y.

SOURCE: Reprinted from E. S. Pearson and H. O. Hartley, eds., Table 29 of *Biometrika Tables for Statisticians*, *Vol. I*, 3rd ed., 1966, by permission of the *Biometrika* Trustees, London.

TABLE B.7 Arcsin Root Transformation (Angles Corresponding to Percentages Are Measured in Degrees)[a]

%	Angles	%	Angles	%	Angles	%	Angles
.0	.00	25.0	30.00	50.0	45.00	75.0	60.00
1.0	5.74	26.0	30.66	51.0	45.57	76.0	60.67
2.0	8.13	27.0	31.31	52.0	46.15	77.0	61.34
3.0	9.98	28.0	31.95	53.0	46.72	78.0	62.03
4.0	11.54	29.0	32.58	54.0	47.29	79.0	62.72
5.0	12.92	30.0	33.21	55.0	47.87	80.0	63.44
6.0	14.18	31.0	33.83	56.0	48.45	81.0	64.16
7.0	15.34	32.0	34.45	57.0	49.02	82.0	64.90
8.0	16.43	33.0	35.06	58.0	49.60	83.0	65.65
9.0	17.46	34.0	35.67	59.0	50.18	84.0	66.42
10.0	18.44	35.0	36.27	60.0	50.77	85.0	67.21
11.0	19.37	36.0	36.87	61.0	51.35	86.0	68.03
12.0	20.27	37.0	37.47	62.0	51.94	87.0	68.87
13.0	21.13	38.0	38.06	63.0	52.53	88.0	69.73
14.0	21.97	39.0	38.65	64.0	53.13	89.0	70.63
15.0	22.79	40.0	39.23	65.0	53.73	90.0	71.56
16.0	23.58	41.0	39.82	66.0	54.33	91.0	72.54
17.0	24.35	42.0	40.40	67.0	54.94	92.0	73.57
18.0	25.10	43.0	40.98	68.0	55.55	93.0	74.66
19.0	25.84	44.0	41.55	69.0	56.17	94.0	75.82
20.0	26.56	45.0	42.13	70.0	56.79	95.0	77.08
21.0	27.28	46.0	42.71	71.0	57.42	96.0	78.46
22.0	27.97	47.0	43.28	72.0	58.05	97.0	80.02
23.0	28.66	48.0	43.85	73.0	58.69	98.0	81.87
24.0	29.33	49.0	44.43	74.0	59.34	99.0	84.26
						100.0	90.00

[a]Angles $= \arcsin \sqrt{\%}$.

TABLE B.8 The Fisher Z Transformation[a]

r	Z_{F_r}	r	Z_{F_r}	r	Z_{F_r}	r	Z_{F_r}	r	Z_{F_r}
.00	.000	.20	.203	.40	.424	.60	.693	.80	1.099
.01	.010	.21	.213	.41	.436	.61	.709	.81	1.127
.02	.020	.22	.224	.42	.448	.62	.725	.82	1.157
.03	.030	.23	.234	.43	.460	.63	.741	.83	1.188
.04	.040	.24	.245	.44	.472	.64	.758	.84	1.221
.05	.050	.25	.255	.45	.485	.65	.775	.85	1.256
.06	.060	.26	.266	.46	.497	.66	.793	.86	1.293
.07	.070	.27	.277	.47	.510	.67	.811	.87	1.333
.08	.080	.28	.288	.48	.523	.68	.829	.88	1.376
.09	.090	.29	.299	.49	.536	.69	.848	.89	1.422
.10	.100	.30	.310	.50	.549	.70	.867	.90	1.472
.11	.110	.31	.320	.51	.563	.71	.887	.91	1.528
.12	.121	.32	.332	.52	.576	.72	.908	.92	1.589
.13	.131	.33	.343	.53	.590	.73	.929	.93	1.658
.14	.141	.34	.354	.54	.604	.74	.950	.94	1.738
.15	.151	.35	.365	.55	.618	.75	.973	.95	1.832
.16	.161	.36	.377	.56	.633	.76	.996	.96	1.946
.17	.172	.37	.388	.57	.648	.77	1.020	.97	2.092
.18	.182	.38	.400	.58	.662	.78	1.045	.98	2.298
.19	.192	.39	.412	.59	.678	.79	1.071	.99	2.647

[a] $Z_{F_r} = \frac{1}{2} \ln[(1 + r)/(1 - r)]$.

554

TABLE B.9 Critical Values of the Durbin-Watson Statistic d (Critical Values Are One-Sided)[a]

$\alpha = .05$

n	$p=1$ d_l	d_u	$p=2$ d_l	d_u	$p=3$ d_l	d_u	$p=4$ d_l	d_u	$p=5$ d_l	d_u
15	1.08	1.36	.95	1.54	.82	1.75	.69	1.97	.56	2.21
16	1.10	1.37	.98	1.54	.86	1.73	.74	1.93	.62	2.15
17	1.13	1.38	1.02	1.54	.90	1.71	.78	1.90	.67	2.10
18	1.16	1.39	1.05	1.53	.93	1.69	.82	1.87	.71	2.06
19	1.18	1.40	1.08	1.53	.97	1.68	.86	1.85	.75	2.02
20	1.20	1.41	1.10	1.54	1.00	1.68	.90	1.83	.79	1.99
21	1.22	1.42	1.13	1.54	1.03	1.67	.93	1.81	.83	1.96
22	1.24	1.43	1.15	1.54	1.05	1.66	.96	1.80	.86	1.94
23	1.26	1.44	1.17	1.54	1.08	1.66	.99	1.79	.90	1.92
24	1.27	1.45	1.19	1.55	1.10	1.66	1.01	1.78	.93	1.90
25	1.29	1.45	1.21	1.55	1.12	1.66	1.04	1.77	.95	1.89
26	1.30	1.46	1.22	1.55	1.14	1.65	1.06	1.76	.98	1.88
27	1.32	1.47	1.24	1.56	1.16	1.65	1.08	1.76	1.01	1.86
28	1.33	1.48	1.26	1.56	1.18	1.65	1.10	1.75	1.03	1.85
29	1.34	1.48	1.27	1.56	1.20	1.65	1.12	1.74	1.05	1.84
30	1.35	1.49	1.28	1.57	1.21	1.65	1.14	1.74	1.07	1.83

$\alpha = .01$

n	$p=1$ d_l	d_u	$p=2$ d_l	d_u	$p=3$ d_l	d_u	$p=4$ d_l	d_u	$p=5$ d_l	d_u
15	.81	1.07	.70	1.25	.59	1.46	.49	1.70	.39	1.96
16	.84	1.09	.74	1.25	.63	1.44	.53	1.66	.44	1.90
17	.87	1.10	.77	1.25	.67	1.43	.57	1.63	.48	1.85
18	.90	1.12	.80	1.26	.71	1.42	.61	1.60	.52	1.80
19	.93	1.13	.83	1.26	.74	1.41	.65	1.58	.56	1.77
20	.95	1.15	.86	1.27	.77	1.41	.68	1.57	.60	1.74
21	.97	1.16	.89	1.27	.80	1.41	.72	1.55	.63	1.71
22	1.00	1.17	.91	1.28	.83	1.40	.75	1.54	.66	1.69
23	1.02	1.19	.94	1.29	.86	1.40	.77	1.53	.70	1.67
24	1.04	1.20	.96	1.30	.88	1.41	.80	1.53	.72	1.66
25	1.05	1.21	.98	1.30	.90	1.41	.83	1.52	.75	1.65
26	1.07	1.22	1.00	1.31	.93	1.41	.85	1.52	.78	1.64
27	1.09	1.23	1.02	1.32	.95	1.41	.88	1.51	.81	1.63
28	1.10	1.24	1.04	1.32	.97	1.41	.90	1.51	.83	1.62
29	1.12	1.25	1.05	1.33	.99	1.42	.92	1.51	.85	1.61
30	1.13	1.26	1.07	1.34	1.01	1.42	.94	1.51	.88	1.61

TABLE B.9 (Continued)

$\alpha = .05$

n	$p=1$ d_l	d_u	$p=2$ d_l	d_u	$p=3$ d_l	d_u	$p=4$ d_l	d_u	$p=5$ d_l	d_u
31	1.36	1.50	1.30	1.57	1.23	1.65	1.16	1.74	1.09	1.83
32	1.37	1.50	1.31	1.57	1.24	1.65	1.18	1.73	1.11	1.82
33	1.38	1.51	1.32	1.58	1.26	1.65	1.19	1.73	1.13	1.81
34	1.39	1.51	1.33	1.58	1.27	1.65	1.21	1.73	1.15	1.81
35	1.40	1.52	1.34	1.58	1.28	1.65	1.22	1.73	1.16	1.80
36	1.41	1.52	1.35	1.59	1.29	1.65	1.24	1.73	1.18	1.80
37	1.42	1.53	1.36	1.59	1.31	1.66	1.25	1.72	1.19	1.80
38	1.43	1.54	1.37	1.59	1.32	1.66	1.26	1.72	1.21	1.79
39	1.43	1.54	1.38	1.60	1.33	1.66	1.27	1.72	1.22	1.79
40	1.44	1.54	1.39	1.60	1.34	1.66	1.29	1.72	1.23	1.79
45	1.48	1.57	1.43	1.62	1.38	1.67	1.34	1.72	1.29	1.78
50	1.50	1.59	1.46	1.63	1.42	1.67	1.38	1.72	1.34	1.77
55	1.53	1.60	1.49	1.64	1.45	1.68	1.41	1.72	1.38	1.77
60	1.55	1.62	1.51	1.65	1.48	1.69	1.44	1.73	1.41	1.77
65	1.57	1.63	1.54	1.66	1.50	1.70	1.47	1.73	1.44	1.77
70	1.58	1.64	1.55	1.67	1.52	1.70	1.49	1.74	1.46	1.77
75	1.60	1.65	1.57	1.68	1.54	1.71	1.51	1.74	1.49	1.77
80	1.61	1.66	1.59	1.69	1.56	1.72	1.53	1.74	1.51	1.77
85	1.62	1.67	1.60	1.70	1.57	1.72	1.55	1.75	1.52	1.77
90	1.63	1.68	1.61	1.70	1.59	1.73	1.57	1.75	1.54	1.78
95	1.64	1.69	1.62	1.71	1.60	1.73	1.58	1.75	1.56	1.78
100	1.65	1.69	1.63	1.72	1.61	1.74	1.59	1.76	1.57	1.78

$\alpha = .01$

n	$p=1$ d_l	d_u	$p=2$ d_l	d_u	$p=3$ d_l	d_u	$p=4$ d_l	d_u	$p=5$ d_l	d_u
31	1.15	1.27	1.08	1.34	1.02	1.42	.96	1.51	.90	1.60
32	1.16	1.28	1.10	1.35	1.04	1.43	.98	1.51	.92	1.60
33	1.17	1.29	1.11	1.36	1.05	1.43	1.00	1.51	.94	1.59
34	1.18	1.30	1.13	1.36	1.07	1.43	1.01	1.51	.95	1.59
35	1.19	1.31	1.14	1.37	1.08	1.44	1.03	1.51	.97	1.59
36	1.21	1.32	1.15	1.38	1.10	1.44	1.04	1.51	.99	1.59
37	1.22	1.32	1.16	1.38	1.11	1.45	1.06	1.51	1.00	1.59
38	1.23	1.33	1.18	1.39	1.12	1.45	1.07	1.52	1.02	1.58
39	1.24	1.34	1.19	1.39	1.14	1.45	1.09	1.52	1.03	1.58
40	1.25	1.34	1.20	1.40	1.15	1.46	1.10	1.52	1.05	1.58
45	1.29	1.38	1.24	1.42	1.20	1.48	1.16	1.53	1.11	1.58
50	1.32	1.40	1.28	1.45	1.24	1.49	1.20	1.54	1.16	1.59
55	1.36	1.43	1.32	1.47	1.28	1.51	1.25	1.55	1.21	1.59
60	1.38	1.45	1.35	1.48	1.32	1.52	1.28	1.56	1.25	1.60
65	1.41	1.47	1.38	1.50	1.35	1.53	1.31	1.57	1.28	1.61
70	1.43	1.49	1.40	1.52	1.37	1.55	1.34	1.58	1.31	1.61
75	1.45	1.50	1.42	1.53	1.39	1.56	1.37	1.59	1.34	1.62
80	1.47	1.52	1.44	1.54	1.42	1.57	1.39	1.60	1.36	1.62
85	1.48	1.53	1.46	1.55	1.43	1.58	1.41	1.60	1.39	1.63
90	1.50	1.54	1.47	1.56	1.45	1.59	1.43	1.61	1.41	1.64
95	1.51	1.55	1.49	1.57	1.47	1.60	1.45	1.62	1.42	1.64
100	1.52	1.56	1.50	1.58	1.48	1.60	1.46	1.63	1.44	1.65

[a] n = number of observations; p = number of independent variables.

SOURCE: This table is reproduced from *Biometrika*, Vol. 41 (1951), pp. 173 and 175, with the permission of the *Biometrika* Trustees.

TABLE B.10 Table of Random Numbers

	Column							
Row	00000 12345	00001 67890	11111 12345	11112 67890	22222 12345	22223 67890	33333 12345	33334 67890
01	49280	88924	35779	00283	81163	07275	89863	02348
02	61870	41657	07468	08612	98083	97349	20775	45091
03	43898	65923	25078	86129	78496	97653	91550	08078
04	62993	93912	30454	84598	56095	20664	12872	64647
05	33850	58555	51438	85507	71865	79488	76783	31708
06	55336	71264	88472	04334	63919	36394	11095	92470
07	70543	29776	10087	10072	55980	64688	68239	20461
08	89382	93809	00796	95945	34101	81277	66090	88872
09	37818	72142	67140	50785	22380	16703	53362	44940
10	60430	22834	14130	96593	23298	56203	92671	15925
11	82975	66158	84731	19436	55790	69229	28661	13675
12	39087	71938	40355	54324	08401	26299	49420	59208
13	55700	24586	93247	32596	11865	63397	44251	43189
14	14756	23997	78643	75912	83832	32768	18928	57070
15	32166	53251	70654	92827	63491	04233	33825	69662
16	23236	73751	31888	81718	96546	83246	47651	04877
17	45794	26926	15130	82455	78305	55058	52551	47182
18	09893	20505	14225	68514	46427	56788	96297	78822
19	54382	74598	91499	14523	68479	27686	46162	83554
20	94750	89923	37089	20048	80336	94598	26940	36858
21	70297	34135	53140	33340	42050	82341	44104	82949
22	85157	47954	32979	26575	57600	40881	12250	73742
23	11100	02340	12860	74697	96644	89439	28707	25815
24	36871	50775	30592	57143	17381	68856	25853	35041
25	23913	48357	63308	16090	51690	54607	72407	55538
26	79348	36085	27973	65157	07456	22255	25626	57054
27	92074	54641	53673	54421	18130	60103	69593	49464
28	06873	21440	75593	41373	49502	17972	82578	16364
29	12478	37622	99659	31065	83613	69889	58869	29571
30	57175	55564	65411	42547	70457	03426	72937	83792
31	91616	11075	80103	07831	59309	13276	26710	73000
32	78025	73539	14621	39044	47450	03197	12787	47709
33	27587	67228	80145	10175	12822	86687	65530	49325
34	16690	20427	04251	64477	73709	73945	92396	68263
3 5	70183	58065	65489	31833	82093	16747	10386	59293
36	90730	35385	15679	99742	50866	78028	75573	67257
37	10934	93242	13431	24590	02770	48582	00906	58595
38	82462	30166	79613	47416	13389	80268	05085	96666
39	27463	10433	07606	16285	93699	60912	94532	95632
40	02979	52997	09079	92709	90110	47506	53693	49892
41	46888	69929	75233	52507	32097	37594	10067	67327
42	53638	83161	08289	12639	08141	12640	28437	09268
43	82433	61427	17239	89160	19666	08814	37841	12847
44	35766	31672	50082	22795	66948	65581	84393	15890
45	10853	42581	08792	13257	61973	24450	52351	16602
46	20341	27398	72906	63955	17276	10646	74692	48438
47	54458	90542	77563	51839	52901	53355	83281	19177
48	26337	66530	16687	35179	46560	00123	44546	79896
49	34314	23729	85264	05575	96855	23820	11091	79821
50	28603	10708	68933	34189	92166	15181	66628	58599

SOURCE: Reprinted from The Rand Corporation, *A Million Random Digits with 100,000 Normal Deviates*, The Free Press, Glencoe, Ill., 1955.

ANSWERS

TO

SELECTED PROBLEMS

3.2. Testing for homoscedasticity:

$$F_{max} = 2.44 < F_{max(.95); 5,5} = 16.3.$$

There is no evidence that the equality of variance assumption is violated. The same conclusion is reached using the L test. Here

$$L = .29 < F_{.95; 4,25} = 2.76.$$

Testing for means:

Source	Degrees of Freedom	Sum of Squares	Mean Square	F	P Value
Among Groups	4	2.123453	.530863	18.03	.0001
Within Groups	25	.736217	.029449		
Total	29	2.859670	—		

Since $F = 18.03 > F_{.95; 4,25} = 2.76$, reject H_0 and conclude that product perceptions differ for various levels of expectations.

4.3. (a) A posteriori analysis using Tukey T-method:

$$Q_{.95; 5,25} = 4.16 \text{ (by interpolation).}$$

$$\text{Critical range} = 4.16\sqrt{\frac{.029449}{6}} = .291$$

95% Confidence Intervals		
A, B	.097 ± .291	not significant
A, C	.623 ± .291	Significant
A, D	.640 ± .291	significant
A, E	.445 ± .291	significant
B, C	.526 ± .291	significant
B, D	.543 ± .291	significant
B, E	.348 ± .291	significant
C, D	.017 ± .291	not significant
C, E	−.178 ± .291	not significant
D, E	−.195 ± .291	not significant

(b) A priori analysis using orthogonal contrasts:

Source	Degrees of Freedom	Sum of Squares	Mean Square	F	
A vs B	1	.028033	.028033	.95	not significant
C vs D	1	.000833	.000833	.03	not significant
A, B vs C, D	1	2.041664	2.041664	69.33	significant
E vs A, B, C, D	1	.052920	.052920	1.80	not significant
Within Groups	25	.736217	.029449		
Total	29	2.859670	—		

5.2. (a)

Source	Degrees of Freedom	Sum of Squares	Mean Square	F
Among Dosage Levels (α)	2	2.040952	1.020476	40.06 significant
Among Blocks (B)	6	3.245714	.540952	(21.23 significant)
Error	12	.305714	.025476	
Total	20	5.592380	—	

Since $F_\alpha = 40.06 > F_{.95; 2, 12} = 3.89$, we may reject H_0 and conclude that there is a difference in the mean reaction times under the three barbiturate dosage levels.

A posteriori analysis:

$$\text{Tukey T-method: } Q_{.95; 3, 12} = 3.77$$

$$\text{Critical range} = 3.77\sqrt{\frac{.025476}{7}} = .227$$

95 % Confidence Intervals		
5 mg. vs 10 mg.	.085 ± .227	not significant
5 mg. vs 25 mg.	.700 ± .227	significant
10 mg. vs 25 mg.	.615 ± .227	significant

(b) 95 % confidence interval estimate:

$$\bar{Y}_{25\,mg.} \pm t_{.975;\,12}\sqrt{\frac{\text{MSE}}{r}}. \qquad 3.555 \leq \mu_{25\,mg.} \leq 3.817$$

6.1.

Source	Degrees of Freedom	Sum of Squares	Mean Square	F
Aisle Location	2	4,844.4444	2,422.2222	51.904
Height	2	1,869.7774	934.8887	20.033
Height x Aisle Location	4	286.2226	71.5557	1.533
Error	9	420	46.6667	
Total	17	7,420.444	—	

(a) $F = 51.904 > F_{.95;\,2,9} = 4.26$. There is an effect due to aisle location.
(b) $F = 20.033 > F_{.95;\,2,9} = 4.26$. There is an effect due to shelf height.
(c) $F = 1.533 < F_{.95;\,4,9} = 3.63$. There is no evidence of an interaction between shelf height and aisle location.
(e) Tukey T-method: $Q_{.95;\,3,9}\sqrt{\dfrac{\text{MSE}}{6}} = 11.016$

95 % Confidence Intervals		
Aisle Location		
Front vs Middle	40.000 ± 11.016	significant
Front vs Rear	23.333 ± 11.016	significant
Middle vs Rear	16.667 ± 11.016	significant
Height of Shelf		
Top vs Middle	17.000 ± 11.016	significant
Top vs Bottom	24.333 ± 11.016	significant
Middle vs Bottom	7.333 ± 11.016	not significant

(f) Sales are highest for a top shelf height and a front aisle location.

7.5. Using the method of unweighted means:

Source	Degrees of Freedom	Sum of Squares	Mean Square	F
Surgery (α)	1	18.354006	18.354006	23.06 significant
Age Group ($\mathcal{B}$)	2	.553024	.276512	.35 not significant
Interaction ($\alpha \times \mathcal{B}$)	2	.016749	.008375	.01 not significant
Error	34	—	.796080	

Since $F_\alpha = 23.06 > F_{.95; 1, 34} \cong 4.13$, we may reject $H_{0\alpha}$ and conclude that there is a significant difference in the mean number of required post-surgical hospitalization days based on the two surgical procedures. On the other hand, there is no evidence of an interaction effect, nor is there reason to suspect the existence of an effect due to age group on required post-surgical hospitalization days.

8.1. $X = $ Programmers $Y = $ Managers.

(a) $b_0 = -.08844$ $b_1 = .44989$ $\hat{Y}_i = -.08844 + .44989 X_i$

(b) For each additional programmer, .45 managers are staffed; the constant portion of the model is $-.09$ managers.

(c) $S_{Y|X} = 1.42$

(d) $r^2 = .702$ 70.2% of the variation in the number of managers can be explained by variation in the number of programmers.

(e) $r = +.838$

(f) $t = 4.34 > t_{.975; 8} = 2.306$. There is a significant linear relationship.

(g) $F = .494 < F_{.95; 6, 2} = 19.33$. There is no evidence of lack of fit in the linear regression model.

(h) $.211 \leq \beta_1 \leq .689$

(i) $-1.627 \leq \beta_0 \leq +1.451$

(j) $2.966 \leq \mu_{Y|X} \leq 5.855$

(k) $.832 \leq \hat{Y}_I \leq 7.988$

(l) $.34 \leq \rho \leq .97$

10.1. (a) $\hat{Y}_i = 16.196 + 2.038 X_{1_i} + .563 X_{2_i}$

For a given shipping mileage, each additional option increases average delivery time by 2.038 days. For a given number of options, each increase in shipping mileage of 100 miles increases average delivery time by .563 days.

(b) $\hat{Y}_i = 41.08$ days

(c) $F = 270.584 > F_{.95; 2, 13} = 3.81$. There is a significant relationship between delivery time and at least one of the independent variables.

(d) 97.654% of the variation in delivery time can be explained by variation in the number of options ordered and shipping mileage.

(e) $F = 509.16$ and $F = 10.53$ each $> F_{.95; 1, 13} = 4.67$. Each independent variable makes a significant contribution and should be included in the model.

(f) $1.844 < \beta_1 < 2.232$

(g) $r^2_{Y1.2} = .9751$ and $r^2_{Y2.1} = .4474$. For a given shipping mileage, 97.51% of the variation in delivery time can be explained by variation in the number of options. For a given number of options, 44.74% of the variation in delivery time can be explained by variation in the shipping mileage.

11.1. (a) $\hat{Y}_i = 729.8665 - 10.887X_{1_i} + .0465X_{1_i}^2$

(b) There is a decrease in sales as price increases. However, this decrease levels off for further price increases.

(c) $\hat{Y}_i = 107.8$

(d) $F = 37.56 > F_{.95; 2, 12} = 3.89$. Reject H_0. There is a significant second degree polynomial relationship.

(e) $r_{Y.12}^2 = .862$ 86.2% of the variation in sales can be explained by the curvilinear relationship between sales and price.

(f) $F = 9.70$ and $F = 6.96$ each $> F_{.95; 1, 12} = 4.75$ Both the linear and curvilinear effects make significant contributions to the model and thus the second degree polynomial model should be utilized.

(g) There is clearly no relationship between the residuals and the price level. Thus the curvilinear model appears to be the best model.

12.2. (a) Neighborhood I: $\hat{Y}_i = 33.945082 + 8.031967X_{1_i}$
Neighborhood II: $\hat{Y}_i = 28.042991 + 10.121184X_{1_i}$

(b) Neighborhood I: $F = 19.08 > F_{.95; 1, 8} = 5.32$. Reject H_0 that $\mathcal{B}_1 = 0$.
Neighborhood II: $F = 88.36 > F_{.95; 1, 8} = 5.32$. Reject H_0 that $\mathcal{B}_1 = 0$.

(c) Neighborhood I: $F = 1.60 < F_{.95; 4, 4} = 6.39$. Don't reject H_0.
Simple linear model seems adequate.
Neighborhood II: $F = 2.94 < F_{.95; 4, 4} = 6.39$. Don't reject H_0.
Simple linear model seems adequate.

(d) To test for parallelism first requires a test for homoscedasticity. $F = 82.5014/37.2162 = 2.22 < F_{.95; 8, 8} = 3.44$. There is no evidence the homoscedasticity assumption is violated. Pool $S_{Y|X}^2$ values. $S_{p|X}^2 = 59.8588$. $H_0: \beta_{1_I} = \beta_{1_{II}}$. Since $t = -1.005 > t_{.025; 16} = -2.120$ do not reject H_0. Combined estimate of common slope is $\hat{\beta}_1 = 9.218938$.

(e) To test for equality of intercepts first requires a test for homoscedasticity. [see part (d)]. After obtaining $S_{p|X}^2 = 59.8588$ test $H_0: \beta_{0_I} = \beta_{0_{II}}$. Since $t = .313 < t_{.975; 16} = 2.120$ do not reject H_0. Combined estimate of common intercept is $\hat{\beta}_0 = 30.994037$.

(f) Since the tests in parts (d) and (e) were separate tests, we cannot conclude that the two sample regression equations are coming from the same or identical populations [see Section 12.5.1].

12.4. (a) ANACOVA Model: $\hat{Y}_i = 23.737133 + 9.218938X_{1_i} + 12.696743D_{1_i}$
Neighborhood I: $\hat{Y}_i = 23.737133 + 9.218938X_{1_i}$
Neighborhood II: $\hat{Y}_i = 36.433876 + 9.218938X_{1_i}$
Having "adjusted" for number of rooms, b_2' represents the incremental difference in selling prices of homes in Neighborhoods I and II. That is, for those homes having the same number of rooms, selling prices are approximately $12,697 higher in Neighborhood II.

(b) To test for significance: $\bar{Y}_{I(ADJ.)} = 106.246628$ (in thousands of dollars) and $\bar{Y}_{II(ADJ.)} = 118.943371$ (in thousands of dollars). $H_0: \beta_2 = 0$. Since $F = 12.90 > F_{.95; 1, 17} = 4.45$ reject H_0. There is evidence that the adjusted mean selling prices are significantly different in the two neighborhoods.

(c) If ANOVA methods are used for these data instead of ANACOVA, similar conclusions would be reached. Testing for homoscedasticity: Since $F_{.025; 9, 9}$ $= .25 < F = .62 < F_{.975; 9, 9} = 4.03$ do not reject H_0. There is no evidence that the population variances are unequal. To test $H_0: \mu_{Y_I} = \mu_{Y_{II}}$ we develop the ANOVA table (see Table 3.4). Since $F = 5.67 > F_{.95; 1, 18}$ $= 4.41$ we reject H_0. There is evidence that the mean selling prices are significantly different in the two neighborhoods.

12.12. Using the generalized least-squares analysis:

Source	Degrees of Freedom	Sum of Squares	Mean Square	F
Surgery (α)	1	116.456185	116.456185	23.06 significant
Age Group (β)	2	3.431935	1.715968	.34 not significant
Interaction ($\alpha \times \beta$)	2	.110901	.055451	.01 not significant
Error	34	171.736905	5.051085	
Total	39	—	—	

The results here are virtually the same as those obtained using the method of unweighted means in Problem 7.5. The conclusions are identical.

13.2. (a) We list the best two models (for each p^*) and their associated $R_{p^*}^2$ and C_{p^*}

p^*	$R_{p^*}^2$	C_{p^*}	Variables
2	.195	63.92	YLD
2	.401	39.63	EPS
3	.518	27.89	DVD, YLD
3	.616	16.32	YLD, EPS
4	.682	10.44	YLD, EPS, INC
4	.725	5.38	DVD, YLD, EPS
5	.738	5.89	DVD, YLD, EPS, INC
5	.742	5.44	DVD, YLD, EPS, ROS
6	.756	5.74	DVD, YLD, EPS, INC, ROS
6	.762	5.05	DVD, YLD, EPS, SALES, ROS
7	.762	7.02	DVD, YLD, EPS, SALES, ROS, ROE
7	.771	6.04	DVD, YLD, EPS, SALES, INC, ROS
8	.771	8.00	DVD, YLD, EPS, SALES, INC, ROS, ROE

(1) Based on the $R_{p^*}^2$ criterion we might choose either the model containing DVD, YLD and EPS or the model containing DVD, YLD, EPS and ROS. In the interest of parsimony the model containing DVD, YLD and EPS seems preferable.

(2) Based on the C_{p^*} criterion, the best models appear to be the DVD, YLD, EPS model, the DVD, YLD, EPS, ROS model, the DVD, YLD, EPS, SALES, ROS model and the DVD, YLD, EPS, SALES, INC,

ROS model. In the interest of both parsimony and minimum C_{p*}, it would appear that the model containing DVD, YLD, EPS and ROS is preferable.

(3) Using the .05 level of significance, the model containing DVD, YLD and EPS has been chosen by the stepwise, forward selection, and backward elimination approaches.

(b) The dummy variable indicating the exchange in which the stock is traded (EXC) has been included in the analysis. We list the best two models (for each $p*$) and their associated R_{p*}^2 and C_{p*}.

$p*$	R_{p*}^2	C_{p*}	Variables
2	.195	70.58	YLD
2	.401	44.59	EPS
3	.518	31.89	DVD, YLD
3	.616	19.51	YLD, EPS
4	.682	13.08	YLD, EPS, INC
4	.725	7.66	DVD, YLD, EPS
5	.742	7.58	DVD, YLD, EPS, ROS
5	.750	6.62	DVD, YLD, EPS, EXC
6	.762	7.02	DVD, YLD, EPS, SALES, ROS
6	.766	6.52	DVD, YLD, EPS, ROS, EXC
7	.778	7.07	DVD, YLD, EPS, INC, ROS, EXC
7	.783	6.40	DVD, YLD, EPS, SALES, ROS, EXC
8	.785	8.19	DVD, YLD, EPS, SALES, ROS, ROE, EXC
8	.792	7.26	DVD, YLD, EPS, SALES, INC, ROS, EXC
9	.794	9.00	DVD, YLD, EPS, SALES, INC, ROS, ROE, EXC

(1) The results are consistent with those of part a. The model containing DVD, YLD, and EPS seems preferable.

(2) Based on the C_{p*} criterion, the best models appear to be the DVD, YLD, EPS, ROS, EXC model, the DVD, YLD, EPS, SALES, ROS, EXC model and the DVD, YLD, EPS, SALES, INC, ROS, EXC model.

(3) The results are consistent with part a. The model containing DVD, YLD, and EPS has been chosen by the stepwise, forward selection, and backward elimination approaches.

14.1. (a) $\hat{Y}_i = -154.71916 + 2.30037 X_{1_t}$

(b) (1) $\hat{\beta}_1$: For each additional 1.0 billion dollars in the stock of money, consumer expenditures increase by 2.30037 billions of dollars.

(2) r^2: 95.73% of the variation in consumer expenditures is explained by variation in the stock of money.

(3) Significance: For this fitted OLS model $F = 403.22 > F_{95; 1, 18} = 4.41$. We may reject $H_0: \beta_1 = 0$ and conclude the model is significant.

(c) (1) A plot of residuals against time strongly suggests the presence of autocorrelation. Clusters of residuals (each having the same sign) are readily identified as in Figure 14.5.

(2) Using the Durbin-Watson test, $d = .3282 < d_l = 1.20$. Thus, at the $\alpha = .05$ level we may reject $H_0: \rho_a = 0$ and conclude there is evidence of significant positive autocorrelation.

(d) (1) Cochrane-Orcutt method: $\hat{\rho}_a = .8745$. $n = 19$.

$$\hat{Y}_i = -245.15088 + 2.79597X_{1_i}$$

$d = 1.6686 > d_u = 1.40$. At the $\alpha = .05$ level we cannot reject H_0: $\rho_a = 0$. There is no longer any evidence of autocorrelation.

(2) SAS PROC AUTOREG: $\hat{\rho}_a = .7506$. $n = 20$.

$$\hat{Y}_i = -157.99192 + 2.32549X_{1_i}$$

(e)

Parameter Being Estimated	Method		
	OLS	Cochrane-Orcutt	SAS PROC AUTOREG
β_0	−154.719	−245.151	−157.992
β_1	2.300	2.796	2.325
Std. Error for β_1	.115	.609	.180
ρ_a	—	.875	.751
Coeff. of Determination	.957	.554	.903

While all three fitted models are significant, the Cochrane-Orcutt and SAS iterative-estimation procedures are more appropriate than OLS because of the initial significant autocorrelation that was present.

15.2. X_1 = number of daily newspapers
X_2 = radios per 1000
X_3 = televisions per 1000
X_4 = theater seats per 1000
X_5 = literacy rate

(a) Principal components:

	Component Weights				
	1	2	3	4	5
X_1	−.404	.894	.188	.028	−.016
X_2	.825	.153	−.136	.496	.174
X_3	.911	.010	.212	.101	−.340
X_4	.752	.309	−.474	−.337	−.022
X_5	.836	−.008	.421	−.283	.210
Variance (Eigenvalue)	2.939	.919	.501	.450	.191
Proportion	.588	.184	.100	.090	.038
Cumulative Proportion	.588	.772	.872	.962	1.000

The first principal component (which explains almost 60% of the variance) appears to be a composite of the number of radios, televisions, theater seats along with the literacy rate. The second principal component (which explains almost 20% of the variance) appears to primarily reflect the number of newspapers.

(b) Varimax rotation

			Factor Loadings		
	1	2	3	4	5
X_1	−.109	.987	−.035	−.075	−.089
X_2	.218	−.090	.266	.902	.243
X_3	.433	−.160	.257	.394	.752
X_4	.220	−.037	.928	.247	.165
X_5	.903	−.134	.235	.216	.256

The first factor primarily reflects the literacy rate, the second factor reflects the number of daily newspapers, the third factor is the number of theater seats, the fourth factor is the number of radios and the fifth factor is primarily the number of televisions.

16.1. (a) (1) The two dimensions may be interpreted as oriental vs. western for the vertical axis and spicy vs. bland for the horizontal axis.

(2) Spanish, Italian, and Greek appear to be moderately spicy western cuisines. Japanese, Cantonese, and Mandarin are bland oriental cuisines. Szechuan is clearly distinguished as a spicy oriental cuisine while American is perceived as a bland western cuisine.

(b) The cluster analysis is consistent with the results obtained from multidimensional scaling. At the two cluster level, the cuisines are grouped into spicy and bland, indicating the importance of this factor. The two most similar cuisines are Mexican and Szechuan followed by Spanish and Greek.

17.4. A = Estrogen B = Bleeding C = Case or Control

	Model	G^2	Degrees of Freedom
M_1	$[A][B][C]$	73.99	4
M_2	$[AB][C]$	63.44	3
M_3	$[AC][B]$	65.38	3
M_4	$[BC][A]$	13.70	3
M_5	$[AB][AC]$	54.84	2
M_6	$[AB][BC]$	3.16	2
M_7	$[AC][BC]$	5.09	2
M_8	$[AB][AC][BC]$	0.003	1

Models M_6, M_7, and M_8 are good fitting models. $G^2(M_7|M_8) = 5.087 > X^2_{.95;1} = 3.841$ while $G^2(M_6|M_8) = 3.157 < X^2_{.95;1} = 3.841$. Therefore, since there is no difference between M_6 and M_8, choose $M_6([AB][BC])$ as the simpler model. There is an association between estrogen and bleeding, and between bleeding and the occurrence of disease; but given these associations, there is no association between estrogen and the occurrence of disease. For the logit model $\frac{\hat{m}_{111}}{\hat{m}_{112}} = 1.596$ and $\frac{\hat{m}_{121}}{\hat{m}_{122}} = .117$ so that those who experience bleeding have a 61.48% chance of getting the disease while those who do not experience bleeding have only a 10.5% chance of getting the disease. These estimates are irrespective of whether estrogen was used since the estrogen-disease association (AC) has been eliminated from the model.

18.1. $Y = -16.21 + .075X_1 + .025X_2$
$\bar{Y}_1 = +.709 \qquad \bar{Y}_2 = -1.418$

(a) High discriminant scores are associated with successful students while low discriminant scores are associated with unsuccessful students.

(b) For $X_1 = 3.33$ and $X_2 = 560$, then $Y = -1.96$.
Therefore this student should be classified as unsuccessful. For these data, 19 of 20 successful students are correctly classified while 8 out of 10 unsuccessful students are correctly classified. We may note from the output of our computer package that if a stepwise discriminant analysis were performed only the GMAT score would be used in the discriminant analysis. In addition, there appears to be a significant difference in the covariance matrices of the two groups.

18.3. Using the full multinomial rule states (1, 1, 2), (1, 2, 1), (1, 2, 2), (2, 2, 1), and (2, 2, 2) are assigned to Brand X and states (1, 1, 1), (2, 1, 1), and (2, 1, 2) are assigned to Brand M. The error rate is 42.3%.
Using the logit rule the best parsimonious model selected is ([AC][AD][BC] [BD][CD]). States (1, 2, 1), (1, 2, 2), (2, 2, 1), and (2, 2, 2) are assigned to Brand X and states (1, 1, 1), (1, 1, 2), (2, 1, 1), and (2, 1, 2) are assigned to Brand M. The error rate is 43.5%.

REFERENCES

AFIFI, A. A., and S. AZEN, *Statistical Analysis: A Computer Oriented Approach*, 2nd Ed. (New York: Academic Press, 1979).

ANDERBERG, M., *Cluster Analysis for Applications* (New York: Academic Press, 1973).

ANSCOMBE, F. J., "Graphs in Statistical Analysis," *American Statistician*, Vol. 27, pp. 17–21, 1973.

APPELBAUM, M. I., and E. M. CRAMER, "Some Problems in the Nonorthogonal Analysis of Variance," *Psychological Bulletin*, Vol. 81, pp. 335–343, 1974.

AWAD, E. M., *Business Data Processing*, 5th Ed. (Englewood Cliffs, N.J.: Prentice-Hall, 1980).

BELSLEY, D. A., E. KUH, and ROY E. WELSCH, *Regression Diagnostics; Identifying Influential Data and Sources of Collinearity* (New York: Wiley, 1980).

BERENSON, M. L., "A Comparison of Several k Sample Tests for Ordered Alternatives in Completely Randomized Designs," *Psychometrika*, Vol. 47, pp. 265–280, 1982.

BERENSON, M. L., and D. M. LEVINE, *Basic Business Statistics: Concepts and Applications*, 2nd Ed. (Englewood Cliffs, N.J.: Prentice-Hall, 1983).

BERKSON, J., "Applications of the Logistic Function to Bio-Assay," *Journal of the American Statistical Association*, Vol. 39, pp. 357–365, 1944.

BIRCH, M. W., "Maximum Likelihood in Three-Way Contingency Tables," *Journal of the Royal Statistical Society*, Series B, Vol. 25, pp. 220–233, 1963.

BISHOP, Y. M. M., S. E. FIENBERG, and P. W. HOLLAND, *Discrete Multivariate Analysis: Theory and Practice* (Cambridge, Mass.: The MIT Press, 1975).

BOWERMAN, B. L., and R. T. O'CONNELL, *Time Series and Forecasting* (North Scituate, Mass.: Duxbury Press, 1979).

BOX, G. E. P., "Some Theorems on Quadratic Forms Applied in the Study of Analysis of Variance Problems I. Effect of Inequality of Variance in the One-Way Classification," *Annals of Mathematical Statistics*, Vol. 25, pp. 290–302, 1954.

BOX, G. E. P., and S. L. ANDERSEN, "Permutation Theory in the Derivation of Robust Criteria and the Study of Departures from Assumption," *Journal of the Royal Statistical Society*, Series B, Vol. 17, pp. 1–26, 1955.

BOX, G. E. P., W. G. HUNTER, and J. S. HUNTER, *Statistics for Experimenters: An Introduction to Design, Data Analysis and Model Building* (New York: Wiley, 1978).

BRADY, J. P., "Studies on the Metronome Effect on Stuttering," *Behaviour Research and Therapy*, Vol. 7, pp. 197–204, 1969.

BROWN, M. B., and A. B. FORSYTHE, "Robust Tests for the Equality of Variances," *Journal of the American Statistical Association*, Vol. 69, pp. 364–367, 1974.

CARLSON, J. E., and N. H. TIMM, "Analysis of Nonorthogonal Fixed-Effects Designs," *Psychological Bulletin*, Vol. 81, pp. 563–570, 1974.

CARMER, S. G., and M. R. SWANSON, "An Evaluation of Ten Pairwise Multiple Comparison Procedures by Monte Carlo Methods," *Journal of the American Statistical Association*, Vol. 68, pp. 66–74, 1973.

CHATTERJEE, S., and B. PRICE, *Regression Analysis by Example* (New York: Wiley, 1977).

COCHRAN, W. G., "Approximate Significance Levels of the Behrens-Fisher Test," *Biometrics*, Vol. 20, pp. 191–195, 1964.

COCHRAN, W. G., and G. M. COX, *Experimental Designs*, 2nd Ed. (New York: Wiley, 1957).

COCHRANE, D., and G. H. ORCUTT, "Application of Least Squares Regression to Relationships Containing Autocorrelated Error Terms," *Journal of the American Statistical Association*, Vol. 44, pp. 32–61, 1949.

COHEN, J., "Multiple Regression as a General Data Analytic System," *Psychological Bulletin*, Vol. 70, pp. 426–443, 1968.

CONOVER, W. J., "Some Reasons for Not Using the Yates Continuity Correction on 2×2 Contingency Tables (with comments and a rejoinder)," *Journal of the American Statistical Association*, Vol. 69, pp. 374–382, 1974.

CONOVER, W. J., *Practical Nonparametric Statistics*, 2nd Ed. (New York: Wiley, 1980).

CONOVER, W. J., M. E. JOHNSON, and M. M. JOHNSON, "A Comparative Study of Tests for Homogeneity of Variances, with Applications to the Outer Continental Shelf Bidding Data," *Technometrics*, Vol. 23, pp. 351–361, 1981.

COX, D. R., *Planning of Experiments* (New York: Wiley, 1958).

CRAMER, E. M., and M. I. APPELBAUM, "Nonorthogonal Analysis of Variance—Once Again," *Psychological Bulletin*, Vol. 87, pp. 51–57, 1980.

CURTIS, G. A., *ESP: Econometric Software Package* (Chicago: Graduate School of Business, University of Chicago, 1976).

DAMASER, E. C., R. E. SHOR, and M. T. ORNE, "Physiological Effects During Hypnotically Requested Emotions," *Psychosomatic Medicine*, Vol. 25, pp. 334–343, 1963.

DANIEL, C. and F. S. WOOD, *Fitting Equations to Data: Computer Analysis of Multifactor Data*, 2nd ed. (New York: Wiley, 1980).

DIXON, W. J., M. B. BROWN, L. ENGELMAN, J. FRANE, M. HILL, R. JENNRICH and

J. TOPOREK Eds., *BMDP Biomedical Computer Programs* (Berkeley, Cal.: University of California Press, 1981).

DIXON, W. J., and F. J. MASSEY, JR., *Introduction to Statistical Analysis*, 3rd Ed. (New York: McGraw-Hill, 1969).

DRAPER, N. R., and H. SMITH, *Applied Regression Analysis*, 2nd Ed. (New York: Wiley, 1981).

DURBIN, J., and G. S. WATSON, "Testing for Serial Correlation in Least Squares Regression, (II)," *Biometrika*, Vol. 38, pp. 159–178, 1951.

FIENBERG, S. *The Analysis of Cross Classified Categorical Data*, 2nd Ed. (Cambridge, Mass.: MIT Press, 1980).

FISHER, R. A., "The Use of Multiple Measurements in Taxonomic Problems," *Annals of Eugenics*, Vol. 7, pp. 179–188, 1936.

FISHER, R. A., *Statistical Methods and Scientific Inference*, 2nd Ed. (New York: Hafner Press 1959).

FRIEDMAN, M., and D. MEISELMAN, "The Relative Stability of Monetary Velocity and the Investment Multiplier in the United States, 1897–1958," in *Commission on Money and Credit Stabilization Policies* (Englewood Cliffs, N.J.: Prentice-Hall, 1963), p. 266.

GINI, C., "Variabilità e Mutabilità Contributo allo Studio delle Distribuzioni; Relazioni Statische." In *Studi Economico-Giuridici della R. Università di Cagliari*, 1912.

GLASS, G. V., and J. C. STANLEY, *Statistical Methods in Education and Psychology* (Englewood Cliffs, N.J.: Prentice-Hall, 1970).

GNANADESIKAN, R., *Methods for Statistical Data Analysis of Multivariate Observations* (New York: Wiley, 1977).

GOLDSTEIN, M., and W. R. DILLON, *Discrete Discriminant Analysis* (New York: Wiley, 1978).

GOLDSTEIN, M., and W. R. DILLON, *Multivariate Analysis: Methods and Applications* (New York: Wiley, 1983).

GOODMAN, L. A., and W. H. KRUSKAL, "Measures of Association for Cross-Classifications," *Journal of the American Statistical Association*, Vol. 49, pp. 732–764, 1954.

GRAYBILL, F. A., *An Introduction to Linear Statistical Models* (New York: McGraw-Hill, 1961).

GREEN, P. E., and V. R. RAO, *Applied Multidimensional Scaling: A Comparison of Approaches and Algorithms* (Hinsdale, Ill.: Dryden Press, 1972).

GRIZZLE, J. E., "Continuity Correction in the χ^2 Test for 2×2 Tables," *American Statistician*, Vol. 21, pp. 28–32, 1967.

HABERMAN, S., *Analysis of Qualitative Data* (New York: Academic Press, 1980).

HADLEY, G., *Linear Algebra* (Reading, Mass.: Addison-Wesley, 1961).

HARMAN, H. H., *Modern Factor Analysis*, 2nd Ed. (Chicago: University of Chicago Press, 1967).

HARTIGAN, J. A., *Clustering Algorithms* (New York: Wiley, 1975).

HARTLEY, H. O., "The Maximum F-Ratio as a Short-Cut Test for Heterogeneity of Variance," *Biometrika*, Vol. 37, pp. 308–312, 1950.

HERR, D. G., and J. GAEBELEIN, "Nonorthogonal Two-Way Analysis of Variance," *Psychological Bulletin*, Vol. 85, pp. 207–216, 1978.

HICKS, C. R., *Fundamental Concepts in the Design of Experiments*, 2nd Ed. (New York: Holt, Rinehart and Winston, 1973).

HORWITZ, R. I., and A. R. FEINSTEIN, "Alternative Analytic Methods for Case-Control Studies of Estrogens and Endometrial Cancer," *The New England Journal of Medicine*, Vol. 299, p. 1091, 1978.

HOSKING, J. D., and R. M. HAMER, "Nonorthogonal Analysis of Variance Programs: An Evaluation," *Journal of Educational Statistics*, Vol. 4, pp. 161–185, 1979.

HUBERT, L., and F. B. BAKER, "Data Analysis by Single-Link and Complete-Link Hierarchical Clustering," *Journal of Educational Statistics*, Vol. 1, pp. 87–111, 1976.

HULL, C. H., and N. H. NIE, *SPSS Update-New Procedures and Facilities for Releases 7–9* (New York: McGraw-Hill, 1981).

HUNDAL, P. S., "Knowledge of Performance as an Incentive in Repetitive Industrial Work," *Journal of Applied Psychology*, Vol. 53, pp. 224–226, 1969.

IMAN, R. L., and W. J. CONOVER, "The Use of the Rank Transform in Regression," *Technometrics*, Vol. 21, pp. 499–509, 1979.

KAISER, H. F., "The Varimax Criterion for Analytic Rotation in Factor Analysis," *Psychometrika*, Vol. 23, pp. 187–200, 1958.

KAISER, H. F., "Computer Programs for Varimax Rotation in Factor Analysis," *Educational and Psychological Measurement*, Vol. 19, pp. 413–420, 1959.

KENDALL, M. G., *Multivariate Analysis* (London: Griffin, 1975).

KLEINBAUM, D. G., and L. L. KUPPER, *Applied Regression Analysis and Other Multivariable Methods* (North Scituate, Mass.: Duxbury Press, 1978).

KLINE, M. V., "Hypnotic Age Regression and Intelligence," *Journal of Genetic Psychology*, Vol. 77, pp. 129–132, 1950.

KOHR, R. L., and P. A. GAMES, "Robustness of the Analysis of Variance, the Welch Procedure and a Box Procedure to Heterogeneous Variances," *The Journal of Experimental Education*, Vol. 43, pp. 61–69, 1974.

KRUSKAL, J. B., and M. WISH, *Multidimensional Scaling*. Sage University Paper series on Quantitative Applications in the Social Sciences, series no. 07–011. (Beverly Hills and London: Sage Publications, 1978).

LACHENBRUCH, P. A., *Discriminant Analysis* (New York: Hafner Press, 1975).

LEVENE, H., "Robust Tests for Equality of Variances," *Contributions to Probability and Statistics*, I. Olkin, Ed. (Palo Alto, Calif.: Stanford University Press, pp. 278–292, 1960).

LEVINE, D. M., S. WACHSPRESS, P. McGUIRE, and M. S. MAYZNER, "Visual Information Processing of Numerical Inputs," *Bulletin of Psychonomic Society*, Vol. 1 (6A), pp. 404–406, 1973.

LEVINE, D. M., "Nonmetric Multidimensional Scaling and Hierarchical Clustering: Procedures for the Investigation of the Perception of Sports," *Research Quarterly*, Vol. 48, pp. 341–348, 1977.

LIGHT, R. J., and B. H. MARGOLIN, "An Analysis of Variance for Categorical Data," *Journal of the American Statistical Association*, Vol. 66, pp. 534–544, 1971.

MALLOWS, C. L., "Choosing Variables in a Linear Regression: A Graphical Aid." Presented at the Central Regional Meeting of the Institute of Mathematical Statistics, Manhattan, Kansas, May 7–9, 1964.

MALLOWS, C. L., "Choosing a Subset Regression." Presented at Annual Meeting of the American Statistical Association, Los Angeles, August 15–19, 1966.

MALLOWS, C. L., "Some Comments on C_p," *Technometrics*, Vol. 15, pp. 661–675, 1973.

MARASCUILO, L., and M. MCSWEENEY, *Nonparametric and Distribution-Free Methods for the Social Sciences* (Belmont, Calif.: Wadsworth, 1977).

MARDIA, K. V., J. T. KENT, and J. M. BIDDY, *Multivariate Analysis* (New York: Academic Press, 1979).

MCNAMARA, A. M., and J. J. BROWNE, "An Interesting Correlation—United States Oil and Gold Prices," *The New York Statistician*, Vol. 32, No. 2, p. 2, 1980.

MILLER, R. G., *Simultaneous Statistical Inference*, 2nd ed. (New York: Springer-Verlag, 1981).

MOSTELLER, F., and J. W. TUKEY, *Data Analysis and Regression: A Second Course in Statistics* (Reading, Mass.: Addison-Wesley, 1977).

MURACH, M., *Business Data Processing and Computer Programming in FORTRAN* (Chicago, Science Research Associates, 1979).

NETER, J., and W. WASSERMAN, *Applied Linear Statistical Models* (Homewood, Ill.: Richard D. Irwin, 1974).

NIE, N. H., C. H. HULL, J. JENKINS, K. STEINBRENNER, and D. H. BENT, *Statistical Package for the Social Sciences*, 2nd Ed. (New York: McGraw-Hill, 1975).

NOBLE, B., and J. DANIEL, *Applied Linear Algebra*, 2nd Ed. (Englewood Cliffs, N.J.: Prentice-Hall, Inc., 1977).

OSTROM, C. W., *Time Series Analysis: Regression Techniques*, Sage University Paper series on Quantitative Applications in the Social Sciences, series no. 07–009 (Beverly Hills and London: Sage Publications, 1978).

OTT, L., *An Introduction to Statistical Methods and Data Analysis* (North Scituate, Mass.: Duxbury Press, 1977).

OVERALL, J. E., and D. K. SPIEGEL, "Concerning Least Squares Analysis of Experimental Data," *Psychological Bulletin*, Vol. 72, pp. 311–322, 1969.

OVERALL, J. E., D. K. SPIEGEL, and J. COHEN, "Equivalence of Orthogonal and Nonorthogonal Analysis of Variance," *Psychological Bulletin*, Vol. 82, pp. 182–186, 1975.

PARKER, J. B., and A. HOLFORD, "Optimum Test Statistics With Particular Reference to a Forensic Science Problem," *Applied Statistics*, Vol. 17, p. 246, 1968.

PETRE, PETER D., "The Fortune Directory of the Largest Non-Industrial Companies," *Fortune*, July 14, 1980, pp. 146–159.

PRESS, S. J., and S. WILSON, "Choosing Between Logistic Regression and Discriminant Analysis," *Journal of the American Statistical Association*, Vol. 73, pp. 699–705, 1978.

RAO, C. R., *Linear Statistical Inference and its Applications*, 2nd Ed. (New York: Wiley, 1973).

RAO, V. R., "The Salience of Price in the Perception and Evaluation of Product Quality: A Multidimensional Model and Experimental Test," Doctoral Thesis, University of Pennsylvania, 1970.

REIS, T. N., and H. T. SMITH, "The Use of Chi-Square for Preference Testing in

Multidimensional Problems," *Chemical Engineering Progress*, Vol. 59, pp. 39–43, 1963.

ROMNEY, A. K., R. N. SHEPARD, and S. B. NERLOVE, (Eds.) *Multidimensional Scaling: Theory and Applications in the Behavioral Sciences, Vol. II—Applications* (New York: Seminar Press, 1972).

RYAN, T. A., B. L. JOINER, and B. R. RYAN, *MINITAB Student Handbook* (North Scituate, Mass.: Duxbury Press, 1976).

SAS User's Guide, 1979 Edition (Raleigh, N.C.: SAS Institute, 1979).

SASS, C. J., *FORTRAN IV Programming and Applications* (San Francisco: Holden-Day, 1974).

SCHEFFÉ, H., "A Method for Judging All Contrasts in an Analysis of Variance," *Biometrika*, Vol. 40, pp. 87–104, 1953.

SCHEFFÉ, H., *The Analysis of Variance* (New York: Wiley, 1959).

SCHIFFMAN, S. S., M. L. REYNOLDS, and F. W. YOUNG, *Handbook of Multidimensional Scaling* (New York: Academic Press, 1981).

SHAPIRO, S. S., M. B. WILK, and H. J. CHEN, "A Comparative Study of Various Tests for Normality," *Journal of the American Statistical Association*, Vol. 63, pp. 1343–1372, 1968.

SHEPARD, R. N., "The Analysis of Proximities: Multidimensional Scaling with an Unknown Distance Function—I and II," *Psychometrika*, Vol. 27, pp. 125–140 and 219–246, 1962.

SMITH, C. A. B., "Some Examples in Discrimination," *Annals of Eugenics*, Vol. 18, pp. 272–283, 1947.

SNEDECOR, G. W., and W. G. COCHRAN, *Statistical Methods*, 7th Ed. (Ames, Iowa: Iowa State University Press, 1980).

TAKANE, Y., F. W. YOUNG, and J. DeLEEUW, "Nonmetric Individual Differences Multidimensional Scaling: An Alternating Least Squares Method with Optimal Scaling Features," *Psychometrika*, Vol. 42, pp. 7–67, 1977.

THURSTONE, L. L., *Multiple Factor Analysis* (Chicago: University of Chicago Press, 1947).

TUKEY, J. W., "One Degree of Freedom for Non-Additivity," *Biometrics*, Vol. 5, pp. 232–242, 1949.

TUKEY, J. W., "The Problem of Multiple Comparisons," Mimeographed Notes (Princeton, N. J.: Princeton University, 1953).

TUKEY, J.W. *Exploratory Data Analysis* (Reading, Mass.: Addison-Wesley, 1977).

VELLEMAN, P. F., and D. C. HOAGLIN, *Applications, Basics, and Computing of Exploratory Data Analysis* (Boston: Duxbury Press, 1981).

WAGNER, E. H., R. A. GREENBERG, P. B. IMREY, C. A. WILLIAMS, S. H. WOLF and M. A. IBRAHIM, "Influence of Training and Experience on Selecting Criteria to Evaluate Medical Care," *The New England Journal of Medicine*, Vol. 294, pp. 871–876, 1976.

WELCH, B. L., "On the Comparison of Several Mean Values: An Alternative Approach," *Biometrika*, Vol. 38, pp. 330–336, 1951.

WESTIN, S.A., and W.R. DILLON, "CANDI: A Program to Provide a Canonical Representation of Distance Scores," *Journal of Marketing Research*, Vol. 16, pp. 559–560, 1979.

WILCOXON, F. and R. A. WILCOX, *Some Rapid Approximate Statistical Procedures*, revised pamphlet (Pearl River, N.Y.: Lederle Laboratories of the American Cyanamid Company, 1964).

WINER, B. J., *Statistical Principles in Experimental Design*, 2nd Ed. (New York: McGraw-Hill, 1971).

WOODWARD, W. F., "A Comparison of Base Running Methods in Baseball," *M. Sc. Thesis*, Florida State University, 1970, in Hollander, M., and D. A. Wolfe, *Nonparametric Statistical Methods* (New York: Wiley, 1973).

YATES, F., "Contingency Tables Involving Small Numbers and the χ^2 Test," *Journal of the Royal Statistical Society* (*Supplements*) Vol. 1, pp. 217–235, 1934.

"The Fortune Directory of the 500 Largest Industrial Corporations," *Fortune*, May 5, 1980, pp. 276–299.

"The Dividend Achievers," *Dun's Review*, Vol. 116, No. 6, Dec. 1980, pp. 77–93.

INDEX